Lecture Notes in C 4040

Commenced Publication in 1973
Founding and Former Series Edit
Gerhard Goos, Juris Hartmanis, a

Ralf Reulke   Ulrich Eckardt
Boris Flach   Uwe Knauer
Konrad Polthier (Eds.)

Combinatorial Image Analysis

11th International Workshop, IWCIA 2006
Berlin, Germany, June 19-21, 2006
Proceedings

 Springer

Volume Editors

Ralf Reulke
Uwe Knauer
Humboldt-Universität zu Berlin
Institut für Informatik
Unter den Linden 6, 10099 Berlin, Germany
E-mail: {reulke,knauer}@informatik.hu-berlin.de

Ulrich Eckardt
Universität Hamburg
Fachbereich Mathematik
Bundesstr. 55, 20146 Hamburg
E-mail: Eckhardt@math.uni-hamburg.de

Boris Flach
Technische Universität Dresden
Fakultät Informatik
01062 Dresden, Germany
E-mail: bflach@inf.tu-dresden.de

Konrad Polthier
Freie Universität Berlin
Institut für Mathematik und Informatik
Arnimallee 3, 14195 Berlin, Germany
E-mail: polthier@mi.fu-berlin.de

Library of Congress Control Number: 2006927111

CR Subject Classification (1998): I.4, I.5, I.3.5, F.2.2, G.2.1, G.1.6

LNCS Sublibrary: SL 6 – Image Processing, Computer Vision, Pattern Recognition, and Graphics

ISSN   0302-9743
ISBN-10   3-540-35153-1 Springer Berlin Heidelberg New York
ISBN-13   978-3-540-35153-5 Springer Berlin Heidelberg New York

Springer is a part of Springer Science+Business Media

springer.com

© Springer-Verlag Berlin Heidelberg 2006
Printed in Germany

Typesetting: Camera-ready by author, data conversion by Scientific Publishing Services, Chennai, India
Printed on acid-free paper   SPIN: 11774938   06/3142   5 4 3 2 1 0

Preface

This volume presents the proceedings of the 11th International Workshop on Combinatorial Image Analysis. IWCIA 2006 was the 11th in a series of international workshops devoted to combinatorial image analysis. Prior meetings took place in Paris (France 1991), Ube (Japan 1992), Washington DC (USA 1994), Lyon (France 1995), Hiroshima (Japan 1997), Madras (India 1999), Philadelphia (USA 2001), Palermo (Italy 2003) and Auckland (New Zealand 2004). For this workshop we received 59 papers from all over the world. Each paper was assigned to three independent referees and carefully revised. Finally, we selected 34 papers for the conference based on content, significance, relevance, and presentation.

Conference papers are presented in this volume in the order they were presented at the conference. The topics of the conference covered combinatorial image analysis, grammars and models for analysis and recognition of scenes or images, combinatorial topology and geometry for images, digital geometry of curves or surfaces, algebraic approaches to image processing, image, point-clouds or surface registration as well as fuzzy and probabilistic image analysis.

The program followed a single-track format with presentations of all published conference papers. Non-overlapping oral and poster sessions ensured that all attendees had opportunities to interact personally with presenters. Among the highlights of the meeting were the talks of our two invited speakers, renowned experts in the field of discrete geometry, digital topology, and image analysis:

- David Coeurjolly (University of Lyon, France):
  Computational Aspects of Digital Plane and Hyperplane Recognition
- Longin Jan Latecki (Temple University, Philadelphia, USA):
  Polygonal Approximation of Point Sets.

The editors thank all the referees for their big effort in reviewing the submissions and maintaining the high standard of IWCIA conferences. We are also thankful to the sponsors of IWCIA 2006: Humboldt University for hosting the workshop, IAPR for advertising the event, and the German Aerospace Center for financial support. Finally, the organizers wish to thank all contributing authors and our sponsors. Their support was essential for realizing this workshop. In addition, we like to express our appreciation to the people whose efforts made this conference a success.

June 2006   Ulrich Eckardt, Boris Flach, Uwe Knauer,
                                                    Konrad Polthier and Ralf Reulke

Organization

IWCIA 2006 was organized by the department of Computer Science, Humboldt-Universität zu Berlin. The workshop was endorsed by the IAPR.

Conference Chair

Ralf Reulke   Humboldt-Universität zu Berlin, Germany

Organizing Committee

Ralf Reulke	Humboldt-Universität zu Berlin, Germany
Konrad Polthier	Freie Universität Berlin, Germany
Boris Flach	Technische Universität Dresden, Germany
Ulrich Eckardt	University of Hamburg, Germany

Steering Committee

Ralf Reulke	Humboldt-Universität zu Berlin, Germany
Reinhard Klette	University of Auckland, New Zealand
Gabor T. Hermann	City University of New York, USA

Scientific Secretariat

Uwe Knauer	Humboldt-Universität zu Berlin, Germany
Anko Boerner	DLR Berlin, Germany

Program Committee and Referees

Tetsuo Asano	JAIST, Japan
Reneta Barneva	SUNY Fredonia, USA
Gilles Bertrand	ESIEE, France
Gunilla Borgefors	Swedish University of Agricultural Sciences
Anko Börner	DLR Berlin, Germany
Valentin Brimkov	SUNY Buffalo State College, USA
Hans-Dieter Burkhard	Humboldt-Universität zu Berlin, Germany

Table of Contents

Combinatorics and Counting

Thinning and Watersheds

Distances

Image Representation and Segmentation

Invited Paper

Approximations I

Digital Topology

Shape and Matching

Invited Paper

Approximations II

Combinatorics and Grammars

Tomography

Poster Session

Topological Map: An Efficient Tool to Compute Incrementally Topological Features on 3D Images

Guillaume Damiand, Samuel Peltier, Laurent Fuchs, and Pascal Lienhardt

SIC - bât. SP2MI, Bvd M. et P. Curie
BP 30179, 86962 Futuroscope Chasseneuil Cedex - France
{damiand, peltier, fuchs, lienhardt}@sic.univ-poitiers.fr

Abstract. In this paper, we show how to use the three dimensional *topological map* in order to compute efficiently topological features on objects contained in a 3D image. These features are useful for example in image processing to control operations or in computer vision to characterize objects. Topological map is a combinatorial model which represents both topological and geometrical information of a three dimensional labeled image. This model can be computed incrementally by using only two basic operations: the removal and the fictive edge shifting. In this work, we show that Euler characteristic can be computed incrementally during the topological map construction. This involves an efficient algorithm and open interesting perspectives for other features.

Keywords: topological features, model for image representation, intervoxel boundaries, combinatorial map.

1 Introduction

In this paper, we show how to use the three dimensional *topological map* [1, 2] in order to compute efficiently topological features on objects contained in a 3D image. Topological map is a combinatorial model which represents both topological and geometrical information of a three dimensional labeled image with particular properties that makes it a good model for features extraction. Indeed, it represents the topology of 3D labeled images with a minimal number of cells, while conserving all the region adjacencies and incidences.

More precisely, the topological map is incrementally built from a 3D image by using simple removal operations of subdivision cells that verify particular properties. Moreover, removal operations are controlled in order to preserve topological information. After all removals, the topological map represents the regions of a 3D labeled image by their boundaries, which are closed orientable subdivided surfaces.

The main idea of this work is to incrementally compute topological features on regions of a 3D image during the topological map construction. We present here the case of Euler characteristic computation; this is a first example and our

U. Eckardt et al. (Eds.): IWCIA 2006, LNCS 4040, pp. 1–15, 2006.

approach can be extended to several topological features (as canonical polygonal schema or homology classes computation [3]).

Euler characteristic χ of a subdivided object is the alternating sum of numbers of cells (vertices, edges, faces, etc). Let S be a closed orientable subdivided surface and let $\#V$ (resp. $\#E$, $\#F$ and g) be its number of vertices (resp. edges, faces and tunnels[1]). In this case, it is well known that $\chi(S) = \#V - \#E + \#F = 2(1-g)$ [4] gives the complete classification of surfaces.

Euler characteristic and its variants have several applications to image analysis and digital geometry [5]. For example, it can be used to prevent topological alterations in a transformation process or to validate a given segmentation.

Usually, Euler characteristic is computed from a given subdivision, see [6, 7, 8, 9] and references therein. Indeed, it is difficult to analyze the consequences of local changes (adding or removing cells) for topological features. However in our approach, thanks to image scanning and to topological map, consequences of adding cells to the subdivision can be translated into local cases analysis and allows us to obtain the variation of the topological features. Hence the Euler characteristic is computed during the topological map construction with only a small additional cost.

To the authors knowledge such incremental approach had not been yet proposed. In the general context of pavings, an incremental algorithm can be deduced from some results of [10] but this general approach is not well suited for 3D digital imagery.

The paper is organized as follows: Section 2 gives some recalls on topological map. Section 3 presents our incremental method to compute incrementally Euler Characteristic and Section 4 concludes and gives some perspectives.

2   Recalls on Topological Maps

2.1   Combinatorial Maps

A subdivision of a 3D topological space is a partition of the space into 4 subsets whose elements are 0D, 1D, 2D and 3D *cells* (respectively called vertices, edges, faces and volumes, and noted i-cell, $i = 0 \ldots 3$). Border relations are defined between these cells, where the border of an i-cell is a set of *($j < i$)-cells*. Two cells are *incident* when one belongs to the border of the second, and two i-cells are *adjacent* if they are both incident to a common *($j < i$)-cell*.

The topology of nD subdivision of orientable spaces without boundary can be represented by n-dimensional combinatorial maps, or *n-maps* [11, 12, 13, 14, 15]. Intuitively, a 3D combinatorial map can be obtained by successive decompositions of an orientable 3D object. We first distinguish the volumes of this object, then the faces of these volumes, and then the edges of these faces. The elements resulting from the last decomposition are called *darts* and are the basic elements of the combinatorial map definition. To obtain the map, adjacency relations between i-cells are reported onto darts (denoted β_i). These β_i have to verify some

---

[1] Or holes in more general topological context.

particular properties in order to ensure the validity of the represented subdivision (for example β_1 is a permutation and other β_i are involutions, see for example [16] for the formal definition).

We present an example of combinatorial map in Fig. 1B, and the corresponding represented object in Fig. 1A. β_1 connects an oriented edge and the following oriented edge incident to the same face and the same volume, β_2 connects the two faces incident to the same edge and the same volume, and β_3 connects the two volumes incident to the same edge and the same face. In order to simplify the figures, β_i are not explicitly drawn but can be (generally) deduced from the shape of objects.

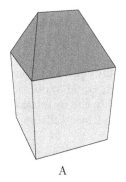 

A   B

Fig. 1. Usual representation of a 3D combinatorial map. (A) A 3D object. (B) Implicit representation of the corresponding combinatorial map, where β_i applications are not explicitly drawn.

Within the combinatorial map framework, all cells are implicitly represented through the notion of *orbit*. Intuitively, an orbit $< \beta_{i_1}, \ldots, \beta_{i_j} > (d)$ is the set of darts that can be reached with a breadth-first search algorithm, starting with d, and using all combinations of all β_{i_k} or $\beta_{i_k}^{-1}$ permutations $\forall k, 1 \leq k \leq j$. With this notion, each cell is defined as a particular orbit. Based on the cells definition, we can retrieve the classical *cell degree* notion. The degree of an i-cell c is the number of distinct $(i+1)$-cells incident to c. Note that in a n-dimensional space, the degree is not defined for n-cells, since $(n+1)$-cells do not exists in such a space.

2.2 Removal Operations

Topological maps are constructed mainly by using removal operations. The i-dimensional removal operation (denoted i-removal) consists in removing an i-cell. This leads to the merging of the two $(i+1)$-cells incident to the removed cell. For 3D subdivisions, we can remove a face (2-removal, e.g. Fig. 2), an edge (1-removal) or a vertex (0-removal). We only present here the main notions about these operations. A more complete description can be found in [17] where general definitions of removal and contraction operations are provided for any dimension.

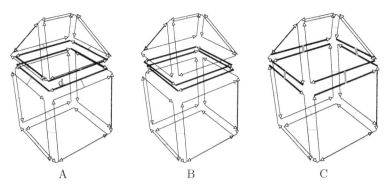

Fig. 2. 2-removal of the face incident to dart d. (A) Initial configuration with two adjacent volumes. (B) The removed face is isolated. (C) Adjacent faces of the initial removed face are joined by modifying β_2 relations.

Any face of a 3-map can be removed without any constraint (e.g. Fig. 2), since the degree of a face, in a 3D subdivision, is always equal to one or two. The face removal operation consists mainly to locally modify the β_2 relation for each dart that belongs to the neighborhood of the removed face (all removal operations are based on similar principle).

The 1-removal (removal of an edge) can be applied only for edges whose degree is one or two. Otherwise it is not possible to automatically decide how to connect the faces incident to the removed edge. This operation is achieved in a similar way than for face removal, but here by modifying β_1 relation. Vertex removal can only be applied for vertices whose degree is one or two. This operation is performed in a similar way than for edge removal, but with different cases to take into account, due to the un-homogeneous definition of combinatorial maps (β_1 is a permutation while others β_i are involutions).

Validity of removal operations can be proved whatever the initial configuration and the cell to remove (even for degenerated cases, as for example removal of a dangling face adjacent to an unique volume, see [17]).

2.3   Topological Map

Combinatorial maps can be used to represent labeled images [18,19,20,21,22,23, 2,24] where cells correspond to interpixel or intervoxel elements (pointels, linels, surfels or voxels). For representing 3D labeled images, the main idea of our approach is first to build a complete combinatorial map, that represents all the intervoxel cells of the image, and then to progressively simplify it with removal operations, as long as no topological information is lost. The minimal map obtained by this construction scheme, called *topological map*, represents all the adjacency and incidence relations between regions of the image.

This is the main property of topological map: to be minimal according to the number of cells, while conserving all the adjacency and incidence relations. To avoid losses of information, we control the operations used during the construction. There are two cases to consider:

- the first case is volume disconnection, when a region is completely included into another one. In this case, we obtain in our model two connected components, one which represents the external surface, and a second which represents the inner surface. We add an inclusion tree on the regions of the image, that allows us to keep relations between these two surfaces[2];
- the second case is face disconnection, when a face has different borders. Here, we add a constraint on the 1-removal operation in order to avoid this type of disconnection. Indeed, this case only occurs when we remove a degree one edge, which is not a dangling edge. By avoiding to remove such an edge, we keep each face connected, and thus homeomorphic to a topological disk. We call *fictive edges* the particular edges kept by this additional constraint, since they do not represent an adjacency relation between regions. By opposition, other edges are called *real edges*. We introduce the notion of *real degree* of a vertex, which is the vertex degree but without considering incident fictive edges.

The topological map construction is made through 5 steps, each one being a simplification of the map obtained by the previous step:

Step 1: Initialization. Given a 3D labeled image, build a 3-map representing a 3D grid made of cubic volumes, plus an enclosing volume which represents the infinite region.

Step 2: Remove each face shared by two voxels having the same label. This step merge volumes that belong to the same region. After this step, each boundary between two regions is represented by a unique surface made of square faces (corresponding to surfels).

Step 3: Remove each degree two edge, and each dangling edge, except isolated edges. This step simplifies the boundaries of each region by merging its faces. We can classify each edge e depending on its degree d:

- $d > 2$: e is not removed due to the precondition of the 1-removal operation. This type of edge belongs to a junction of different boundaries;
- $d = 2$: e is removed because the two incident faces belong to the same boundary and can thus be merged into a unique face. Moreover, this removal can not involves a disconnection since the two faces are different;
- $d = 1$ and e is an isolated edge: this case corresponds to the minimal representation of a sphere with two vertices, one isolated edge and one face. Thus, e is not removed otherwise we remove a surface that represents an adjacency relation;
- $d = 1$ and e is a dangling edge: e is removed because it not represents an adjacency relation and its removal can not involves a disconnection;
- $d = 1$ and e is not a dangling edge: e is not removed to avoid the disconnection of the face. This is the unique case which involves the creation of a fictive edge.

---

[2] Nevertheless the problem of interlaced rings is not take into account by inclusion tree but this is a main drawback of all topological structures. This could be eventually avoided by adding fictive faces to keep cells homeomorphic to balls.

Step 4: Remove each real degree two vertex incident to two non-loop edges, after shifting all fictive edges incident to this vertex. This step simplifies the boundaries of each region by merging real edges. Since this step and the following are both concerned by the real degree vertex, explanations are merged and presented after the last step.

Step 5: Remove each real degree zero vertex incident to at least two edges, including one non-loop edge, after shifting all fictive edges incident to this vertex, except one non-loop edge. This step simplifies the boundaries of each region by grouping fictive edges on same vertices.

We can classify each vertex v depending on its real degree d. We consider the real degree and not the degree since fictive edges are not take into account during this simplification of boundaries. But they are necessary to keep each face connected and for that, they are shifted (pushed along one incident edge) before the vertex removal. If the real degree d is:

- $d > 2$: v is not removed since it belongs to a junction of different boundaries with more than two real edges;
- $d = 2$: if at least one real edge is a loop, v is not removed since the loop corresponds to a face and thus represents an adjacency information. Otherwise, v is removed since the two incident real edges belong to the same boundary;
- $d = 1$: v is not removed because the real edge is a loop (same reason than the previous case);
- $d = 0$: if v is incident to an unique edge, it is not removed since this is the case of the sphere. If v is only incident to $2k$ loops, v is not removed since this case corresponds to the minimal representation of a torus with k holes. Otherwise, there are at least 2 edges and at least one non-loop edge. In this case, v is removed in order to regroup fictive edges on a same vertex.

We can see a first example in Fig. 3 which shows a 3D image and the corresponding topological map. The second example given in Fig. 4 shows a case where a region R_1 is totally included into another one R_2 without other adjacency regions. In such a case, the representation obtained in topological map

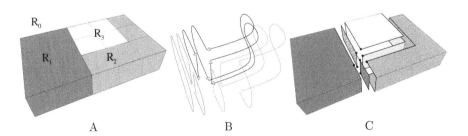

A   B   C

Fig. 3. (A) A 3D image. (B) The corresponding topological map (partial representation without the infinite volume). (C) The represented subdivision in intervoxel elements.

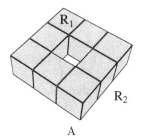

A   B   C

Fig. 4. (A) A region R_1 totally included into another one R_2. This is a partial representation of a 3D image with other regions around R_2 but that are not adjacent to R_1. (B) The corresponding topological map (partial representation without the infinite volume) with one face, two edges and one vertex. (C) The surfels that compose the boundary surface between R_1 and R_2.

corresponds to classical canonical representation of surfaces (in our example we obtain the torus canonical representation with 1 vertex, 2 edges and 1 face).

We have presented here the construction of topological map by successive steps. But in practice, we use an incremental extraction algorithm (presented for example in [25, 2]) which extract topological map in a single scan of the image. The image is scanned from left to right, from behind to front and from up to bottom. For each voxel, a cube is added to the combinatorial map already built. Then, we remove some faces, edges and vertices, depending on the local current configuration.

3   Incremental Euler Characteristic Computation

Euler characteristic χ of each surface of topological map can be computed using the alternated sum of numbers of i-cells, for each $i = 0 \ldots 3$. We propose here to compute incrementally the Euler characteristic during the topological map construction, with only a small additional cost. This can be achieved just by studying the different removal operations and their effect on the number of cells.

The incremental Euler computation given here is only valid for regions represented by a unique surface in topological map. An orientable surface without boundary is completely characterized by χ (this also can be done with its genus). Euler characteristic of a set of surfaces does not give any characteristic information on surfaces. To extend this work in order to obtain topological features for region made of many surfaces, we need to study other characteristics. For example, we are currently interesting on homology groups and the way they can be computed incrementally by using topological map.

In the following, we note $\#F$ the number of faces, $\#E$ the number of edges, and $\#V$ the number of vertices of a region, and χ the Euler characteristic before each operation, and we use the same notation with the prefix n ($\#nF$, $\#nE$, $\#nV$ and $n\chi$) for the same numbers after the operation.

3.1   Cube Creation

The first step of the incremental extraction algorithm consists in creating a cube and add it to the combinatorial map already built. Thus, we just increase the number of cells of the region that contains this voxel $\#nV = \#V + 8$, $\#nE = \#E + 12$ and $\#nF = \#F + 6$ in order to count all the cells of the new cube. After this step, some cells are eventually counted twice. Moreover we can have temporally an invalid Euler characteristic since it corresponds to several surfaces. But the following simplifications are going to eventually decrease these numbers, depending on the current configuration, and finally, one connected component is re-obtained and thus the valid Euler characteristic.

3.2   2-Removal

The face removal is used directly after the cube creation, and uniquely on faces of the new cube in the case of the incremental algorithm. For this reason, faces considered here are only square faces, made of 4 edges. Faces are only removed between volumes that belong to the same region, and thus there are only a region which is concerned by this removal and for which we need to update its topological characteristics.

$\#nF = \#F - 2$, since faces are counted twice in the initial subdivision and the both half faces are removed during the 2-removal (see example in Fig. 5).

$\#nE = \#E - 4$. There are two cases depending on the degree of the edges incident to the removed face. If the degree of each edge is greater than two (e.g. Fig. 5), each edge is counted twice in the initial subdivision and grouped after the removal. Thus, there are 8 edges before, and 4 after which gives the difference -4. The second case is when some edges incident to the removed face are degree two edges (e.g. Fig. 6). In this case, degree one edges are counted only once in the initial subdivision, but are completely removed after the 2-removal

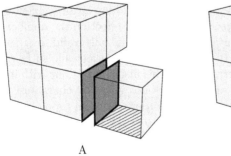
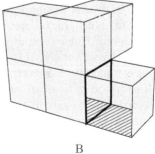

A   B

Fig. 5. Face removal where no edge incident to the face is a degree 2 edge. (A) Before the 2-removal The new cube is drawn on the right of the current region (we do not have represented two faces to see the interior of the volume). (B) After the removal: the 2 dark grey faces are removed, and the 8 bold edges are merged into 4 edges.

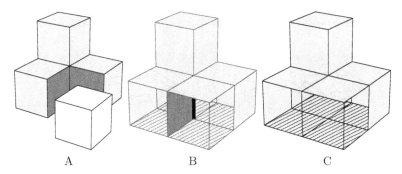

Fig. 6. Face removal where some edges incident to the face are degree 2 edges. (A) Before added the new cube (drawn in white) in a region made of 4 voxels. This addition will be done by 2-removing both dark grey faces. (B) Subdivision obtained after the first 2-removal. 3 faces are not drawn in order to see the interior of the volume. The black thick edge incident to the second face to remove is a degree one edge (i.e. incident twice to the same face). (C) After the second face removal, the degree one edge has completely disappeared.

operation. For this reason, both cases involve exactly the same evolution on the number of edges.

$\#nV =?$. Concerning the number of vertices, the problem is more complicated. Indeed, there are many different cases, depending on the number of vertices counted twice in the initial subdivision, and depending also on the number vertices that are grouped or not after the face removal. Since the number of cases seems to be too much important, we use the topological map in order to update the number of vertices.

We just count the number of vertices incident to the removed face before its removal, and count again the same number after this removal. The difference gives immediately the new number of vertices depending on the old one. Of course, this solution involves a small additional cost. But this cost is very small since we are in a 3D discrete grid and thus we are sure that at most 6 edges are incident to a given vertex.

3.3   1-Removal

The third step of the construction of topological map consists in removing each degree two edge, and each dangling edge (except isolated edges). Now, the initial combinatorial map can have different kind of volumes, since we have already merged some of them during the first step. But the map is already closed and when we process an edge, we are sure that this edge is incident to two volumes. For this reason, number of cells need to be updated in a same way for both regions incident to the removed edge.

When we remove an edge, two possible different cases can be obtained:

– when a degree two edge is removed;
– when a degree one dangling edge is removed.

Other cases are avoided by definition of topological map.

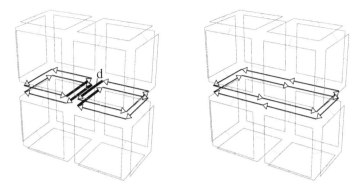

Fig. 7. 1-removal of the degree two edge incident to dart d. (A) Before the 1-removal. (B) After.

First case: Degree two edge. This case is shown in Fig. 7. The new characteristics are: $\#nF = \#F - 1$: two faces are merged into one; $\#nE = \#E - 1$: one edge is removed; $\#nV = \#V$: the number of vertices is still unchanged; and thus $n\chi = \chi$: there are no topological modification.

Second case: Degree one dangling edge. This case is shown in Fig. 8. The new characteristics are: $\#nF = \#F$: since the removed edge is inside a face, no face are merged; $\#nE = \#E - 1$: the edge is removed; $\#nV = \#V - 1$: due to the removal of the edge, the degree one vertex incident to the removed edge vanishes; and thus $n\chi = \chi$: the Euler characteristic remains unchanged.

Thus we can conclude that the 1-removal does not change Euler characteristic of concerned regions, whatever the configuration of the removed edge.

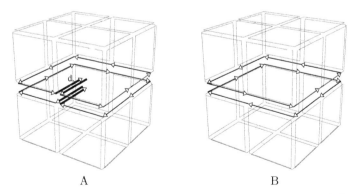

A   B

Fig. 8. 1-removal of the degree one dangling edge incident to dart d. (A) Before the 1-removal. (B) After. The removal of the edge involves the disappearance of the degree one vertex.

3.4   0-Removal

The last step of the construction of topological map consists in removing each vertex which is either:

- a real degree two vertex incident to two non-loop edges, after shifting all fictive edges incident to this vertex;
- a real degree zero vertex incident to at least two edges, including one non-loop edge, after shifting all fictive edges incident to this vertex, except one non-loop edge.

The additional conditions (concerning the non-loop edges) ensure that the 0-removal does not involves the disappearance of a face. Thus, there are only two cases to consider:

- when the degree of the vertex is two;
- when the degree of the vertex is one and the edge is dangling.

Indeed, there are the two unique possible configurations obtained after the fictive edges shifting starting from both cases (given above) of the topological map construction.

Actually, these two cases involve the same modifications: the disappearance of one edge and one vertex. Thus, the new characteristics are: $\#nF = \#F$, $\#nE = \#E - 1$, $\#nV = \#V - 1$ and thus $n\chi = \chi$. We can conclude as for the 1-removal: the 0-removal does not change Euler characteristic of concerned regions.

Note that for 0-removal, there are many regions that are concerned by these modifications: each region which is incident to the removed vertex. Thus, topological characteristics need to be updated for each such regions.

3.5   Fictive Edge Shifting

We also need to study the possible evolutions during the fictive edge shifting. This can be done immediately since there is no modification, neither for the number of volumes, nor for faces, edges and vertices. Obviously, the Euler characteristic remains unchanged after this operation.

3.6   Experimentations

We have implemented the incremental Euler characteristic computation in our computer software which computes the topological map incrementally. This program is developed in C++ without particular optimization. All our experiments were made on a classical personal computer with a Athlon 2000MHz CPU and 512Mb of memory and a Linux Debian System.

Our experiments are made on random artificial images in order to be able to test easily many different images. We have generated images of size range $4 \times 4 \times 4$ to $160 \times 160 \times 160$, and for each size we have generated 10 random images in order to compute an average of the obtained results. For each image, the number of generated regions is a random number between 1 and the size of the image.

We compared the time needed to compute Euler characteristic with the classical method (counting the number of cells and compute the altering sum) and with our

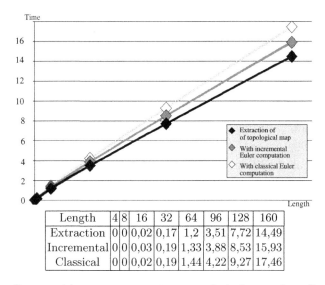

Length	4	8	16	32	64	96	128	160
Extraction	0	0	0,02	0,17	1,2	3,51	7,72	14,49
Incremental	0	0	0,03	0,19	1,33	3,88	8,53	15,93
Classical	0	0	0,02	0,19	1,44	4,22	9,27	17,46

Fig. 9. Time (in seconds) necessary to extract topological map alone (in black in the figure and the first line of the array), with the incremental Euler computation (in dark grey in the figure and the second line of the array), and with the classical algorithm (in white and dash line in the figure and the last line of the array). Each time is the average of 10 extractions for image of size $Length \times Length \times Length$.

incremental method. We can see in Fig. 9 the results obtained by our experiments. Moreover, we have also verified that both methods give the same result.

We can first observe that the additional cost taken by our incremental Euler characteristic computation is small compared to the time necessary to extract the topological map alone. Since we do not have optimized our software, this additional time can be reduced by using some programming techniques. Second, we can observe that our incremental algorithm is faster than the classical algorithm (about 10%), what shows the interest to use the incremental solution.

4   Conclusion and Perspectives

In this paper, we show how to compute Euler Characteristic "on the fly" during the topological map construction. This computation is efficient since topological map construction is efficient and only a small additional cost is needed to compute Euler Characteristic.

The proposed algorithm is incremental as it uses computations from one step to determine the result for the next step. Our experiments show that the additional time necessary to compute incrementally Euler characteristic is very small. Moreover, this solution is more efficient than the classical algorithm which consists in counting the number of cells of the final subdivision and using the altering sum.

This first result is interesting since we are able to compute efficiently during the topological map construction, a topological feature. Moreover, this computation can be optimized, either by using some programming techniques, or by studying the different cases that can occur during face removal in order to remove the counting of vertices. For that, it could be possible to study directly the possible cases by considering directly the added voxel and not only face removal.

Now, we are working on the computation, in a similar way, of other topological features. We have first results for the canonical polygonal schema computation. Intuitively, a canonical polygonal schema of a given surface is the minimal polygon such that when each edges are correctly identified two by two, we obtain the initial surface. We have shown that this notion can be directly obtained from topological map (for regions that are composed by a unique face since the polygonal schema is not defined for other cases).

The next step is to extend computations of topological features for higher dimensions in order to be able to characterize any type of regions and not only those made of an unique surface (which is the case for Euler characteristic). For that, we are interesting on the calculation of homology groups and generators of these groups.

Homology groups are topological invariants that deal with holes in a topological spaces. These invariants can be computed into each dimension and concrete interpretation can be given for low dimensions. In dimension 0, homology groups characterizes connected components, in dimension 1, homology groups characterizes holes, and cavities are described by dimension 2 homology groups[3].

In order to represent homological informations directly on the image, an algorithm that computes generators of homology groups can be used [26]. Such generators are cycles (i.e. closed paths) that surround holes, see examples given on figure 10. Our goal is now to compute these generators incrementally during topological map construction.

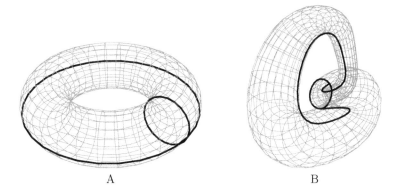

| A | B |

Fig. 10. Illustration of homology generators. (A) For a torus, two paths that surround the two holes are highlighted. (B) For a Klein bottle, two paths are highlighted, one surround the hole and the other indicates the torsion part of the Klein bottle.

---

[3] Roughly speaking for each dimension p, the p-th homology group H_p is isomorphic to a direct sum $\underbrace{\mathbb{Z} \oplus \cdots \oplus \mathbb{Z}}_{\beta_p} \oplus \mathbb{Z}/t_1^p \mathbb{Z} \oplus \cdots \oplus \mathbb{Z}/t_n^p \mathbb{Z}$. where β_p is called the *p-th Betti number*, the integers $t_1^p, \ldots, t_n^p$ are called the *torsion coefficients* of H_p.

References

1. Bertrand, Y., Damiand, G., Fiorio, C.: Topological map: Minimal encoding of 3d segmented images. In: Workshop on Graph-Based Representations in Pattern Recognition, Ischia, Italy, IAPR-TC15 (2001) 64–73
2. Damiand, G.: Définition et étude d'un modèle topologique minimal de représentation d'images 2D et 3D. Thèse de doctorat, Université Montpellier II (2001)
3. Hatcher, A.: Algebraic Topology. Cambridge University Press (2002) available on http://www.math.cornell.edu/~hatcher/AT/ATpage.html.
4. Agoston, M.K.: Algebraic Topology, a first course. Pure and applied mathematics. Marcel Dekker Ed. (1976)
5. Klette, R., Rosenfeld, A.: Digital Geometry - Geometrics Methods for Digital Pictures Analysis. Morgan Kaufmann, San Francisco (2004)
6. Lee, C., Poston, T., Rosenfeld, A.: Holes and genus of 2D and 3D digital images. CVGIP: Graphical Models and Image Processing **55** (1993) 20–47
7. Imiya, A., Eckhardt, U.: The Euler Characteristics of discrete objects and discrete quasi-objects. Computer Vision and Image Understanding **75** (1999) 307–318
8. Desbarats, P., Domenger, J.P.: Retrieving and using topological characteristics from 3d discrete images. In: Proceedings of the Computer Vision Winter Workshop. (2002) 130–139
9. Brimkov, V., Maimone, A., Nordo, G.: An explicit formula for the number of tunnels in digital objects. ArXiv Computer Science e-prints (2005)
10. Spehner, J.: Merging in maps and in pavings. Theoretical Computer Science **86** (1991) 205–232
11. Edmonds, J.: A combinatorial representation for polyhedral surfaces. Notices of the American Mathematical Society **7** (1960)
12. Tutte, W.: A census of planar maps. Canad. J. Math. **15** (1963) 249–271
13. Jacques, A.: Constellations et graphes topologiques. In: Combinatorial Theory and Applications. Volume 2. (1970) 657–673
14. Cori, R.: Un code pour les graphes planaires et ses applications. PhD thesis, Universit Paris VII (1973)
15. Cori, R.: Un code pour les graphes planaires et ses applications. In: Astérisque. Volume 27. Soc. Math. de France, Paris, Francc (1975)
16. Lienhardt, P.: Topological models for boundary representation: a comparison with n-dimensional generalized maps. Commputer Aided Design **23** (1991)
17. Damiand, G., Lienhardt, P.: Removal and contraction for n-dimensional generalized maps. In: Discrete Geometry for Computer Imagery. Number 2886 in Lecture Notes in Computer Science, Naples, Italy (2003) 408–419
18. Fiorio, C.: A topologically consistent representation for image analysis: the frontiers topological graph. In: Discrete Geometry for Computer Imagery. Number 1176 in Lecture Notes in Computer Science, Lyon, France (1996) 151–162
19. Brun, L., Domenger, J.P., Braquelaire, J.P.: Discrete maps : a framework for region segmentation algorithms. In: Workshop on Graph-Based Representations in Pattern Recognition, Lyon, IAPR-TC15 (1997) published in Advances in Computing (Springer).
20. Pailloncy, J.G., Jolion, J.M.: The frontier-region graph. In: Workshop on Graph-Based Representations in Pattern Recognition. Volume 12 of Computing Supplementum., Springer (1997) 123–134

21. Braquelaire, J.P., Brun, L.: Image segmentation with topological maps and inter-pixel representation. Journal of Visual Communication and Image Representation **9** (1998) 62–79
22. Braquelaire, J.P., Desbarats, P., Domenger, J.P., Wüthrich, C.: A topological structuring for aggregates of 3d discrete objects. In: Workshop on Graph-Based Representations in Pattern Recognition, Austria, IAPR-TC15 (1999) 193–202
23. Bertrand, Y., Fiorio, C., Pennaneach, Y.: Border map: a topological representation for nd image analysis. In: Discrete Geometry for Computer Imagery. Number 1568 in Lecture Notes in Computer Science, Marne-la-Vallée, France (1999) 242–257
24. Damiand, G., Bertrand, Y., Fiorio, C.: Topological model for two-dimensional image representation: definition and optimal extraction algorithm. Computer Vision and Image Understanding **93** (2004) 111–154
25. Bertrand, Y., Damiand, G., Fiorio, C.: Topological encoding of 3d segmented images. In: Discrete Geometry for Computer Imagery. Number 1953 in Lecture Notes in Computer Science, Uppsala, Sweden (2000) 311–324
26. Peltier, S., Alayrangues, S., Fuchs, L., Lachaud, J.O.: Computation of homology groups and generators. In: Discrete Geometry for Computer Imagery. Volume 3429 of LNCS., Springer (2005) 195–205

Counting Gaps in Binary Pictures

Valentin E. Brimkov[1], Angelo Maimone[2], and Giorgio Nordo[2]

[1] Mathematics Department, SUNY Buffalo State College, Buffalo, NY 14222, USA
brimkove@buffalostate.edu
[2] Dipartimento di Matematica, Università di Messina, 98166 Messina, Italy
amaimone@dipmat.unime.it, giorgio.nordo@unime.it

Abstract. An important concept in combinatorial image analysis is that of gap. In this paper we derive a simple formula for the number of gaps in a 2D binary picture. Our approach is based on introducing the notions of free vertex and free edge and studying their properties from point of view of combinatorial topology. The number of gaps characterizes the topological structure of a binary picture and is of potential interest in property-based image analysis.

Keywords: digital geometry, 2D binary picture, gap, gap-freeness.

1 Introduction

An important concept in combinatorial image analysis is that of gap. Intuitively, gaps are locations in a digital picture (that is any finite set of pixels/voxels in 2D/3D) through which a "discrete path" can penetrate. Gaps play an important role in rendering pixelized/voxelized scenes by casting digital rays from the image to the scene [8, 9]. Thus it is useful to know if a digital picture is gap-free or it has gaps of certain type. This is particularly interesting when dealing with digital curves or surfaces. It may also be helpful to have an estimation for the number of gaps (if any) in the considered object, possibly as a function of other object characteristics. Such kind of information may help better understand the topological structure of a binary picture and is of potential interest in property-based image analysis.

Results of this sort belong to combinatorial topology, but are also of interest in several other disciplines, such as digital geometry, combinatorial image analysis, and theory of computer graphics. A classical result is the famous Descartes-Euler formula $v - e + f = 2$ that relates the number of vertices (v), edges (e), and facets (f) of a polytope. For various applications of this last formula and other similar results to image analysis and digital geometry, see Chapters 4 and 6 of [10]. In particular, digital picture gap-freeness appears to be equivalent to the notion of well-composedness of a set of pixels proposed by Latecki, Eckhardt, and Rosenfeld [13]. This last paper demonstrates the wealth of using well-composed (i.e., gap-free) sets in image analysis.

A recent work [7] provided the formula $g = v - 2(p + c - h) + b$, where g is the number of gaps, v the number of vertices, p the number of pixels, h the number of holes, c the number of connected components, and b the number of

U. Eckardt et al. (Eds.): IWCIA 2006, LNCS 4040, pp. 16–24, 2006.

2×2 grid squares in a digital picture. In the present paper we obtain a simpler (and computationally more relevant) formula that expresses the number of gaps in a generic 2D digital picture in terms of the new notions of *free vertex* and *free edge*. We achieve this by certain combinatorial considerations within a digital topology framework.

In the next section we introduce some basic notions and notations of digital topology. In Section 3 we present our main results, and we conclude in Section 4.

2 Preliminaries

2.1 Some Basic Notions of Digital Topology

In this section we introduce some basic notions of digital topology to be used in the sequel. We conform to the terminology used in [10]. See also [11, 15, 18] for further details.

All considerations take place in the *grid cell model* that consists of the grid cells of $\mathbb{Z}^2$, together with the related topology. In the grid cell model we represent pixels as squares, called 2-*cells*. Their edges and vertices are called 1-*cells* and 0-*cells*, respectively.

For every $i = 0, 1, 2$ the set of all cells of size i (i-cells) is denoted by $\mathbb{C}_2^{(i)}$. Further, we define the space $\mathbb{C}_2 = \bigcup_{i=0}^{2} \mathbb{C}_2^{(i)}$. We say that two 2-cells are 0-adjacent (1-adjacent) if $e \cap e' \in \mathbb{C}_2^{(0)}$ ($e \cap e' \in \mathbb{C}_2^{(1)}$). The relation of 0-adjacency (resp., 1-adjacency) is denoted by A_0 (resp., A_1). Given a 2-cell p, by $A_0(p)$ and $A_1(p)$ we denote the A_0 and A_1 *neighborhoods* of p, respectively, that are the sets of all 2-cells which are 0-adjacent (resp. 1-adjacent) to p. (These are also called 0/1-neighbors of p.)

We can also consider the grid cell model as an *incidence structure*, i.e., as a triple $(\mathbb{C}_2, I, dim)$ where I is an *incidence relation* defined as follows. For every pair of cells $e, e' \in \mathbb{C}_2$, we have eIe' if and only if e is adjacent to e' or e' is adjacent to e and dim is a mapping from $\mathbb{C}_2$ to the set $\{0, 1, 2\}$. Note that the incidence relation I is reflexive and symmetric while the adjacency relations A_0 and A_1 are irreflexive and symmetric. The grid cell model can also be considered as an *abstract cell complex* $(\mathbb{C}_2, <, dim)$ (see [12]). Here $<$ is a *bounding relation*, that is antisymmetric, irreflexive, and transitive, and such that for every $e, e' \in \mathbb{C}_2$, $e < e'$ if and only if eIe' and $dim(e) < dim(e')$. Hence $<$ is a partial order on $\mathbb{C}_2$ and the corresponding order topology $\tau(<)$ is called the *grid cell topology*. In this topology the *open sets* are precisely the sets $U \subseteq \mathbb{C}_2$ such that for every $u \in U$ and every $v \in \mathbb{C}_2$ with $u < v$, we have $v \in U$.

Since this topology forms a T_0 Alexandroff space, for every point $c \in \mathbb{C}_2$ it is possible to define its *minimal neighborhood* $\eta(c)$, i.e., the smallest open set containing c. For more details about Alexandroff spaces see [10]. For related recent results the reader is referred to [3].

For the sake of completeness, we recall that for any subset A of $\mathbb{C}_2$ its *boundary* $\partial(A)$ is defined as the set of all points x of $\mathbb{C}_2$ such that every open neighborhood of x meets A and $\mathbb{C}_2 \setminus A$, while its *interior* $int(A)$ is the set of all points

x of $\mathbb{C}_2$ such that there exists some open neighborhood of x contained in A. The points of $int(A)$ will be called *internal points* of A. In the rest of the paper, we will assume that the abstract cell complex $(\mathbb{C}_2, <, dim)$ is equipped with the topology $\tau(<)$.

2.2 Gaps in Digital Picturess

Several definitions of a gap are available in the literature (see, e.g., [1, 2, 5]). In what follows, we will refer to the following one, that is believed to fit best our purposes.

Definition 1. *Let v be a 0-cell (vertex) of a digital picture D. We say that D has a* gap *at v if there are two 2-cells (pixels) p_1 and p_2, such that:*

1. $v < p_1$ and $v < p_2$,
2. $p_1 \in A_0(p_2) \setminus A_1(p_2)$, and
3. $A_1(p_1) \cap A_1(p_2) \cap D = \emptyset$.

Figure 1 illustrates the notion of gap. Let us note that Condition 2 of the above definition is equivalent to $p_2 \in A_0(p_1) \setminus A_1(p_1)$, that is, the relation "to have a gap in a common vertex", defined in the sets of 2-cells $\mathbb{C}_2^{(2)}$, is symmetric.

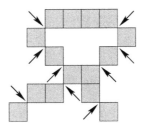

Fig. 1. Gaps in a digital picture

3 Main Results

We start this section by introducing a new definition which will play an important role in obtaining our main results.

Definition 2. *Let c be a cell (vertex or edge) of a digital picture D. We say that c is* free *if $\eta(c) \nsubseteq D$. A cell that does not satisfy this condition will be called non-free.*

In the following, the number of free vertices and free edges of a digital picture D will be respectively denoted by $v^*(D)$ and $e^*(D)$, or simply by v^* and e^*, if no confusion arises. By $v'(D)$ and $e'(D)$ (or simply by v' and e') we will denote the number of non-free vertices and non-free edges, respectively.

Proposition 1. *Let c be a vertex or edge of a digital picture D. Then c is free iff it belongs to the topological boundary $\partial(D)$.*

Proof. Let $c \in D$ be free. Suppose by contradiction that $c \notin \partial(D)$. Then $c \in D \setminus \partial(D) = int(D)$ and so there is an open neighborhood N of c such that $N \subseteq D$. Since $\eta(c) \subseteq N$ is the smallest open set that contains c, it follows that $\eta(c) \subseteq D$, which contradicts the hypothesis that c is free. Conversely, let $c \in \partial(D)$. Then D and its complement have non-empty intersection with every neighborhood of c. In particular, we have $\eta(c) \cap \mathbb{C}^2 \setminus D \neq \emptyset$. This implies that $\eta(c) \nsubseteq D$. Hence c is free. □

Corollary 1. *A vertex or an edge c of a digital picture D is non-free iff it is internal for D.*

Let us note that a vertex v is non-free if and only if it is the unique common vertex of the four pixels in a 2×2-block $B(v)$ centered at v. Similarly, an edge e is non-free if and only if it is the common edge of the two pixels in a 2×1-block $B(e)$ (see Figure 2).

(a)   (b)

Fig. 2. a) Non-free vertex. b) Non-free edge.

Example 1. Recall that a closed digital k-curve C ($k = 0$ or 1) is a k-connected set of pixels such that every its pixel has exactly two k-neighbors in C. It is easy to see that all vertices of C are free.

Given a digital picture D, we will denote by $B(D)$ and $b(D)$ (or, simply, by B and b) the number of 2×2- and 2×1-blocks of D, respectively. The following are easy facts.

Proposition 2. *Let D be a digital picture. Then $B(D)$ equals the number v' of non-free vertices, and $b(D)$ equals the number e' of non-free edges.*

The above proposition suggests that the study of the gaps of a digital picture can be based on the analysis of the type of adjacency in a block. As usual, by $|D|$ we will denote the cardinality of a set of pixels D. With this preparation, we are ready to prove our main result.

Theorem 1. *Let D be a digital picture. Then*

$$g = e^* - v^*,$$

where g is the number of gaps of D and e^* and v^* denote the number of free edges and free vertices of D, respectively.

Proof. We use induction on the number of pixels of D. Let $|D| = 1$. Then we have $e^* = 4$, $v^* = 4$, and $g = 4 - 4 = 0$, so the basis of induction is proved. Assume that the theorem holds for any digital picture D with $|D| = n-1$, where n is an integer greater than or equal to 2. Now let $D' = D \cup \bar{p}$ for a pixel $\bar{p} \notin D$. Then $|D'| = n$. We will examine how adding $\bar{p}$ to D influences the number of gaps, free edges, and free vertices. Since obviously these last parameters do not change outside $A_0(\bar{p})$, it suffices to pay attention only to changes that occur in $A_0(\bar{p})$.

For convenience, we will denote by $v_-, e_-, g_-, v_+, e_+, g_+$ the number of vertices, edges, and gaps, respectively, that may vanish or occur when adding $\bar{p}$ to D. We analyze different cases classified with respect to $|A(\bar{p}) \cap D| = i$ for $i = 0, 1, \ldots, 8$ (i.e., the cardinality of the set of 0-neighbors of $\bar{p}$). So, we have $2^8 = 256$ different feasible cases.

Table 1. Values of parameters v_-^*, v_+^*, e_-^*, e_+^*, g_-, g_+, Σ_-, and Σ_+ for $|A(\bar{p}) \cap D| = 0, 1, 2,$ and 3

$|A(\bar{p}) \cap D|$	code	symmetric	v_-^*	v_+^*	e_-^*	e_+^*	g_-	g_+	Σ_-	Σ_+
0	#0	none	0	4	0	4	0	0	0	0
1	#1	3	0	2	1	3	0	0	-1	-1
	#2	3	0	3	0	4	0	1	0	0
2	#3	7	0	2	1	3	0	0	-1	-1
	#5	3	0	1	2	2	1	0	-1	-1
	#9	7	0	1	1	3	0	1	-1	-1
	#10	3	0	2	0	4	0	2	0	0
	#17	1	0	0	2	2	0	0	-2	-2
	#34	1	0	2	0	4	0	2	0	0
3	#7	3	1	1	2	2	0	0	-1	-1
	#11	7	0	1	1	3	0	1	-1	-1
	#13	7	0	1	2	2	1	0	-1	-1
	#14	3	0	2	1	3	0	0	-1	-1
	#19	7	0	0	2	2	0	0	-2	-2
	#21	3	0	0	3	1	2	0	-1	-1
	#35	7	0	1	1	3	0	1	-1	-1
	#37	3	0	0	2	2	1	1	-1	-1
	#41	3	0	0	1	3	0	2	-3	-3
	#42	3	0	1	0	4	0	3	0	0

To facilitate the further description, we encode each configuration of pixels in a possible neighborhood of $\bar{p}$, as follows. To any pixel of $A_0(\bar{p})$ we assign a position in a binary string W of length 8 (remember that $A_0(\bar{p})$ has 8 elements). Pixels of $A_0(\bar{p})$ are counted clockwise, starting from the top-left corner of the neighborhood. Accordingly, to each configuration $A_0(\bar{p}) \cap D$, there corresponds an 8-bit string W, in a way that positions in W corresponding to pixels from $A_0(\bar{p})$ contain 1's, while the others are 0's. For brevity, we encode a configuration by the decimal number corresponding to its binary string. See Figure 3 for illustration.

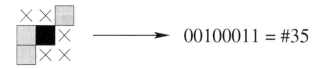

$$00100011 = \#35$$

Fig. 3. An example of a configuration and its binary and decimal labels

Given a configuration $A_0(\bar{p}) \cap D$, its complement $A_0(\bar{p}) \setminus (A_0(\bar{p}) \cap D)$ to $A_0(\bar{p})$ is called *dual* to $A_0(\bar{p}) \cap D$. It can be viewed as obtained from $A_0(\bar{p}) \cap D$ by exchanging v_-, e_-, g_- and v_+, e_+, g_+. Clearly, if $|A_0(\bar{p}) \cap D| = i$, then $|A_0(\bar{p}) \setminus (A_0(\bar{p}) \cap D)| = 8 - i$. Therefore, we can restrict ourselves to configurations with $|A(\bar{p}) \cap D| = i$, for $i = 0, 1, 2, 3, 4$. Moreover, every case, except for four of them (see the samples labeled by #0 (and its dual), #85, and #170) admits symmetric ones (at least one and at most seven) that may also be disregarded. Thus finally we obtain that it is enough to consider 32 essentially different cases displayed in Figure 4, the rest being either symmetric or dual to them with essentially analogous gap characterization.[1]

Table 2. Values of parameters v_-^*, v_+^*, e_-^*, e_+^*, g_-, g_+, Σ_-, and Σ_+ for $|A(\bar{p}) \cap D| = 4$

$|A(\bar{p}) \cap D|$	code	symmetric	v_-^*	v_+^*	e_-^*	e_+^*	g_-	g_+	Σ_-	Σ_+
4	#15	7	1	1	2	2	0	0	-1	-1
	#23	7	1	0	3	1	1	0	-1	-1
	#27	3	0	0	2	2	0	0	-2	-2
	#39	3	1	0	2	2	0	1	-1	-1
	#43	7	0	0	1	3	0	2	-1	-1
	#45	7	0	0	2	2	1	1	-1	-1
	#46	7	0	1	1	3	0	1	-1	-1
	#51	3	0	0	2	2	0	0	-2	-2
	#53	7	0	0	3	1	2	0	-1	-1
	#54	3	0	1	2	2	1	0	-1	-1
	#57	7	0	0	2	2	0	0	-2	-2
	#85	none	0	0	4	0	4	0	0	0
	#170	none	0	0	0	4	0	4	0	0

Let us denote $\Sigma_- = v_- - e_- + g_-$ and $\Sigma_+ = v_+ - e_+ + g_+$. Keeping in mind the definition of gaps, Proposition 2, and Corollary 1, one can easily conclude that,

---

[1] A simple computer program allowed us to easily generate the 32 configurations. Since we are interested in finding only configurations up to symmetries and duality, for a current configuration under consideration we have to check if it can be obtained by one of the seven possible symmetries of the square or if it is the dual to a configuration already found. If this is the case, we do not count the configuration as a new one. Note that at any step, in order to quickly verify if such a configuration has to be added to the list, we do not need to compare it with the whole list of configurations already found: the representative of each class of symmetries is the configuration with the minimal label.

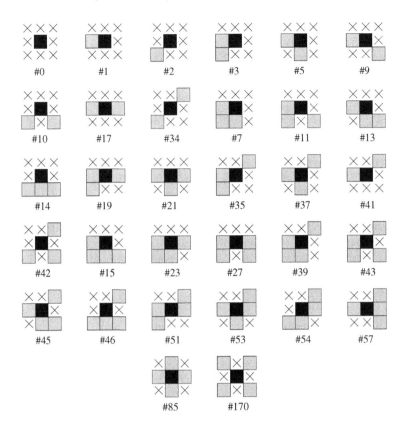

Fig. 4. The 32 essentially different configurations $A_0(\bar{p}) \cap D$

for a given configuration the statement of the theorem holds iff $\Sigma_- = \Sigma_+$. Thus it remains to show that this last equality is verified for all the 32 configurations. As Tables 1 and 2 demonstrate, this condiction holds in all these cases, which completes the proof.   □

Corollary 2. *Let D be a digital picture. Then $g = e - v + B - b$, where B is the number of 2×2-block and b is the number of 2×1-block.*

Proof. By Theorem 1 we have that $g = e^* - v^*$. Moreover, we have $v = v^* + v'$, $e = e^* + e'$, $v' = B$, and $e' = b$, from where the thesis holds.   □

4   Concluding Remarks

In this paper we proposed a formula for the total number of gaps in a binary picture in terms of its free vertices and edges. It can be used to test a binary picture for existence of gaps. As mentioned, the number of gaps characterizes the topological structure of a binary picture. This may be of practical importance, e.g., when trying to define the borders of a digital picture in a way to ensure gap-freeness (see [10]).

Let us mention that the obtained formula may allow a shorter proof if one uses certain classical results, e.g., from combinatorial topology or graph theory. The advantage of the proof presented in this paper is that it is a direct one and does not resort to previous theoretical developments.

Work in progress is aimed at extending the results to higher dimensions.

Acknowledgments

The authors are indebted to the three anonymous referees for their useful remarks and suggestions. The first author thanks Reneta Barneva for proofreading and for some useful remarks.

References

1. E. Andres, R. Acharya, and C. Sibata. Discrete analytical hyperplanes. *Graphical Models and Image Processing* **59** (1997) 302–309
2. E. Andres, Ph. Nehlig and J. Françon, Tunnel-free supercover 3D polygons and polyhedra, In: D. Fellner and L. Szirmay-Kalos (Guest Eds.), *EUROGRAPH-ICS'97*, 1997, pp. C3-C13
3. Arenas, F.G., Alexandroff spaces, *Acta Mathematica*, University of Comenianae, Vol. LXVIII-1 (1999) 17–25
4. Barneva, R.P, V.E. Brimkov, and Ph. Nehlig, Thin discrete triangular meshes, *Theoretical Computer Science* **246** (1-2) (2000) 73–105
5. Brimkov, V.E., E. Andres, and R.P. Barneva, Object discretizations in higher dimensions, *Pattern Recognition Letters* **23** (2002) 623–636
6. Brimkov, V.E., R.P. Barneva, and Ph. Nehlig, Minimally thin discrete triangulations, In: *Volume Graphics*, A. Kaufman, R. Yagel, M. Chen (Eds.), Chapter 3, Springer Verlag, 2000, pp. 51-70
7. Brimkov, V.E., A. Maimone, G. Nordo, R. Barneva, and R. Klette, The number of gaps in binary pictures, In: Bebis et al. (Eds.): *International Symposium of Visual Computing* (Lake Tahoe, Nevada), *LNCS* 3804 (2005) 35–42
8. Cohen-Or, D. and A. Kaufman, 3D line voxelization and connectivity control, *IEEE Computer Graphics and Applications* **17** (6) (1997) 80–87
9. Kaufman, A., D. Cohen, and R. Yagel, Volume graphics, *IEEE Computer* **26**(7) (1993) 51–64
10. Klette, R. and A. Rosenfeld, *Digital Geometry - Geometric Methods for Digital Picture Analysis*, Morgan Kaufmann, San Francisco, 2004
11. Kong, T.Y., Digital topology, In: Davis, L.S., editor. *Foundations of Image Understanding*, Kluwer, Boston, Massachusetts, 2001, pp. 33–71
12. Kovalevsky, V.A., Finite topology as applied to image analysis, *Computer Vision, Graphics and Image Processing*, **46**(2) 141–161
13. Latecki, L., U. Eckhardt, A. Rosenfeld, Well-composed sets, *Computer Vision and Vision Understanding* **61** (1995) 70–83

14. Mylopoulos, J.P. and T. Pavlidis, On the topological properties of quantized spaces. I. The notion of dimension, *J. ACM* **18** (1971) 239–246
15. Pavlidis, T, *Algorithms for Graphics and Image Processing*, Computer Science Press, Rockville, MD, 1982
16. Rosenfeld, A., Arcs and curves in digital pictures, *Journal of the ACM* **18** (1973) 81–87
17. Rosenfeld, A., Adjacency in digital pictures, *Information and Control* **26** (1974) 24–33
18. Voss, K., *Discrete Images, Objects, and Functions in* $\mathbf{Z}^n$, Springer Verlag, Berlin, 1993

The Exact Lattice Width of Planar Sets and Minimal Arithmetical Thickness

F. Feschet

LLAIC1 - IUT Clermont-Ferrand,
BP 86, 63172 Aubière, France
feschet@llaic3.u-clermont1.fr

Abstract. We provide in this paper an algorithm for the exact computation of the lattice width of an integral polygon K with n vertices in $O(n \log s)$ arithmetic operations where s is a bound on all integers defining vertices and edges. We also provide an incremental version of the algorithm whose update complexity is shown to be $O(\log n + \log s)$. We apply this algorithm to construct the arithmetical line with minimal thickness, which contains a given set of integer points.

1 Introduction

Integer Linear Programming is a fundamental tool in optimization, operation research, economics... Moreover it is interesting in itself since the problem is NP-hard in the general case. Several results were known for the planar case [1, 2, 3] before Lenstra [4] proved that Integer Linear Programming can be solved in polynomial time when the dimension is fixed. Faster and faster algorithms are nowadays developed and available making the use of Integer Linear Programming reliable even for high dimensional problems. The approach of Lenstra used the notion of *lattice width* (see section 2.1 for precise lattice definition) to detect directions for which the polyhedron of solutions is thin. In polynomial time, the problem is then reduced to a feasibility question: given a polyhedron P, determine whether P contains an integer point in it. To solve it, Lenstra approximated the width of the polyhedron and gave a recursive solution using problems of smaller dimension. The approximate lattice width is also used in the recent algorithms of Eisenbrand and Rote [5] and Eisenbrand and Laue [6] for the 2-variables problem.

In the present paper, we are interested in the computation of the exact lattice width as well as the computation of the whole set of directions for which the width is the lattice width. To do this, we propose a partitioning of the set of directions in cones where the problem is shown to be solvable in $O(\log s)$ time where s is a bound on the integers appearing in the problem. For computing the complexity we use the arithmetic model where each arithmetic operation $+$, $-$, $\times$ and $/$ are unit-cost operations. In this model the complexity of our solution is $O(n \log s)$ arithmetic operations on any integer polygon with n vertices.

U. Eckardt et al. (Eds.): IWCIA 2006, LNCS 4040, pp. 25–33, 2006.

2   Preliminaries

2.1   The Lattice Width

In this section, we review some definitions from algorithmic number theory and provide a precise formulation of the problem we solve. Definitions are taken from [5, 7, 8].

Let K be a set of points of $\mathbb{R}^d$ and Λ a lattice included in $\mathbb{R}^d$. Without loss of generality, we suppose both structures to be full-dimensional. The *width* of K along a direction $c \neq 0$ in $\mathbb{R}^d$ is defined as

$$\omega_c(K) = \max\left\{c^T x \mid x \in K\right\} - \min\left\{c^T x \mid x \in K\right\} \tag{1}$$

The *lattice width* of K is the minimal value of the width of K among the directions belonging to the dual lattice Λ^*,

$$\omega_\Lambda(K) = \min\left\{\omega_c(K) \mid c \in \Lambda^* \setminus \{0\}\right\} \tag{2}$$

where Λ^* is given by $\Lambda^* = \{x \in \mathbb{R}^d \mid x^T v \in \mathbb{Z}, \forall v \in \Lambda\}$. When $\Lambda = \mathbb{Z}^d$, we have $\Lambda^* = \mathbb{Z}^d$ and we write $\omega(K)$ instead of $\omega_{\mathbb{Z}^d}(K)$. Geometrically, if a set K has a lattice width of l then K can be covered by at most $\lfloor l \rfloor + 1$ parallel hyperplanes given by $c^T x = \text{cst}$ with cst in the integer interval from $\min\{c^T x \mid x \in K\}$ to $\max\{c^T x \mid x \in K\}$.

It is straightforward to see that $\omega(K) = \omega(\text{conv}(K))$ where $\text{conv}(K)$ denotes the convex hull of K. Here, we also suppose that K is a polyhedral convex set. K can be given either in H- or in V-representation [9]. In the H-representation, K is defined as $K = \{x \in \mathbb{R}^d \mid Ax \leq b\}$ where $A \in \mathbb{Z}^{m \times d}$ and $b \in \mathbb{Z}^m$. In the V-representation, we suppose that K is given by a list of vertices and edges. In this paper, we focus on the V-representation of K. A pre-processing must be done when K is given by an H-representation. All numbers appearing in the coordinates of vertices or in the components of vectors are supposed to have bit size bounded above by $\log s$.

In the present paper, we suppose that $d = 2$ otherwise stated. Moreover, we suppose that K is an integer polygon. We denote by n the number of vertices of K. The lattice Λ is also supposed to be $\mathbb{Z}^2$. In this case, the lattice width is an integer. The problem we would like to solve is the following one:

Problem (Lattice Width)
Given an integer polygon $K \in \mathbb{Z}^2$ in a V-representation, find its lattice width $\omega(K)$ as well as all vectors $c \in \mathbb{Z}^2$ such that $\omega_c(K) = \omega(K)$.

It should be noted that we are not interested in *approximate* solutions of the problem. An approximate solution of the problem might be found by the following algorithm suggested by one of the referees of a preliminary version of this paper. It is known [4] that the lattice width of a convex set K is obtained for the shortest vector with respect to the *dual norm* whose unit ball is the polar set of the set $\frac{1}{2}(K + (-K))$. In the general case, computing the shortest vector is NP-hard. In the case of a polygon, the dual norm is a polyhedral norm whose unit ball is a polygon with $O(n)$ vertices. The standard approach [7] is to apply a linear transform which makes the unit ball approximately round. Following the method of Kaib and Schnörr [10] or a reduction similar to the one of Rote [11], the shortest vector with respect to the Euclidean norm is a good approximation of the shortest vector according to the dual norm. This leads to an algorithm with $O(\log s)$ arithmetic operations, to solve the approximate version of the problem of computing the lattice width.

2.2   Relations with Discrete Geometry

Arithmetical discrete lines have been introduced by Reveillès in [12]. They are defined by the set of integer points $(x, y) \in \mathbb{Z}^2$ such that

$$\mu \leq ax - by < \mu + \omega \tag{3}$$

where (a, b) is the direction of the discrete line, μ is its shift at the origin and ω is its arithmetical thickness. Special values of ω lead to classical 8-connected lines ($\omega = \max(|a|, |b|)$) or 4-connected lines ($\omega = |a| + |b|$). After the introduction of this definition, a linear time recognition algorithm has been published by Debled and Reveillès [13].

The arithmetical thickness corresponds merely to the number of Bezoutian's lines ($ax - by = k$) which are necessary to cover the integer points inside the discrete lines. When it increases, discrete lines become thick. However, no good recognition algorithm have been published for arbitrary thick lines. The problem can be solved using Integer Linear Programming [7] as a black-box, using the Fourier-Motzkin elimination [14] when the thickness is a linear function of (a, b), or partially using ad-hoc techniques [15] where optimality could sometimes be reached.

It is easy to see that computing the arithmetical thickness is exactly the same problem than the one of computing the lattice width of the points of a discrete line. In the present paper, we use the lattice width to extract the directions of lines minimizing the arithmetical thickness. This permits to list all the directions which are solutions. As depicted in Fig. 1, arithmetical thickness can be minimized while geometrical thickness is not. Indeed, if d is the geometrical thickness of a set along the direction (a, b) and if ω is its arithmetical thickness along (a, b) then simple geometrical relations lead to the equation: $1 + d\sqrt{a^2 + b^2} = \omega$.

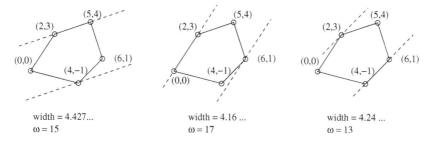

Fig. 1. Minimizing geometrical thickness does not minimize arithmetical thickness

3   Computing the Exact Lattice Width

3.1   Algorithm

We now propose an algorithm to solve our problem. The idea is based on the principle that the lattice width of K is necessarily reached for two *opposite* vertices of K. The meaning of *opposite* is given later in the description of the method. To solve the problem it is then sufficient to have both an efficient solution for two opposite vertices and an efficient way of considering vertices such that not all n^2 pairs are examined. We will show that we can examine $O(n)$ pairs of vertices and for each pair, the cost of computing all directions with minimal width is $O(\log s)$. This permits us to provide an algorithm with $O(n \log s)$ arithmetic operations for a convex polygon K. If the input of the algorithm is a set of points not necessarily convex, the computation cost becomes $O(n \log n + n \log s)$ arithmetic operations.

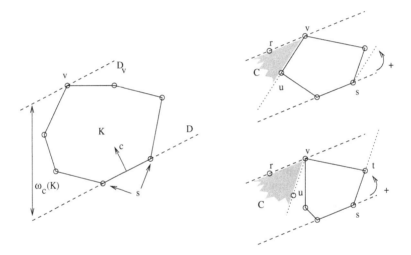

Fig. 2. (left) Supporting lines and width $\omega_c(K)$ (right) Cone of rotations

To define the notion of *opposite*, we rely on the notion of *supporting lines* well known in computational geometry [9]. A *supporting line* of K is a line D such that $D \cap K \neq \emptyset$ and K is contained entirely in one of the half-planes bounded by D. For each supporting line D, there exist at least one vertex v of K such the parallel line D_v to D passing through v is such that K entirely leads in the strip bounded by D and D_v. If s denotes a vertex of K belonging to D then s and v are called *opposite* (see Fig. 2, left). Opposite pairs are also called *antipodal* pairs. Note that in general, a supporting line intersects K at only one point. The supporting line D intersecting D along an edge is called *principal* supporting lines.

We now suppose K to be oriented counter clockwise. As in the classical Rotating Calipers algorithm of Toussaint [16], we can rotate the principal supporting lines D around the right vertex of $D \cap K$. D_v is also rotated around v to keep it parallel to D. This rotation can be pursued until D or D_v becomes another principal supporting line. Note that D and D_v are simply supporting lines during the rotation. At each position of the rotation s and v form an opposite pair of points which exactly define $w_c(K)$ where c is the normal direction to D. Hence, as depicted in Fig. 2 (right), s and v exactly define $w_c(K)$ for all D in a cone whose apex is v. The point r is such that the segment from v to r has exactly the same length than the opposite edge of K and the point u is either the next vertex of K after v or the point on the parallel of the line (st) such that the length of $[st]$ equals the length of $[vu]$.

After one turn around K, we have constructed at most $2n$ opposite pairs and associated cones. Hence, the number of cones is $O(n)$. Moreover, the series of cones forms a partition of all possible directions of computation for the lattice width taking into account that $w_c(K) = w_{-c}(K)$. Hence, as previously announced, the computation of the lattice width is reduced to the computation of the minimal value of $w_c(K)$ for each cone.

In each cone C, the computation of the minimal value of $w_c(K)$ is the computation of the shortest vectors for the dual norm. They are thus located at the vertices of the border of the convex hull of integer points except v inside the cone [10]. This set is also known as Klein's sail [17, 18, 19]. Note that to allow the possibility to find all solutions, repetitions in the convex envelope must be kept. To compute Klein's sail, we use an adapted version of the algorithm of Harvey [20] whose complexity is $O(\log s)$ arithmetic operations for constructing the sail. To bound the complexity of the search, we could also rely on the general theorem of Cook et al [21] which says that there exists at most $O((\log s)^{d-1})$ vertices in dimension d, a result also shown by Harvey [20] with an explicit example of the worst case in two dimension.

We now detail the adapted version of Harvey's algorithm [20] to our problem. First, if the vectors vr and vu are reducible via gcd division then we perform the reduction. Hence, for each vector, we can suppose that their x and y components are relatively prime. Suppose that $r = (a, b)$ and $u = (c, d)$.

Klein's sail is perfectly known [19] when considering points in the cone v, r, u when $v = (0, 0)$, $r = (q, p)$ and $u = (1, 0)$. Let $x = p/q$. The continued fraction [22] of x is the expansion,

$$a_0 + \cfrac{1}{a_1 + \cfrac{1}{a_2 + \cfrac{1}{a_3 + \dots}}} \tag{4}$$

where a_i's are called the partial quotients. The principal convergents of x are the rational approximations p_k/q_k obtained by truncating the continued fraction expansion of x after the k's term. There are recurrence relations to obtain all of them in $O(\log s)$ arithmetic operations and it is well known that the points of the sail are exactly the principal convergents with even index [22]. Let us introduce the matrix

$$T = \begin{bmatrix} \alpha + kd & \beta - kc \\ -d & c \end{bmatrix} \tag{5}$$

with α and β such that $\alpha c + \beta d = 1$ and $k \in \mathbb{Z}$. Then it is straightforward that T is unimodular with determinant $+1$ and $Tu = (1,0)$. We also have $Tr = (a\alpha + b\beta - k\Delta, \Delta)$ with $\Delta = bc - ad$. Hence, we can enforce Tr to have a positive x coordinates by setting, $k = \lfloor (a\alpha + b\beta - 1)/\Delta \rfloor$. The cone v, Tu and Tr leads to a sail computed from the principal convergents of the fraction defined by Tr. To get the original sails, we apply the transformation $T^{(-1)}$.

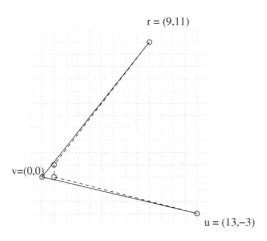

Fig. 3. A sail of a cone with irreducible support

To illustrate an example, let us consider the cone of Fig. 3. α and β are solution of $13\alpha - 3\beta = 1$. This leads to $\alpha = 1$ and $\beta = 4$. We also have $\Delta = 170$. Thus, $k = 0$. The image of u is $(1,0)$ has requested and the image of r is $(53, 170)$. The even principal convergents are respectively: $0/1$, $3/1$, $16/5$, $170/53$. The matrix $T^{(-1)}$ is the matrix,

$$\begin{bmatrix} 13 & -4 \\ -3 & 1 \end{bmatrix} \tag{6}$$

The image of the previous points by $T^{(-1)}$ are respectively: $(13, -3)$, $(1,0)$, $(1,1)$ and $(9, 11)$. All those points define the sail of the original cone.

To end our algorithm, we store for each cone its list of directions for which the minimum is reached. Then, we output all lists whose value is the global minimum. Hence, we obtain the exact lattice width as well as all directions for which it is attained. The computation of the global minimum obviously costs $O(n)$ arithmetic operations. Hence, the claim of a complexity in $O(n \log s)$ arithmetic operations is valid. We have solved the problem when $\Lambda = \mathbb{Z}^2$. This result naturally extends to the case where K is still an integer polygon and Λ is any full-dimensional lattice of $\mathbb{Z}^2$. Indeed, there exists a linear invertible transform M of $\mathbb{Z}^2$ such that $M\Lambda = \mathbb{Z}^2$. If we consider $K_M = M^{-1}K$ then $\omega_\Lambda(K) = \omega_{\mathbb{Z}^2}(K_M)$.

3.2   Example

In this part, we consider the example of Fig. 4. All points have been placed such that the coefficients of all the edges correspond to irreducible couple (a, b). We also provide on Fig. 4 all the sails obtained for one positive turn around the convex set. In each cone, we have depicted the sail and shown the arithmetical thickness associated to the shortest vector. Several comment could be done. First, the arithmetical thickness of each edge is very high such that the optimal solution could not correspond to direction given by an edge of the convex set (see the values of ω inside the convex set). Second, the minimal thickness corresponds to an horizontal discrete line followed by the solution corresponding to a vertical one. If we do not consider these specific cases, then all other cones provide solutions with the same arithmetical thickness. We hence obtained three vectors of direction leading to the same thickness. It should be noticed that the solution given by the sail on the lower right correspond to a solution that is only 14 percent higher in terms of geometrical thickness but more than three times lower in terms of arithmetical thickness than the solution only minimizing the geometrical thickness. This particular solution could then be considered as a compromise between both measures.

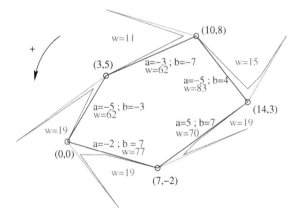

Fig. 4. A complete execution of the algorithm with all sails and the arithmetical thickness corresponding to the shortest vector

4   Incremental Construction

We now study an incremental version of the algorithm. The incremental construction of the convex hull can be done in $O(\log n)$ update complexity [23]. In this algorithm, the convex hull is stored in a concatenable queue. We now consider the adding of the new point p (see Fig. 5).

   The addition of p implies that two cones will be generated by the edges f_u and f_l adjacent to p in the new convex hull and other cones will be added when p is

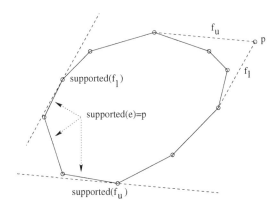

Fig. 5. Incremental construction of the cones

a supporting point of an edge e of the convex hull. The set of edges for which p is a supporting point, is a connected portion of the convex hull. This portion can be easily determined when the supporting points of f_u and f_l are known. Only the extremal edges need to be considered. Indeed if we consider the set of slopes of these edges, it is monotonous. Hence the corresponding cones at the vertex p are included one in the other. So to obtain the largest cones, it is sufficient to consider the extremal positions.

Only two new cones at most will be generated at the point p. Hence, we finally obtain that at most four new cones must be computed to determine the new lattice width. Since each computation costs $O(\log s)$, we obtain an update of the cones in $O(\log s)$. Now, to find the supporting points of f_u anf f_l, we use a binary search over the convex hull. This leads to a $O(\log n)$ complexity. At the end, the total complexity of the modification becomes $O(\log n + \log s)$ per update including the complexity of maintaining the convex hull.

5　Conclusion

Our paper raises the problem of the exact computation of the lattice width of any integer polygon K by a geometric partitioning of the space of possible directions. Moreover, all directions leading to the lattice width can be computed. In the incremental case, we provide an extension of the algorithm to maintain the lattice width. The whole set of solutions could also be obtained by enumerating the new cones introduced during a modification. An extension to $\mathbb{Z}^d$ is currently a work in progress.

References

1. Kannan, R.: A polynomial algorithm for the two variable integer programming problem. J. Assoc. Comput. Mach. **27** (1980) 118–122
2. Scarf, H.: Production sets with indivisibilities part i and part ii. Econometrica **49** (1981) 1–32, 395–423

3. Hirschberg, D., Wong, C.: A polynomial-time algorithm for the knapsack problem with two variables. J. Assoc. Comput. Mach. **23** (1976) 147–154
4. Lenstra, H.: Integer Programming with a Fixed Number of Variables. Math. Oper. Research **8** (1983) 535–548
5. Eisenbrand, F., Rote, G.: Fast 2-variable integer programming. In Aardal, K., Gerards, B., eds.: Integer Programming and Combinatorial Optimization. Volume 2081 of LNCS., Springer-Verlag (2001) 78–89
6. Eisenbrand, F., Laue, S.: A linear algorithm for integer programming in the plane. Math. Program. Ser. A **102** (2005) 249–259
7. Schrijver, A.: Theory of Linear and Integer Programming. John Wiley and Sons (1998)
8. Barvinok, A.: A Course in Convexity. Volume 54 of Graduates Studies in Mathematics. Amer. Math. Soc. (2002)
9. de Berg, M., Schwarzkopf, O., van Kreveld, M., Overmars, M.: Computational Geometry: Algorithms and Applications. Springer-Verlag (2000)
10. Kaib, M., Schnörr, C.P.: The Generalized Gauss Reduction Algorithm. Journal of Algorithms **21**(3) (1996) 565–578
11. Rote, G.: Finding a shortest vector in a two-dimensional lattice modulo m. Theoretical Computer Science **172**(1-2) (1997) 303–308
12. Reveillès, J.P.: Géométrie discrète, calcul en nombres entiers et algorithmique. Thèse d'etat, Université Louis Pasteur, Strasbourg, France (1991)
13. Debled-Rennesson, I., Reveillès, J.P.: A linear algorithm for segmentation of digital curves. In: International Journal on Pattern Recognition and Artificial Intelligence. Volume 9. (1995) 635–662
14. Françon, J., Schramm, J.M., Tajine, M.: Recognizing artimethic straight lines and planes. In Miguet, S., Montanvert, A., Ubéda, S., eds.: Discrete Geometry for Computer Imagery. Volume 1176 of LNCS., Berlin: Springer-Verlag (1996) 141–150
15. Gérard, Y., Debled-Rennesson, I., Zimmermann, P.: An elementary digital plane recognition algorithm. Discrete Applied Mathematics **151** (2005) 169–183
16. Houle, M., Toussaint, G.: Computing the width of a set. IEEE Trans. on Pattern Analysis and Machine Intelligence **10**(5) (1988) 761–765
17. Lachaud, G.: Klein polygons and geometric diagrams. Contemporary Math. **210** (1998) 365–372
18. Lachaud, G.: Sails and klein polyhedra. Contemporary Math. **210** (1998) 373–385
19. Arnold, V.: Higher dimensional continued fractions. Regular and chaotic dynamics **3** (1998) 10–17
20. Harvey, W.: Computing two-dimensional Integer Hulls. SIAM Journal on Computing **28**(6) (1999) 2285–2299
21. Cook, W., Hartman, M., Kannan, R., McDiarmid, C.: On integer points in polyhedra. Combinatorica **12** (1992) 27–37
22. Hardy, G., Wright, E.: An Introduction to the Theory of Numbers. Oxford University Press (1996)
23. Preparata, F.P., Shamos, M.I.: Computational Geometry : An Introduction. Springer-Verlag (1985)

Branch Voxels and Junctions in 3D Skeletons

Gisela Klette

CITR, University of Auckland, Tamaki Campus, Building 731
Auckland, New Zealand

Abstract. Branch indices of points on curves (introduced by Urysohn and Menger) are of basic importance in the mathematical theory of curves, defined in Euclidean space. This paper applies the concept of branch points in the 3D orthogonal grid, motivated by the need to analyze curve-like structures in digital images. These curve-like structures have been derived as 3D skeletons (by means of thinning). This paper discusses approaches of defining branch indices for voxels on 3D skeletons, where the notion of a junction will play a crucial role. We illustrate the potentials of using junctions in 3D image analysis based on a recent project of analyzing the distribution of astrocytes in human brain tissue.

Keywords: 3D skeletons, 3D curve analysis, branch nodes, branch index, thinning, medical image analysis, astrocytes.

1 Introduction and Basic Notions

Our theoretical studies on discrete versions of 3D branch indices have been initiated within a recent project about the analysis of confocal microscope images of human brain tissue. Those images are taken layer by layer and constitute a volume, which we assume to be defined within a regular orthogonal grid in 3D space. Figure 1 shows such a volume where 3D rendering has been used.

This 3D view clearly shows some type of "curve-like structures", which can be analyzed after segmentation, skeletonization, and property calculations for voxels on skeletons. Similar curve-like structures appear in other biomedical images such as, for example, in 3D scans of blood vessels, or in 3D ultrasound images. Long term observations in the School of Medicine at The University of Auckland produced the hypothesis that the number, distribution and "complexity" of astrocytes (i.e., brain cells whose shape resembles that of a star) are related to brain normality or defined types of abnormality (e.g., epilepsy). However, the intuitive concept of "complexity" requires a definition and quantitative studies. Following discussion with colleagues from the School of Medicine we decided to focus on the number, distribution and complexity of "junctions" of curve-like structures. This paper will provide a definition of such junctions, and discusses consequences of the chosen definition. The definition is an adaptation of basic concepts in 3D curve theory of Euclidean space.

We will point out at first that 3D topological thinning algorithms (see [10, 4]) deliver skeletons which consist of different "types" of voxels, characterized by branching index. Branching indices will then allow to cluster specific voxels into

U. Eckardt et al. (Eds.): IWCIA 2006, LNCS 4040, pp. 34–44, 2006.

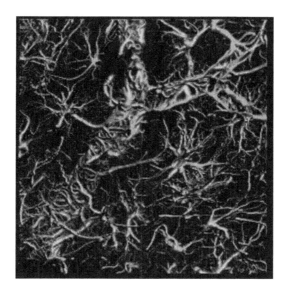

Fig. 1. Example of an input data set of 42 slices of 256×256 density images, all generated by confocal microscopy from a sample of human brain tissue: The astrocytes are partially located around a blood vessel which has approximately "Y-shape" (from lower left to upper right) in this sample

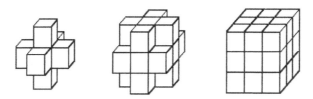

Fig. 2. Neighborhoods (left) $N_2(p)$, (middle) $N_1(p)$, and (right) $N_0(p)$

"junctions". We then map the 3D skeleton into an undirected graph where nodes are defined by junctions and endvoxels. We also propose ways of labeling this graph for supporting the quantitative analysis of curve-like structures. This will be illustrated for the shown sample (in Figure 1) of brain tissue.

We use common adjacency concepts: 4-, 8- (in 2D), 6-, 18-, 26- (in 3D) for the grid point model, and 0-, 1- (in 2D), 0-, 1-, 2- (in 3D) for the grid cell model (see Figure 2 for an illustration in the grid cell model), with notations as in [6]. Any of these adjacency relations A_α, $\alpha \in \{0, 1, 2, 4, 6, 8, 18, 26\}$, are irreflexive and symmetric. The α-*neighborhood* $N_\alpha(p)$ of a pixel or voxel location p includes p and its α-adjacent pixel or voxel locations.

Concepts for describing curve points in a continuous space are known for more than 80 years (see [6] for a review). P. Urysohn in 1923 and K. Menger in 1932 proposed (independently) equivalent definitions for simple curves (arcs) based on the notion of the branching index of a point on a curve (arc). The *branching index* of a point on a curve was defined as follows:

Definition 1. *Let p be a point, ε be a positive real, $U_\varepsilon(p)$ be the ε-neighborhood of p and $F(U_\varepsilon(p))$ be the frontier of $U_\varepsilon(p)$. A curve γ has branching index m $(m \geq 0)$ at $p \in \gamma$ if and only if for any $r > 0$, there is an $\varepsilon < r$ such that the cardinality of $F(U_\varepsilon(p)) \cap \gamma$ equals m.*

Figure 3 shows two examples where q has branching index 4 and p has branching index 2. It is obvious that the branching index of a curve point $p \in \gamma$ in the Euclidean space is the number of crossings of a circle (with radius $\varepsilon < r$ and middle point p) and curve γ. For all circles close enough to p this number has to be constant for defining a branching index.

A *simple curve* in the Euclidean space is a curve γ in which every point $p \in \gamma$ has branching index 2. A *simple arc* is either a curve in which every point p has branching index 2 except for two endpoints, which have branching index 1, or a simple curve with one of its points labeled as an endpoint.

An application of those concepts in the 3D digital space (based on 0-adjacency in the cell model, read if and only if for iff) may lead to the following definitions:

- digital curve ρ has *branching index* $m > 0$ at voxel $p \in \rho$ iff exactly m 0-adjacent voxels are elements of ρ;
- voxel $p \in \rho$ is *regular* iff p has branching index 2;
- $p \in \rho$ is a *branch voxel* iff p has a branching index of at least 3;
- $p \in \rho$ is an *endvoxel* iff p has branching index 1;
- $p \in \rho$ is a *singular voxel* iff p is either a branch voxel or an endvoxel;
- the digital curve ρ in 3D space is *simple* iff every voxel in ρ is regular; and
- ρ is a *simple arc* iff it is either a curve in which every voxel p is regular except for two endvoxels, or a simple curve where one of its voxels is labeled to be a (double) endvoxel.

This results into limitations of branching indices at voxels which restricts the generality of the concept.

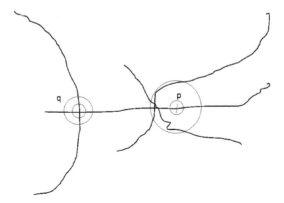

Fig. 3. $U_\varepsilon(q) \cap \gamma = 4$ and $U_\varepsilon(p) \cap \gamma = 2$, assuming that ε is sufficiently small

2   Junctions and Abstract Curve Graphs

For ensuring unlimited branching indices we introduce specific clusters of branch voxels. The need for unlimited branching indices occurred when studying 3D skeletons of curve-like structures, produced by a 3D topological thinning algorithm. See, for example, [10, 4] for a discussion of 3D skeletonization. (Those algorithms iteratively delete simple voxels until only non-simple voxels or endvoxels remain.) A 3D curve skeleton ρ is a digital curve, which we consider with respect to 0-adjacency.

Definition 2. *A 0-region of branch voxels of a digital curve ρ is called a junction. The branching index of a junction J in ρ is the number of regular voxels or end voxels in ρ being 0-adjacent to any one of the branch voxels in J.*

Figure 4 illustrates a junction which consists of three branch voxels. Note that a junction is a non empty 0-connected set of branch voxels. A single branch voxel also represents a junction (with cardinality one).

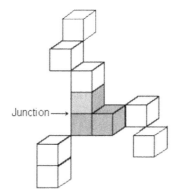

Fig. 4. Example of a junction containing three voxels

It follows that a junction has a branching index greater than 2. For example, the branching index of the junction shown in Figure 4 is 3. The complexity of a junction is measured by its branching index.

The following definition is useful for determining the geometric location of a junction. Let J be a junction, n be the number of branch voxels p_i constituting J, with $p_i = (x_i, y_i)$, $1 \leq i \leq n$. The *centroid* $c(J)$ of J is a 3D point with coordinates:

$$x = \frac{\sum_{i=1}^{n} x_{p_i}}{n}, \quad y = \frac{\sum_{i=1}^{n} y_{p_i}}{n}, \quad z = \frac{\sum_{i=1}^{n} z_{p_i}}{n} \quad (1)$$

We identify the geometric position of a junction with that of its centroid.

Definition 3. *A digital curve ρ is mapped into an undirected graph G, where a node of G is either a junction or an endvoxel of ρ. Two nodes in G are connected*

by an edge iff the corresponding junctions or endvoxels are 0-connected in ρ. G
is the abstract curve graph of ρ.

G is uniquely defined by the chosen adjacency (0-adjacency in our case). The
geometric positions of a junction or of an endvoxel define the geometric positions
of the nodes of G.

In experiments we assign indices to all nodes of the abstract curve graph.
All branch voxels of one junction obtain the same label this way. Edges of the
abstract curve graph correspond to digital arcs between junctions or endvoxels
of the curve. See Figure 5 for an example where the skeleton has been calculated
for a 3D brain tissue scan. Due to using only one label for all branch voxels of one
junction, different arcs may start from different branch voxels which all have the
same label. For example, B_{14} in Figure 5 consists of three branch voxels. Each of
them is an endvoxel of a digital arc. We use the geometric position (i.e., centroid)
of B_{14} as endpoint for all of those three arcs, for example for the calculation of
the Euclidean distance to other nodes (i.e., junctions or endvoxels).

A straightforward application of this convention allows the calculation of
Euclidean distances between nodes. For a more accurate estimation of distances
between nodes we apply length estimation based on connecting digital arcs.
First we identify the two endvoxels for each arc. This can be done by a second labeling process where all arcs are uniquely labeled. All voxels in one arc
obtain the same label; each branch voxel is mapped (say randomly) to exactly
one arc [5]. After these assignments, we apply a (global) DSS-based length measurement (see [6]), where we decided for the DSS-algorithm as published in [1].
The algorithm cuts an arc into a set of digital straight segments, and the total
length is the sum of the lengths of those segments. We have chosen the way of
DSS-based length estimations because (besides a general theoretical benefit of
being multigrid convergent to the true length) it also proved to be adequate for
characterizing the complexity of distributions of astrocytes.

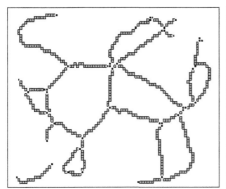 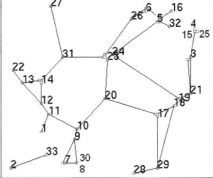

Fig. 5. Left: a skeleton (junctions are shown as black voxels). Right: abstract curve
graph for this skeleton (nodes are labeled by indices).

The distances between pairs of singular voxels can now be used for calculating weights for all edges in the abstract curve graph (using means in case of multiple arcs). Based on the cost matrix for this weighted graph together with the coordinates of all nodes we then applied traditional algorithms from graph theory (such as algorithms for calculating the minimum path between any two nodes, algorithms for determining the total weight of the minimum spanning tree, or algorithms for finding the diameter of the graph and so on) for further analysis of the curve-like structure in the given 3D image.

We determined the *uniformity* of junctions as follows: The volume data are divided into a set of subcubes (small cubes of identical size). For a fixed branching index j, we count the number of junctions in each cube having branching index j. If the number of junctions with branching index j is equal in every subcube then we say that junctions of branching index j are *uniformly distributed* in the whole volume. The deviation from this ideal case characterizes the degree of non-uniformity.

The division into subcubes can be fixed (a segmentation into pairwise disjoint subcubes), or there can be a sliding subcube of varying size. In the examples below we only illustrate the case of a fixed segmentation using pairwise disjoint, uniformly sized cubes of voxels. (Sliding subcubes experiments are not reported in this paper.)

For the description of the *density* of junctions we calculate the shortest path between (unordered) pairs of distinct junctions with the same branching index j. The shorter the path, the more *dense* are the junctions positioned in 3D space. The total number of junctions in a subcube is a (simple) expression of density of junctions in this subcube.

The data set shown in Figure 1 is divided into 36 subcubes, all of size 42^3. (This also generates some excessive data.) We have chosen this subdivision based on the sizes of given data sets and we had in mind that experts in the school of medicine have the hypothesis that there is a relationship between the number of astrocytes close to the main blood vessels and stages of epilepsy. See Figure 6 for the resulting curve (3D skeleton). We illustrate the approach by results for this example data set.

All identified junctions have branching indices between 3 and 7. The shaded cubes in Figure 7 correspond to the location of the main blood vessel, and they contain in total more than 50% of all junctions, for each branching index between three and seven. Table 1 presents the total number of junctions per branching index for all the gray cubes (volume V_1) and for the whole volume V_2.

We counted the number of junctions of equal types per cube to find out how they are distributed in the volume. Obviously, they are not (ideally) uniformly distributed (see Table 2) in the whole volume. Most of them are located close to the blood vessel.

The cardinality of junctions in this experiment did not exceed four and the maximum branching index did not exceed seven. The original structure of the image (elongated parts) and a range of preprocessing steps (segmentation and

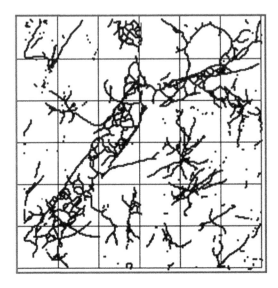

Fig. 6. 3D skeleton of the binarized volume shown in Figure 1

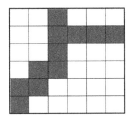

Fig. 7. Location of a main blood vessel (shown as gray cubes) detected by analyzing the 3D skeleton shown in Figure 6

Table 1. Number of junctions per branching index in the gray cubes (volume V_1) and in the total volume (volume V_2)

Branching index j	Junctions in V_1	Junctions in V_2	Ratio between V_1 and V_2
$j = 3$	150	276	54.3%
$j = 4$	53	85	62.4%
$j = 5$	16	21	76.2%
$j = 6$	5	7	71.4%
$j = 7$	2	2	100%
$3 \leq j \leq 7$	226	391	57.8%

noise reduction by a sequence of morphological operations) are reasons for keeping the cardinality and the branching index at low values. Theoretically, this is not always the case.

Table 2. Distribution of junctions in subcubes: The horizontal axis represents the numbers of subcubes, and the vertical axis represents the number of all junctions in a subcube

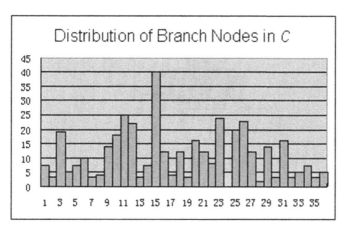

3   Properties of Junctions in 3D Curves

The branching index in the continuous space is defined for a single point $p \in \gamma$ of a curve γ. With above definitions we merge a set (i.e., a 0-connected region) of branch voxels into a single node in the abstract curve graph G. Interestingly, the size of this region can grow behind any limit.

The cardinality of a junction can grow if the image size or the grid resolution grows. Let us consider Figure 8.

 - All black voxels are non-regular with a branch index $m \geq 3$, and all white voxels are endvoxels (if the white voxels would be regular then the junction would not change).
 - If endvoxel q (as a grid cube) would share two more vertices with two more voxels, then q would change into a branch voxel. We could continue this process of adding two more voxels to one of the new endvoxels. As a consequence the junction would grow and the branching index could increase behind any limit.

The maximum branching index for a junction with cardinality one is eight; see Figure 9. We recall the concept of an attachment set to separate branch voxels into two types. The *frontier* of a voxel is the union of its six faces. A face of a voxel includes its four edges, and each edge includes its two vertices. Let p be an n-cell, $0 \leq n \leq 3$. The frontier of an n-cell p is a union of i-cells with $0 \leq i < n$ (i.e., excluding p itself). For example, if p is a voxel (a 3-cell) then the frontier consists of eight 0-cells, twelve 1-cells and six 2-cells. Kong [7] defined the I-attachment set of a cell p for the grid cell model as follows, where I is an image:

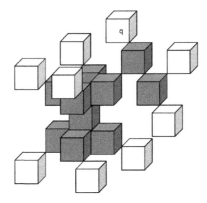

Fig. 8. A junction with cardinality 10 and $m = 9$

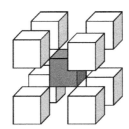

Fig. 9. A junction with cardinality 1 and $m = 8$

Definition 4. *Let p and q are grid cells. The I-attachment set of a n-cell p in I is the union of all i-cells, $0 \leq i < n$, on the frontier of p that also lie on frontiers of other grid cells q with $I(p) = I(q), p \neq q$.*

Let m be the number of voxels in $N_0(p) \bigcap S$ and n the number of components in the I-attachment set of p. A branch voxel p is called:

– a *proper branch voxel* if $m = n$,
– a *normal branch voxel* if $m > n$.

A junction is either a 0-region of normal branch voxels, or a proper branch voxel. It follows that a proper branch voxel is a junction of cardinality one. This definition splits the large junction in Figure 8 (for example) into three disjoint junctions. The black voxels (see Figure 10) represent a new junction with cardinality eight and $m = 7$. Voxels p and q are disjoint junctions with $m = 3$ each.

This approach increases the number and the density of junctions and it prevents junctions from growing in a certain direction. For the application of the DSS algorithm we use the centroid of each junction J as a node in the abstract curve graph G, and each 0-connected regular voxel or end voxel to J is a start (or end) voxel for the length measurement of a digital arc between two nodes. We do not use voxels in junctions as start or end voxel for the DSS algorithm.

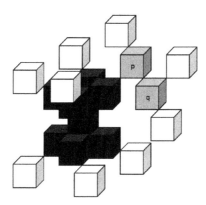

Fig. 10. Three 0-connected junctions, one formed by the dark gray voxels, and two defined by single voxels each (voxels p and q)

The length between two nodes is the calculated length of the arc between two nodes plus the Euclidean distance from the start voxel of the arc to the centroid of the 0-connected junction (if it is connected to a junction) plus the Euclidean distance from the end voxel of the arc to the centroid of the 0-connected junction (if it is connected to a junction).

4   Conclusions

In this paper we propose a classification of voxels in 3D skeletons for subsequent length measurements of digital arcs. The definition of branch voxels follows curve theory for the Euclidean space. Junctions are defined as 0-connected regions of branch voxels. These junctions and their properties are useful for the analysis of curve-like structures in biomedical images. An adjustment for the definition of junctions is introduced to prevent arbitrary growth and to improve the accuracy of length measurements using the DSS algorithm. The dependency between grid resolution and the discussed features is an interesting subject of future research.

Acknowledgment. The author acknowledges the work of her students, in particular of Mian Pan, who contributed to this work in his Master thesis project.

References

1. I. Debled-Rennesson and J.-P. Reveilles. A linear algorithm for segmentation of digital curves. *Int. J. Pattern Recognition Artificial Intelligence*, **9**:635–662, 1995.
2. G. Klette: A comparative discussion of distance transformations and simple deformations in digital image processing. *Machine Graphics & Vision*, **12**:235–256, 2003.
3. G. Klette: Simple points in 2D and 3D binary images. In Proc. *Computer Analysis Images Patterns*, LNCS 2756, pages 57–64, Springer, Berlin, 2003.

4. G. Klette and M. Pan: 3D topological thinning by identifying non-simple voxels. In Proc. *Int. Workshop Combinatorial Image Analysis*, LNCS 3322, pages 164–175, Springer, Berlin, 2004.
5. G. Klette and M. Pan: Characterization of curve-like structures in 3D medical images. In Proc. *Image Vision Computing New Zealand*, pages 164-175, 2005.
6. R. Klette and A. Rosenfeld. *Digital Geometry – Geometric Methods for Digital Picture Analysis*. Morgan Kaufmann, San Francisco, 2004.
7. T. Y. Kong: On topology preservation in 2-D and 3-D thinning. *Int. J. Pattern Recognition Artificial Intelligence*, **9**:813–844, 1995.
8. K. Palagyi, E. Sorantin, E. Balogh, A. Kuba, C. Halmai, B. Erdohelyi, and K. Hausegger: A sequential 3D thinning algorithm and its medical applications. In Proc. *Information Processing Medical Imaging*, LNCS 2082, pages 409–415, Springer, Berlin, 2001.
9. K. Palagyi and A. Kuba: Directional 3D thinning using 8 subiterations. In Proc. *Discrete Geometry Computational Imaging*, LNCS 1568, pages 325–336, Springer, Berlin, 2003.
10. K. Palagyi and A. Kuba: A 3D 6-subiteration thinning algorithm for extracting medial lines. *Pattern Recognition Letters*, **19**: 613–627, 1998.

New 2D Parallel Thinning Algorithms Based on Critical Kernels

Gilles Bertrand and Michel Couprie

Institut Gaspard-Monge
Laboratoire A2SI, Groupe ESIEE
Cité Descartes, BP 99
93162 Noisy-le-Grand Cedex France
g.bertrand@esiee.fr, m.couprie@esiee.fr

Abstract. Critical kernels constitute a general framework settled in the category of abstract complexes for the study of parallel thinning in any dimension. In this context, we propose several new parallel algorithms, which are both fast and simple to implement, to obtain symmetrical skeletons of 2D objects in 2D or 3D grids. We prove some properties of these skeletons, related to topology preservation, and to the inclusion of the topological axis which may be seen as a generalization of the medial axis.

1 Introduction

Forty years ago, in 1966, D. Rutovitz proposed an algorithm which is certainly the first parallel thinning algorithm [23]. Since then, many 2D parallel thinning algorithms have been proposed, see for example [25, 1, 19, 7, 11, 13, 9, 18]. A fundamental property required for such algorithms is that they do preserve the topology of the original objects. In fact, such a guarantee is not obvious to obtain, even for the 2D case, see [8] where some counter-examples are given.

In [3], one of the authors introduces a general framework for the study of parallel thinning in any dimension in the context of abstract complexes. A new definition of a simple point (a point which may be deleted without "changing the topology of the object") has been proposed, this definition is based on the collapse operation which is a classical tool in algebraic topology and which guarantees topology preservation. The most fundamental result proved in [3] is that, if a subset Y of X contains the critical kernel of X, then Y has the same topology as X.

In this paper, we focus on 2D structures in 2D and 3D spaces. We introduce the notions of *crucial faces and pixels* (Sec. 5, Sec. 6) which permit to make a link with the framework of digital topology [16]. Thanks to simple local characterizations, we are able to express thinning algorithms by the way of sets of masks, as in most papers related to parallel thinning. We introduce the formal definition of a *minimal symmetric skeleton*, and we propose an algorithm to compute it (Sec. 7). The quality of a curvilinear skeleton is often assessed by the fact that it contains, approximately or completely, the medial axis of the shape. We introduce the *topological axis* (Sec. 8), a generalization of the medial axis (which is not defined for the case of two-dimensional structures in discrete n-dimensional spaces, $n > 2$). In 2D, we propose a new parallel algorithm (Sec. 9)

U. Eckardt et al. (Eds.): IWCIA 2006, LNCS 4040, pp. 45–59, 2006.

to compute skeletons which are guaranteed to include the medial axis. We extend our algorithms to the 3D case by proposing a new algorithm to compute minimal symmetric skeletons of 2D objects in 3D grids, and also a new algorithm to compute skeletons of 2D objects in 3D grids which are guaranteed to contain the topological axis (Sec. 10).

For the sake of space, proofs are not given in this paper, most of them may be found in [3] or [5].

2 Cubical Complexes

In this section, we give some basic definitions for cubical complexes, see also [17]. We consider here only the two-dimensional case. The reader is invited to check that many of the notions introduced in the first sections make sense in arbitrary n-dimensional cubical spaces.

If T is a subset of S, we write $T \subseteq S$, we also write $T \subset S$ if $T \subseteq S$ and $T \neq S$.

Let $\mathbb{Z}$ be the set of integers. We consider the families of sets $\mathbb{F}_0^1$, $\mathbb{F}_1^1$, such that $\mathbb{F}_0^1 = \{\{a\} \mid a \in \mathbb{Z}\}$, $\mathbb{F}_1^1 = \{\{a, a+1\} \mid a \in \mathbb{Z}\}$. A subset f of $\mathbb{Z}^n$, $n \geq 2$, which is the Cartesian product of exactly m elements of $\mathbb{F}_1^1$ and $(n - m)$ elements of $\mathbb{F}_0^1$ is called a *face* or an *m-face* of $\mathbb{Z}^n$, m is the *dimension of f*, we write $dim(f) = m$.

We denote by $\mathbb{F}_2^n$ the set composed of all m-faces of $\mathbb{Z}^n$, $m = 0, 1, 2$ and $n \geq 2$. An m-face of $\mathbb{Z}^n$ is called a *point* if $m = 0$, a *(unit) interval* if $m = 1$, a *(unit) square* if $m = 2$.

In this paper, we will consider only 2D objects which are in 2D or 3D spaces. Thus, in the following, we suppose that $n = 2$ or $n = 3$.

Let f be a face in $\mathbb{F}_2^n$. We set $\hat{f} = \{g \in \mathbb{F}_2^n \mid g \subseteq f\}$ and $\hat{f}^* = \hat{f} \setminus \{f\}$.

Any $g \in \hat{f}$ is a *face of f*, and any $g \in \hat{f}^*$ is a *proper face of f*.

If X is a finite set of faces in $\mathbb{F}_2^n$, we write $X^- = \cup\{\hat{f} \mid f \in X\}$, X^- is the *closure of X*.

A set X of faces in $\mathbb{F}_2^n$ is a *cell* or an *m-cell* if there exists an m-face $f \in X$, such that $X = \hat{f}$. The *boundary of a cell* $\hat{f}$ is the set $\hat{f}^*$.

A finite set X of faces in $\mathbb{F}_2^n$ is a *complex (in $\mathbb{F}_2^n$)* if $X = X^-$. Any subset Y of a complex X which is also a complex is a *subcomplex of X*. If Y is a subcomplex of X, we write $Y \preceq X$. If X is a complex in $\mathbb{F}_2^n$, we also write $X \preceq \mathbb{F}_2^n$.

Let $X \preceq \mathbb{F}_2^n$. A face $f \in X$ is *principal for X* if there is no $g \in X$ such that $f \in \hat{g}^*$. We denote by X^+ the set composed of all principal faces of X.

Observe that, in general, X^+ is not a complex, and that $[X^+]^- = X$.

Let $X \preceq \mathbb{F}_2^n$, $dim(X) = \max\{dim(f) \mid f \in X^+\}$ is the *dimension of X*. We say that X is an *m-complex* if $dim(X) = m$.

We say that X is *pure* if, for each $f \in X^+$, we have $dim(f) = dim(X)$.

Let $X \preceq \mathbb{F}_2^n$ and $Y \preceq X$. If $Y^+ \subseteq X^+$, we say that Y is a *principal subcomplex* of X and we write $Y \sqsubseteq X$. Observe that, for any $X \preceq \mathbb{F}_2^n$, $\emptyset \sqsubseteq X$.

If $X \preceq \mathbb{F}_2^n$ and if X is a pure 2-complex, we also write $X \sqsubseteq \mathbb{F}_2^n$.

Let $X \preceq \mathbb{F}_2^n$ and let $Y \preceq X$. We set $X \oslash Y = [X^+ \setminus Y^+]^-$. The set $X \oslash Y$ is a complex which is the *detachment of Y from X*.

Two distinct faces f and g of $\mathbb{F}_2^n$ are *adjacent* if $f \cap g \neq \emptyset$. Two complexes X, Y in $\mathbb{F}_2^n$ are *adjacent* if there exist $f \in X$ and $g \in Y$ which are adjacent.

Let $X \preceq \mathbb{F}_2^n$. A sequence $\pi = \langle f_0, ..., f_l \rangle$ of faces in X is a *path in X (from f_0 to f_l)* if f_i and f_{i+1} are adjacent for each $i = 0, ..., l-1$; the number l is the *length of π*. We say that X is *connected* if, for any pair of faces (f, g) in X, there is a path in X from f to g. We say that $Y \preceq X$ is a *connected component of X* if $Y \subseteq X$, Y is connected, and if Y is maximal for these two properties (*i.e.*, we have $Z = Y$ whenever $Y \preceq Z \preceq X$ and Z connected).

Two 2-faces f and g of $\mathbb{F}_2^n$ are *strongly adjacent* if $f \cap g$ is a 1-face.

Let $X \sqsubseteq \mathbb{F}_2^n$. A sequence $\pi = \langle f_0, ..., f_l \rangle$ of 2-faces in X is a *strong path in X (from f_0 to f_l)* if f_i and f_{i+1} are strongly adjacent for each $i = 0, ..., l-1$; the number l is *the length of π*. We say that X is *strongly connected* if, for any pair of 2-faces (f, g) in X, there is a strong path in X from f to g.

If f is a 2-face of $\mathbb{F}_2^n$, we set:

$\Gamma^*(f) = \{g \in \mathbb{F}_2^n \mid g$ is a 2-face adjacent to $f\}$, $\Gamma(f) = \Gamma^*(f) \cup \{f\}$; and
$\Gamma_{\mathcal{S}}^*(f) = \{g \in \mathbb{F}_2^n \mid g$ is strongly adjacent to $f\}$, $\Gamma_{\mathcal{S}}(f) = \Gamma_{\mathcal{S}}^*(f) \cup \{f\}$.

3 Simple Cells

Intuitively a cell $\hat{f}$ of a complex X is simple if its removal from X "does not change the topology of X". In this section we propose a definition of a simple cell based on the operation of collapse [10], which is a discrete analogue of a continuous deformation (a homotopy). Note that this definition is a rather general one, in particular, it may be directly extended to n-dimensional cubical complexes [3].

Let X be a complex in $\mathbb{F}_2^n$ and let $f \in X^+$. The face f is a *border face for X* if there exists one face $g \in \hat{f}^*$ such that f is the only face of X which contains g. Such a face g is said to be *free for X* and the pair (f, g) is said to be a *free pair for X*. We say that $f \in X^+$ is an *interior face for X* if f is not a border face. In Fig. 1 (a), the pair (f, j) is a free pair for X, and the complex X has no interior face.

Let X be a complex, and let (f, g) be a free pair for X. The complex $X \setminus \{f, g\}$ is an *elementary collapse of X*.

Let X, Y be two complexes. We say that X *collapses onto Y* if there exists a *collapse sequence from X to Y*, *i.e.*, a sequence of complexes $\langle X_0, ..., X_l \rangle$ such that $X_0 = X$, $X_l = Y$, and X_i is an elementary collapse of X_{i-1}, $i = 1, ..., l$; the number l is the *length* of the collapse sequence. If X collapses onto Y, we also say that *Y is a retraction of X*. See illustration Fig. 1 (a), (b), (c).

We give now a definition of a simple point, it may be seen as a discrete analogue of the one given by T.Y. Kong in [15] which lies on continuous deformations in the n-dimensional Euclidean space.

Definition 1. Let $X \preceq \mathbb{F}_2^n$. Let $f \in X^+$.
We say that $\hat{f}$ and f are *simple for X* if X collapses onto $X \oslash \hat{f}$.

The notion of attachment, as introduced by T.Y. Kong [14, 15], leads to a local characterization of simple cells.

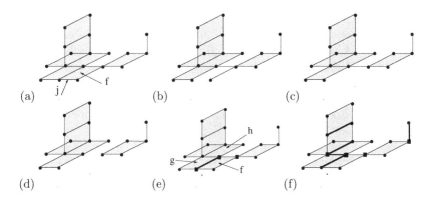

Fig. 1. (a) A complex X, (b) and (c) two steps of elementary collapse of X, (d) the detachment of $\hat{f}$ from X, (e) the attachment of the 2-face f is highlighted, the face f is not simple, whereas g and h are simple, (f) the essential 0- and 1-faces for X are highlighted

Definition 2. Let $X \preceq \mathbb{F}_2^n$ and let $f \in X^+$. The *attachment of* $\hat{f}$ *for* X is the complex $Attach(\hat{f}, X) = \hat{f}^* \cap [X \oslash \hat{f}]$.

In other words, a face g is in $Attach(\hat{f}, X)$ if g is in $\hat{f}^*$ and if g is a (proper) face of a principal face h distinct from f.

The following proposition is an easy consequence of the above definitions.

Proposition 3. *Let* $X \preceq \mathbb{F}_2^n$, *and let* $f \in X^+$.
The cell $\hat{f}$ *is simple for* X *if and only if* $\hat{f}$ *collapses onto* $Attach(\hat{f}, X)$.

The attachment of a 2-face f of a complex X is highlighted Fig. 1 (e) and $X \oslash \hat{f}$ is depicted in (d). It may be seen that f is not simple: there is no collapse sequence from X (a) to $X \oslash \hat{f}$ (d). On the other hand the faces g and h are simple. The next property may be directly derived from Prop. 3.

Proposition 4. *Let* $X \preceq \mathbb{F}_2^n$, *and let* $f \in X^+$.

1) If $\hat{f}$ *is a 0-cell, then* $\hat{f}$ *is not simple for* X;
2) If $\hat{f}$ *is a 1-cell, then* $\hat{f}$ *is simple for* X *if and only if* $Attach(\hat{f}, X)$ *is made of a single point;*
3) If $\hat{f}$ *is a 2-cell, then* $\hat{f}$ *is simple for* X *if and only if* f *is a border face and* $Attach(\hat{f}, X)$ *is non-empty and connected.*

From Prop. 4, we easily derive a characterization of simple 2-faces which is an equivalent, in the framework of 2D complexes in $\mathbb{F}_2^n$, of the well-known characterization of simple pixels in the square grid given by A. Rosenfeld [21].

Proposition 5. *Let* $X \sqsubseteq \mathbb{F}_2^n$, *and let* f *be a 2-face for* X. *The face* f *is simple for* X *if and only if:*

i) f *is a border face; and*
ii) $\Gamma^*(f) \cap X$ *is non-empty and connected.*

4   Critical Kernels

Let X be a complex in $\mathbb{F}_2^n$. We observe that, if we remove simultaneously simple cells from X, we may obtain a set Y such that X does not collapse onto Y. In other words, if we remove simple cells in parallel, we may "change the topology" of the original object X. For example, in Fig. 1 (e), g and h are simple for X, but the complexes X and $X \oslash [\hat{g} \cup \hat{h}]$ have not "the same topology" (here, the same number of connected components). Thus, it is not possible to use directly the notion of simple cell for thinning discrete objects in a symmetrical manner.

In this section, we introduce a new framework for thinning in parallel discrete objects with the warranty that we do not alter the topology of these objects. This method may be extended for complexes of arbitrary dimension [3]. As far as we know, this is the first method which allows to thin arbitrary complexes in a symmetric way.

This method is based solely on three notions, the notion of an *essential face* which allows to define the *core of a face*, and the notion of a *critical face*.

Definition 6. Let $X \preceq \mathbb{F}_2^n$ and let $f \in X$. We say that f is an *essential face* for X if f is precisely the intersection of all principal faces of X which contain f, i.e., if $f = \cap \{g \in X^+ \mid f \subseteq g\}$. We denote by $Ess(X)$ the set composed of all essential faces of X. If f is an essential face for X, we say that $\hat{f}$ is an *essential cell for X*.

Observe that a principal face for X is necessarily an essential face for X, i.e., $X^+ \subseteq Ess(X)$. The essential 0- and 1-faces of the complex X of Fig. 1 (a) are highlighted Fig. 1 (f).

Definition 7. Let $X \preceq \mathbb{F}_2^n$ and let $f \in Ess(X)$. The *core of $\hat{f}$ for X* is the complex, denoted by $Core(\hat{f}, X)$, which is the union of all essential cells for X which are in $\hat{f}^*$, i.e., $Core(\hat{f}, X) = \cup \{\hat{g} \mid g \in Ess(X) \cap \hat{f}^*\}$.

The preceding definition may be seen as a generalization of the notion of attachment for arbitrary essential cells (not necessarily principal).

Proposition 8. Let $X \preceq \mathbb{F}_2^n$ and let $f \in X^+$. The attachment of $\hat{f}$ for X is precisely the core of $\hat{f}$ for X, i.e, we have $Attach(\hat{f}, X) = Core(\hat{f}, X)$.

Definition 9. Let $X \preceq \mathbb{F}_2^n$ and let $f \in X$. We say that f and $\hat{f}$ are *regular for X* if $f \in Ess(X)$ and if $\hat{f}$ collapses onto $Core(\hat{f}, X)$. We say that f and $\hat{f}$ are *critical* for X if $f \in Ess(X)$ and if f is not regular for X.

We set $Critic(X) = \cup \{\hat{f} \mid f$ is critical for $X\}$, $Critic(X)$ is the *critical kernel* of X. A face f in X is a *maximal critical face*, or an *M-critical face* for X, if f is a principal face of $Critic(X)$.

Again, the preceding definition of a regular cell is a generalization of the notion of a simple cell. As a corollary of Prop. 8, a principal face of a complex $X \preceq \mathbb{F}_2^n$ is regular for X if and only if is simple for X.

We propose the following classification of critical faces which is specific to the 2D case.

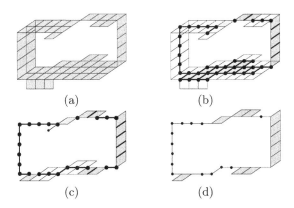

Fig. 2. (a): a complex X_0 in $\mathbb{F}_2^3$. (b): highlighted, $X_1 = Critic(X_0)$. (c): highlighted, $X_2 = Critic(X_1)$. (d): X_2 is such that $Critic(X_2) = X_2$.

Definition 10. Let $X \preceq \mathbb{F}_2^n$, and let $f \in Ess(X)$.

i) f is *0-critical for* X if $Core(\hat{f}, X) = \emptyset$;
ii) f is *1-critical for* X if $Core(\hat{f}, X)$ is not connected;
iii) f is *2-critical for* X if f is an interior 2-face.

Note that a face f is critical for $X \preceq \mathbb{F}_2^n$ if and only if f is k-critical for some $k \in \{0, 1, 2\}$.

The following theorem holds for complexes of arbitrary dimensions (see [3]), it may be proved quite in a simple manner in the 2D case (first, we collapse regular 2-faces onto their core, then we collapse regular 1-faces onto their core). This is our basic result in this framework. See Fig. 2 where the successive critical kernels of a complex are depicted.

Theorem 11. Let $X \preceq \mathbb{F}_2^n$. The critical kernel of X is a retraction of X. Furthermore, if $Y \sqsubseteq X$ is such that Y contains the critical kernel of X, then Y is a retraction of X.

5   Crucial Kernels

If X is a complex in $\mathbb{F}_2^n$, the subcomplex $Critic(X)$ is not necessarily a principal subcomplex of X as illustrated Fig. 2. In this paper we investigate thinning algorithms which take as input a pure 2-complex and which return a principal subcomplex of the input (thus also a pure 2-complex). In this section, we propose some notions which allow to recover a principal subcomplex Y of an arbitrary complex X, with the constraint that Y is a retraction of X.

Definition 12. Let $X \preceq \mathbb{F}_2^n$, and let $f \in X^+$ be a simple face for X.
We say that f is *crucial for* X, if $\hat{f}^*$ contains a face which is M-critical for X.

We say that f is *k-crucial for* X, if $\hat{f}^*$ contains an M-critical face which is k-critical for X, $k = 0, 1$.

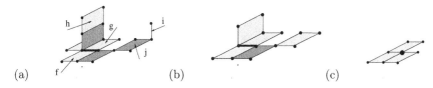

Fig. 3. (a): A complex X_0 and its M-critical faces (highlighted). (b): $X_1 = Cruc(X_0)$ and its M-critical faces. (c): The complex $X_2 = Cruc(X_1)$ contains only one M-critical face (highlighted), and $X_2 = Cruc(X_2)$.

Thus, a critical face for X is either a principal face (which is not simple) or is included in a crucial face (which is also simple and principal).

In Fig. 3 (a), the M-critical faces of a complex are highlighted. The faces f and g are crucial (1-crucial), the faces i and h are simple but not crucial (the critical faces included in i and h are not M-critical), the face j is not simple (it is M-critical), thus j is not crucial.

Definition 13. Let $X \preceq \mathbb{F}_2^n$, and let K be a set of crucial faces for X.
We say that K is *a (k-) crucial clique for X*, if there exists a (k-critical) face f which is M-critical for X and such that K is precisely the set of principal faces of X which contain f. We also say that K is the *crucial clique induced by f*.

In Fig. 3 (a), the set of faces $K = \{f, g\}$ is a 1-crucial clique, in (c) the set K' composed of the three 2-faces is a 0-crucial clique.

Definition 14. Let $X \preceq \mathbb{F}_2^n$ and let $Y \sqsubseteq X$.
We say that Y is a *crucial retraction of X* if:

 i) Y contains each principal face of X which is critical; and
 ii) Y contains at least one face of each crucial clique for X.

From the above definitions, we immediately derive the following property.

Proposition 15. *Let $X \preceq \mathbb{F}_2^n$ and let $Y \sqsubseteq X$.*
We have $Critic(X) \subseteq Y$ if and only if Y is a crucial retraction of X.

Thus, by Th. 11, if Y is a crucial retraction of X, then Y is a retraction of X. All algorithms proposed in this paper will iteratively compute crucial retractions.

Let us define the *crucial kernel of X* as the set $Cruc(X)$ which is the union of all cells of X which are either not simple for X or crucial for X. In Fig. 3 (a), a complex X_0 and its M-critical faces (three 2-faces and one 1-face) are depicted. The complex $X_1 = Cruc(X_0)$ is given in (b) also with its M-critical faces (one 2-face and one 1-face, which are both 1-critical). Finally, in (c), the complex $X_2 = Cruc(X_1)$ contains only one M-critical face (which is 0-critical), and it may be seen that $X_2 = Cruc(X_2)$.

For thinning objects, we often want to keep other faces than the ones which are either not simple or crucial. That is why we introduce the following definition.

Definition 16. Let $X \preceq \mathbb{F}_2^n$. Let P be a set of faces which are simple for X, and let $f \in P$. We say that f is *(k-) crucial for* $\langle X, P \rangle$, if f belongs to a *(k-)* crucial clique which is included in P $(k = 0, 1)$.

Intuitively, the set P corresponds to a set of faces which are candidate for deletion in parallel. The following definition may be seen as a "template" for our thinning algorithms (see the expression of all the algorithms proposed in the next sections). Here, the set K corresponds to a set which is preserved by a thinning algorithm (like extremities of curves, if we want to obtain a curvilinear skeleton).

Definition 17. Let $X \preceq \mathbb{F}_2^n$. Let K be a set of principal faces of X, let P be the set of faces in $X \setminus K$ which are simple for X, and let R be the set composed of all faces which are crucial for $\langle X, P \rangle$. The set $[X^+ \setminus P]^- \cup R^-$ is the *crucial kernel of X constrained by K*.

From the previous definitions, we immediately deduce the following proposition which ensures that any constrained crucial kernel preserves topology.

Proposition 18. *Let $X \preceq \mathbb{F}_2^n$, and let K be a set of principal faces of X. The crucial kernel of X constrained by K is a crucial retraction of X.*

6   Crucial Pixels in the Square Grid

We introduce the following definitions in order to establish a link between planar pure complexes (*i.e.*, pure 2-complexes in $\mathbb{F}_2^2$) and the square grid as considered in image processing.

We define the *square grid* as the set $\mathbb{G}^2$ composed of all 2-faces of $\mathbb{F}_2^2$. A 2-face of $\mathbb{G}^2$ is also called a *pixel*. In the sequel, we consider only finite subsets of $\mathbb{G}^2$.

For any pure 2-complex in $\mathbb{F}_2^2$, *i.e.*, for any $X \sqsubseteq \mathbb{F}_2^2$, we associate the subset X^+ of $\mathbb{G}^2$. In return, to each finite subset S of $\mathbb{G}^2$, we associate the complex S^- of $\mathbb{F}_2^2$. This will be our basic methodology to "interpret" a set of pixels. In particular, all definitions given for a principal face in X^+ have their counterparts for a pixel in $\mathbb{G}^2$. For example if $S \subseteq \mathbb{G}^2$ and $p \in S$, we will say that the pixel p is *simple for S* if p is simple for S^-. Border, interior, (k-) critical, and (k-) crucial pixels are defined in the same manner. Observe that, if $p \in \mathbb{G}^2$, $\Gamma^*(p)$ and $\Gamma_S^*(p)$ correspond to the so-called 8- and 4-neighborhood of p, respectively.

We give now some simple local conditions, in the square grid, for crucial pixels. We express these local conditions by a set of masks, as in most papers related to parallel thinning in the digital topology framework. The definition of the masks $C, C_1, ..., C_4$ is given Fig. 4.

Proposition 19. *Let $S \subseteq \mathbb{G}^2$, $p \in S$, and let P be a set of simple pixels of S.*

i) The pixel p is 1-crucial for $\langle S, P \rangle$ if and only if p is matched by pattern C;
ii) The pixel p is 0-crucial for $\langle S, P \rangle$ if and only if p is matched by one of the patterns $C_1, ...C_4$.

Using the terminology of section 5, the mask C is a mask for 1-crucial cliques, and $C_1, ..., C_4$ are masks for 0-crucial cliques. For each of these masks, the crucial

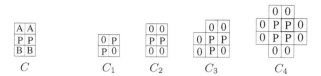

Fig. 4. Patterns and masks for crucial pixels. The 11 masks corresponding to these 5 patterns are obtained from them by applying any series of $\pi/2$ rotations. The label 0 indicates pixels that must belong to the complement of S. The label P indicates pixels that must belong to the set P which is a set composed of simple pixels of S. For mask C, at least one of the pixels marked A and at least one of the pixels marked B must be in S. If one of these masks matches the sets S, P, then all the pixels which correspond to a label P in the mask are recorded as "matched".

clique is the set composed of P's. In fact, it can be shown [5] that these masks also characterize the minimal non-simple sets introduced by C. Ronse [20], see also [12, 14]. We observe that, since P is composed of simple pixels of S, the set of P's of each mask $C_1, ..., C_4$ is necessarily surrounded by 0's. Thus, we have:

Proposition 20. *Let $S \subseteq \mathbb{G}^2$, and let K be a 0-crucial clique for S. Then K is a connected component of S.*

7   Minimal $\mathcal{K}$-Skeletons

A minimal symmetric skeleton of an object may be obtained by deleting iteratively, in parallel, all pixels which are neither critical nor crucial.

Definition 21. *Let $S \subseteq \mathbb{G}^2$. The* crucial kernel of S *is the set $Cruc(S)$ which is composed of all critical pixels and all crucial pixels of S.*

Let $\langle S_0, S_1, ..., S_k \rangle$ *be the unique sequence such that $S_0 = S$, $Cruc(S_k) = S_k$ and $S_i = Cruc(S_{i-1})$, $i = 1, ..., k$. The set S_k is the* minimal $\mathcal{K}$-skeleton of S.

By Prop. 18 (here $K = \emptyset$), the minimal $\mathcal{K}$-skeleton of a set S is a retraction of S. The following algorithm computes a minimal $\mathcal{K}$-skeleton. The pixels of S which are kept at each step (04) of the algorithm correspond precisely to the pixels which are either critical (the set $S \setminus P$) or crucial (the set R).

> **Algorithm** MK_a^2 (**Input** /**Output** : a set $S \subseteq \mathbb{G}^2$)
> 01. **Repeat Until Stability**
> 02. $P \leftarrow$ set of pixels which are simple for S
> 03. $R \leftarrow$ set of pixels in P which are 0- or 1-crucial for S
> 04. $S \leftarrow [S \setminus P] \cup R$

From Prop. 19, we may check if a pixel is 1-crucial by using the pattern C. Considering all possible rotations, there are in fact only two masks corresponding to C. On the other hand it may be seen that the checking of a 0-crucial pixel with the patterns $C_1, ..., C_4$ involves 9 masks. In the following, we propose an algorithm which avoids the use of these 9 masks. This algorithm is based on

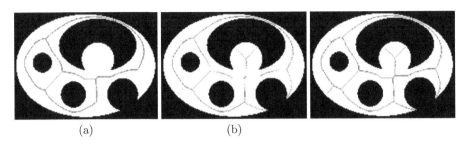

(a)   (b)

Fig. 5. (a): A subset S of $\mathbb{G}^2$ (in white) and its minimal $\mathcal{K}$-skeleton (in gray). (b): The medial axis of S (in gray). (c): in gray, $AK^2(S)$.

a technic used for computing the so-called *ultimate erosions* in the context of mathematical morphology (see [24]).

Let $S \subseteq \mathbb{G}^2$, we denote by $S \ominus \Gamma^* = \{p \in S \mid \Gamma^*(p) \subseteq S\}$, *the erosion of S by Γ^**, and by $S \oplus \Gamma^* = \cup\{\Gamma^*(p) \mid p \in S\}$, *the dilation of S by Γ^**.

> **Algorithm MK^2 (Input /Output** : a set $S \subseteq \mathbb{G}^2$)
> **01. Repeat Until Stability**
> 02. $P \leftarrow$ set of pixels which are simple for S
> 03. $R \leftarrow$ set of pixels in P which are 1-crucial for S
> 04. $T \leftarrow [S \setminus P] \cup R$
> 05. $S \leftarrow T \cup [S \setminus (T \oplus \Gamma^*)]$

The correctness of the algorithm lies on the following property.

Proposition 22. *Let $S \subseteq \mathbb{G}^2$, and let $p \in S$ be a simple pixel.*

i) If p is not crucial for S, then there exists $q \in \Gamma^*(p) \cap S$ such that q is either critical or 1-crucial for S.

ii) If p is 0-crucial for S, then any $q \in \Gamma^*(p) \cap S$ is neither critical, nor 1-crucial.

Let us denote by $MK^2(S)$ the result obtained by algorithm MK^2 from the input S. The set T (line 04) is the set of pixels which are either critical or 1-crucial. From Prop. 22, the pixels which are added to the set T at step 05 of MK^2 are precisely 0-crucial pixels. Thus, we have the following property.

Proposition 23. *Let $S \subseteq \mathbb{G}^2$. The set $MK^2(S)$ is the minimal $\mathcal{K}$-skeleton of S.*

An example of a minimal $\mathcal{K}$-skeleton is given Fig. 5 (a). As far as we know, MK^2 is the first algorithm for a minimal symmetric skeleton. Furthermore, the result of MK^2 is an object which is well-defined. To our best knowledge, this is also the first attempt to give a precise definition of such a notion.

8   Topological Axis and Medial Axis

The quality of a curvilinear skeleton is often assessed by the fact that it contains, approximately or completely, the medial axis of the shape. We introduce the following definitions in order to generalize the medial axis for pure 2-complexes in $\mathbb{F}_2^n$, for arbitrary n.

Definition 24. Let $X \sqsubseteq \mathbb{F}_2^n$, and let $f \in X^+$. We set $\rho(f, X)$ as the minimum length of a collapse sequence of X necessary to remove f from X, if such a sequence exists, and $\rho(f, X) = \infty$ otherwise. We define the *topological axis of X* as the set of faces f in X^+ such that $\rho(f, X) = \infty$ or $\rho(f, X) \geq \max\{\rho(g, X) \mid g \in \Gamma_S^*(f) \text{ and } \rho(g, X) \neq \infty\}$.

Note that we have $\rho(f, X) = 1$ if and only if f is a border face for X.

Let $X \sqsubseteq \mathbb{F}_2^n$, and let $f \in X^+$. We denote by $\pi'(f, X)$ the length of a shortest strong path, in X, from f to a border face of X, if such a path exists, and $\pi'(f, X) = \infty$ otherwise. We denote by $\pi(f, X)$ the length of a shortest strong path, in $\mathbb{F}_2^n$, from f to a border face of X. We observe that $\rho(f, X) = \pi'(f, X) + 1$.

Now we focus our attention on the case $n = 2$. Let $X \sqsubseteq \mathbb{F}_2^2$, and let $f \in X^+$. We have necessarily $\rho(f, X) \neq \infty$. Furthermore, since any 1-face in $\mathbb{F}_2^2$ is included in precisely two 2-faces, it may be seen that $\pi(f, X) = \pi'(f, X)$, thus $\rho(f, X) = \pi(f, X) + 1$.

In [22], A. Rosenfeld and J.L. Pfaltz have proved that, for the city-block and the chessboard distance, the medial axis of a shape can be obtained by detecting the local maxima of its distance transform, the medial axis being defined as the set of the centers of all the maximal balls for S. From the definition of the topological axis, and from the preceding remarks, we may deduce that the medial axis of S with the city-block distance is precisely the topological axis of S^-. This shows that the notion of topological axis indeed generalizes the one of medial axis (which is not defined for the case of two dimensional structures in discrete n-dimensional spaces, $n > 2$).

9 $\mathcal{K}$-Skeletons and Medial Axis

For obtaining a skeleton which includes the medial axis of an object, we define the following notion of $\mathcal{K}$-skeleton which is constrained to include a given set K.

Definition 25. Let $S \subseteq \mathbb{G}^2$ and let $K \subseteq S$. Let P be the set composed of all simple pixels for S which are not in K. We denote by $Cruc(S, K)$ the set composed of all pixels in $S \setminus P$ and all pixels which are crucial for $\langle S, P \rangle$.

Let $\langle S_0, S_1, ..., S_k \rangle$ be the unique sequence such that $S_0 = S$, $S_k = Cruc(S_k, K)$ and $S_i = Cruc(S_{i-1}, K)$, $i = 1, ..., k$. The set S_k is the *$\mathcal{K}$-skeleton of S constrained by K*.

Again, by Prop 18, the $\mathcal{K}$-skeleton of a set S constrained by a set K is a retraction of S. We give now a general result on constrained thinning which permits, under some conditions, to avoid the checking of the 9 masks (corresponding to $C_1, ..., C_4$) for the detection of 0-crucial pixels. This result is a direct consequence of Prop. 20.

Proposition 26. *Let $S \subseteq \mathbb{G}^2$. Let $K \subseteq S$, such that each connected component of S contains at least one pixel of K, and let P be the set composed of all simple pixels for S which are not in K. Then, any $p \in P$ is necessarily not 0-crucial for $\langle S, P \rangle$.*

For computing a $\mathcal{K}$-skeleton constrained by the medial axis, we could first extract the medial axis, and then compute the constrained skeleton, this method is followed by B.K. Jang and R.T. Chin [13]. We present here an algorithm which computes at the same time the medial axis and the skeleton.

> **Algorithm** AK^2 (**Input /Output** : set $S \subseteq \mathbb{G}^2$)
> 00. $K \leftarrow \emptyset$; $T \leftarrow S$
> 01. **Repeat Until Stability**
> 02. $E \leftarrow T \ominus \Gamma_S$; $D \leftarrow T \setminus [E \oplus \Gamma_S]$; $T \leftarrow E$; $K \leftarrow K \cup D$
> 03. $P \leftarrow$ set of pixels of $S \setminus K$ which are simple for S
> 04. $R \leftarrow$ set of pixels in P which are 1-crucial for $\langle S, P \rangle$
> 05. $S \leftarrow [S \setminus P] \cup R$

If we denote by $AK^2(S)$ the result obtained by algorithm AK^2, we have:

Proposition 27. *Let $S \subseteq \mathbb{G}^2$. The set $AK^2(S)$ is the $\mathcal{K}$-skeleton of S constrained by the topological axis of S.*

In Fig. 5, we show a subset S of $\mathbb{G}^2$ together with its topological (medial) axis (b) and its medial $\mathcal{K}$-skeleton (c). As far as we know, AK^2 is the first algorithm for a symmetric skeleton which contains the medial axis.

10 $\mathcal{K}$-Skeletons of 2D Objects in 3D Grids

We consider in this section objects which are pure 2-complexes in $\mathbb{F}_2^3$. We denote by $\mathbb{G}_2^3$ the set composed of all 2-faces of $\mathbb{F}_2^3$. A 2-face of $\mathbb{G}_2^3$ is also called a *surfel*. In the sequel, we consider only finite subsets of $\mathbb{G}_2^3$.

As for the square grid, definitions of principal faces of $\mathbb{F}_2^3$ have their counterparts in $\mathbb{G}_2^3$. For example, if $S \subseteq \mathbb{G}_2^3$ and $p \in S$, we say that the surfel p *is simple for S* if p is simple for S^-.

In the square grid, we were able to give a combinatorial characterization of 0- and 1-crucial pixels. In fact, the number of configurations for 0-crucial surfels is too high for being directly exhibited. Fortunately, such a characterization is not mandatory to implement parallel thinning operators based on crucial kernels. It is possible to have a characterization for 1-crucial surfels which is based solely on the pattern D given Fig. 6.

Proposition 28. *Let $S \subseteq \mathbb{G}_2^3$, $p \in S$. Let P be a set of simple surfels of S. The surfel p is 1-crucial for $\langle S, P \rangle$ if and only if p is matched by the pattern D.*

The following algorithm computes a minimal $\mathcal{K}$-skeleton, it has exactly the same structure as algorithm MK^2 for a square grid, but here, the checking of 1-crucial elements is made with the mask D.

> **Algorithm** MK_2^3 (**Input /Output** : a set $S \subseteq \mathbb{G}_3^2$)
> 01. **Repeat Until Stability**
> 02. $P \leftarrow$ set of surfels which are simple for S
> 03. $R \leftarrow$ set of surfels in P which are 1-crucial for S
> 04. $T \leftarrow [S \setminus P] \cup R$
> 05. $S \leftarrow T \cup [S \setminus (T \oplus \Gamma^*)]$

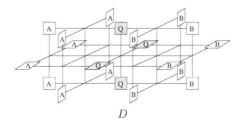

D

Fig. 6. Pattern and masks for 1-crucial surfels. The masks corresponding to this pattern are obtained by applying any series of $\pi/2$ rotations. The label Q indicates surfels that must either be in P or in the complement of S; at least two surfels labeled Q must be in P. At least one of the surfels marked A and at least one of the surfels marked B must be in S. If one of these masks matches the sets S, P, then all the surfels of P which correspond to a label Q are recorded as "matched".

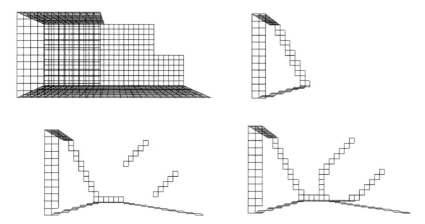

Fig. 7. Top left: A set of surfels S in $\mathbb{F}_2^3$. Top right: The minimal $\mathcal{K}$-skeleton of S. Bottom left: The topological axis of S. Bottom right: The result of algorithm BK_2^3.

The topological soundness of the algorithm may be proved by establishing the analogue of Prop. 22 in $\mathbb{G}_2^3$. An example of a minimal $\mathcal{K}$-skeleton is given Fig. 7. As far as we know, MK_2^3 is the first algorithm for a minimal symmetric skeleton for an object made of surfels.

In a similar way, algorithm AK^2 may be transposed to design an algorithm which produces a skeleton containing the topological axis of an object which is made of surfels. We give here another example of an algorithm which has such a property.

Algorithm BK_2^3 (Input /Output : set $S \subseteq \mathbb{G}_2^3$)
00. $T \leftarrow S$
01. **Repeat Until Stability**
02. $T \leftarrow \{s \in T \mid s$ is an interior surfel of $T\}$
03. $P \leftarrow$ set of simple surfels for S such that $\Gamma_S^*(p) \cap T \neq \emptyset$
04. $R \leftarrow$ set of surfels in P which are 1-crucial for $\langle S, P \rangle$
05. $S \leftarrow [S \setminus P] \cup R$

It may be seen that $BK_2^3(S)$, the result obtained by BK_2^3 from the input S, contains the topological axis of S. The topological soundness may be proved by establishing the equivalent of Prop. 26 in $\mathbb{G}_2^3$.

An example of a skeleton obtained with BK_2^3 is given Fig. 7. To our best knowledge, there is only one other algorithm for symmetric curvilinear skeletons of 2D objects in 3D spaces which is the one given by J. Burguet and R. Malgouyres [6]. This algorithm is based on the technic of P-simple points [2]. The 2D objects which are considered are the sets of surfels which constitute the boundary of 3D objects, or subsets of such boundaries. In this context, surfels which share a point or an interval are not necessarily considered as adjacent which makes a difference with the notion of adjacency used in this section. Another difference is that our algorithm always produce a skeleton which contains the topological axis of the original object.

11   Conclusion

Based on the framework of critical kernels [3], we studied the case of 2D structures in 2D and 3D grids. The salient outcomes of this article are the following:

- the definition and some characterizations of crucial faces, allowing for fast and simple implementations,
- the definition and an algorithm for a minimal symmetric skeleton (MK^2),
- the introduction of the topological axis, which generalizes the medial axis,
- a parallel algorithm for a symmetric skeleton which contains the medial axis,
- a parallel algorithm for a minimal symmetric skeleton of an object made of surfels,
- a parallel algorithm for a symmetric skeleton, which contains the topological axis of an object made of surfels.

As far as we know, all the above algorithms have no equivalent.

In future works, we will study the case of general skeletons (*i.e.*, which are not necessarily principal subcomplexes), and the important case of parallel thinning of 3D objects [4].

References

[1] C. Arcelli, L.P. Cordella, and S. Levialdi. Parallel thinning of binary pictures. *Electronic Letters*, 11(7):148–149, 1975.
[2] G. Bertrand. On P-simple points. *Comptes Rendus de l'Académie des Sciences, Série Math.*, I(321):1077–1084, 1995.
[3] G. Bertrand. On critical kernels. *Internal Report*, Université de Marne-la-Vallée, IGM2005-05, 2005. Also submitted for publication.
[4] G. Bertrand and M. Couprie. Three-dimensional parallel thinning algorithms based on critical kernels. *In preparation*.
[5] G. Bertrand and M. Couprie. Two-dimensional parallel thinning algorithms based on critical kernels. *Internal Report*, Université de Marne-la-Vallée, IGM2006-02, 2006. Also submitted for publication.

 [6] J. Burguet and R. Malgouyres. Strong thinning and polyhedric approximation of the surface of a voxel object. *Discrete Applied Mathematics*, 125:93–114, 2003.

 [7] R.T. Chin, H.K. Wan, D.L. Stover, and R.D. Iverson. A one-pass thinning algorithm and its parallel implementation. *Computer Vision, Graphics, and Image Processing*, 40(1):30–40, October 1987.

 [8] M. Couprie. Note on fifteen 2d parallel thinning algorithms. *Internal Report*, Université de Marne-la-Vallée, IGM2006-01, 2005.

 [9] U. Eckhardt and G. Maderlechner. Invariant thinning. *International Journal of Pattern Recognition and Artificial Intelligence*, 7(5):1115–1144, 1993.

[10] P. Giblin. *Graphs, surfaces and homology*. Chapman and Hall, 1981.

[11] R.W. Hall. Fast parallel thinning algorithms: Parallel speed and connectivity preservation. *Communication of the ACM*, 32(1):124–131, January 1989.

[12] R.W. Hall. Tests for connectivity preservation for parallel reduction operators. *Topology and its Applications*, 46(3):199–217, 1992.

[13] B.K. Jang and R.T. Chin. Reconstructable parallel thinning. *Pattern Recognition and Artificial Intelligence*, 7:1145–1181, 1993.

[14] T. Y. Kong. On topology preservation in 2-d and 3-d thinning. *International Journal on Pattern Recognition and Artificial Intelligence*, 9:813–844, 1995.

[15] T. Y. Kong. Topology-preserving deletion of 1's from 2-, 3- and 4-dimensional binary images. In *Lecture Notes in Computer Science*, volume 1347, pages 3–18, 1997.

[16] T. Y. Kong and A. Rosenfeld. Digital topology: introduction and survey. *Comp. Vision, Graphics and Image Proc.*, 48:357–393, 1989.

[17] V.A. Kovalevsky. Finite topology as applied to image analysis. *Computer Vision, Graphics and Image Processing*, 46:141–161, 1989.

[18] A. Manzanera and T.M. Bernard. Metrical properties of a collection of 2d parallel thinning algorithms. In *Electronic Notes on Discrete Mathematics, Proc. 9th IWCIA*, volume 12, 2003.

[19] T. Pavlidis. An asynchronous thinning algorithm. *Computer Graphics and Image Processing*, 20(2):133–157, October 1982.

[20] C. Ronse. Minimal test patterns for connectivity preservation in parallel thinning algorithms for binary digital images. *Discrete Applied Mathematics*, 21(1):67–79, 1988.

[21] A. Rosenfeld. Connectivity in digital pictures. *Journal of the Association for Computer Machinery*, 17:146–160, 1970.

[22] A. Rosenfeld and J.L. Pfaltz. Sequential operations in digital picture processing. *Journal of the Association for Computer Machinery*, 13:471–494, 1966.

[23] D. Rutovitz. Pattern recognition. *Journal of the Royal Statistical Society*, 129:504–530, 1966.

[24] J. Serra. *Image analysis and mathematical morphology*. Academic Press, 1982.

[25] R. Stefanelli and A. Rosendeld. Some parallel thinning algorithms for digital pictures. *Journal of the Association for Computing Machinery*, 18(2):255–264, 1971.

Grayscale Watersheds on Perfect Fusion Graphs

Jean Cousty, Michel Couprie, Laurent Najman, and Gilles Bertrand

Institut Gaspard-Monge
Laboratoire A2SI, Groupe ESIEE
Cité Descartes, BP99 93162 Noisy-le-Grand Cedex France
{j.cousty, m.couprie, l.najman, g.bertrand}@esiee.fr

Abstract. In this paper, we study topological watersheds on perfect fusion graphs, an ideal framework for region merging. An important result is that contrarily to the general case, in this framework, any topological watershed is thin.

Then we investigate a new image transformation called C-watershed and we show that, on perfect fusion graphs, the segmentations obtained by C-watershed correspond to segmentations obtained by topological watersheds. Compared to topological watershed, a major advantage of this transformation is that, on perfect fusion graph, it can be computed thanks to a simple linear-time immersion-like algorithm. Finally, we derive characterizations of perfect fusion graphs based on thinness properties of both topological watersheds and C-watersheds.

1 Introduction

Region merging methods [1] consist of improving an initial segmentation by merging some pairs of neighboring regions. The watershed transform [2, 3, 4, 5] produces a set of connected regions separated by a divide. Therefore it has long been used as an entry point for region merging methods [6]. In [7], we developed a theoretical framework for the study of merging properties in graphs. A *(binary) watershed set* is a set of vertices which cannot be reduced without changing the number of connected components of its complementary. It models a frontier in a graph. In the general case such a watershed set can be thick and thus the induced region neighboring relationship, used by further merging procedures, can lack important properties.

An original approach to grayscale watershed [5, 8, 9, 10] consists of modifying the original image by lowering some points while preserving the connectivity of each lower section. Such a transformation (and its result) is called a W-thinning, a *topological (grayscale) watershed* being an "ultimate" W-thinning. In [8, 10], the authors prove that the only lowering transformation which preserves the connection value (a notion of contrast) between the minima of the original image is precisely the W-thinning. Due to this contrast preservation property, the divide (*e.g.*, the points not in any minimum) of a topological watershed is an interesting segmentation of the original image. Furthermore, this contrast preservation property is necessary for the correctness of many region merging methods based on watersheds (see [11, 12] for examples of such methods).

U. Eckardt et al. (Eds.): IWCIA 2006, LNCS 4040, pp. 60–73, 2006.

An important result in [7] is that the class of all graphs in which any binary watershed set is thin is precisely the class of graphs in which any region can always be merged. Any element in this class is called a fusion graph. Surprisingly, the divides produced by watershed algorithms [2, 3, 4] and in particular by topological watershed algorithms [9], are not always binary watershed sets and can sometimes be thick, even on fusion graphs.

Therefore, in this paper, we consider a more restricted class of graphs called perfect fusion graph [7] which constitutes an ideal framework for region merging. An important result is that, on perfect fusion graphs, the divide of any topological watershed is a thin binary watershed set. The algorithms to compute topological watershed are not linear and require the computation of an auxiliary data structure called component tree [13]. Therefore, we investigate a new grayscale transformation: the *C-watershed*. Our main contributions concerning C-watersheds are the following:

1) We prove that, contrarily to the general case, on perfect fusion graphs, any C-watershed of a map is indeed a W-thinning and thus possesses the contrast preservation property, needed by morphological region merging methods.
2) On these graphs, the divide of any C-watershed is a thin binary watershed set. Consequently, we derive characterizations of perfect fusion graphs based on thinness properties of both C-watersheds and topological watershed functions.
3) We propose and prove the correctness of a new simple and linear-time algorithm to compute C-watershed on these graphs, while the correctness of such an algorithm cannot be guaranteed in the general case [10].

The proofs of the properties presented in this paper will be given in a forthcoming extended version.

2   Watersheds and Fusion Graphs

2.1   Basic Notions and Notations

Let E be a finite set, we denote by 2^E the set composed of all the subsets of E. We denote by $|E|$ the number of elements of E.

We define a graph as a pair (E, Γ) where E is a finite set and Γ is a binary relation on E (*i.e.*, $\Gamma \subseteq E \times E$), which is reflexive (for all x in E, $(x, x) \in \Gamma$) and symmetric (for all x, y in E, $(y, x) \in \Gamma$ whenever $(x, y) \in \Gamma$). Each element of E (resp. Γ) is called a *vertex* or a *point* (resp. *an edge*). We will also denote by Γ the map from E to 2^E such that, for all $x \in E$, $\Gamma(x) = \{y \in E \mid (x, y) \in \Gamma\}$. If $y \in \Gamma(x)$, we say that y is *adjacent to* x. Let $X \subseteq E$, we define $\Gamma(X) = \cup_{x \in X} \Gamma(x)$, and $\Gamma^\star(X) = \Gamma(X) \setminus X$. If $y \in \Gamma(X)$, we say that y *is adjacent to* X. If $X, Y \subseteq E$ and $\Gamma(X) \cap Y \neq \emptyset$, we say that Y *is adjacent to* X.

Let (E, Γ) be a graph, let $X \subseteq E$, a *path in* X is a sequence $\pi = \langle x_0, ..., x_l \rangle$ such that $x_i \in X$, $i \in [0, l]$, and $x_i \in \Gamma(x_{i-1})$, $i \in [1, l]$. We also say that π is a *path from* x_0 *to* x_l *in* X and that x_0 and x_l are *linked for* X. We say that X *is connected* if any x and y in X are linked for X. In the sequel we will consider that (E, Γ) is a graph and we will assume that E is connected.

Let $X \subseteq E$ and $Y \subseteq X$. We say that Y *is a connected component of X*, or simply *a component of X*, if Y is connected and if Y is maximal for this property, *i.e.*, if $Z = Y$ whenever $Y \subseteq Z \subseteq X$ and Z connected. We denote by $\mathcal{C}(X)$ the set of all connected components of X.

Let $k_{\min}$ and $k_{\max}$ be two elements of $\mathbb{Z}$ such that $k_{\min} < k_{\max}$. We set $\mathbb{K} = \{k \in \mathbb{Z}; k_{\min} \leq k < k_{\max}\}$ and $\mathbb{K}^+ = \mathbb{K} \cup \{k_{\max}\}$. We denote by $\mathcal{F}(E)$ the set composed of all functions from E to $\mathbb{K}$. Let $F \in \mathcal{F}(E)$, let $k \in \mathbb{K}^+$. We denote by $F[k]$ the set $\{x \in E; F(x) \geq k\}$ and by $\overline{F}[k]$ its complementary set; $F[k]$ is called an *upper section* of F and $\overline{F}[k]$, a *lower section* of F. A connected component of $\overline{F}[k]$ which does not contain a connected component of $\overline{F}[k-1]$ is a *(regional) minimum* of F. We denote by $M(F) \subseteq E$ the set of all points which are in a minimum of F. We say that $\overline{M}(F) = \overline{M(F)}$ is the *divide* of F.

2.2   Watershed Set and Fusion Graphs

The notion of (binary) watershed set may be seen as a model of frontier in a graph. Many segmentation algorithms expect to compute such watershed sets.

In the following definitions "W-" stands for watershed.

Definition 1. *Let $X \subseteq E$ and let $p \in X$. We say that p is* W-simple *for X if p is adjacent to exactly one component of $\overline{X}$.*
The set X is a watershed *if there is no W-simple point for X.*

In Fig. 1a, y is W-simple for the set constituted by the black vertices. Observe that the set X of black points in Fig. 1b is a watershed set since it contains no W-simple point for X.

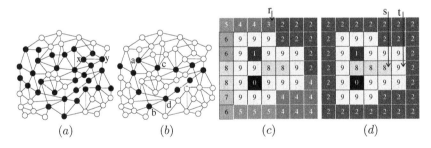

(a)	(b)	(c)	(d)

Fig. 1. Illustration of watersheds. (a): A graph (E, Γ) and a subset X (black points) of E; (b): the set of black points is a watershed of X; (c): a graph corresponding to the 8-adjacency relation and a function F; (d): a topological watershed of F.

Let $X \subseteq E$ and let $p \in X$. We say that p is *an inner point for X* if p is not adjacent to $\overline{X}$. The interior of X is the set of all inner points for X, denoted $int(X)$. If $int(X) = \emptyset$, we say that X is *thin*.

For example, the point x in Fig. 1a is an inner point for the set of black vertices. In Fig. 1b, the set of black vertices is thin. The sets made of black and gray points in Fig. 3a and b are not thin: their interior, depicted in gray, are not empty. Observe also that they are watershed sets since they do not contain any W-simple points.

The theoretical framework set up in [7] allows to study the properties of region merging methods in graphs. In particular, one of the most striking theorems, allows to link the region merging properties with the thinness properties of watershed set.

In the following definition the prefix "F-" stands for fusion.

Let $X \subseteq E$. Let $x \in X$, we say that x *is F-simple (for X)*, if x is adjacent to exactly two components of $\overline{X}$. Let $S \subseteq X$. We say that S *is F-simple (for X)* if S is adjacent to exactly two components $A, B \in \mathcal{C}(\overline{X})$ such that $A \cup B \cup S$ is connected.

Let us look at Fig. 1b. The set X made of the black vertices separates its complementary set into four components. The points a and c are F-simple for X whereas b and d are not. The set $S = \{a, c\}$ is F-simple for X and $\{b, d\}$ is not. If we remove from X an F-simple set, S for instance, we obtain a set which separates its complementary into three components: we "merged two components of $\overline{X}$ through S". This operation may be seen as an elementary merging in the sense that only two components were merged.

Let $X \subset E$ and let A and B be two elements of $\mathcal{C}(\overline{X})$ with $A \neq B$. We say that *A and B can be merged (for X) through S* if S is F-simple and A and B are precisely the two components of $\overline{X}$ adjacent to S. We say that *A can be merged (for X)* if there exists $B \in \mathcal{C}(\overline{X})$ and $S \subseteq X$ such that A and B can be merged through S.

We say that (E, Γ) is a *fusion graph* if for any subset of vertices $X \subseteq E$ such that $|\mathcal{C}(\overline{X})| \geq 2$, any component of $\overline{X}$ can be merged.

Notice that all graphs are not fusion graphs. For instance, the graphs induced by the 4-adjacency relation depicted on Fig. 3a is not a fusion graph. On the other hand, the graph induced by the 8-adjacency depicted on Fig. 3c is an example of fusion graph.

The most striking theorem (33) in [7] states that the class of fusion graphs is precisely the class of graphs in which any watershed set is thin.

We set $\Gamma^{\star}(A, B) = \Gamma^{\star}(A) \cap \Gamma^{\star}(B)$ and if $\Gamma^{\star}(A, B) \neq \emptyset$, we say that *$A$ and B are neighbors*.

Definition 2. *We say that (E, Γ) is a perfect fusion graph if, for any $X \subseteq E$, any neighbors A and B in $\mathcal{C}(\overline{X})$ can be merged through $\Gamma^{\star}(A, B)$.*

In other words, the perfect fusion graphs are the graphs in which merging two neighboring regions can always be performed by removing from the frontier set all the points which are adjacent to both regions. This class of graphs allows, in particular, to rigorously define hierarchical schemes based on region merging and to implement them in a straightforward manner. It has been shown [7] that any perfect fusion graph is a fusion graph and that the converse is not true. For instance, the graphs induced by the 8-adjacency relation are not, in general, perfect fusion graphs (see counter-examples in Appendix A) whereas they are fusion graphs. In [7] the authors introduce a family of adjacency relations on $\mathbb{Z}^n$

that can be used in image processing and that induce perfect fusion graphs. See Appendix B for an illustration of these relations on $\mathbb{Z}^2$ and $\mathbb{Z}^3$.

The two following necessary and sufficient conditions for perfect fusion graphs show show the deep relation existing between perfect fusion graphs and thin watershed set.

Theorem 1 (from 41 in [7]). *The three following statements are equivalent:*

i) (E, Γ) is a perfect fusion graph;
ii) for any $x \in E$, any $X \subseteq \Gamma(x)$ contains at most two connected components;
iii) for any watershed Y in E such that $\mathcal{C}(\overline{Y}) \geq 2$, each point x in Y is F-simple.

To conclude this section, we recall the definition of line graphs. This class of graphs allows to make a strong link between the framework developed in this paper and the approaches of watershed and region merging based on edges rather than vertices.

Let (E, Γ) be a graph. The *line graph* of (E, Γ) is the graph (E', Γ') such that $E' = \Gamma$ and (u, v) belongs to Γ' whenever $u \in \Gamma$, $v \in \Gamma$ and u, v share a common vertex of E.

We say that the graph (E', Γ') is a line graph if there exists a graph (E, Γ) such that (E', Γ') is isomorphic to the line graph of (E, Γ).

It has been proved [7] that any line graph is a perfect fusion graph and that the converse is not true. We point out that the definitions, properties and algorithm for watershed on perfect fusion graph developed in Section 3 also holds for watershed approaches based on edges rather than vertices.

2.3   W-Thinnings and Topological Grayscale Watersheds

We now recall the notions of W-thinning and topological grayscale watershed which have been introduced and studied in [5, 8, 9, 10].

Let $F \in \mathcal{F}(E)$. We denote by $[F \setminus x]$ the map in $\mathcal{F}(E)$ such that $[F \setminus x](x) = F(x) - 1$, and $[F \setminus x](y) = F(y)$ for any $y \in E$, $y \neq x$.

Definition 3. *Let $x \in E$. Let $F \in \mathcal{F}(E)$ and let $k - \Gamma(p)$. We say that p is W-destructible for F if p is W-simple for $F[k]$.*
If there is no W-destructible point for F we say that F is a (topological) watershed.
Let $G \in \mathcal{F}(E)$.
We say that G is a W-thinning of F if $G = F$ or if there exists a W-thinning $H \in \mathcal{F}(E)$ of F and a point $x \in E$, which is W-destructible for H, such that $G = [H \setminus x]$.
If G is both a W-thinning of F and a watershed we say that G is a (topological) watershed of F.

In Fig. 1c and d, assume that the graph is the one corresponding to the 8-adjacency relation. In both Fig. 1c and d, it may be seen that there are three minima which are the components with levels 0,1 and 2. In Fig. 1c, the point labeled r is W-destructible. In Fig. 1d, no point is W-destructible. The function depicted in Fig. 1d is a watershed of the function in Fig. 1c. Observe that in Fig. 1d the minima of Fig. 1c have been extended as much as possible while preserving the number of components of all the lower sections of Fig. 1c. The divide of a topological watershed constitutes an interesting segmentation [8, 10] which possesses important properties not guaranteed by most watershed algorithms [2, 3]. In particular, it preserves the connection value between the minima of the original function; the connection value (see [8, 10, 14, 15]) between two minima is the minimal altitude at which one need to climb in order to reach one minimum from the other. This contrast preservation property is a requirement for region merging method based on watershed [11, 12].

3   C-Watersheds: Definitions, Properties and Algorithm

In [7], we have shown that any subset of $\mathbb{Z}^2$ equipped with the 8-adjacency forms a fusion graph but not, in general, a perfect fusion graph. In particular, the graphs considered in Fig. 1c and d are fusion graphs but not perfect fusion graphs. Let us consider the function F depicted in Fig. 1d. We have seen that F is a topological watershed. If we examine the divide of F, it may be seen that the point labeled s is inner (in the binary sense) for the divide. Thus, on fusion graphs, there exist topological watersheds whose divides are not thin.

On the same figure, remark also that the point labeled t is W-simple for $\overline{M}(F)$, thus $\overline{M}(F)$ is not a binary watershed set. Thus, on fusion graphs, there exist topological grayscale watersheds whose divides are not binary watershed sets.

In the remaining of this paper, we study W-thinnings and topological grayscale watersheds on perfect fusion graphs and we show, among other properties, that the divide of any topological watershed is necessarily a thin watershed set.

Let us first define a type of points that we call M-cliff. Given a graph and a function, these points are the lowest points adjacent to a single minimum. We will show that if the graph is a perfect fusion graph, any M-cliff point is W-destructible (Th. 2).

Definition 4. *Let $F \in \mathcal{F}(E)$ and let $x \in E$. We say that x is a cliff point (for F) if $x \in \overline{M}(F)$ and if it is adjacent to a single minimum of F. We say that x is M-cliff (for F) if x is a cliff point with minimal altitude (i.e., $F(x) = \min\{F(y) \mid y \in E$ is a cliff point for $F\}$).*

Let us look at Fig. 2. Thanks to Th. 1.ii, it may be seen that the depicted graphs are perfect fusion graphs. In Fig. 2a, the points with level 3 are cliff points and the bold circled point is the only M-cliff point. In figure 2b and c, it can be seen that there is no M-cliff point and no cliff point.

Let $F \in \mathcal{F}(E)$ and $j \in \mathbb{K}$. The point x *is W-destructible with lowest value j (for F)* if for any $h \in \mathbb{K}$ such that $j < h \le F(x)$, x is W-simple for $F[h]$ and if x is not W-simple for $F[j]$.

Let $h \in \mathbb{K}$ such that $h < F(x)$, we denote by $[F \setminus x \downarrow h]$ the function of $\mathcal{F}(E)$ such that $[F \setminus x \downarrow h](x) = h$ and $[F \setminus x \downarrow h](y) = F(y)$ for all $y \in E \setminus \{x\}$.

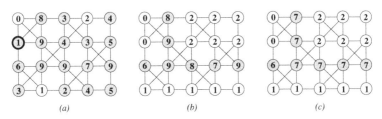

(a) (b) (c)

Fig. 2. Example of function on perfect fusion graphs, the minima are in white; (a): the bold circled vertex is M-cliff; (b): a C-watershed of (a); (c): a topological watershed of both (a) and (b)

Theorem 2. *Let $F \in \mathcal{F}(E)$. Let $x \in E$ be M-cliff for F and let $l \in \mathbb{K}$ be the level of the only minimum adjacent to x. If (E, Γ) is a perfect fusion graph then x is W-destructible with lowest value l for F.*

Remark that on non-perfect fusion graphs, the points which are M-cliff are not necessarily W-destructible. For example, the point labeled t in Fig. 1d is M-cliff whereas it is not W-destructible.

Definition 5. *Let $F \in \mathcal{F}(E)$, we say that $G \in \mathcal{F}(E)$ is a C-thinning of F if*

i) $G = F$, or if
ii) there exists a function H which is a C-thinning of F and there exists a point x M-cliff for H, with lowest value k such that $G = [H \setminus x \downarrow k]$.

We say that F is a C-watershed if there is no M-cliff point for F. If G is both a C-thinning of F and a C-watershed we say that G is a C-watershed of F.

The following property follows immediately from definition 5 and Th. 2.

Property 3. *Let (E, Γ) be a perfect fusion graph and let $F \in \mathcal{F}(E)$. If F is a topological watershed then F is a C-watershed. If G is a C-thinning of F then G is a W-thinning of F.*

The converses of the two propositions in Prop. 3 are not true. The function of Fig. 2b is a C-watershed of Fig. 2a but is not a topological watershed. Indeed, the points at altitude 9 are W-destructible. The function depicted in Fig. 2c is a W-thinning of Fig. 2a but not a C-thinning of 2a. Indeed some points at level 9 have been lowered down to 7, and 7 is not the altitude of any minimum.

Observe that, on perfect fusion graphs, since any C-thinning is a W-thinning, from the contrast preservation theorem presented in the introduction, we can immediately deduce that C-thinnings, and hence C-watershed, preserves the connection value between the minima of the original map.

It can be easily seen, that in a C-thinning sequence the points which are in a minimum at a given step become neither M-cliff, nor W-destructible further in the sequence. This observation leads us to the definition of Algorithm 1, a very simple algorithm for computing C-watersheds.

At each iteration of the main loop (line 6) of Algorithm 1, F is a C-thinning and a W-thinning of the input function.

Algorithm 1. C-watershed

Data: a perfect fusion graph (E, Γ), a function $F \in \mathcal{F}(E)$
Result: F

1    $L := \emptyset; K := \emptyset;$
2   Attribute distinct labels to all minima of F and label the points of $M(F)$ with the corresponding labels;
3   **foreach** $x \in E$ **do**
4   **if** $x \in M(F)$ **then** $K := K \cup \{x\};$
5   **else if** x *is adjacent to* $M(F)$ **then** $L := L \cup \{x\}; K := K \cup \{x\};$

6   **while** $L \neq \emptyset$ **do**
7        $x :=$ an element with minimal altitude for F in L;
8        $L := L \setminus \{x\};$
9   **if** x *is adjacent to exactly one minimum of* F **then**
10   Set $F[x]$ to the altitude of the only minimum of F adjacent to x;
11   Label x with the corresponding label;
12   **foreach** $y \in \Gamma^*(x) \cap \overline{K}$ **do** $L := L \cup \{y\}; K := K \cup \{y\};$

At the end of Algorithm 1, F is a C-watershed of the input function.

In Algorithm 1, the operations performed on the set L are the insertion of an element and the extraction of an element with minimal altitude. Thus L may be managed as a priority queue.

Lemma 4. *Let* $F \in \mathcal{F}(E)$. *Let* $x \in E$ *be M-cliff for* F *and let* $k = F(x)$. *If* (E, Γ) *is a perfect fusion graph, any* $y \in E$ *which is inner for* $\overline{M}(F)$ *is such that* $F(y) \geq k$.

On non-perfect fusion graphs, the previous lemma is in general not true.

From Lem. 4, we deduce that in Algorithm 1, when the function F is lowered at a point x with altitude k, any point inserted further in the set L has a level greater than or equal to k. Thus the set L may be managed by a monotone priority queue. Recently, M. Thorup [16] proved that if we can sort n-keys in time $n.s(n)$ then and only then there is a monotone queue with capacity n, supporting the *insert* and *extract-min* operations in $s(n)$ amortized time.

Property 5. *If the elements of* E *can be sorted according to* F *in* $o(|E|)$, *then Algorithm 1 terminates in linear time with respect to* $(|E| + |\Gamma|)$.

Since Algorithm 1 possesses the monotone property discussed above, it can be classified in the group of immersion algorithms (see [2, 3, 10] for examples). Moreover, it is the first immersion algorithm proved to compute W-thinnings in linear time with respect to the size of the graph.

Notice that computing a topological grayscale watershed from a C-watershed is not straightforward. For more details we refer to [9].

Let us now state some properties of C-watersheds on perfect fusion graphs. Let $G \in \mathcal{F}(E)$ and let $k = G(x)$. If x is F-simple for $G[k]$, we say that x is *F-simple for* G.

Theorem 6 (Grayscale characterizations of perfect fusion graphs). *The three following statements are equivalent:*

   i) (E, Γ) is a perfect fusion graph;
   ii) for any C-watershed $G \in \mathcal{F}(E)$, any point of $\overline{M}(G)$ is F-simple for $\overline{M}(G)$;
   iii) for any topological grayscale watershed $G \in \mathcal{F}(E)$, any point in $\overline{M}(G)$ is F-simple for G.

Thanks to Th. 6, we immediately deduce the following theorem.

Theorem 7. *Let (E, Γ) be a perfect fusion graph and let $F \in \mathcal{F}(E)$. If F is a C-watershed then $\overline{M}(F)$ is a watershed.*

In other words, on a perfect fusion graph, the minima of a C-watershed cannot be further extended.

   In this section, we have seen that:

1) on perfect fusion graphs, the C-watersheds preserves the connection value between the minima of the original map; and
2) in this framework, the divide of any C-watersheds is a thin binary watershed set.

Since perfect fusion graphs allow to rigorously define region merging procedure, the divide of C-watersheds on perfect fusion graphs is an ideal entry point for hierarchical methods based on watersheds.

   On perfect fusion graph, any topological watershed is a C-watershed (Prop. 3), thus we may easily deduce from Th. 6 and 7 that:

   i) a graph is a perfect fusion graph if and only if, for any topological watershed F, any point of the divide of F is adjacent to exactly two minima of F; and
   ii) on a perfect fusion graph, the divide of any topological grayscale watershed is a binary watershed set.

4   Perspectives: Perfect Fusion Grids and Hierarchical Schemes

Following some properties given in [7] and the examples depicted in this paper (see Fig. 1cd, Fig. 6), it may be seen that there exist topological watersheds whose divides are not thin in 2D on the 4-, 6- and 8-connected grids, and in 3D on the 6- and 26-connected grids. In this paper, we have shown that, on perfect fusion graphs, the divide of any topological grayscale watershed is a thin binary watershed set. On these graphs, region merging schemes are easy to rigorously define and straightforward to implement. Thus, the framework of perfect fusion graph is adapted for region merging methods based on topological watersheds.

   In [7], we introduced the family of perfect fusion grids over $\mathbb{Z}^n$, for any $n \in \mathbb{N}$. Any element of this family is indeed a perfect fusion graph. We proved that any of these grids is "between" the direct adjacency graph (which generalizes the 4-adjacency to $\mathbb{Z}^n$) and the indirect adjacency graph (which generalizes the

8-adjacency to $\mathbb{Z}^n$). These n-dimensional grids are all equivalent (up to a translation) and, in a forthcoming paper, we intend to prove that they are the only graphs that possess these two properties. Examples of (restrictions of) 2 and 3-dimensional perfect fusion grids are presented in Appendix B.

Perfect fusion grids constitute an interesting alternative for region merging methods based on watersheds. Future work will include revisiting hierarchical segmentation methods [11, 12] on perfect fusion grids.

References

1. Pavlidis, T. In: Structural Pattern Recognition. Volume 1 of Springer Series in Electrophysics. Springer (1977) 90–123 segmentation techniques, chapter 4–5.
2. Vincent, L., Soille, P.: Watersheds in digital spaces: An efficient algorithm based on immersion simulations. PAMI **13**(6) (1991) 583–598
3. Meyer, F.: Un algorithme optimal de ligne de partage des eaux. In: Actes du 8ème Congrès AFCET, Lyon-Villeurbanne, France (1991) 847–859
4. Beucher, S., Meyer, F.: The morphological approach to segmentation: the watershed transformation. E. Dougherty (Ed.), Mathematical Morphology in Image Processing, Marcel Decker (1993) 443–481
5. Couprie, M., Bertrand, G.: Topological grayscale watershed transform. In: SPIE Vision Geometry V Proceedings. Volume 3168. (1997) 136–146
6. Jasiobedzki, P., Taylor, C., Brunt, J.: Automated analysis of retinal images. IVC **1**(3) (1993) 139–144
7. Cousty, J., Bertrand, G., Couprie, M., Najman, L.: Fusion graphs: merging properties and watershed (2006) submitted. Also in technical report IGM2005-04, http://igm.univ-mlv.fr/LabInfo/rapportsInternes/2005/04.pdf.
8. Bertrand, G.: On topological watersheds. JMIV **22**(2-3) (2005) 217–230
9. Couprie, M., Najman, L., Bertrand, G.: Quasi-linear algorithms for the topological watershed. JMIV **22**(2-3) (2005) 231–249
10. Najman, L., Couprie, M., Bertrand, G.: Watersheds, mosaics and the emergence paradigm. DAM **147**(2-3) (2005) 301–324 Special issue on DGCI.
11. Beucher, S.: Watershed, hierarchical segmentation and waterfall algorithm. In: Mathematical Morphology and its Applications to Image Processing. (1994) 69–76
12. Najman, L., Schmitt, M.: Geodesic saliency of watershed contours and hierarchical segmentation. PAMI **18**(12) (1996) 1163–1173
13. Najman, L., Couprie, M.: Building the component tree in quasi-linear time. IEEE Trans. on Image Processing (2006) To appear.
14. Bertrand, G.: A new definition for the dynamics. In: 7th International Symposium on Mathematical Morphology Proceedings. (2005) 197–206
15. Rosenfeld, A.: The fuzzy geometry of image subsets. Pattern Recognition Letters **2** (1984) 311–317
16. Thorup, M.: On ram priority queues. In: 7th ACM-SIAM Symposium on Discrete Algorithms. (1996) 59–67

Appendix A: Counter-Examples of Merging Properties in Usual Grids

Let us first consider the 4-connected graph depicted in Fig. 3a. Since none of the components of the complementary of the black vertices can be merged, the depicted graph is not a fusion graph. As an illustration of the fusion graphs

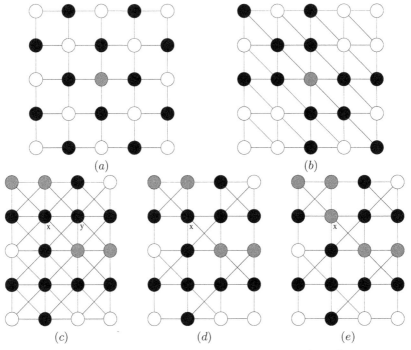

Fig. 3. Counter example of merging properties in usual grids $(a - c)$, and illustration of merging properties in a perfect fusion grid (d, e)

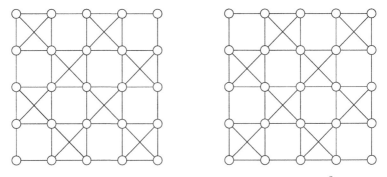

Fig. 4. Samples of the two perfect fusion grids on $\mathbb{Z}^2$

fundamental theorem (33 in [7]) recalled in Section 2, we can observe that the set of black and gray points is a non-thin binary watershed.

The graph of Fig. 3b, which is a 2D 6-connected graph, is not a fusion graph since the gray point which is a component of the complementary of the black vertices cannot be merged.

Since the graph, depicted on Fig. 3c, induced by the 8-adjacency relation is a fusion graph, any binary watershed on this graph is thin. This property can be verified, in particular, for the watershed made of the black points on Fig. 3c. Observe on the same figure that the two neighboring gray components cannot be merged through $\{x, y\}$ their common neighborhood. The black vertices is thus a counter-example of the perfect fusion property for the depicted graph.

In Fig. 3d, the same sets of black and gray points are considered on a perfect fusion grid. Observe that the two gray components can now be merged through their common neighborhood $\{x\}$. Remark also that the set obtained by removing $\{x\}$ from the black points (Fig. 3e) is still a watershed. This desirable property,

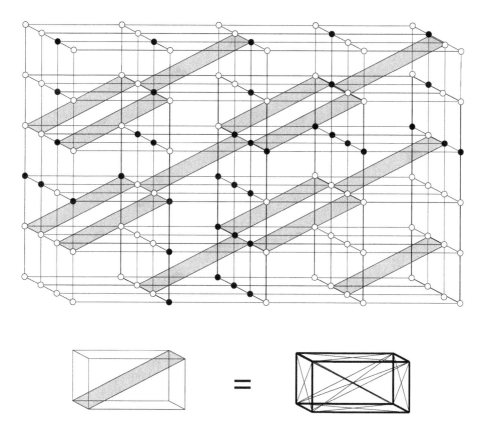

Fig. 5. A 3-dimensional perfect fusion grid. Black points constitute a set which is a watershed.

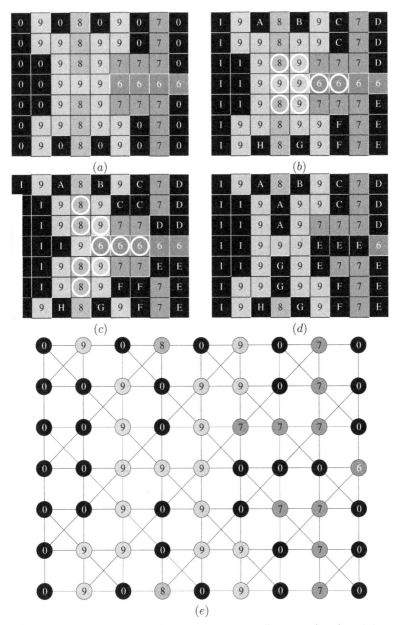

Fig. 6. Comparison of topological watershed using different grids. The minima, labeled by letters, are supposed to be at altitude 0; the circled points are inner for the divide of the depicted function with respect to the assumed adjacency; (a), an image; (b), a topological watershed of (a) when the 8-adjacency graph is assumed; (c), a topological watershed of (a) when the 4-adjacency graph is assumed; (d), a topological watershed of (a) when one of the perfect fusion grids is assumed; (e) same as (d) showing the assumed adjacency relation.

which does not hold in the general case, can be easily proved on perfect fusion graphs.

We finish this appendix section, with a table that sums up the status of the different graphs used in 2D and 3D image processing. See [7] for more details. For non trivial images, we have:

	fusion graph	perfect fusion graph
$2D$: 4-connected graph	is not a	is not a
$3D$: 6-connected graph	is not a	is not a
$2D$: 8-connected graph	is a	is not a
$3D$: 26-connected graph	is not a	is not a
$2D$: 6-connected graph	is not a	is not a

Appendix B: Perfect Fusion Grids: 2D and 3D Cases

A formal definition of perfect fusion grids can be found in [7]. In $\mathbb{Z}^2$, there are two distinct perfect fusion grids, in $\mathbb{Z}^3$ there are four. Actually it has been proved that, for any strictly positive integer n, there are exactly 2^{n-1} perfect fusion grids over $\mathbb{Z}^n$ which are all equivalent (up to a unit translation). Samples of the two perfect fusion grids on $\mathbb{Z}^2$ are depicted in Fig. 4. Fig. 5 shows a binary watershed (black points) on one of the 3D perfect fusion grids. To clarify the figure, we use the following convention: any two points belonging to a same cube marked by a gray stripe are adjacent to each other.

Matching of the Multi-channel Images with Improved Nonparametric Transformations and Weighted Binary Distance Measures

Bogusław Cyganek

AGH - University of Science and Technology
Al. Mickiewicza 30, 30-059 Kraków, Poland
`cyganek@uci.agh.edu.pl`

Abstract. This paper extends the concepts of image matching in the non-parametric space and binary distance measures. Matching in the nonparametric domain exhibits many desirable properties at relatively small computation complexity: It concentrates on capturing mutual relation among pixels in a small neighbourhoods rather than bare intensity values, thus improving matching discrimination. It is also more resistive against noise and uneven lighting conditions of the matched images. Last but not least, the matching algorithms operate in the integer domain and can be easily implemented in hardware what benefits in dramatic improvement of their run times. In this paper we extend the concept of nonparametric image transformation into the realm of colour images taking into consideration different colour spaces and different distances defined in these spaces. We propose significant bit reduction for aggregated block matching in the Census domain. We propose also the sparse sampling model for the Census transformation that increase the discriminative power of this representation and allows even further reduction of bits necessary for matching. The presented techniques have been applied to matching of the stereo images but can be employed in any computer vision task that requires comparison of images, such as image registration, object detection and recognition, etc. Presented experiments exhibit interesting properties of the described techniques.

Keywords: image matching, Census, binary distance.

1 Introduction

This paper is a continuation of the work comparing matching strategies for images in different nonparametric representations and with various similarity measures. The basic concepts for grey valued (or: one channel) images were reported in [5]. In this paper we extend the idea of nonparametric image representation into multi-channel images with the colour images as the most frequent examples of such spaces.

The general idea behind different transformations of data (images in our case) is to change their properties into representations which are more suitable for given applications. For example, the Karhunen-Loève transformation allows detection of the

U. Eckardt et al. (Eds.): IWCIA 2006, LNCS 4040, pp. 74–88, 2006.

dominating directions in the multi dimensional data, which further allows data compression [7]. The wavelet transform allows time-frequency decomposition of the compound signals what enables e.g. a scale space representation of images [10]. Similarly, the nonparametric transformations change statistical properties of signals by changing in any data sample an absolute value by its rank among the other values in this sample. For images the data samples usually are defined as small compact rectangular regions around pixels of that image. By this token the nonparametric representation exhibits the uniform distribution compared to an unknown distribution of the input signal. For the matching task the most interesting showed to be the Census transformation, introduced by Zabih and Woodfill [17]. This transformation is very helpful in image matching since it conveys information on *mutual relations* among pixels in their local neighbourhoods. This, in turn, allows disambiguation of the matches. Its other desirable feature in the light of image matching is the resistance to the noise and non-uniform lighting conditions among images. In [5] we presented the Census and its modification called the Detailed-Neighbourhood-Relation, as well as we compared different binary comparison metrics for image correlation. The main conclusion of this work is that, for the same bit-rate, the better matching quality is obtained when extending range of local neighbourhoods instead of increasing number of bits for precise description of relations among pixels.

In this paper we focus on matching of the colour images in the nonparametric representation. The motivation comes from our experiments of comparing matching strategies for the grey-valued (one-channel) with the matching of colour (multi-channel) images by means of the common measures, such as SAD, SSD, ZSSD, etc. For many tested colour images they did not show any, or showed very small, improvement in the matching quality, although the amount of data, and computational effort, in that case were tripled [3]. This complies as well with results obtained by other researchers [12], although they reported a little bit higher degree of quality improvement. Thus for image matching the common practice is to convert colour into grey (one channel) representation and perform faster matching. Our idea is similar but before matching we propose and check some non-linear colour conversions that capture the mutual relations in local topological spaces of the multi-channel images. As a result this method allows better signal conditioning for the matching stage, on the one hand.

The further contributions of this papers are as follows: Based on the observed redundancy in the aggregated block matching we propose a 50% reduction of bits resulting in relaxed demands on memory and improved speed of computations. We also propose the sparse sampling model for the Census transformation which further improves matching with low bit rate.

The paper is organized as follows: section (2.1) describes nonparametric representation for scalar images. Then the techniques for data reduction in aggregated blocks (2.2) and sparse Census sampling models (2.3) are introduced and explained. Then we analyze different methods of the nonparametric representation for multi-channel images (2.4). Section (2.5) discusses the distance measures for nonparametric representation. The paper closes with the experimental results (4) and conclusions (5).

2 Nonparametric Representations of Images

2.1 Nonparametric Census Representation for a Single Pixel with Scalar Value

The nonparametric measures rely on the mutual relations among pixel values in a local neighbourhood defined around pixels of an image [1][2].

The Census transform maps the local pixel neighbourhoods, located around a certain central pixel P, to a bit string. In this series each bit conveys a binary 0/1 information, indicating whether a given pixel is less or not from the central one. The Census transform for a pixel P in the image $\mathbf{I}$ is defined as follows [17]:

$$T[\mathbf{I},P]= \underset{P'\in W(P,\beta)}{\otimes} \xi(I,P,P') \tag{1}$$

where I stands for intensity, P is a central pixel, $\otimes$ denotes concatenation, $W(P,\beta)$ is a local pixel neighbourhood around a pixel P with a radius β, P' denotes pixels belonging to W, and ξ is given by the following formula:

$$\xi(I,P,P') = \begin{cases} 0 & if \quad I(P) \le I(P') \\ 1 & otherwise \end{cases} . \tag{2}$$

To find correspondence of images we first apply (1) to the images and then usually compute the Hamming distance between bit strings, although some other measures are also possible [5]. The separate question is choice of the window W [4][5].

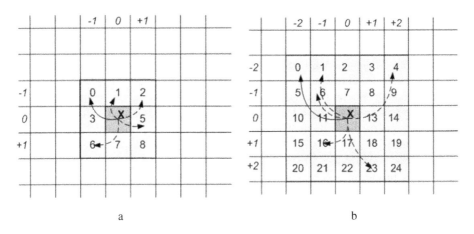

Fig. 1. The Census transformation for a single pixel in a 3×3 (a) and 5×5 (b) neighbourhoods

Computation of the Census transformation for a single pixel "x" in a 3×3 neighbourhood depicts Fig. 1a, whereas for the 5×5 Fig. 1b (shown selected relations only).

2.2   Reduced Match Aggregation Schemes for the Census Representation

The intention of image matching process is unambiguous selection of the best fit among pixels. Although this can be done solely on a pixel-by-pixel basis, in practice such an approach leads to many errors due to ambiguous matches, which come from the limited dynamics of the values assigned to pixels and ubiquitous noise. As was shown in [5], binary matching in the Census representation has many advantageous to the other measures in this respect and performs much better in the pixel-by-pixel scheme since each pixel in Census representation conveys information on its closest neighbourhood. Thus with sufficiently large windows W in (1) it is even possible to perform reliable pixel-by-pixel matches.

However, in many cases matching based on single pixels is not sufficient (e.g. for large baseline stereo) and larger support region is necessary [11]. If so, then the matching usually is done in the corresponding rectangular windows placed in the source and destination images. Then for each pair of pixels from the corresponding windows the matching measure is computed and added to the total result. If one against many windows is checked, then the window with the minimal (or maximal, depending on the used measure) score is chosen as the corresponding one.

This scheme works fine for many different windows and comparison metrics. However, for the Census representation of many neighbouring pixels we encounter some data redundancy. To see this let us analyse the case presented in Fig. 2 of a 3×3 match window with each pixel already converted to the 3×3 Census representation (i.e. although having different meaning, the two windows are of the same size).

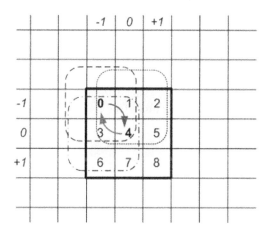

Fig. 2. Explanation of data redundancy for the block of pixels in the Census domain

We see that the relation between the pixels no. 0 and 4 is computed and stored twice: once when computing the 3×3 Census representation (2) for no. 0, then its negated value is stored in the 3×3 Census representation, this time computed for no. 4 (see Fig. 2). Such repeated bits do not convey useful information and one of the two can be simply omitted. So, if computing Census representations for pixels that will be treated as aggregated blocks (e.g. very common in stereo matching) we need only to compute *half* the number of comparisons (2). Thus, the computation schemes from Fig. 1a and Fig. 1b have to be modified to the ones in Fig. 3a and Fig. 3b, respectively.

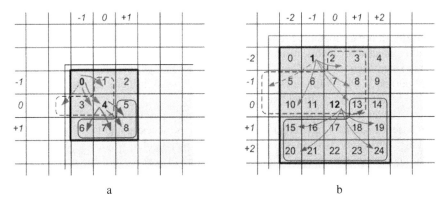

a   b

Fig. 3. Reduced Census coding for block of pixels (light grey). An example with Census for a 3×3 (a) and 5×5 (b) neighbourhoods (darker grey). Larger neighbourhoods are encoded in the same way.

For other Census representations (i.e. the windows W in (1)) the computation scheme for the reduced representation is the same: With the top-down and left-right bit numbering in W, having selected a central pixel with a number n_X, (e.g. in Fig. 3a: $n_X = 4$) only pixels with numbers greater than n_X are taken into the representation.

This simple observation leads to significant improvement of the image matching in terms of computation time and memory occupation. This is also a good feature for hardware realizations, as well [1][16].

The proposed reduction of bits for the Census representation for blocks of pixels can be interpreted as taking each k-*th* sample from that block and with the Census representation not reduced at the same time. This concept is illustrated in Fig. 4.

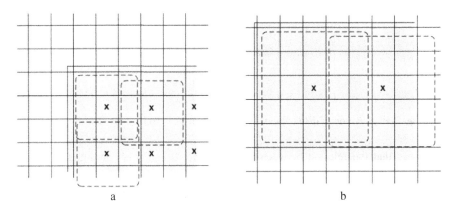

a   b

Fig. 4. Taking each k-*th* pixel in the matching blocks in the not reduced Census representation. An example for the 3×3 (a) and 5×5 (b) Census neighbourhoods. Central pixels denoted by 'x'.

However, the drawback of the second approach is its more spread representation of the matching block. Let us consider, for instance, the 5×5 Census transformation and a 3×3 matching block, which in this method would take 11×11 pixels. In the case of stereo matching this can produce excessive smearing in the resulting disparity map.

2.3 Sparse Sampling Model for the Census Transformation

An increase of the size of the Census window (Fig. 1) leads to an increase of the discriminating power of such feature representation. To some extent this improves quality of the image matching [5]. However, too big neighbourhoods do not lead to further improvements, since we encounter local deformations in the matching neighbourhoods, which are due to different (projective) transformations of the matched images. Also, the polynomial growth of a number of bits inhibits practical realizations of Census windows larger than say 7×7 (6 bytes per pixel). During experiments we noticed that very high discriminating power can be achieved if we *sparsely sample* a Census neighbourhood, computing mutual relations among the central pixel and its neighbours *separated* by a certain distance. Fig. 5 depicts this idea. Notice that we employ also the reduced aggregation scheme (2.2), so only the neighbours to the right and down are taken into relations. This can be seen as special definition of the window $W(P,\beta)$ in (1). In Fig. 5 we consider two Census windows: the inner 5×5 and the outer 9×9, respectively. For the inner representation "x" is compared with only four neighbours n_{11}-n_{14}, separated from each other by d_1 pixels.

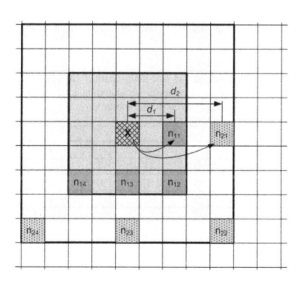

Our experiments showed that this four bits representtation produces almost the same matching results as the full 5×5 representation (24 bits). However, we can increase this property even more by addition of the next four bits from the outer window with pixels n_{21}-n_{22} which are separated by d_2. In total we get 8 bits (i.e. very practical size of 1 byte) for repre-sentation of the whole 9×9 neighbourhood. Let us remind that such sparse sampling is done for *each* pixel in the matching window, i.e. at x_i, x_{i+1}, etc.

Fig. 5. Sparse relations for Census matching. The inner window size is 5×5 but we compute relations with only 4 neighbours distant by d_1. In the outer window we also compute only 4 relations among pixels distant by d_2.

For many images this feature could be explained by the probabilistic dependence among pixels and their nearest neighbours. This comes also from some physical phenomena encountered in CCD cameras, e.g. charge leaking in neighbouring cells.

2.4 Nonparametric Representation for Multi-channel Images

For scalar valued images a definition of a certain order among pixels does not pose a problem. Things become more complicated for non-scalar valued images, such as colour or in general matrix or tensor valued. Basically the following schemes can be used in the latter case:

1. Compute the nonparametric representation in each channel separately resulting in the nonparametric multi-channel representation;
2. Convert the multi-channel signal into scalar valued, then compute the nonparametric representation;
3. Compute the nonparametric representation directly from the multi-channel representation.

A direct application of the first option of computing the nonparametric representation for each channel separately leads to the number of bits multiplied by a number of channels. This showed up to be an over-representation which due to the usual strong correlation among the channels does not improve the matching results [3]. Needless to say that such a method requires more memory and computation time. Therefore the novel idea is to combine the nonparametric representations from separate channels into one channel, however in a manner that conveys some additional information. One of the methods relies on a dominating-bit voting scheme – called a Dominating-Census (DC). It is defined for multi channel images with odd number of channels. For a given pixel position all bits are checked separately in each of the (binary) channels. Then the most dominating value of bits is chosen – see Table 1. It is easy to verify that this nonlinear method is different from the second option of converting a colour image into a scalar one.

Table 1. The Dominating-Census (DC) transformation for odd-multi-channel images (three colour channels assumed for simplicity). At first the Census is computed in each channel separately, then the dominating value across the channels is chosen as an output.

Channel	Bit values							
r	0	0	0	0	1	1	1	1
g	0	0	1	1	0	0	1	1
b	0	1	0	1	0	1	0	1
DC	0	0	0	1	0	1	1	1

It is easy to derive the logical expression describing the Dominating-Census (DC) majority voting rule presented in Table 1 – it is as follows:

$$DC = (r \wedge g) \vee (r \wedge b) \vee (g \wedge b) = r \wedge (g \vee b) \vee (g \wedge b) \qquad (3)$$

The last expression requires four logical operations per one output bit.

The second of the aforementioned options is a conversion from the multi-channel to scalar representation. In the next part we limit our considerations to the three-channel colour images, although the concepts can be easily extended into more channels.

There are many ways to convert multi-channel colour images into grey valued ones by means of the linear combination of the colour components [10]. For example the very common IHS coordinate system provides a quantitative means of specifying the intensity I, as well as saturation S and colour H:

$$I = \begin{bmatrix} \dfrac{1}{3} & \dfrac{1}{3} & \dfrac{1}{3} \end{bmatrix} \begin{bmatrix} R & G & B \end{bmatrix}^T \tag{4}$$

where $[R\ G\ B]^T$ is a three-channel colour vector for a given pixel, I a grey value (intensity) of that pixel. The other colour space YC_rC_b provides us with the following:

$$Y = \begin{bmatrix} 0.299 & 0.587 & 0.114 \end{bmatrix} \begin{bmatrix} R & G & B \end{bmatrix}^T \tag{5}$$

where Y represents a grey value.

The last option is computation of the nonparametric representation directly from the multi-channel representation. The three schemes presented in Fig. 6 –Fig. 8 were tested for purpose of this research. In all of them we assumed the same number of bits as for the uncompressed representation in a single channel (e.g. 8 bits for 3×3 neighbourhood). However the 'cube of influence' (i.e. a space around a central pixel) was of size $n{\times}n{\times}n$ (the best results were obtained for $n=4\text{-}6$).

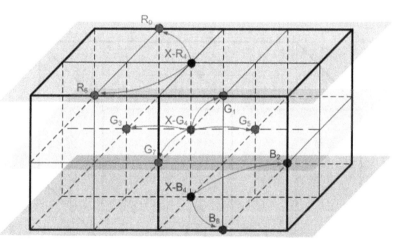

Fig. 6. The Census transformation computed directly in a $n{\times}n{\times}n$ neighbourhood of pixels. Three reference values of a single pixel (X) are used in this scheme.

Fig. 6 presents the star-like representation with one reference pixel. However, each component of the reference is used as a scalar reference in the corresponding channel, i.e. red to red, etc. This scheme was motivated by the fact of usual strong correlation among channels. The relations X-R0 and X-R6, as well as X-B2 and X-B8, convey information among the most distant pixels in the cube.

Fig. 7 presents a mutation of the scheme presented in Fig. 6. Only one component value of a central pixel (X-G4) is used. The other pixels are set in accordance with the compressed scheme from Fig. 3.

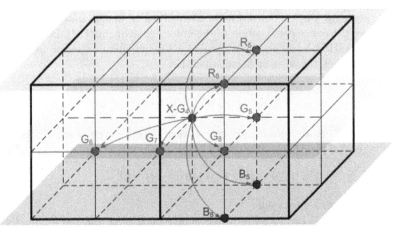

Fig. 7. The Census transformation in a $n\times n\times n$ neighbourhood. One reference pixel (X-G4) used, however other pixels set in accordance with the compressed scheme from Fig. 3.

Fig. 8 depicts another mutation of the scheme from Fig. 7. This time the three reference values (X) are used, each is located in a different channel as in Fig. 6.

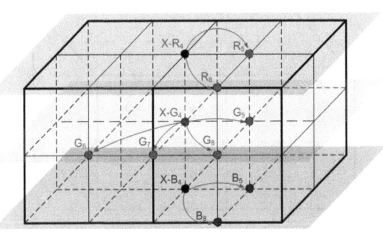

Fig. 8. A mutation of the scheme from Fig. 7. Three components of a single pixel (X) used.

In all direct schemes only 8 bits are computed for a single pixel. In experiments the best results were obtained if the pixels were distant by 4 pixels.

3   Distance Measures for Nonparametric Representation

The most popular measure for comparison of binary strings of the Census representation is the Hamming D_H measure. It treats all bits (0 or 1) with the same weight. However, other metrics, such as Tanimoto D_T or Dixon-Koehler D_{DK} are also possible, which stress more matches on '1s' than on '0s' in a specific way. The latter can be of some advantage if bit '0' is defined in (2) to convey information for weak order (i.e. for the equality) in (1). In such situations the potential areas of the same intensity, which usually cause problems for matching, are treated with a littlelower weight. For completeness we provide the formulas for the mentioned measures [5][6]:

$$D_H\left(\mathbf{a},\mathbf{b}\right) = \frac{1}{N}\sum_{i=1}^{N} a_i \otimes b_i ,\qquad(6)$$

$$D_T\left(\mathbf{a},\mathbf{b}\right) = \begin{cases} 1 & if \quad \mathbf{a} = \mathbf{b} = \mathbf{0} \\ 1 - \dfrac{\mathbf{a}^{\mathrm{T}}\mathbf{b}}{\mathbf{a}^{\mathrm{T}}\mathbf{a} + \mathbf{b}^{\mathrm{T}}\mathbf{b} - \mathbf{a}^{\mathrm{T}}\mathbf{b}} & otherwise \end{cases},\qquad(7)$$

$$D_{DK}\left(\mathbf{a},\mathbf{b}\right) = D_H\left(\mathbf{a},\mathbf{b}\right)D_T\left(\mathbf{a},\mathbf{b}\right).\qquad(8)$$

where a, b are the compared vectors of the same length N, $\otimes$ denotes the exclusive-or operation.

In this paper we used also the weighted Tanimoto D_{WT} measure [15], which originates from the biological and chemical sciences, and is defined as follows [8]:

$$D_{WT}\left(\mathbf{a},\mathbf{b}\right) = \eta D_T\left(\mathbf{a},\mathbf{b}\right) + \left(1-\eta\right)D_T\left(\neg\mathbf{a},\neg\mathbf{b}\right),\qquad(9)$$

where $\neg$ denotes bit negation. The second term in the above is called a complement of D_T. The weight parameter η stabilizes situations of strong correlations only on '1s' or only on '0s'. In [8] its value is proposed as follows:

$$\eta = \frac{2-p}{3}, \text{ where } p = \frac{\mathbf{a}^{\mathrm{T}}\mathbf{a} + \mathbf{b}^{\mathrm{T}}\mathbf{b}}{2N}.\qquad(10)$$

Certainly, $p \in [0,1]$ and $\mathbf{a}^{\mathrm{T}}\mathbf{a}$ is a number of '1s' in $\mathbf{a}$, while $\mathbf{b}^{\mathrm{T}}\mathbf{b}$ in $\mathbf{b}$. However, in our experiments we found that better results are obtained for $\eta=(3-p)/4$, since it always favours matches on '1s', which is preferable.

4   Experimental Results

The presented system was implemented with the Microsoft® Visual C++ 6.0 on the IBM PC with Pentium 4 3.4G and 2 GB RAM. It was built upon the image matching platform presented in [5]. The test images provided by the Middlebury University were used since they are colour versions and are endowed with the ground-truth disparity maps [14][13]. The following experimental results are organised to verify the consecutive concepts presented in this paper. Then the final conclusions are drawn that can lead further research and implementations.

The first tested concept is comparison of the full and compressed Census representations (2.2) and different distance measures (3).

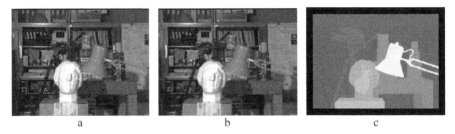

a   b   c

Fig. 9. The "Tsukuba" 384×288 colour image used for experiments: left image (a), right (b), ground-truth map (max. disparity 15) (c)

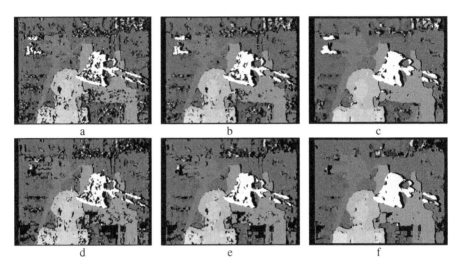

a   b   c

d   e   f

Fig. 10. Comparison of "Tsukuba" disparities for the reduced (a,b,c) and full (d,e,f) 5×5 Census representations. The aggregation windows: 5×5 (a,d), 7×7 (b,e), 11×11 (c,f). The Hamming measure used for comparisons; the colour pixels were linearly transformed to grey values.

a   b   c

Fig. 11. The "Venus" 434×383 colour image used for experiments: left image (a), right (b), ground-truth map (max. disparity 19) (c)

The tested images are presented in Fig. 9 for 384×288 "Tsukuba", and in Fig. 11 for 434×383 "Venus". In these experiments the colour pixels were linearly transformed into grey values in accordance with (5). The other two test images are: 434×380 "Sawtooth" and 284×216 "Map" (grey valued).

Fig. 10 depicts comparisons of the disparity computation with the full (d,e,f) and reduced (a,b,c) 5×5 *Census* representations for the "Tsukuba". The aggregation windows were: 5×5 (a,d), 7×7 (b,e), 11×11 (c,f). The Hamming measure was used for comparisons and the colour pixels were linearly transformed to grey values (5).

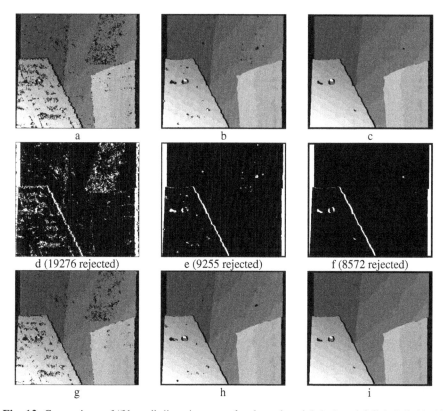

Fig. 12. Comparison of "Venus" disparity maps for the reduced (a,b,c) and full (g,h,i) 11×11 *Census*. The rejected points (white) after the cross-checking of the reduced representations (d,e,f). The aggregation windows: 3×3 (a,d,g), 9×9 (b,e,h), 13×13 (c,f,i). The Weighted-Tanimoto D_{WT} measure used; the colour pixels were linearly transformed to grey values.

The similar tests for the "Venus" test pair and the Weighted-Tanimoto comparison measures presents Fig. 12. The maps in Fig. 12d,e,f show pixels removed by the cross-checking process which detects the inconsistencies in disparity maps [9]. These and many other tests validate our concept of half bit reduction for the aggregated blocks of pixels in the Census domain. These important results allowed *reduction* of the computation time from 25-45%, depending on the size of the matching blocks. It also shows that better results are obtained for considerably larger *Census* represent-tation and possibly smaller aggregation blocks, what is consistent with the results presented in [5]. Quantitative accuracy assessments of the methods can be found in

Table 2. The "Gr-Tr" column contains the overall error rate of the computed disparities and the ground-truth. This is the ratio of the total number of bad matches and total number of pixels. The second column denotes ratio of inconsistent pixels that were rejected by the cross-checking process to the total number of pixels. The last column contains an average execution time in seconds.

Table 2. Accuracy and computation time for different methods, images and settings ($n \times nC$ stands for *Census* size, $k \times kM$ denotes matching block size, *Lin* – linear colour conversion, time in seconds). Values in parenthesis concern the redundant (full) *Census* representations.

Method	Tsukuba (384×288)			Venus (434×383)			Sawtooth (434×380)		
	Gr-T	Mis	Tme	Gr-T	Mis	Tme	Gr-T	Mis	Tme
5×5C, 5×5M Lin, D_H	19.4	0.21 (0.2)	0.7 (1.2)	18.2	0.21 (0.2)	1.1 (1.9)	16	0.18	1
5×5C, 7×7M Lin, D_H	17	0.17 (0.2)	1.2 (2.3)	17.4	0.13 (0.15)	1.8 (3.5)	15.8	0.19	1.8
5×5C, 11×11M Lin, D_H	13.3	0.12 (0.1)	2.4 (4.1)	16.9	0.2 (0.23)	3.8 (6.2)	13.1	0.1	3.8
11×11C, 3×3M Lin, D_{WT}	11.4	0.08	3.8	11.4	0.116	4.1	9.14	0.07	4
11×11C, 9×9M Lin, D_{WT}	10.2	0.05	4.7	10.2	0.055	6.6	7.2	0.06	6.2
11×11C, 13×13 Lin, D_{WT}	13.7	0.06	6.4	13.7	0.05	9.1	6.23	0.06	9.7

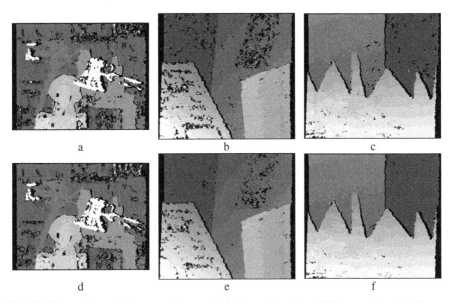

a   b   c

d   e   f

Fig. 13. Comparison of the conversion methods from the colour RGB space into the Census representation: the Dominating-Census (a,b,c) vs. linear (d,e,f). "Tsukuba" (a,d), "Venus" (b,e), and "Sawtooth" (c,f). All disparity maps were obtained in the 11×11 Census representation, 5×5 matching blocks, and with the Hamming measure.

It is interesting to notice that the reduced bit version of the aggregated matches shows even greater accuracy (Table 2), although a visual results seem to be worse (compare Fig. 10 abc vs. def).

The next group of experiments was conducted to assess differences between the linear conversions of the colour images and the nonlinear Dominating-Census method presented in (2.4). Fig. 13 and present results of comparison of the conversion methods from the colour RGB space into the Census representation for the three test images. The Dominating-Census (a,b,c) vs. linear (d,e,f). All disparity maps were obtained in the 11×11 Census, 5×5 matching blocks, and with the Hamming measure. These results show no noticeable difference between the two conversion methods.

Table 3. Comparison of different matching methods (Maj.- majority voting, nb-C – Census representation with n-bits, DH – Hamming distance, Tme – time [s]). The best values in bold.

Method		Tsukuba (384×288)			Venus (434×383)			Sawtooth (434×380)		
		Gr-T	Mis	Tme	Gr-T	Mis	Tme	Gr-T	Mis	Tme
Max.disp.16, Cross check, 5x5 Match window, D_H	Lin, 12b-C	20.67	0.22	2.2	22.21	0.13	3.6	26.75	0.09	3.2
	Maj., 12b-C	20.64	0.23	2	23.52	0.17	3.3	27.44	0.12	3.0
	Fig. 6, 8b-C	**15.70**	**0.14**	1.3	22.53	0.14	2.4	25.22	0.1	2.3
	Fig. 7, 8b-C	17.27	0.15	1.3	23.29	0.16	2.3	26.28	0.12	2.3
	Fig. 8, 8b-C	17.32	0.17	1.4	23.08	0.15	2.3	27.66	0.12	2.3
	Sparse, 8b-C	16.97	0.18	1.4	**21.22**	**0.09**	2.4	**25.7**	**0.08**	2.4
Max.disp.16, Cross check, 9x9 Match window, D_H	Lin, 12b-C	15.87	0.15	6	20.34	0.07	9.9	24.46	0.07	9.9
	Maj., 12b-C	15.74	0.15	6	21.04	0.08	9.7	24.75	0.07	9.8
	Fig.6, 8b-C	**12.45**	**0.09**	4.4	21.18	0.09	6.8	23.37	0.07	6.6
	Fig. 7, 8b-C	13.69	0.1	4.4	21.36	0.1	6.9	23.76	0.07	6.9
	Fig.8, 8b-C	13.55	0.11	4.3	21.47	0.09	6.8	24.25	0.07	6.8
	Sparse, 8b-C	14.22	0.12	4.4	**19.94**	**0.05**	6.7	**23.40**	**0.06**	6.7

Table 3 presents comparison among different methods for three colour images. For methods that convert colour into scalars we use 12 bits of the Census representation, for colour we use 8 bits only. The best methods are: from Fig. 6 operating directly in the RGB space, and the sparse method in Fig. 5 that converts to scalar.

The HSI colour space [10], as well as the normalized HSI space, were also tested but no improvements were detected compared to the already presented methods that operate in the RGB space (2.4).

5 Conclusions

This paper presents different matching techniques that operate in the nonparametric representation of multi-channel (non scalar) images. It extends the work presented in [5][3]. The conclusions drawn from this research can be summarized as follows:

1. The best matching methods are: directly in the RGB space – the method operating according to the scheme in Fig. 6, for colour and grey valued images – the method with *sparse Census sampling model*, presented in Fig. 5.
2. The bit reduction method in the aggregated block matching in the Census domain (Fig. 3) allows a 50% reduction of bits without noticeable loss of accuracy.

3. The sparse Census sampling model allows increase of the discriminative properties of the Census representation and further reduction of bits (Fig. 5).
4. The type of a binary matching measure does not influence much results, although the Weighted-Tanimoto allows control of the balance between match solely on '1' bits vs. '0' bits. However, the preferable (due to shortest computation) is the Hamming measure.

The presented techniques showed to be robust against noise and local image imbalances. The software implementation does not require other than integer arithmetic. The version with the fixed 11x11 Census and 5x5 matching window was also implemented in FPGA and successfully operates in real-time.

Acknowledgement

This work was sponsored by the scientific grant no. KBN 3T11C 045 26 of the Polish Committee for Scientific Research.

References

1. Banks J., Bennamoun M., Corke P.: Non-Parametric Techniques for Fast and Robust Stereo Matching. CSIRO Manufacturing Science and Technology, Australia (1997)
2. Bhat D.N., Nayar S.K.: Ordinal Measures for Image Correspondence. IEEE Transaction on Pattern Analysis and Machine Intelligence Vol. 20 No. 4 (1998)
3. Cyganek, B., Socha Ł.: Comparison Of Matching Strategies For Colour Images. Int. Conf. on Computer Vision Theory and Applications – VISAPP, Setúbal, Portugal (2006)
4. Cyganek, B.: Adaptive Window Growing Technique for Efficient Image Matching, Springer LNCS 3522, (2005) 308-315
5. Cyganek, B.: Comparison of Nonparametric Transformations and Bit Vector Matching for Stereo Correlation. Springer LNCS 3322 (2004) 534-547
6. Dixon, S.L., Koehler, R.T. J. Med Chem. 42 (1999) 2887–2900
7. Duda, R.O., Hart, P.E., Stork, D.G.: Pattern Classification. Wiley (2001)
8. Fligner, M., Verducci, J., Bjoraker, J., Blower, P.: A new association coefficient for molecular dissimilarity. Conf. on Chemoinformatics. Sheffield, England (2001)
9. Fua P.: A Parallel Stereo Algorithm that Produces Dense Depth Maps and Preserves Image Features, INRIA Technical Report No 1369 (1991)
10. Gonzalez, R.C., Woods, R.E.: Digital Image Processing. Prentice-Hall (2002)
11. Hartley, R.I., Zisserman A.: Multiple View Geometry in Computer Vision. CUP (2000)
12. Koshan, A.: Improving Robot Vision by Color Information. Proc. 7th International Conference on Artificial Intelligence and Information-Control Systems of Robots. Smolenice, Slovakia, 10-14 September, 1997
13. Scharstein, D., Szeliski, R.: A Taxonomy and Evaluation of Dense Two-Frame Stereo Correspondence Algorithms. Int. Journal of Computer Vision 47(1/2/3) (2002) 7-42
14. Scharstein, D., Szeliski, R.: www.middlebury.edu/stereo (2005)
15. Sloan Jr., K. R., Tanimoto, S. L.: Progressive Refinement of Raster Images, IEEE Transactions on Computers, Vol. 28, No. 11 (1979) 871-874
16. Yang, R., Pollefeys, M.: A versatile stereo implementation on commodity graphics hardware. Real-Time Imaging, 11 (2005) 7-18
17. Zabih, R., Woodfill, J.: Non-Parametric Local Transforms for Computing Visual Correspondence. Proc. Third European Conf. Computer Vision (1994) 150-158

Approximating Euclidean Distance Using Distances Based on Neighbourhood Sequences in Non-standard Three-Dimensional Grids

Benedek Nagy[1] and Robin Strand[2]

[1] Department of Computer Science, Faculty of Informatics, University of Debrecen,
PO Box 12, 4010, Debrecen, Hungary and
Research Group on Mathematical Linguistics,
Rovira i Virgili University, Tarragona, Spain
nbenedek@inf.unideb.hu
[2] Centre for Image Analysis, Uppsala University,
Lägerhyddsvägen 3, SE-75237 Uppsala, Sweden
robin@cb.uu.se

Abstract. In image processing, it is often of great importance to have small rotational dependency for distance functions. We present an optimization for distances based on neighbourhood sequences for the face-centered cubic (fcc) and body-centered cubic (bcc) grids. In the optimization, several error functions are used measuring different geometrical properties of the balls obtained when using these distances.

1 Introduction

When computing the distance transform of a segmented image, each object grid point is assigned the value to the closest background grid point. In [1], the distance transform is defined using the classical city block and chessboard distances, defined as the shortest path between two grid points using only 4- and 8-neighbours, respectively. Seen as approximations of the Euclidean distance, these distances are very rough. On the other hand, they have other advantages. For example, when computing a reversible skeleton, the centres of maximal balls (the grid points needed for the skeleton to be reversible) are very easy to extract; they are local maxima in the distance transform, [1]. In [2], it is noted that by mixing the city block and chessboard distances, the Euclidean distance is better approximated; the authors state that the approximation obtained when the ratio between the number of steps using the different neighbourhood relations in $\mathbb{Z}^2$ is equal to $1 : \sqrt{2}$ is optimal. These distances can also be used for computing reversible skeletons that include the centres of maximal balls, [3].

There is another very common way of modifying the city block and chessboard distances in order to obtain a less rotational dependent distance, the *weighted* distances, [4, 5]. With these distances, each local step is given a weight, which is considered when computing the distance, i.e., when finding the shortest path. The calculation of optimal local weights have been the subject of many papers, see for example [5, 6].

U. Eckardt et al. (Eds.): IWCIA 2006, LNCS 4040, pp. 89–100, 2006.

Non-standard grids in 2D and 3D have also been considered. Weighted distances for the two-dimensional hexagonal grid is examined in [7, 8] and a skeletonization algorithm in [9]. The generalizations of the hexagonal grid in three dimensions are the face-centered cubic (fcc) and body-centered cubic (bcc) grids. Weighted distances on these grids have also been examined, [8] and a skeletonization algorithm for the fcc and bcc grids is found in [10].

There are several reasons for using the fcc and bcc grids. For example, since these grids are reciprocal and both have higher packing density than the cubic grid (the fcc grid is a densest packing lattice in 3D), less samples can be used without affecting the image representation/reconstruction quality, [11]. The high number of neighbours at approximately the same distance on these grids (12 neighbours at distance $\sqrt{2}$ for the fcc grid and 14 neighbours at distance $\sqrt{3}$ and 2 for the bcc grid) implies that the rotational dependency for these grids is lower than for the cubic grid. Many image reconstruction techniques for computed tomography images can easily be adjusted to work on the fcc and bcc grids. For example, the filtered backprojection method use 1D Fourier transforms corresponding to projections of the original object. The dependence on the grid on which the reconstructed image will be on comes first in the last step, which is an interpolation from the filtered 1D projections to the grid points. The algebraic reconstruction technique is applied to the bcc grid in [11].

The distances obtained by mixing steps corresponding to 4- and 8-neighbours suggested in 1968 by Rosenfeld [2] are called *distances based on neighbourhood sequences*. The literature on distances based on neighbourhood sequences is rich; a theory for periodic neighbourhood sequences not connected to any specific neighbourhood relations in $\mathbb{Z}^n$ is presented in [12, 13] and further developed for the natural neighbourhood structure, by the so-called octagonal neighbourhood sequences in [14, 15]. Results for general (not necessarily periodic) neighbourhood sequences are presented in [16, 17].

Many approaches where the deviation from the Euclidean distance is minimized in order to find the optimal neighbourhood sequence have been proposed for $\mathbb{Z}^2$ and $\mathbb{Z}^3$. Several error functions minimizing the asymptotic maximum difference of two balls of equal radius using a distance based on periodic neighbourhood sequences and a Euclidean sphere, respectively, is presented in [13, 18, 19] ($\mathbb{Z}^2$), [20] ($\mathbb{Z}^3$). An investigation of optimal non-periodic neighbourhood sequences in $\mathbb{Z}^2$ with optimal sequences also for finite distances is found in [21]. In [22], an optimization for $\mathbb{Z}^3$ is carried out using a geometric approach.

All above approaches are based on the difference between a Euclidean ball and a ball generated by distances based on neighbourhood sequences of the same radius. Such error functions are natural in $\mathbb{Z}^n$, because a ball generated by a distance obtained by only considering first-order/n-th order neighbours will always be an underestimation/overestimation of the Euclidean ball of the same radius. Using non-standard grids as in this paper, there is in general no order of neighbours generating a distance that is an overestimation of the Euclidean distance. Instead, error functions that are independent of the radius of the Euclidean ball are used in the optimization. In [23], optimal neighbourhood sequences for the

2D hexagonal and triangular grids are found using a "non-compactness ratio" – the ratio between the squared perimeter and the area of the convex hull of the disks obtained by using neighbourhood sequences. In this paper, we calculate the neighbourhood sequences best approximating the Euclidean distance by using different error functions for the fcc and bcc grids. We use the formulas for distances based on neighbourhood sequences derived in [24] to find formulas that describe the geometry of the balls in these grids. The formulas are then used for the error functions.

2   Notation and Preliminary Results

The following definitions of the fcc and bcc grids are used:

$$\mathbb{F} = \{(x, y, z) : x, y, z \in \mathbb{Z} \text{ and } x + y + z \equiv 0 \pmod{2}\}. \tag{1}$$

$$\mathbb{B} = \{(x, y, z) : x, y, z \in \mathbb{Z} \text{ and } x \equiv y \equiv z \pmod{2}\}. \tag{2}$$

When the two grids are handled in parallel, $\mathbb{G}$ is used to denote either $\mathbb{F}$ or $\mathbb{B}$. Observe that, using these definitions, each grid point has integer coordinates.

Two grid points $p, q \in \mathbb{G}$ are r-neighbours, $1 \leq r \leq 2$ if

1. $\sum_{i=1}^{3} |p(i) - q(i)| \leq 3$ and
2. $\max_{i \in \{1,2,3\}} |p(i) - q(i)| \leq r$

The points $p, q \in \mathbb{G}$ are *adjacent* if p and q are r-neighbours for some r. The neighbourhood relations are visualized in Figure 1 by showing the Voronoi regions (the voxels) corresponding to some grid points.

Fig. 1. The light grey voxel is 1-neighbour to the dark grey voxel. The white and light grey voxels are 2-neighbours to the dark grey voxel. Left: fcc, right: bcc.

The r-neighbours which are not $(r-1)$-neighbours are called *strict* r-neighbours. In addition, we will use path-generated distances, therefore, the terms 1- and 2-steps will also be used insted of step to a 1-neighbour and step to a 2-neighbour, respectively.

A neighbourhood sequence B is a sequence of integers $b(i)$, $B = (b(i))_{i=1}^{\infty}$. If B is periodic, i.e., if for some fixed $l \in \mathbb{N}$ ($l > 0$), $b(i) = b(i + l)$ is valid for all $i \in \mathbb{N}$, then B is written $B = (b(1), b(2), \ldots, b(l))$.

A *path* in a grid is a sequence $(p = p_0), p_1, \ldots, (p_m = q)$ of adjacent grid points. A path is a *B-path* of length m if, for all $i \in \{1, 2, \ldots, m\}$, p_{i-1} and p_i are $b(i)$-neighbours. The *B-distance* $d(p, q; B)$ is defined as the length of the shortest B-path(s) between p and q. Usually in digital geometry the shortest path is not unique. The distance function generated by B is denoted $d(B)$.

Let

$$\mathbf{1}_B^k = |\{i : b(i) = 1, 1 \leq i \leq k\}| \text{ and}$$

$$\mathbf{2}_B^k = |\{i : b(i) = 2, 1 \leq i \leq k\}|.$$

When B is clear from the context, we will exclude the subscript and write $\mathbf{1}^k$ and $\mathbf{2}^k$ for short. Note that, for any B and any k, $\mathbf{1}_B^k + \mathbf{2}_B^k = k$.

Definition 1. *Let A and B be two neighbourhood sequences in $\mathbb{G}$. The relation $A \sqsupseteq^* B$ (A is faster than B) is defined as*

$$d(p, q; A) \leq d(p, q; B) \ \forall p, q \in \mathbb{G}. \tag{3}$$

Definition 2. *For any neighbourhood sequence $B = (b(i))_{i=1}^{\infty}$, the sequence $B(j) = (b(i))_{i=j}^{\infty}$ is the j-shifted sequence of B.*

The following theorems are from [24]. Observe that Theorem 2 gives a computationally efficient way of deciding if a distance generated by a neighbourhood sequence is a metric or not.

Theorem 1. *The distance function based on a neighbourhood sequence B is a metric on $\mathbb{G}$ if and only if $B(i) \sqsupseteq^* B$ for all $i \in \mathbb{N}$.*

Theorem 2. *For the fcc and bcc grids, a neighbourhood sequence A is faster than a neighbourhood sequence B if and only if*

$$\sum_{i=1}^{j} a(i) \geq \sum_{i=1}^{j} b(i) \quad \text{for all } j \in \mathbb{N}.$$

This can be written as the following equivalent condition:

$$\mathbf{2}_A^j \geq \mathbf{2}_B^j \quad \text{for all } j \in \mathbb{N}.$$

3   Balls in $\mathbb{F}$ and $\mathbb{B}$ Generated by Distances Based on Neighbourhood Sequences

A ball $\mathcal{B}_{\mathbb{G}}$ and a sphere $\mathcal{S}_{\mathbb{G}}$ in $\mathbb{G}$ are defined as

$$\mathcal{B}_{\mathbb{G}}(B, k) = \{q \in \mathbb{G} : d(0, q; B) \leq k\} \text{ and}$$

$$\mathcal{S}_{\mathbb{G}}(B, k) = \{q \in \mathbb{G} : d(0, q; B) = k\}, \text{ respectively.}$$

The vector (x, y, z) is called the sorted absolute difference vector of p and q, if it is a permutation of the values $d(i) = |p(i) - q(i)|$ such that $x \geq y \geq z$. Theorem 3 and 4 are proven in [24].

Theorem 3 (*B*-distance in $\mathbb{F}$). *Let a neighbourhood sequence B be given. Let d be the distance of the two grid points $(0,0,0)$ and (x, y, z) in $\mathbb{F}$ (having sorted absolute difference vector (x, y, z)). Then*

$$d = \min\left\{ k \;\middle|\; k = \max\left\{ \frac{x+y+z}{2}, x - 2^k \right\} \right\}.$$

Corollary 1. *Let the neighbourhood sequence B and the positive integer k be given. The value of $D = \left(x^2 + y^2 + z^2 : (x, y, z) \in S_{\mathbb{F}}(B, k)\right)$ is*

– *maximized by*

$$\left(1_B^k + 2 \cdot 2_B^k, 1_B^k, 0\right). \tag{4}$$

– *minimized by either*

$$\left\lfloor \frac{k}{3} \right\rfloor \cdot (2, 2, 2) + \left(k \bmod 3, \left\lceil \frac{k \bmod 3}{2} \right\rceil, \left\lfloor \frac{k \bmod 3}{2} \right\rfloor \right) \tag{5}$$

or

$$\left(2 \cdot 2^k + 1^k, 0, 0\right) \; \text{if } 1^k \text{ is even} \tag{6}$$

$$\left(2 \cdot 2^k + 1^k, 1, 0\right) \; \text{otherwise} \tag{7}$$

Proof. For (4), see [24].

The minimum of D is reached when the value of $\max\left\{ \frac{x+y+z}{2}, x - 2^k \right\}$ is maximal. This occurs when either $\frac{x+y+z}{2}$ or $x - 2^k$ is maximal (independent of the other argument).

Obviously, the value of $\frac{x+y+z}{2}$ is maximal when $x + y + z = 2k$. This is the case for a set of grid points. The grid point satisfying $d((x, y, z), (0, 0, 0); B) = k$ that minimizes D can be written as either $(2l, 2l, 2l)$, $(2l + 1, 2l + 1, 2l)$, or $(2l + 2, 2l + 1, 2l + 1)$ for some l. This is the case even if $2^k = k$, i.e. if all steps are 2-steps. Since any 1-neighbour is also a 2-neighbour and since, for any $k \geq 0$,

$$(2k + 2)^2 + (2k)^2 + (2k)^2 \geq (2k + 1)^2 + (2k + 1)^2 + (2k)^2 \text{ and}$$

$$(2k + 2)^2 + (2k + 2)^2 + (2k)^2 \geq (2k + 2)^2 + (2k + 1)^2 + (2k + 1)^2,$$

we conclude that using 1-steps gives shorter distance and thus, the optimal grid point can be written as a sum of 1-steps. Equation (5) follows.

The second argument $x - 2^k$ is maximized when x is maximized. Equation (6) and (7) are obviously the grid points best approximating this. $\square$

Theorem 4 (*B*-distance in $\mathbb{B}$). *Let a neighbourhood sequence B be given. Let d be the distance of the two grid points $(0,0,0)$ and (x, y, z) in $\mathbb{B}$ (having sorted absolute difference vector (x, y, z)). Then*

$$d = \min\left\{ k \;\middle|\; k = \max\left\{ \frac{x+y}{2}, x - 2^k \right\} \right\}.$$

Corollary 2. *Given a neighbourhood sequence B and a positive integer k, the value of $D = \left(x^2 + y^2 + z^2 : (x, y, z) \in S_\mathbb{B}(B, k) \right)$ is*

– *maximized by*

$$\left(1_B^k + 2 \cdot 2_B^k, 1_B^k, 1_B^k \right) \qquad if \qquad 2 \cdot 1_B^k \leq 2_B^k \tag{8}$$
$$(k, k, k) \quad otherwise. \tag{9}$$

– *minimized by either*

$$(k, k, 0) \qquad if \qquad k \ is \ even \tag{10}$$
$$(k, k, 1) \quad otherwise \tag{11}$$

or

$$\left(1^k + 2 \cdot 2^k, 0, 0 \right) \qquad if \qquad 1^k \ is \ even \tag{12}$$
$$\left(1^k + 2 \cdot 2^k, 1, 1 \right) \quad otherwise \tag{13}$$

Proof. For (8) and (9), see [24].

The minimum of D is reached when the value of $\max \left\{ \frac{x+y}{2}, x - 2^k \right\}$ is maximal. This occurs for all x, y, and z such that either $\frac{x+y}{2}$ or $x - 2^k$ is maximal. Among the grid points satisfying this, $x + y = 2k$ and z as close to zero as possible minimize D. By noting that

$$(k + 2)^2 + k^2 + 0^2 \geq (k + 1)^2 + (k + 1)^2 + 1^2$$

for $k \geq 0$, and thus, using only 1-steps give minimal D, the optimal grid points are obtained by using combinations of only $(1, 1, 1)$, and $(1, 1, -1)$ as in (10) and (11).

By setting $x = k$, the value of $x - 2^k$ is maximal. With this satisfied, values of y, z as close to zero as possible minimize D. In $\mathbb{B}$, with this restriction, (12) and (13) are obtained. $\qquad \square$

4   Balls Generated by Distances Based on Neighbourhood Sequences in $\mathbb{R}^3$

In this section, the convex hull of $\mathcal{B}_\mathbb{G}(B, k)$ in $\mathbb{R}^3$, denoted $\mathcal{H}_\mathbb{G}(B, k)$, is considered. The shape of $\mathcal{H}_\mathbb{G}(B, k)$ for different B and k is shown in Figure 2.

Observe that when $2^k = 0$, $\overrightarrow{QR} = (0, 0, 0)$ and when $1^k = 0$, then $\overrightarrow{PQ} = (0, 0, 0)$.

Given an integer k and a neighbourhood sequence B, the grid points in $\mathcal{B}_\mathbb{G}(B, k)$ located at maximum Euclidean distance from the origin are given by Corollary 1 and 2. Since these points also are in the convex hull, we have the following result:

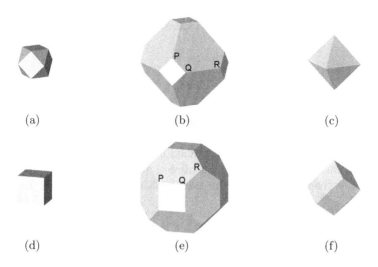

Fig. 2. The shape of $\mathcal{H}_{\mathbb{G}}(B, k)$, where $B = (1), k = 1$ (left), $B = (1, 2), k = 2$ (middle), and $B = (2), k = 1$ (right). Up: fcc, down: bcc.

Lemma 1. *The vertices of $\mathcal{H}_{\mathbb{G}}(B, k)$ with $x \geq y \geq z \geq 0$ are:*

$$\mathbb{F}: \left(\mathbf{1}^k + 2 \cdot \mathbf{2}^k, \mathbf{1}^k, 0\right) \tag{14}$$

$$\mathbb{B}: \left(\mathbf{1}^k + 2 \cdot \mathbf{2}^k, \mathbf{1}^k, \mathbf{1}^k\right) \quad and \quad (k, k, k). \tag{15}$$

Proof. A direct consequence of Corollary 1 and 2. □

Using the vertices of the polyhedra given in Lemma 1, the surface area and the volume of the polyhedra are computed.

Lemma 2. *Given a neighbourhood sequence B and a positive integer k, the surface area A and the volume V of $\mathcal{H}_{\mathbb{G}}(B, k)$ are*

$$A_{\mathbb{F}} = 16 \left(\mathbf{1}^k + \mathbf{2}^k\right)^2 \sqrt{3} - 12 \left(\mathbf{1}^k\right)^2 \sqrt{3} + 12 \left(\mathbf{1}^k\right)^2,$$
$$A_{\mathbb{B}} = 24 \left(\mathbf{1}^k\right)^2 + 24 \left(\mathbf{2}^k\right)^2 \sqrt{2} + 48 \cdot \mathbf{2}^k \mathbf{1}^k \sqrt{2},$$
$$V_{\mathbb{F}} = \tfrac{32}{3} \left(\mathbf{2}^k\right)^3 + 32 \left(\mathbf{2}^k\right)^2 \mathbf{1}^k + 32 \cdot \mathbf{2}^k \left(\mathbf{1}^k\right)^2 + \tfrac{20}{3} \left(\mathbf{1}^k\right)^3, \quad and$$
$$V_{\mathbb{B}} = 16 \left(\mathbf{2}^k\right)^3 + 48 \left(\mathbf{2}^k\right)^2 \mathbf{1}^k + 48 \cdot \mathbf{2}^k \left(\mathbf{1}^k\right)^2 + 8 \left(\mathbf{1}^k\right)^3.$$

The proof of Lemma 2 consists entirely of geometric calculations and is omitted.

Lemma 1 can also be used to compute the length of the sides of the polyhedra $\mathcal{H}_{\mathbb{G}}(B, k)$, see Figure 2:

$$|\overrightarrow{PQ}| = \sqrt{2} \, \mathbf{1}^k \quad |\overrightarrow{QR}| = 2\sqrt{2} \, \mathbf{2}^k \text{ (fcc)}$$
$$|\overrightarrow{PQ}| = 2 \, \mathbf{1}^k \quad |\overrightarrow{QR}| = \sqrt{3} \, \mathbf{2}^k \text{ (bcc)}$$

In the optimization, we need the grid points in $\mathcal{B}_{\mathbb{G}}(B, k)$ located at minimum Euclidean distance from $(0, 0, 0)$. The points in $\mathcal{H}_{\mathbb{G}}(B, k)$ located at minimum Euclidean distance are easy to find by the geometry of the polyhedra in $\mathbb{R}^3$ – they are given by the following lemma.

Lemma 3. *The points on $\partial\mathcal{H}_{\mathbb{G}}(B, k)$ such that $x \geq y \geq z \geq 0$ located at minimal Euclidean distance from $(0, 0, 0)$ are:*

$$\mathbb{F}: \qquad \left(\tfrac{2k}{3}, \tfrac{2k}{3}, \tfrac{2k}{3}\right) \qquad \text{or} \tag{16}$$

$$\left(1^k + 2 \cdot 2^k, 0, 0\right) \tag{17}$$

$$\mathbb{B}: \qquad (k, k, 0) \qquad \text{or} \tag{18}$$

$$\left(1^k + 2 \cdot 2^k, 0, 0\right). \tag{19}$$

Proof. These values of (x, y, z) are obtained by finding the point in $\mathcal{H}_{\mathbb{G}}(B, k)$ at maximal distance from the origin in the directions $(1, 1, 1)$, $(1, 0, 0)$, $(1, 1, 0)$, and $(1, 0, 0)$, respectively, as concluded in the proofs of Corollary 1 and 2.   □

The grid points closest to the origin satisfying $d((x, y, z), (0, 0, 0); B) = k$ as stated in Corollary 1 and 2 are in general not equal to the continuous counterpart. The difference is, however, bounded as is shown by the following theorem.

Theorem 5. *Considering vectors (x, y, z) such that $x \geq y \geq z \geq 0$, the (Euclidean) length of the difference vector between*

i *(5) and (16) is bounded by $\sqrt{\tfrac{2}{3}}$,*
ii *(6), (7) and (17) is bounded by 1,*
iii *(10), (11) and (18) is bounded by 1, and*
iv *(12), (13) and (19) is bounded by $\sqrt{2}$.*

Proof. For i, we note that

$$\left(\frac{2k}{3}, \frac{2k}{3}, \frac{2k}{3}\right) = \left\lfloor \frac{k}{3} \right\rfloor (2, 2, 2) + (k \bmod 3) \left(\frac{2}{3}, \frac{2}{3}, \frac{2}{3}\right)$$

The difference vector is thus

$$(k \bmod 3) \left(\frac{2}{3}, \frac{2}{3}, \frac{2}{3}\right) - \left(k \bmod 3, \left\lceil \frac{k \bmod 3}{2} \right\rceil, \left\lfloor \frac{k \bmod 3}{2} \right\rfloor \right),$$

which is $(0, 0, 0)$ for $k = 0$, $\left(-\tfrac{1}{3}, -\tfrac{1}{3}, \tfrac{2}{3}\right)$ for $k = 1$, and $\left(-\tfrac{2}{3}, \tfrac{1}{3}, \tfrac{1}{3}\right)$ for $k = 2$. The proofs of ii − iv are trivial.   □

5   Best Approximating Neighbourhood Sequences

In this section, the sequences B that give the best approximations of the Euclidean distance are calculated. Let B_r denote the Euclidean ball of radius r. The following error functions are used:

$$E_{\mathbb{G}}^1 = \max_{x,y \in \mathcal{S}_{\mathbb{G}}(B,k)} (|x| - |y|) \quad \text{(absolute error)} \tag{20}$$

$$E_{\mathbb{G}}^2 = \max_{x,y \in \mathcal{S}_{\mathbb{G}}(B,k)} \left(\frac{|x| - |y|}{|y|}\right) \quad \text{(relative error)} \tag{21}$$

$$E_{\mathbb{G}}^3 = \frac{\frac{A_{\mathbb{G}}^3}{V_{\mathbb{G}}^2} - 36\pi}{36\pi} \qquad \text{(compactness ratio)} \tag{22}$$

$$E_{\mathbb{G}}^4 = \min_{B_r \subset \mathcal{H}_{\mathbb{G}}(B,k)} (V_{\mathbb{G}} - V_{B_r}) \quad \text{(maximal inscribed ball)} \tag{23}$$

$$E_{\mathbb{G}}^5 = \min_{\mathcal{H}_{\mathbb{G}}(B,k) \subset B_r} (V_{B_r} - V_{\mathbb{G}}) \quad \text{(minimal covering ball)} \tag{24}$$

To calculate (20) and (21) for the fcc grid, equations (4), (5), (6), and (7) from Corollary 1 are needed. For the bcc grid, we need (8), (9), (10), (11), (12) and (13) from Corollary 2. Formulas for area and volume of the balls given by neighbourhood seqences are needed to calculate (22). These are given in Lemma 2. The radius of the maximal inscribed balls, (23), are given by (16), (17), (18), and (19) in Lemma 3. To calculate (24), we can use (14) and (15) from Lemma 1 since these equations give the radii of the minimal covering balls.

Observe that $E_{\mathbb{G}}^1$ and $E_{\mathbb{G}}^2$ are calculated using the discrete spheres $\mathcal{S}_{\mathbb{G}}(B,k)$ and not the border of the convex hull in $\mathbb{R}^3$, $\mathcal{H}_{\mathbb{G}}(B,k)$. Since the points in Lemma 1 and 3 are located at further distance from the origin than the corresponding values in Corollary 1 and 2, the values attained using $\mathcal{S}_{\mathbb{G}}(B,k)$ are less than or equal the values attained in $\mathcal{H}_{\mathbb{G}}(B,k)$.

The values of $\mathbf{1}^k$ and $\mathbf{2}^k$ are integers by definition, so the optimal ratio is in general not possible to achieve – since it is irrational it would require neighbourhood sequences of infinite length. By using a sufficiently long initial part of the neighbourhood sequence, the ratio can be approximated as close as needed. Among the neighbourhood sequences of length k, the ones with closest approximations of the above ratios result in the distances with the least deviation from the Euclidean distance. The convergence of the error functions are shown in Figure 3. For each $0 < k \leq 1000$, the values of $\mathbf{1}^k$ and $\mathbf{2}^k (= k - \mathbf{1}^k)$ that minimize $E_{\mathbb{G}}$ are used to calculate $E_{\mathbb{G}}$. These values of $E_{\mathbb{G}}$ are plotted in Figure 3.

The asymptotic optima are easy to calculate using the equations derived in this paper. They are:

$$\mathbf{1}^k = \frac{6 - 2\sqrt{3}}{3} k \ \text{ gives } \ E_{\mathbb{F}}^1 = \frac{2}{3}\left(\sqrt{15 - 6\sqrt{3}} - \sqrt{3}\right)k \approx 0.2763k$$

$$\mathbf{1}^k = \frac{6 - 2\sqrt{3}}{3} k \ \text{ gives } \ E_{\mathbb{F}}^2 = \sqrt{5 - 2\sqrt{3}} - 1 \approx 0.2393$$

$$\mathbf{1}^k = \frac{6 - 2\sqrt{3}}{3} k \ \text{ gives } \ E_{\mathbb{F}}^3 \approx 0.2794$$

$$\mathbf{1}^k = \frac{6 - 2\sqrt{3}}{3} k \ \text{ gives } \ E_{\mathbb{F}}^4 \approx 1.8016k^3$$

$$\mathbf{1}^k \approx 0.7924k \ \text{ gives } \ E_{\mathbb{F}}^5 \approx 3.9453k^3$$

for the fcc grid and

$$\frac{1}{3}k \leq \mathbf{1}^k \leq \left(2 - \sqrt{2}\right)k \ \text{ gives } \ E_{\mathbb{B}}^1 = \left(\sqrt{3} - \sqrt{2}\right)k \approx 0.3178k$$

$$\frac{1}{3}k \leq \mathbf{1}^k \leq \left(2 - \sqrt{2}\right)k \ \text{ gives } \ E_{\mathbb{B}}^2 = \sqrt{\frac{3}{2}} - 1 \approx 0.2247$$

$$1^k = \left(2 - \sqrt{2}\right) k \quad \text{gives} \quad E^3_{\mathbb{B}} \approx 0.2147$$

$$1^k = \left(2 - \sqrt{2}\right) k \quad \text{gives} \quad E^4_{\mathbb{B}} \approx 2.5442 k^3$$

$$1^k = \frac{1}{3} k \quad \text{gives} \quad E^5_{\mathbb{B}} = \left(4\pi\sqrt{3} - \frac{424}{27}\right) k^3 \approx 6.0619 k^3$$

for the bcc grid.

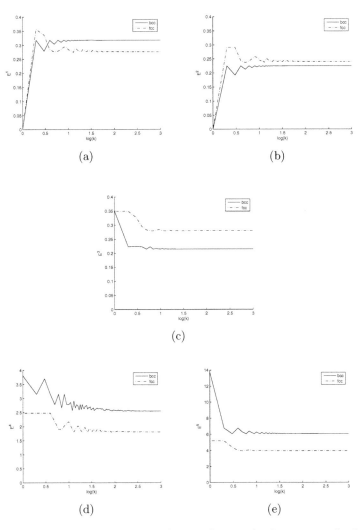

Fig. 3. The performance of $E^1_{\mathbb{G}}/k$ (a), $E^2_{\mathbb{G}}$ (b), $E^3_{\mathbb{G}}$ (c), $E^4_{\mathbb{G}}/k^3$ (d), and $E^5_{\mathbb{G}}/k^3$ (e). For each k, the value of the error function given by the values of 1^k and 2^k that gives the smallest value of the error functions are plotted. The x-axis shows $\log_{10}(k)$.

6   Conclusions

We have presented an optimization for distances based on neighbourhood sequences for the fcc and bcc grids. Using fcc or bcc there are two kinds of neighbours, therefore the 3D space can be described in a more simpler way than using the cubic grid $\mathbb{Z}^3$. For four of the five proposed error functions, the balls on the fcc grid are optimal for $1^k = \frac{6-2\sqrt{3}}{3}k$, so we conculde that this value should be as closely approximated as possible when using distances based on neighbourhood sequences. For the bcc grid, $1^k = (2 - \sqrt{2})k$ seems to be optimal, but any value between $k/3$ and $(2 - \sqrt{2})k$ is near optimal, since the absolute and the relative error are constant on this interval.

In this paper, the asymptotic error is given in closed form when possible. From the plots in Figure 3, we see that the error functions are bounded also for neighbourhood sequences of finite length. This is an important point, since in practice, neighbourhood sequences of infinite length are never used. Figure 3 shows that after the initial fluctuation the optimal value occurs rapidly. This is due to the fact that a new element of the neighbourhood sequence changes drastically the characteristics at short (periodic) sequences. So, usually a 10-length peridic sequence approximates the optimum very well. For $k = 10$, we have $E_{\mathbb{F}}^1 \approx 0.2846k$, $E_{\mathbb{F}}^2 \approx 0.2459$, $E_{\mathbb{F}}^3 \approx 0.2832$, $E_{\mathbb{F}}^4 \approx 2.1696k^3$, and $E_{\mathbb{F}}^5 \approx 3.9469k^3$ for the fcc grid and $E_{\mathbb{B}}^1 \approx 0.3178k$, $E_{\mathbb{B}}^2 \approx 0.2247$, $E_{\mathbb{B}}^3 \approx 0.2150$, $E_{\mathbb{F}}^4 \approx 2.7780k^3$, and $E_{\mathbb{F}}^5 \approx 6.2776k^3$ for the bcc grid. These values are all close to the asymptotic optima and we conclude that longer neighbourhood sequences is superfluous in most applications, but might be useful when higher exactness is needed or the aim is more complex.

Acknowledgement

Many thanks to prof. Gunilla Borgefors for valuable help during the preparation of this manuscript.

Benedek Nagy acknowledges the financial support provided through the European Community's Human Potential Programme under contract HPRN-CT-2002-00275, SegraVis. The work is partially supported by grants of the Hungarian National Foundation for Scientific Research (OTKA F043090 and OTKA T049409) and by the Universitas Foundation.

References

1. Rosenfeld, A., Pfaltz, J.L.: Sequential operations in digital picture processing. J. ACM **13**(4) (1966) 471–494
2. Rosenfeld, A., Pfaltz, J.L.: Distance functions on digital pictures. Pattern Recognition **1** (1968) 33–61
3. Kumar, M.A., Chatterji, B.N., Mukherjee, J., Das, P.P.: Representation of 2D and 3D binary images using medial circles and spheres. International Journal of Pattern Recognition and Artificial Intelligence **10**(4) (1996) 365–387

4. Borgefors, G.: Distance transformations in arbitrary dimensions. Computer Vision, Graphics, and Image Processing **27** (1984) 321–345
5. Borgefors, G.: Distance transformations in digital images. Computer Vision, Graphics, and Image Processing **34** (1986) 344–371
6. Verwer, B.J.H.: Local distances for distance transformations in two and three dimensions. Pattern Recognition Letters **12**(11) (1991) 671–682
7. Borgefors, G.: Distance transformations on hexagonal grids. Pattern Recognition Letters **9** (1989) 97–105
8. Strand, R., Borgefors, G.: Distance transforms for three-dimensional grids with non-cubic voxels. Computer Vision and Image Understanding **100**(3) (2005) 294–311
9. Borgefors, G., Sanniti di Baja, G.: Skeletonizing the distance transform on the hexagonal grid. In: Proceedings of 9th International Conference on Pattern Recognition, Rome, Italy. (1988) 504–507
10. Strand, R.: Surface skeletons in grids with non-cubic voxels. In: Proceedings of 17th International Conference on Pattern Recognition (ICPR'04), Cambridge, UK. Volume I. (2004) 548–551
11. Matej, S., Lewitt, R.M.: Efficient 3D grids for image reconstruction using spherically-symmetric volume elements. IEEE Transactions on Nuclear Science **42**(4) (1995) 1361–1370
12. Yamashita, M., Honda, N.: Distance functions defined by variable neighbourhood sequences. Pattern Recognition **17**(5) (1984) 509–513
13. Yamashita, M., Ibaraki, T.: Distances defined by neighbourhood sequences. Pattern Recognition **19**(3) (1986) 237–246
14. Das, P.P., Chakrabarti, P.P.: Distance functions in digital geometry. Information Sciences **42** (1987) 113–136
15. Das, P.P., Chakrabarti, P.P., Chatterji, B.N.: Generalized distances in digital geometry. Information Sciences **42** (1987) 51–67
16. Fazekas, A., Hajdu, A., Hajdu, L.: Lattice of generalized neighbourhood sequences in nD and ∞D. Publicationes Mathematicae Debrecen **60** (2002) 405–427
17. Nagy, B.: Distance functions based on neighbourhood sequences. Publicationes Mathematicae Debrecen **63**(3) (2003) 483–493
18. Das, P.P.: Best simple octagonal distances in digital geometry. Journal of Approximation Theory **68** (1992) 155–174
19. Das, P.P., Chatterji, B.N.: Octagonal distances for digital pictures. Information Sciences **50** (1990) 123–150
20. Mukherjee, J., Das, P.P., Kumar, M.A., Chatterji, B.N.: On approximating Euclidean metrics by digital distances in 2D and 3D. Pattern Recognition Letters **21** (2000) 573–582
21. Hajdu, A., Hajdu, L.: Approximating the Euclidean distance using non-periodic neighbourhood sequences. Discrete Mathematics **283** (2004) 101–111
22. Danielsson, P.: 3D octagonal metrics. In: Proceedings of 8th Scandinavian Conference on Image Analysis, Tromsø, Norway. (1993) 727–736
23. Hajdu, A., Nagy, B.: Approximating the Euclidean circle using neighbourhood sequences. In: Proceedings of 3rd Hungarian Conference on Image Processing, Domaszék. (2002) 260–271
24. Strand, R., Nagy, B.: Some properties for distances based on neighbourhood sequences in the face-centered cubic grid and the body-centered cubic grid. Technical report, Centre for Image Analysis, Uppsala University, Sweden (2005) Internal report 39.

Fuzzy Distance Based Hierarchical Clustering Calculated Using the A* Algorithm

Magnus Gedda and Stina Svensson

Centre for Image Analysis
Uppsala University and Swedish University of Agricultural Sciences
Uppsala, Sweden
{magnusg, stina}@cb.uu.se

Abstract. We present a method for calculating fuzzy distances between pairs of points in an image using the A* algorithm and, furthermore, apply this method for fuzzy distance based hierarchical clustering. The method is general and can be of use in numerous applications. In our case we intend to use the clustering in an algorithm for delineation of objects corresponding to parts of proteins in 3D images. The image is defined as a fuzzy object and represented as a graph, enabling a path finding approach for distance calculations. The fuzzy distance between two adjacent points is used as edge weight and a heuristic is defined for fuzzy sets. A* is applied to the calculation of fuzzy distance between pair of points and hierarchical clustering is used to group the points. The normalised Hubert's statistic is used as validity index to determine the number of clusters. The method is tested on three 2D images; two synthetic images and one fuzzy distance transformed microscopy image of stem cells. All experiments show promising initial results.

1 Introduction

In this manuscript we introduce a content based clustering framework for automatic clustering of points in an image. The framework should contain content based distance calculations between pairs of points in the image, clustering and *cluster validation*, i.e., determining the number of clusters present in a data set. The application in mind is automatic clustering of seed points to aid decomposition of fuzzy objects, corresponding to proteins, into parts [1]. The proteins are imaged using Cryo-Electron Tomography (Cryo-ET) [2], which is a technique used to produce 3D density images of proteins in solution.

In many applications, e.g., grouping of distance map maxima in segmented cell images, classification is desired to group points in an image based on their spatial proximity, i.e., how close the points are in space. If neither information about probabilities nor stochastic distributions of the points is available, classification has to rely on minimisation of criteria functions or clustering techniques. It is often the case that the desired number of groups is not known on beforehand. In such cases *hierarchical clustering* [3] is the preferred method. The *proximity*

U. Eckardt et al. (Eds.): IWCIA 2006, LNCS 4040, pp. 101–115, 2006.

measure, i.e., how "similar" two points are, used for the clustering can be based on a number of different metrics, e.g., Euclidean distance.

The spatial closeness of two points can often depend on the image contents and cannot always be measured as being the shortest path in the Euclidean sense, i.e., the shortest path between two points deviates from a straight line. Imagine a grey-level image as a topographic map where the intensity of each point is equal to the height value in that point, then the shortest path between two points, on a map of a hilly terrain, is often considered to circle the hills instead of going across. In such cases it is of importance to take into account both spatial information and intensity information when traversing a path.

For the application in mind, the point proximities are affected by the density of an imaged object. Like in many other image acquisition systems, e.g., most modalities in medical imaging, the density of the object is reflected by the intensities in the image. Such images often contain uncertainties which make crisp representations not the most suitable technique. A more natural approach is to assign each image element a degree of membership to the object(s) in the scene, thus keeping the uncertainty in the representation. To overcome this problem, *fuzzy sets* were introduced in [4]. A fuzzy set adds to each point of the set a membership value. A *fuzzy digital object* is a fuzzy set defined on the digital space. In an image corresponding to a fuzzy set (or several fuzzy sets), the intensity in each image point is related to its fuzzy membership value.

Fuzzy objects are widely used as representations of objects in medical images, see [5, 6] and following articles by the same groups. Fuzzy sets can be used as a framework for formulating the properties of images with content based distances. In [7], a grey-weighted distance measure was introduced, and in [8] it was given a theoretical framework and got the name *fuzzy distance transform* (FDT). The FDT calculates distances by taking both the spatial information as well as fuzzy membership into account. More specifically, it is shown that for any fuzzy object, the fuzzy distance is a metric for the support of the fuzzy object. This makes the fuzzy distance suitable as a content based proximity measure for images where the proximity is dependent on the density, e.g., the images for the application in mind.

In [7], the FDT of a fuzzy set is calculated by repeated raster scans similarly to the method for computing the chamfer distance transform (CDT) of binary images [9, 10]. However, the number of scans required to calculate the FDT is dependent on the image contents (unlike the CDT). In [8], it is suggested to use dynamic programming for efficiently computing the FDT of fuzzy sets. This approach is also utilised by the image foresting transform (IFT). See [11] for a recent survey. The IFT represents the image as a directed graph, with edge weights corresponding to some application specific weight function. To calculate the FDT of an image by using the IFT, edge weight is set to the length between two neighbouring points on a fuzzy path. The resulting IFT distance map will then correspond to the FDT.

When the FDT of an image is calculated to measure fuzzy distances between N points, a total of $N-1$ fuzzy distance maps are required. For complex images, the raster scan method will result in many traversals of the image, hence, be computationally inefficient. The IFT will also be computationally inefficient as it expands the entire image for each FDT, which means that a large number of unnecessary computations will be performed to calculate distances to uninteresting parts of the image. This is especially the case when the points are closely situated compared to the image size. An alternate, more efficient, approach in cases where not the whole fuzzy distance map is of interest but only the fuzzy distances between a limited set of points, is to view the problem as a minimum cost path finding problem between each pair of points.

In [12], the A* pathfinding algorithm is presented. A* is an algorithm which uses heuristic information to guide the search. Note that there exists a degraded version of A*, i.e., the *uniform cost algorithm*, which is used in [13] to calculate the distance transform. However, since the uniform cost algorithm does not incorporate heuristics, it is basically a single-source case of the IFT. Hence, it gives similar results, and have the same drawbacks, as the IFT if applied to the fuzzy distance transform.

We present a method to calculate the fuzzy distance between pairs of points in an image using the A* algorithm, including heuristics. The method can be viewed as a one-source version of the IFT, where a heuristic is used to guide the path and only expands the tree until the destination node is found. This enables us to use the general concept of the IFT, but imposes information and constraints to suit the specific situation. The method is used as a first step in a framework for automated fuzzy distance based clustering of points in an image. The other steps are hierarchical clustering and *cluster validation*, i.e., determining the number of clusters present in a data set. Many methods have been developed for handling the validation problem, see [14] for a recent survey. A cluster validation method suited for the situation is the relative criteria method using the *normalised Hubert statistic* validity index [14]. This technique is used as the last step in the clustering framework.

The method is presented for nD images, but the experiments are done on 2D images with properties similar to the application in mind, i.e., delineation of parts of proteins in 3D Cryo-ET images. Experimenting on 2D images instead of 3D images, as an initial evaluation of the method, makes the method behaviour both easier to analyse and easier to illustrate. The images used in the experiments are two synthetic images and one bright field microscopy image of stem cells.

Distance calculations on binary sets using A* is covered in Section 2. In Section 3 the concept is expanded to include fuzzy distance calculations. Section 4 focuses on clustering and cluster validation. Section 5 contains the experiments, and in Section 6, the conclusions are discussed.

2   Distance Calculations on Binary Sets Using A*

In this Section we recall the A* algorithm [12]. Furthermore, a method for using A* for point-to-point distance calculations on binary sets is presented.

2.1 The A* Algorithm

A* is a pathfinding algorithm which can compute the distance (cost) from a start node to a goal node in a graph. For A* to find the shortest path, it needs to be *admissible*, i.e., able to find the minimal cost path if a path to the goal exists. To meet the admissibility criterion the algorithm uses an *evaluation function* $\hat{f}(n)$, where n is a graph node, such that the available node having the smallest value of $\hat{f}$ is the node to be expanded next. $\hat{f}(n)$ is defined as

$$\hat{f}(n) = \hat{g}(n) + \hat{h}(n), \tag{1}$$

where $\hat{g}(n)$ is the cost of the path to n with minimum cost so far found by A* and $\hat{h}(n)$ is the heuristic function, i.e., any estimate of the cost of an optimal path from n to the goal node. To consecutively expand the available node closest to the goal, $\hat{h}(n)$ needs to be a lower bound on the true cost of an optimal path from n to the goal node. If $\hat{h}(n)$ is not a lower bound, A* cannot be guaranteed to find an optimal path if such a path exists.

A* uses two sets to keep track of available and visited nodes. Available nodes are put in the OPEN set, and visited nodes are put in the CLOSED set. Since a node is visited only once, $\hat{f}(n)$ does not need to be recalculated once the node is in the CLOSED set.

A* ALGORITHM

1. Initialisation
   (a) Start node: Calculate $\hat{f}(n)$ and mark n OPEN
2. Propagation
   while nodes left OPEN
      (a) Select node n with smallest $\hat{f}(n)$ from OPEN and mark n CLOSED
      (b) If n is the goal node, terminate algorithm
      (c) For each successor of n not CLOSED, calculate $\hat{f}(n)$ and mark OPEN

The algorithm outputs the minimum cost path from the start node to the goal node, and the associated cost, i.e., the distance from start to goal.

2.2 A* Applied to Binary Sets

The distance from a point p to a point q, where $(p, q) \in \mathbf{Z}^n$, can be defined as a function of the number of steps between adjacent points in a minimal path between p and q. In the function, each step is suitably weighted according to the neighbourhood relation between the adjacent points. E.g., steps in 3D images can be weighted according to the chamfer weights $w_{3D} = \langle 3, 4, 5 \rangle$ for steps to a face, edge and vertex sharing voxel respectively [10].

In accordance with the graph concept of the IFT, the length of a link between two adjacent points $\langle p_i, p_{i+1} \rangle \in \mathbf{Z}^n$ can be used as the edge weight. The edge cost function is, hence, defined as

$$W_d(p_i, p_{i+1}) = \|p_i - p_{i+1}\|_{w_{nD}}, \tag{2}$$

where $\|\cdot\|_w$ denotes the weight of a step from point p_i to point p_{i+1}. Consider the length $\Pi(\pi)$ of a path $\pi = \langle p = p_1, \ldots, p_m = q \rangle$ in $\mathbf{Z}^n$ as

$$\Pi(\pi) = \sum_{i=1}^{m-1} W_d(p_i, p_{i+1}), \tag{3}$$

and the minimum distance between p and q as

$$\omega(p, q) = \min_{\pi \in \mathbf{Z}^n} \Pi(\pi). \tag{4}$$

For A* to find the shortest distance, a heuristic which is a lower bound on ω, is needed. The Euclidean distance to the goal node can be used as a heuristic if it is properly weighted to become a lower bound on ω. The distance obtained with chamfer coefficients cannot be compared directly with the actual Euclidean distance, because the distance is scaled. E.g., in 3D the Euclidean distance between two face-neighbours is 1, while the distance we obtain with chamfer coefficients is 3. To use Euclidean distance in the heuristic, a multiplication with a coefficient divided by an Euclidean norm is needed. For simplicity, the coefficients of the chamfer mask are chosen. By taking the minimum fraction, the heuristic is ensured to be a lower bound on ω. Hence, the heuristic is defined as

$$\hat{h}_d(p) = k_d \cdot \|p - q\|, \tag{5}$$

where $\|\cdot\|$ denotes the Euclidean norm, and k_d is a constant which will ensure $\hat{h}_d$ to be a lower bound on the true minimum distance, ω. According to the above, the constant k_d needs to be defined differently depending on the step weights. A lower bound of the edge cost for a single step, W_d, is acquired when

$$k_d = \min_{i=1\ldots n} \left(\frac{w_i}{\|\mathbf{v}_i\|} \right), \tag{6}$$

where n is the number of dimensions, $\mathbf{v}_i$ is the chamfer vector corresponding to the chamfer weight w_i (e.g. in $\mathbf{Z}^3$: $\mathbf{v}_1 = [1, 0, 0]$, $\mathbf{v}_2 = [1, 1, 0]$ and $\mathbf{v}_3 = [1, 1, 1]$), and $\|\cdot\|$ denotes the Euclidean norm. E.g., considering the chamfer weights $w_{3D} = \langle 3, 4, 5 \rangle$ for 3D images,

$$k_d = \min \left(\frac{3}{\sqrt{1}}, \frac{4}{\sqrt{2}}, \frac{5}{\sqrt{3}} \right) = \frac{4}{\sqrt{2}} \approx 2.8, \tag{7}$$

where the rounding is done downwards to ensure the lower bound. Using $k_d = 2.8$ ensures that the distance to the goal node will always be greater than the Euclidean distance multiplied by k_d, thus, with $k_d = 2.8$, the heuristic in Eq. 5 will be a lower bound on ω.

With W_d as edge weight and $\hat{h}_d$ as heuristic, A* can be used to perform effective point-to-point Euclidean distance calculations. The distance is calculated by applying the A* algorithm from the first point, i.e., the start node, to the second point, i.e., the goal node. As covered in Section 2.1, $\hat{h}(n)$ is the estimated cost (distance) from the node n to the goal node, and $\hat{g}(n)$ is the cost of the path from the start node to n.

A* ALGORITHM APPLIED TO BINARY SETS

1. Initialisation
     all nodes: $\hat{g}(n) = \infty$
     start node: $\hat{g}(n) = 0$; $\hat{f}(n) = \hat{g}(n) + \hat{h}_d(n)$; Push($n$, OPEN)
2. Propagation
     while nodes left on OPEN
          $n = $ Pop(OPEN); Push(n, CLOSED)
          if $n == $ goal node: terminate algorithm
          for each p successor of n, where $p \notin$ CLOSED
               $\hat{f}(p) = W_d(n,p) + \hat{g}(n) + \hat{h}_d(p)$
               $\hat{g}(p) = \min\{W_d(n,p) + \hat{g}(n), \hat{g}(p)\}$
               for each $p \in$ OPEN
                    if new $\hat{f}(p) <$ old $\hat{f}(p)$: update p
               for each $p \notin$ OPEN
                    Push(p, OPEN)

By the above algorithm the distance between two points is obtained. A binary heap priority queue [15] is used for the OPEN list, and a lookup table is used for the CLOSED list. A pointer array is used for effective access of nodes in the OPEN set. Extracting the highest priority node is done in $\mathcal{O}(\log n)$ time, where n is the number of nodes in the image, and there are at most n such operations. Changing the priority of a node is done in $\mathcal{O}(\log n)$ time (worst case), and there are at most m such operations, where m is the number of edges in the image. The computational complexity thus becomes $\mathcal{O}((n + m) \log n)$ for the entire image. However, the number of nodes expanded (n) and edges examined (m) before reaching the goal node is highly dependent on the effectiveness of the heuristic.

3   Distance Calculations on Fuzzy Sets Using A*

In this Section, some background theory on fuzzy sets and fuzzy distance is covered. Furthermore, the concept in Section 2.2 is expanded to apply to fuzzy distance calculations.

3.1   Fuzzy Distance

We recall from fuzzy set theory [4] the following definitions: Let X be a reference set, then a *fuzzy set* $\mathcal{A}$ in X is defined as a set of ordered pairs $\mathcal{A} = \{(x, \mu_{\mathcal{A}}(x)) \,|\, x \in X\}$, where $\mu_{\mathcal{A}} : X \to [0,1]$ is the *membership function* of $\mathcal{A}$ in X. An n-dimensional *fuzzy digital object* $\mathcal{O}$ is a fuzzy subset defined on $\mathbf{Z}^n$, i.e., $\mathcal{O} = \{(p, \mu_{\mathcal{O}}(p)) \,|\, p \in \mathbf{Z}^n\}$, where $\mu_{\mathcal{O}} : \mathbf{Z}^n \to [0,1]$.

In [8], a fuzzy distance transform (FDT) is presented. The notion of fuzzy distance between two points p and q is defined as being the shortest length of a path between p and q. The length $\Pi_{\mathcal{O}}(\pi)$ of a path $\pi = \langle p = p_1, p_2, \ldots, q = p_m \rangle$ is

$$\Pi_{\mathcal{O}}(\pi) = \sum_{i=1}^{m-1} \frac{1}{2} \left(\mu_{\mathcal{O}}(p_i) + \mu_{\mathcal{O}}(p_{i+1}) \right) \times \| p_i - p_{i+1} \|, \tag{8}$$

where the norm $\|\cdot\|$ denotes the Euclidean norm. The fuzzy distance, from $p \in \mathbf{Z}^n$ to $q \in \mathbf{Z}^n$ in $\mathcal{O}$, denoted as $w_\mathcal{O}(p,q)$, is the length of any shortest path in $\mathcal{O}$ from p to q [8], i.e.,

$$w_\mathcal{O}(p,q) = \min_{\pi \in P(p,q)} \Pi_\mathcal{O}(\pi), \qquad (9)$$

where $P(p,q)$ is the set of all *paths* from p to q in $\mathcal{O}$.

3.2 A* Applied to Fuzzy Sets

The edge weight is now defined as the length of a step between two adjacent points $\langle p_i, p_{i+1} \rangle$ in a fuzzy object $\mathcal{O}$, instead of two adjacent points in $\mathbf{Z}^n$. The step length can be defined in many different ways, but the definition from Eq. (8) is used for simplicity. Hence, the edge cost function for fuzzy sets is

$$W_{fd}(p_i, p_{i+1}) = \frac{1}{2}(\mu_\mathcal{O}(p_i) + \mu_\mathcal{O}(p_{i+1})) \times W_d(p_i, p_{i+1}). \qquad (10)$$

If the definition of the membership function $\mu_\mathcal{O}$ allows mappings to zero, the shortest possible edge cost will be zero and, hence, the shortest possible fuzzy distance between two points in $\mathcal{O}$ will be zero. This makes the definition of a positive non-zero heuristic impossible since the lower bound on $w_\mathcal{O}$ (Eq. (9)) will be zero and the A* algorithm will degrade to the uniform cost algorithm. This is avoided by using a definition which prevents zero-valued memberships in the range of input values. Consider a fuzzy object $\mathcal{O}$, where each point p in the fuzzy object is an image element. If $I(p)$ is the image intensity of point p, then a non-zero minimal edge cost between any two points p and q in $\mathcal{O}$ can be ensured by defining the membership function as

$$\mu_\mathcal{O}(p) = \begin{cases} 0 & \text{if } I(p) < 0 \\ (1 - g_f)\frac{I(p)}{I_{max}} + g_f & \text{if } I(p) \in [0, I_{max}] \end{cases}, \qquad (11)$$

where I_{max} is the maximum image intensity, and $g_f \in (0,1)$. This will ensure a non-zero minimal edge cost since an image intensity of zero will result in a membership of g_f. Since the membership function of a point with the maximum image intensity will map to one, the membership function in practice will map to $[g_f, 1]$ for all grey-level images. This gives us the possibility of defining a non-zero heuristic, and still use a large portion of the membership range.

To find a heuristic which is a lower bound on $w_\mathcal{O}$, a constant, corresponding to k_d for binary sets, is needed. The heuristic for fuzzy sets is defined as

$$\hat{h}_{fd}(p) = k_{fd} \cdot \|p - q\|, \qquad (12)$$

where $\|\cdot\|$ denotes the Euclidean norm. Since k_d was deduced from the lower bound on W_d and k_{fd} is deduced from the lower bound on W_{fd} (see Eq. 10), k_{fd} is defined as

$$k_{fd} = \min\left(\frac{1}{2}(\mu_\mathcal{O}(p_i) + \mu_\mathcal{O}(p_{i+1}))\right) \cdot k_d = g_f \cdot k_d, \qquad (13)$$

where the constant g_f is the lowest possible value of the term $\frac{1}{2} \left(\mu_{\mathcal{O}} \left(p_i \right) + \mu_{\mathcal{O}} \left(p_{i+1} \right) \right)$ in Eq. 10. Thus, k_{fd} ensures the heuristic $\hat{h}_{fd}$ (Eq. 12) to be a lower bound on $\omega_{\mathcal{O}}$. E.g., for 3D images using the chamfer weights $w_{3D} = \langle 3, 4, 5 \rangle$, k_{fd} is defined as

$$k_{fd} = g_f \cdot k_d = 2.8 \cdot g_f \tag{14}$$

to ensure that the fuzzy distance to the goal node always will be greater than the Euclidean distance multiplied by k_{fd}. Thus, with $k_{fd} = g_f \cdot k_d$, the heuristic in Eq. 12 will be a lower bound on $\omega_{\mathcal{O}}$.

An admissible A* algorithm, for effective point-to-point fuzzy distance calculations, is obtained by using W_{fd} as edge weight, and $\hat{h}_{fd}$ as a heuristic which is a lower bound on $\omega_{\mathcal{O}}$. The algorithm is the same as for binary sets (see Section 2.2), but all instances of W_d are substituted with W_{fd}, and all instances of $\hat{h}_d$ are substituted with $\hat{h}_{fd}$.

For a fuzzy object containing N points, the fuzzy distance between each pair of points can be put in a *proximity matrix*

$$\mathbf{P} = \begin{pmatrix} 0 & ||p_1 - p_2||_f & \cdots & ||p_1 - p_{N-1}||_f & ||p_1 - p_N||_f \\ ||p_1 - p_2||_f & 0 & \cdots & ||p_2 - p_{N-1}||_f & ||p_2 - p_N||_f \\ \vdots & \vdots & \ddots & \vdots & \vdots \\ ||p_1 - p_{N-1}||_f & ||p_2 - p_{N-1}||_f \cdots & 0 & ||p_{N-1} - p_N||_f \\ ||p_1 - p_N||_f & ||p_2 - p_N||_f & \cdots ||p_{N-1} - p_N||_f & 0 \end{pmatrix},$$

$$\tag{15}$$

where $|| \cdot ||_f$ denotes the fuzzy distance. The proximity matrix will be used in Section 4 when clustering the points and validating the results.

4 Clustering and Validation

This Section covers the clustering of the points from the fuzzy distance calculations. Furthermore, cluster validation is covered for determining the number of clusters in the final result.

As mentioned in Section 1, cluster validation is a frequently occurring problem in cluster analysis. Without any probabilistic information, there is no reliable method for solving this. Some methods have been proposed to solve this problem. None of the solutions are above suspicion, but they can aid a more rigorous manual validation process, or provide automation in cases of well behaved data where the result is assessed by an expert.

For the application in mind there is no information about probabilities, stochastic distributions, or even the number of desired clusters. This means that the only information available for grouping the points in the fuzzy object are the fuzzy distances calculated in Section 3. In this case the preferred method for grouping the points is hierarchical clustering [3]. Hierarchical clustering starts by assigning each point to a unique cluster, and then consecutively merges the two most similar clusters according to some proximity measure, until there is only one cluster left. The cluster similarities can be calculated in different ways, e.g. using *single linkage* or *complete linkage* [3]. Single linkage generates stretched out clusters, while complete linkage generates compact clusters. Consider a *clustering scheme* (partition) with n_c clusters $c_1, \ldots, c_{n_c}$, and a proximity measure d, then the cluster similarity is

$$d_s\left(c_i, c_j\right) = \min_{x \in c_i, y \in c_j} d\left(x, y\right), \tag{16}$$

$$d_c\left(c_i, c_j\right) = \max_{x \in c_i, y \in c_j} d\left(x, y\right), \tag{17}$$

for single linkage and complete linkage respectively.

For the application in mind, compact well separated clusters should be favoured. Hence, hierarchical clustering is used with complete linkage, and fuzzy distances as proximity measures. This results in N different clustering schemes, where N is the number of points in the fuzzy object.

To overcome the cluster validity problem, i.e., determining the number of clusters in a data set, various validity indices have been proposed [14]. A validity index gives an indication of the quality of a partition. Validation of hierarchical clustering partitions through validity indices can be done by *external, internal* or *relative criteria* methods [14]. Since external and internal methods are based on statistical testing, which demand large test sets for probability density estimations, focus is put on the relative criteria method. The relative criteria method chooses the best clustering scheme, from a set of defined schemes, according to a pre-defined criterion. In hierarchical clustering, each merge of two clusters result in a new clustering scheme. With the relative criteria method, the clustering scheme which best fits the data can be chosen by evaluating a validity index for each of the partitions. Then the clustering scheme which, according to the criterion, best fits the data set, is chosen.

Few validity indices can be used when the only information available is the pairwise proximity of the points. One of them is the normalised Hubert statistic [14],

$$\hat{\Gamma} = \frac{1}{N_T} \frac{1}{\sigma_P \sigma_M} \sum_{i=1}^{N-1} \sum_{j=i+1}^{N} \left(\mathbf{P}\left(i, j\right) - m_P\right)\left(\mathbf{M}\left(i, j\right) - m_M\right), \tag{18}$$

where N is the total number of points in the data set, $\mathbf{P}(i, j)$ and $\mathbf{M}(i, j)$ are the (i, j) element of matrices $\mathbf{P}$, $\mathbf{M}$ respectively, and $N_T = N(N-1)/2$. Also, m_P, m_M, σ_P, σ_M are the respective means and variances of $\mathbf{P}$, $\mathbf{M}$ matrices. High values of this index indicate a strong similarity between $\mathbf{P}$ and $\mathbf{M}$. Other possible indices are, e.g., Dunn and Dunn-like indices. However, the Dunn and Dunn-like indices are more sensitive to noise and more computationally complex than the Hubert $\hat{\Gamma}$ statistic [14]. Since cluster validation analysis is out of the scope of this article, the cluster validation is chosen to rely on the Hubert $\hat{\Gamma}$ statistic.

To use Hubert $\hat{\Gamma}$ statistic for relative criteria cluster validation, the proximity matrix $\mathbf{P}$, and an additional matrix $\mathbf{M}$, are needed. As mentioned in Section 3, the proximity matrix holds the proximity measures calculated with the A* algorithm, i.e., the pairwise fuzzy distance values. The additional matrix explains which samples reside in the same cluster. Consider a partition of a data set $\mathcal{X}$ as a mapping, $g : \mathcal{X} \rightarrow \{1 \cdots n_c\}$, where n_c is the number of clusters in the resulting clustering scheme. The additional matrix needed, $\mathbf{M}$, is then defined as

$$\mathbf{M}(i,j) = \begin{cases} 1 & \text{if } g(x_i) \neq g(x_j) \\ 0 & \text{otherwise} \end{cases} \quad i,j = 1, \ldots, N, \tag{19}$$

where N is the total number of points in the fuzzy object.

RELATIVE CRITERIA CLUSTER VALIDATION ALGORITHM USING NORMALISED HUBERT'S STATISTIC [14]

1. Create proximity matrix $\mathbf{P}$
2. Run hierarchical clustering algorithm
3. **For** each of the values of n_c
   (a) Create the matrix $\mathbf{M}$
   (b) Calculate the index $\hat{\Gamma}$ using the matrices $\mathbf{P}$ and $\mathbf{M}$
4. Plot values of the index $\hat{\Gamma}$ as a function of n_c

Since the fuzzy distance proximity measures increase as they get further from each other, i.e., low value for high proximity, the best clustering scheme is identified by choosing the n_c where $\hat{\Gamma}$ has a global minimum.

Many validity indices for compact well separated clusters favour cluster schemes where n_c is close to N. In validation methods this is usually handled by assuming $n_c << N$. For the application in mind, this assumption is seldom true. However, by using fuzzy distances in the clustering instead of Euclidean distances, internals of the contents in the image will be enhanced and the clusters will be both further separated and more compact than for Euclidean based distances, thus loosening the need for the assumption.

5   Experiments

Three experiments were carried out. Each experiment used $g_f = 0.01$ for fuzzy membership calculations. Since the application in mind consider points in a high density area to be closely situated, high density should result in low proximity. Therefore, the intensities of each image was inverted before the fuzzy distances were calculated in order to make original low intensities costly to traverse. This led to low fuzzy distance values in high intensity areas, and high fuzzy distance values in low intensity areas. Furthermore, the parts of the objects for the application in mind are approximately spherical. Therefore, experimental images, which aim to mimic the properties of the application images, with approximately circular density regions were chosen. The aim of all experiments was to cluster the points into the natural clusters represented by the high density areas in the images.

The first experiment was done on a synthetic image of three large crisp circles containing 30 points each, and is presented in Section 5.1. The experiment was chosen to examine how the method behaves for large, fairly close, sets of points in crisp circles separated by low density areas.

The second experiment was done on a synthetic image of six small fuzzy discs containing four to five points each, and is presented in Section 5.2. The experiment was chosen to examine how the method behaves for small, closely situated, sets of points in discs with low gradient borders. This experiment resembles cases with cells; fuzzy areas around some points belonging to the same cluster if inside the same fuzzy area, where the points can be, e.g., maxima of the distance map or fluorescent markers.

The last experiment was done on local maxima points of a fuzzy distance transformed segmented microscopic cell image, and is presented in Section 5.3. The experiment was chosen to examine how the method behaves for extremely small, closely situated, sets of points in non-spherical density areas with low gradient borders. The situation highly resembles the application in mind, where delineation is done on proteins imaged by Cryo-ET [1]. However, in the images for the application in mind, the protein parts are slightly more spherical than the cells in the experiment.

The computational time for the experiments was a few minutes for the first experiment, and a matter of seconds for the other two experiments, on an ordinary desktop computer.

5.1   Experiment 1: Points in Crisp Circles

The image in the first experiment was a 100×80 binary image containing three filled circles, each having intensity values 255, a diameter 26 pixels, and centres $\mathbf{c}_1 = [29.5 \ 29.5]$, $\mathbf{c}_2 = [65.5 \ 29.5]$, and $\mathbf{c}_3 = [50.5 \ 55.5]$. Intensity values of the background was zero. The points clustered were 90 points taken from three normal distributions, $N_i(\mathbf{m}_i, 6\mathbf{I})$, $i = 1, 2, 3$, where $\mathbf{m}_1 = [30 \ 30]$, $\mathbf{m}_2 = [70 \ 30]$, and $\mathbf{m}_3 = [50 \ 60]$, 30 points from each distribution. The three circles delineated the points almost completely, and the image along with the points is shown in Fig. 1(a).

The points were first clustered using Euclidean distance. The resulting validity index diagram is shown in Fig. 1(b). Since the number of clusters is determined by the minimum of the diagram, the cluster validity fails for this case. The method might have resulted in a representative number of clusters under the assumption $n_c << N$, but that would depend on the range allowed for n_c.

Secondly, the points were clustered using fuzzy distance. The resulting validity index diagram is shown in Fig. 1(c). It is clear from the diagram that two clustering schemes, three and four clusters, fit the data better than any other partition. The minimum of the diagram is found for four clusters, and the point in the top left corner is, hence, considered a cluster of its own. This is due to the large area of low intensity between the point and the nearest circle which creates a large fuzzy distance.

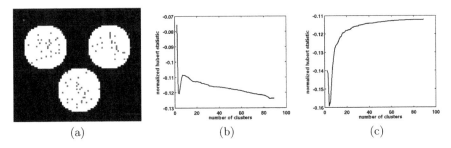

Fig. 1. (a) Synthetic image with random points. (b) Validity index plot for regular distance based clustering. (c) Validity index plot for fuzzy distance based clustering.

5.2   Experiment 2: Points in Fuzzy Discs

The image in the second experiment was a 61×61 grey-level image containing seven filled circles, a diameter of 15 pixels, and centres $c_1 = [30\ 12]$, $c_2 = [13\ 21]$, $c_3 = [47\ 19]$, $c_4 = [30\ 30]$, $c_5 = [13\ 39]$, $c_6 = [47\ 38]$, and $c_7 = [31\ 48]$. The image was Gaussian blurred, with a blur radius of five pixels, to obtain low gradient borders. The intensity values of the circle centres were 255, and the background zero. 32 points were positioned manually inside the circles, with four or five points in each circle. The image along with the points is shown in Fig. 2(a).

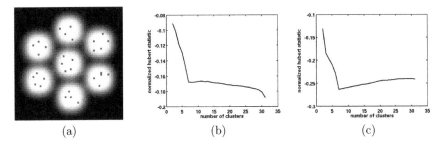

Fig. 2. (a) Synthetic image with random points. (b) Validity index plot for regular distance based clustering. (c) Validity index plot for fuzzy distance based clustering.

The points were first clustered using Euclidean distance. The resulting validity index diagram is shown in Fig. 2(b). Since the resulting number of clusters is chosen as the minimum of the diagram, the cluster validation for this case fails in this experiment well. Since the local minimum at $n_c = 7$ is shallow, a representative result would be difficult to achieve even when assuming $n_c << N$.

Secondly, the points were clustered using fuzzy distance. The resulting validity index diagram is shown in Fig. 2(c). It is clear from the diagram that the clustering scheme containing seven clusters fits the data better than any other partition, and corresponds to the natural clusters in the image. Examination of the clustering scheme for $n_c = 7$ shows that the clusters correspond to the fuzzy discs in the image.

5.3   Experiment 3: Real Image

The third experiment was done on a fuzzy distance transformed image of segmented cells, with the points in the image being the local maxima. The original images are 634×504 bright field microscopy images of in vitro stem cells. The image used in the experiment is a fuzzy distance transformed part (86×72) of one image containing a segmented group of cells. The intensities range from zero (background) to 187. The 18 local maxima of the image were used as points in the clustering. The image along with the local maxima is shown in Fig. 3(a).

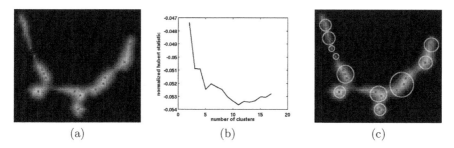

(a)   (b)   (c)

Fig. 3. (a) FDT of segmented cells with local maxima overlayed. (b) Validity index diagram. (c) Clustering result.

The local maxima were clustered using fuzzy distance, and the resulting validity index diagram is shown in Fig. 3(b). The minimum of the diagram is found for $n_c = 11$, and the resulting clustering scheme is shown in Fig. 3(c). Since the method was developed for compact well separated sets in circular density areas, the resulting clusters do not correspond to the seven cells in the image, but to the clusters which best fit the properties inherent in the method. For the five bottom cells, the clusters coincide with the respective density areas of the cells. For the two single point clusters in the upper left part of the image, the low density surrounding of each point, along with the spatial separations from the other local maxima, result in a natural cluster for each of the points, like the outlier in the first experiment. For the four remaining clusters, in the top left and top right part of the image, the spatial separation inside the elongated density areas result in two natural clusters in each cell. Thus, the clustering result correspond to the expected behaviour of the method.

6   Conclusions

We proposed a new method for calculating point-to-point fuzzy distances using the A* algorithm, and used the method in a clustering framework to cluster points in images with content based proximity measures. The fuzzy distance enhance the properties of the image contents, and thus, emphasise the separation and compactness of clusters in density areas approximately spherical. Cluster validation was thus more reliable in these cases, and was not in need of the assumption $n_c << N$.

The examples in this manuscript cover various cases and show results superior to Euclidean based clustering using the same validation technique. Future work consist of a quantitative analysis for determining the robustness of the framework.

One unexplored topic is the effectiveness of the fuzzy distance calculation between two points. In the worst case scenario, the method performs as well as the IFT when the IFT is halted after reaching the goal node. The improvement compared to the IFT, and how it behaves for different values of g_f, need further investigation.

The heuristic is based on the Euclidean distance alone, and does not take the intensities into account. What effect this has on the efficiency of the fuzzy distance calculations is left to be examined. Defining a heuristic based on both distance and intensity would most likely be very difficult, if at all possible.

The performance of the fuzzy distance calculation using different implementations need to be assessed. E.g., using Dial's bucket queue as priority queue, instead of a binary heap, for the OPEN list might increase the speed, and hence, enable fuzzy distance based clustering of larger data sets.

The presented method, or derivatives using other validity indices, is likely to be of use in a number of applications where image contents can be used to emphasise cluster membership when clustering points in an image.

Acknowledgements

Dr. Joakim Lindblad, Dr. Celine Fouard and M.Sc. Erik Vidholm, all at the Centre for Image Analysis, Uppsala University, Sweden, are acknowledged for their scientific support. This project was financially supported by the Swedish Research Council (project 621-2005-5540).

References

1. Svensson, S., Gedda, M., Fanelli, D., Skoglund, U., Sandin, S., Öfverstedt, L.G.: Using a fuzzy framework for delineation and decomposition of immunoglobulin G in cryo electron tomographic images. Submitted for publication in an international conference proceedings (2006)
2. Sandin, S., Öfverstedt, L., Wikström, A., Wrange, O., Skoglund, U.: Structure and flexibility of individual immunoglobulin G molecules in solution. Structure **12** (2004) 409–415
3. Duda, R.O., Hart, P.E., Stork, D.G.: Pattern Classification. John Wiley & Sons, New York, USA (2001)
4. Zadeh, L.A.: Fuzzy sets. Information and Control **8** (1965) 338–353
5. Udupa, J.K., Samarasekera, S.: Fuzzy connectedness and object definition: Theory, algorithms, and applications in image segmentation. Graphical Models and Image Processing **58** (1996) 246–261

6. Carvalho, B.M., Gau, C.J., Herman, G.T., Kong, T.Y.: Algorithms for fuzzy segmentation. Pattern Analysis & Applications **2** (1999) 73–81

7. Levi, G., Montanari, U.: A grey-weighted skeleton. Information and Control **17** (1970) 62–91

8. Saha, P.K., Wehrli, F.W., Gomberg, B.R.: Fuzzy distance transform: theory, algorithms, and applications. Comput. Vis. Image Underst. **86**(3) (2002) 171–190

9. Rosenfeld, A., Pfaltz, J.L.: Sequential operations in digital picture processing. Journal of the Association for Computing Machinery **13**(4) (1966) 471–494

10. Borgefors, G.: On digital distance transforms in three dimensions. Computer Vision and Image Understanding **64**(3) (1996) 368–376

11. Falcao, A.X., Stolfi, J., de Alencar Lotufo, R.: The image foresting transform: theory, algorithms, and applications. IEEE Transactions of Pattern Analysis and Machine Intelligence **26**(1) (2004) 19–29

12. Hart, P.E., Nilsson, N.J., Raphael, B.: A formal basis for the heuristic determination of minimum cost paths. IEEE Transactions on Systems Science and Cybernetics **4**(2) (1968) 100–107

13. Verwer, B.J.H., Verbeek, P.W., Dekker, S.T.: An efficient uniform cost algorithm applied to distance transforms. IEEE Transactions of Pattern Analysis and Machine Intelligence **11**(4) (1989) 425–429

14. Halkidi, M., Batistakis, Y., Vazirgiannis, M.: On clustering validation techniques. Journal of Intelligent Information Systems **17**(2-3) (2001) 107–145

15. Cormen, T.H., Leiserson, C.E., Rivest, R.L.: Introduction to Algorithms. The MIT Press, Cambridge, Massachusetts, USA (1990)

A New Sub-pixel Map for Image Analysis

Hans Meine and Ullrich Köthe

Cognitive Systems Laboratory, University of Hamburg,
Vogt-Kölln-Str. 30, 22527 Hamburg, Germany
{meine, koethe}@informatik.uni-hamburg.de

Abstract. Planar maps have been proposed as a powerful and easy-to-use representation for various kinds of image analysis results, but so far they are restricted to pixel accuracy. This leads to limitations in the representation of complex structures (such as junctions, triangulations, and skeletons) and discards the sub-pixel information available in grayvalue and color images. We extend the planar map formalism to sub-pixel accuracy and introduce various algorithms to create such a map, thereby demonstrating significant gains over the existing approaches.

1 Introduction

When information is extracted from an image's raw pixel data, the results must be stored in a well-defined way. Still, many image analysis approaches use their own representations (labeled images, region adjacency graphs, regular, or irregular pyramids, edgel chains, polygons, etc.). This is not only highly confusing, but also prevents algorithms that perfectly complement each other from actually being used together – their representation are simply incompatible. During the last decade, several researchers have worked on powerful unified representations.

The most promising approach is based on the notion of *planar map*s [1, 2, 3]. Planar maps encode the topological entities (regions, edges, vertices) of a partitioning of the (image) plane, their relations (neighborhood, boundary, containment, etc.) and their geometry. Basic modification operations support well-defined manipulations of an existing map structure. Similar concepts have been used in computer graphics for a long time [4]. Two key problems must be solved to enable their adaptation to image analysis: first, image analysis algorithms must create valid map structures. This requires the establishment of a formal correspondence between the initial pixel data and the map's entities. Second, the map must be realized in an efficient and easy-to-use way due to the huge amount of data and the complexity of the image analysis problem in itself.

So far, these goals have only been achieved with grid-based planar maps. Here, regions, edges and vertices correspond directly to sets of pixels and/or inter-pixel boundaries, i.e. can be accessed and manipulated by fast array operations. The map entities can be derived from labeled images, watershed segmentations, and pixel-based edge detectors (see Sect. 2.2). However, the gray values or colors of real images contain a considerable amount of sub-pixel information. For example, in real images step edges are always blurred by the camera's point spread

U. Eckardt et al. (Eds.): IWCIA 2006, LNCS 4040, pp. 116–130, 2006.
© Springer-Verlag Berlin Heidelberg 2006

function (before sampling) and by the edge detection filters (after sampling). It is well known that the location of the ideal step can be recovered to at least 1/10 of a pixel by careful analysis of the blurred step's shape. This information is discarded when the representation is restricted to pixel accuracy.

Another limitation of grid-based maps is the representation of junctions. In an inter-pixel boundary map, at most four edges at 90° of each other can ever meet at a vertex. A pixel-based map can in principle represent more complex junctions, but these junctions are no longer single Euclidean points [2]. In real images, the corner and junction geometry is often much more complicated. This is one of the reasons why vectorial data structures are preferred for the representation of object geometry in computer graphics. Moreover, grid-based representations are harder to refine as new information arrives, whereas vectorial representations can be refined ad infinitum.

In this paper, we extend the existing grid-based map formalism to sub-pixel accuracy. We show that the map can still be efficiently realized by means of polygonal lines. Finally, we demonstrate various algorithms to create our new representation from image data, not only covering boundary detection, but also the creation of Delaunay triangulations and skeletons. Comparisons of our new results with their pixel-accurate counterparts reveal a significant gain.

2   A Unified Representation for Topology and Geometry

Before we discuss our new sub-pixel accurate GeoMap, let us summarize previous efforts for finding a suitable representation for image segmentation purposes. Segmentation methods impose the following requirements on such a structure [5, 6, 2]:

1. **Topology Inspection.** Algorithms need to access topological properties like the neighborhoods of regions and/or edges, the number of holes, etc. Thus, a sound topological formalism is required.
2. **Geometry Inspection.** During the segmentation process, photometric / geometric properties of regions and / or boundaries are to be derived (e.g. mean color, variance, size, etc.); typical subtasks include region reconstruction in a given image, region containment queries, or inspecting image properties along boundaries (e.g. the image gradient).
3. **Modifications.** If the representation is to be useful for the segmentation process itself, it must not be static. We need operations (e.g. merging two regions) modifying both the topology and the geometry in a consistent way.

2.1   Topology: Combinatorial Maps

For the representation of topology in image processing, a number of graph-like structures have been used (dating back to the RAG [7]). Nowadays, the more powerful formalism of *combinatorial maps* is commonly used, since it allows to efficiently encode most information on the embedding of a planar graph:

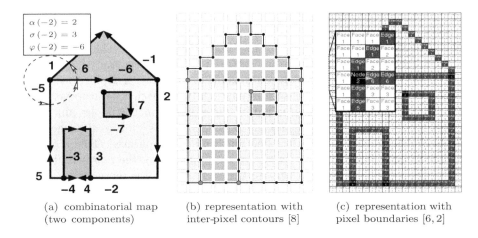

(a) combinatorial map (two components)

(b) representation with inter-pixel contours [8]

(c) representation with pixel boundaries [6, 2]

Fig. 1. Examples of discrete embeddings of combinatorial maps

Definition 1. *A combinatorial map is a triple (D, σ, α) where D is a set of darts (half-edges), and σ, α are permutations defined on D such that all α orbits have length 2 and the map is connected, i.e. there exists a σ-α-path between any two darts:*

$$\forall d_1, d_2 \in D: \exists \pi \in \left\{ \left. \prod_{0 \leq i \leq k} \tau_i \right| \tau_i \in \{\sigma, \alpha\}, k \in \mathbb{N} \right\}: \pi(d_1) = d_2$$

The dual permutation of σ is defined as $\varphi(d) = \sigma^{-1}(\alpha(d))$, where σ^{-1} denotes the σ-predecessor of d.

The orbits of σ, α, and φ are called *vertices*, *edges*, and *faces* respectively, and we use the notation $\sigma^\star(d)$, $\alpha^\star(d)$ and $\varphi^\star(d)$ for the σ-, α-, and φ-orbits which contain d. The orbit $\sigma^\star(d)$ is the start vertex of d, and $\varphi^\star(d)$ is the contour of the face to the left of d. A combinatorial map is *planar*, iff the number of vertices, edges, and faces conforms to Euler's equation:

$$|\sigma| - |\alpha| + |\varphi| = 2 \quad \text{(where } |\alpha| \text{ denotes the number of orbits in } \alpha \text{ etc.)} \quad (1)$$

When one face is designated as the (infinite) *exterior face*, all possible embeddings of a planar combinatorial map become topologically equivalent. By convention, one uses positive and negative integer labels for the darts so that $\alpha(d) = -d$ for each dart labeled d. Since φ is determined through α and σ, a single lookup table for the permutation σ is sufficient to represent a combinatorial map.

Definition 1 does not yet allow to represent multiple boundaries which commonly arise in image segmentation (inner contours like the window in Fig. 1(a)). This is usually solved by using one planar combinatorial map with a marked exterior face per connected component, plus an additional *inclusion relation* between the maps which associates the exterior faces with their parent faces [9, 10].

An alternative is to introduce *auxiliary edges* [11] to make the map connected, which we decided against because it spoils the one-to-one correspondence between topological edges and their geometrical counterparts (we do not want to make up geometrical information for the auxiliary edges).

Note that it is perfectly legal that $-d \in \varphi^\star(d)$, which means that edge $\alpha^\star(d)$ has the same region on both its left and right side. Such edges are called *bridges*, since every path between their two end-vertices must contain d or $-d$. In many publications, bridges have been considered illegal [11, 1], but in fact they are required to (a) represent incomplete boundaries (e.g. arising from edge detectors like Canny's [12], which in general does not deliver complete boundaries, or during sketching) or for (b) representing skeletons (see Sect. 4.4 and Fig. 7).

Given a set of combinatorial maps $(D_i, \sigma_i, \alpha_i)$ with $i \neq j \Rightarrow D_i \cap D_j = \emptyset$, it is possible to define $D = \bigcup D_i$ and compose the permutations into a single tuple (D, σ, α) representing all components, such that e.g. $d \in D_i \Rightarrow \alpha(d) = \alpha_i(d)$. In the following, the orbits of σ, α, and φ are meant to represent *all* vertices, edges, and faces respectively. Furthermore, we will occasionally use the general term "cells" for vertices, edges, or faces, which correspond to 0-, 1-, and 2-cells in the related context of cell complexes [13].

2.2 Pixel-Accurate Approaches

Combinatorial maps can be used to represent the topology of planar subdivisions, but they do not define the geometry of a tessellation, which is crucial for image segmentation. Thus, algorithms often employ a *label image* (aka. "region image") to store the geometry of regions. It is straight-forward to extract a consistent topology from the inter-pixel boundaries of such an image, in which each pixel carries the label of the region it belongs to. It has even been shown that the same is possible for *thin 8-connected pixel boundaries* [14], which for example result from watershed algorithms which leave the watersheds unlabelled.

However, from an applications' perspective it is preferable to have just one structure to deal with, not separate ones for the geometry and the topology. Thus, data structures have been developed [8, 5, 6, 2] which encapsulate both the geometrical and topological aspects and offer means to inspect or modify the tessellation in a consistent way. Fig. 1 illustrates two pixel-based representations:

1. *Inter-pixel boundaries*: In the TOGER framework [15, 1], a boundary plane is used to represent the connections between inter-pixel boundaries (at pixel corners, cf. black dots in 1(b)). This is very memory efficient (only three bits / pixel), but requires traversals and hash lookups to find the edges / regions at arbitrary positions. Darts are represented by the vertex position (cf. gray dots) and a direction.
2. *Pixel-based boundaries*: In [2, 6], the internal representation of a GeoMap is based on a *cell image*, where each pixel carries a label and a type (Region / Line / Vertex). All three topological cell types are represented as connected components of pixels carrying the corresponding type and label. All topological information is extracted via a DARTTRAVERSER, which is represented with a position / direction pair (cf. arrow in Fig. 1(c)). For details see [2, 6].

The limited resolution of these approaches is not only a cosmetic problem but also affects the topology: the vertices of inter-pixel boundaries cannot have a degree > 4, while pixel-based vertices as defined in [2] can have higher degrees if they consist of more than one pixel, which reduces the geometrical quality and needs complicated thinning operations after modifications. The new representation which is presented in the following does not have that problem.

3   Representing Sub-pixel Geometry

The representations discussed in the last section serve as powerful frameworks which ease the implementation of automatic and (semi-)interactive segmentation algorithms. However, they are limited to the pixel grid, while many edge detectors deliver *edgels* (edge elements) with sub-pixel accuracy (e.g. [12, 16]) which cannot be represented within these frameworks. We will now present a new approach which overcomes this limitation.

Let us assume we have sub-pixel accurate edgel positions linked into *edgel chains* (Sect. 4 will discuss some algorithms which produce these). These chains are commonly visualized with their *approximating polyline* (by connecting the points in order), and these ordered point lists serve as the main representation of edges in our new sub-pixel GeoMap. This is illustrated in Fig. 2 (left). It should be stressed that the polylines are only an approximation of the edges, and that the actual run of an edge between two support points is not represented (but could be determined on demand). This matters for algorithms analyzing the geometry, like for instance skeletonization or curvature calculation.

3.1   Meeting Algorithm Requirements

This section explains how the requirements listed in Sect. 2 are fulfilled in our implementation of the GeoMap framework.

Topology Inspection. In our object-oriented design, each cell is represented with a CellInfo object which carries its properties. The framework supports the enumeration of all vertices, edges, or faces of a map, and lookups by label. CellInfo objects can be queried for canonical darts (*anchors*) whose σ, α, or φ-orbits represent the cell (a face contains one anchor per contour, the first always belonging to the outer contour). The central tool to inspect the map topology is the DartTraverser [6]. Similar to an iterator, it represents a current position – a dart within the map. It offers methods to move to the successor / predecessor in any of the three permutations, and to get the start-/end-vertices, the edge it belongs to, or the face to the left/right. Many of the methods are only for your convenience, but this interface has proven to make the GeoMap framework very powerful in practice.

Geometry Inspection. The CellInfo objects mentioned above also carry the cells' geometrical properties (as well as application-specific information, cf. Sect. 3.3):

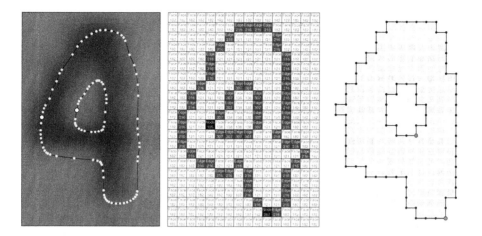

Fig. 2. Comparison of the new sub-pixel representation with approaches restricted by the pixel grid (to integer or half-integer coordinates, respectively)

a vertex simply contains its sub-pixel position, and an edge is represented as a polyline. The geometry of faces is represented implicitly; its anchors can be used to get closed polygons for each contour, and standard polygon techniques can be applied to these for reconstruction of the region, point inclusion tests, or finding the region containing a point. Since these operations are common, but rather slow, we speed them up internally with an additional label image, which Sect. 3.2 describes in more detail.

Note that it is very convenient to have the edge geometry include the vertex positions - in spite of the slight redundancy, this simplifies many algorithms, since all polyline segments can be derived from the edges, without looking at the vertices.

Modifications. We define *Euler operators* to allow the modification of our GeoMap. These are atomic operations which make sure that Euler's equation (here in its form for more than one boundary component) is an invariant:

$$|\sigma| - |\alpha| + |\varphi| - C = 1 \text{ where } C \text{ is the number of connected components} \qquad (2)$$

In contrast to the relatively complex operations used in other approaches (e.g. contraction kernels [11]), we define the following minimal set of simple operations:

merge_edges merge the two edges $\alpha^*(d)$ and $\alpha^*(\sigma(d))$ and the vertex $\sigma^*(d)$ (must have degree 2) into one single edge ($|\sigma'| = |\sigma| - 1$, $|\alpha'| = |\alpha| - 1$)

remove_bridge merge the edge $\alpha^*(d)$ (which must be a bridge) into the surrounding face $\varphi^*(d)$($|\alpha'| = |\alpha| - 1$, $C' = C + 1$)

merge_faces merge the two faces $\varphi^*(d)$ and $\varphi^*(\sigma(d))$ (must not be identical) and their common edge $\alpha^*(d)$ into one face ($|\alpha'| = |\alpha| - 1$, $|\varphi'| = |\varphi| - 1$)

These operations can be composed into more complex ones. For instance, the removal of all edges between two regions[1] is done with the composed operation merge_faces_completely which uses merge_faces to remove the first common edge, after which the rest of the common boundary will consist of bridges which are handled one-by-one with remove_bridge.

Note that after the removal of edges, which reduces the degree of their end-vertices, these vertices may become dispensable. Vertices of degree 2 can be merged into their surrounding contour with merge_edges. However, it may be worthwhile to purposely leave vertices of degree 2 in the structure, if their geometrical counterpart marks a point of interest (e.g. a corner). Singular vertices (degree 0) are discarded in our structure.

In theory, all the mentioned operations have their natural inverses (split_edge, create bridge, split_face respectively). However, we currently restrict ourselves to operations reducing the number of cells. The reasons are manifold: (a) Our Euler Operations can all be parametrized with a single dart, and it is straight-forward to prove their correctness. Their inverses need additional parameters for the geometry of the new cells to be created, which poses a problem when adding edges, since it has to be ensured that the given geometry does not violate the topology. (b) Conventional split and merge algorithms do not split faces into two, but use an implicit description of the split regions which is intrinsically limited to the pixel grid [9]. (c) The bottom-up approach of transforming an initial oversegmentation into the desired result fits well the basic idea of first looking for *any* evidence for boundaries and then applying *relevance filtering* to it.

3.2   Initializing a GeoMap

Assuming that we have already extracted boundaries from an image (examples follow in Sect. 4), this section discusses the remaining task for initializing a complete GeoMap: the determination of the boundary topology from its geometry.

The first problem is the initialization of the permutation σ, which means that we must determine the local cyclic order of edges around vertices. This may be as trivial as calculating the angles of the first segments of the approximating polylines attached to the vertex (see illustration). However, when trying to do this with sub-pixel watersheds (Sect. 4.2), this leads to numerical problems, since watersheds converge tangentially near a maximum, so subgroups of tangential darts have to be followed until they eventually diverge (see [16] for details).

Given the σ-orbits, we still have to determine the exterior faces of each connected boundary component and their parent faces. The exterior faces can be found by calculating the signed area of each contour given by the φ-orbits:

$$A = \frac{1}{2} \sum_i (x_i y_{i+1} - y_i x_{i+1}) \leq 0 \Rightarrow \text{exterior contour} \tag{3}$$

---

[1] Note that merge_faces removes just one edge, whereas the common boundary might consist of several edges (cf. Fig. 1, edges 5 and 2 between wall and background).

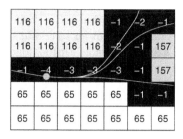
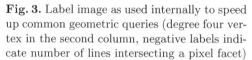

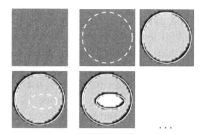

Fig. 3. Label image as used internally to speed up common geometric queries (degree four vertex in the second column, negative labels indicate number of lines intersecting a pixel facet)

Fig. 4. incremental label image initialization and face embedding (from top left to bottom right)

If a contour contains only bridges, it is an exterior contour and A should be zero, but may be a small positive number due to numerical problems. Thus, this case must be checked explicitly.

As mentioned in Sect. 3.1, we make use of an internal label image to speed up geometry queries. For point-in-region tests, we mark pixels whose unit square is not intersected by contours with the corresponding region label. Thus, we can immediately determine which region contains a given point if it's not near the contour, see Fig. 3. Otherwise, the pixel is marked with a negative label, and we must apply a (more expensive) standard point-in-polygon test on all regions whose (cached) axis-parallel bounding box contains the point.

In order to derive the inclusion relation from the geometry, we need to check for polygon inclusion, which corresponds to inclusion of a single point, since the boundaries do not overlap. For efficiency, the following algorithm will do the face embedding in parallel to the initialization of the label image (see Fig. 4):

1. The label image is initialized with the label of the infinite outer face.
2. We sort all contours by decreasing absolute area.
3. For each contour, beginning with the largest:
   (a) If it is an exterior contour, we find the existing face including this hole contour and embed it.
   (b) Else, we add a new face to the map and apply polygon scan conversion techniques to update the label image with the new region and its contour.

In order to facilitate updates of the label image, we store the number of edges intersecting a pixel facet as negative integer (see Fig. 3). Whenever an edge is removed (by merge_faces or remove_bridge), the labels of these pixels are incremented and eventually assigned to the surrounding region if they become zero.

3.3   Maintaining Consistency of Application-Specific Data

A bottom-up image segmentation process can be described as reducing an initial set of candidate boundaries into the final tessellation. We call this reduction process *relevance filtering*. In the context of irregular pyramids, this corresponds to the pyramid bottom containing an initial oversegmentation and a "tapering" stack of levels on top with decreasing numbers of cells. In order to create such a pyramid, automatic segmentation algorithms need to consider (in)homogeneity properties of regions (boundaries) to decide upon insignificant boundaries.

Typical region properties used for relevance filtering are statistics on the regions' colors (mean, variance, ...), area, or circumference. Boundaries are often assessed based on the local image gradient, their length, or curvature. The GeoMap makes it very simple to calculate such information and attach it to the CellInfo objects. During the segmentation process, this information has to be kept up-to-date when removing (parts of) boundaries. It would be possible to re-calculate the information after each change, but for common statistics it is possible (and much more efficient) to incrementally compute it from the cell information before the change.

Our GeoMap representation thus supports to register separate pre- and post-operation callback functions for each Euler operation in order to enable application-specific statistics to be maintained in a consistent way [6, 1]. This ensures that each Euler operation is accompanied by the appropriate updating procedures. The dart which parameterizes the operation is passed to the pre-operation callbacks, to inform them which cells will be merged. The update functions will collect the necessary information from the old cells and wait for the post-operation call, which attaches the updated information to the CellInfo object of the surviving cell, which it gets passed as parameter.

This approach makes it very easy for an application to manage e.g. photometric information on the regions, specific flags needed to perform the segmentation algorithm, or information on the boundary (like the mean gradient or a watersheds' pass value), and it is always guaranteed that this information is up-to-date. The GeoMap itself maintains some meta information on the cells' geometry (lengths, areas, bounding boxes), which is also made available and does not have to be recalculated.

Note that we internally store the partial sum of the signed area (3) for each edge, which allows us to quickly determine the signed area of any contour. (The removal of a bridge leads to a new contour whose area is unknown, and the partial sums efficiently solve the problem that the area is needed to determine the new exterior contour if the bridge belonged to the old exterior contour.)

4 Applications

Now that we have introduced our new sub-pixel precise representation formalism, we will show how it can be used with some image analysis algorithms.

4.1 Preliminaries: Continuous View on Input Images

A key tool to all our sub-pixel resolution experiments is that we can adaptively sample images at any desired (sub-pixel) position. This can be done efficiently by means of spline interpolation.

Splines of order n possess $n - 1$ continuous derivatives and can be efficiently computed at any location $\boldsymbol{x} = (x, y)$ by convolution of discrete spline coefficients c_{ij} with continuous B-spline basis functions β_n:

$$f(x, y) = \sum_{i,j} c_{ij} \, \beta_n(i - x) \, \beta_n(j - y) \tag{4}$$

The coefficients c_{ij} depend on the order n of the spline and can be computed from the sampling values f_{ij} by a cascade of $\lfloor n/2 \rfloor$ first-order recursive filters. Details on these computations can be found in [17, 16]. We use spline interpolation throughout this work for retrieving image values at sub-pixel locations, because of their global continuity across facet borders.

A side effect of the spline reconstruction is that interpolated real images (containing noise) will not have any plateaus in practice (when represented with floating-point accuracy). This is important for methods relying on the gradient vanishing only at isolated points (like the contour following methods described below). Note that it is not necessary to use convolution filters for derivatives, because they can be derived analytically from the spline approximation.

4.2   Sub-pixel Watersheds

When comparing the classical watersheds-by-flooding algorithm [18] with e.g. Canny's edge detector [12], watersheds have the disadvantage of being limited to the pixel grid. On the other hand, they provide closed contours, so that a complete topology can be derived [8, 14]. The advantages of both worlds can be combined by applying a sub-pixel watershed algorithm to the interpolated boundary indicator function [16, 19]. This algorithm is based on a mathematical definition of watersheds given by Maxwell [20]: watersheds are flowlines between maxima and saddles. If the function f is differentiable, a unique flowline exists at every point with non-zero gradient, and flowlines can be traced (upwards, starting at saddle points) by numerically solving their differential equation

$$\frac{\partial \boldsymbol{x}(t)}{\partial t} = \nabla f(\boldsymbol{x}(t)) \tag{5}$$

(e.g. with the Runge-Kutta method). This is stable near a watershed, because all flowlines in a neighborhood converge to the same maximum (for details, see [16]).

The algorithm is significantly slower than pixel-based watershed algorithms, but gives very high resolution (as can be seen in Fig. 5). Since the flowlines connect saddles and maxima, the output of the algorithm naturally forms a graph, which can be turned into a map after determining the σ-order of edges around each vertex (maximum). As mentioned in Sect. 3.2, the cyclic order of edges cannot be determined locally due to numerical problems because watersheds converge tangentially near maxima, see the detailed close-up in Fig. 5. The yellow circles mark the locations where the watersheds diverge (as found by our σ-sorting algorithm [16]). It is advisable to add additional vertices at these

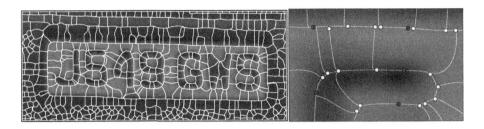

Fig. 5. Sub-pixel watersheds. *left*: initial oversegmentation, *right*: problem of tangential convergence; additional vertices added (yellow) vs. original vertices / maxima (dark red).

positions, since otherwise the statistics of a topological edge may contain mixed-up information from several geometrically unrelated segments. (The exact vertex positions may be refined later.)

4.3 Sub-pixel Level-Set Contour Tracing

An alternative method for finding an initial boundary set is not to look for ridges, but for zero-crossings of an appropriate edge detector (e.g. the Laplacian-of-Gaussian [21]) or of a distance function resulting from variational segmentation in level-set approaches. More generally, this can be used to find any level lines implicitly defined by

$$\phi(x, y) = c \Leftrightarrow \tilde{\phi}(x, y) := \phi(x, y) - c = 0$$

The tangent unit vector t of a level-line is always perpendicular to the gradient direction: $t = \nabla\phi^{\perp}/|\nabla\phi|$. Thus, the points of a level-line fulfill the PDE

$$\frac{\partial \boldsymbol{x}(\tau)}{\partial \tau} = \pm \boldsymbol{t}(\tau) = \pm \frac{\nabla\phi(\tau)^{\perp}}{|\nabla\phi(\tau)|} \tag{6}$$

with initial condition $\phi(\boldsymbol{x}_0) = 0$ and $\nabla\phi(\boldsymbol{x}_0) \neq 0$. In principle, this PDE could be solved with standard methods (like Runge-Kutta's), but this does not take advantage of the fact that the level-line must remain at this particular level. This constraint is used by *predictor-corrector methods* which significantly simplify level-line tracing. They use the tangent to extrapolate the curve towards a new candidate point (predictor step), but these predictions need not be extremely accurate because the level constraint is subsequently used to move the new point's position onto the contour (corrector step). Compared to other methods, this allows simpler predictors or larger steps. The basic algorithm is as follows [22]:

1. Given: a differentiable function $\phi(\boldsymbol{x})$ and a starting point $\boldsymbol{x}_0$ such that $\phi(\boldsymbol{x}_0) = 0$. Select an initial step size h and a bound ϵ_0 that specifies how much $\phi(\boldsymbol{x})$ may deviate from the exact zero level along the line.

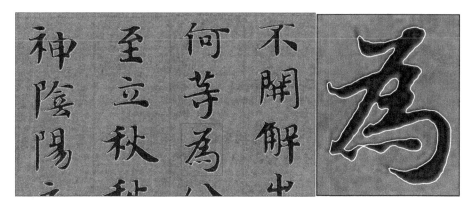

Fig. 6. Level-set contours of an ancient Chinese transcript (*right*: close-up)

2. While stopping criterion not fulfilled:

   (a) Predict candidate point $\hat{\boldsymbol{x}}_{i+1}^{(0)} = h\,\boldsymbol{t}(\boldsymbol{x}_i)$ where $\boldsymbol{t}(\boldsymbol{x}_i) = \frac{\nabla\phi^{\perp}(\boldsymbol{x}_i)}{|\nabla\phi(\boldsymbol{x}_i)|}$ if $\boldsymbol{x}_i$ is not a saddle point of ϕ, and $\boldsymbol{t}(\boldsymbol{x}_i) = \frac{\boldsymbol{x}_i - \boldsymbol{x}_{i-1}}{|\boldsymbol{x}_i - \boldsymbol{x}_{i-1}|}$ otherwise.

   (b) While $\left| \phi\left(\hat{\boldsymbol{x}}_{i+1}^{(k)} \right) \right| > \epsilon_0$:

      i. Correct the candidate point by Newton iterations

$$\hat{\boldsymbol{x}}_{i+1}^{(k+1)} = \hat{\boldsymbol{x}}_{i+1}^{(k)} - \frac{\phi\left(\hat{\boldsymbol{x}}_{i+1}^{(k)} \right)}{\left| \nabla\phi\left(\hat{\boldsymbol{x}}_{i+1}^{(k)} \right) \right|^2} \nabla\phi\left(\hat{\boldsymbol{x}}_{i+1}^{(k)} \right)$$

   (c) If the total correction was small, accept $\hat{\boldsymbol{x}}_{i+1}^{(k+1)}$ as new point $\boldsymbol{x}_{i+1}$, set $i := i + 1$, possibly increase h, and go to 2. Else, reduce h and go to (a).

Since level-lines form closed contours, one wants to stop the algorithm when it returns to the starting point. Detecting this is not trivial, but since we define $\phi(\boldsymbol{x})$ as a spline, there is a simple solution which also solves the problem of detecting starting points: consider the explicit polynomial representation (4) of a spline and the locus of points where $x = i \vee y = j$. We get a set of horizontal and vertical lines through the sampling points, enclosing small unit squares. Along these lines, (4) simplifies to two 1-dimensional polynomials of order n, and the roots of these polynomials can easily be computed by a standard root finder. Each root that lies on the side of the corresponding unit square marks a point where the zero level-line crosses. By iteratively choosing one of the crossings as the starting point and applying the above algorithm to trace the level-line until it leaves this square at another of the known crossings, we get connected level contours.

Finally, we must identify the vertices in order to initialize a GeoMap with the contours (according to Sect. 3.2). Since edges derived from zero-crossings always form closed contours, there are two kinds of vertices: if the curve self-intersects, all intersection points are vertices. Otherwise, an arbitrary point on the curve must be selected as a vertex.

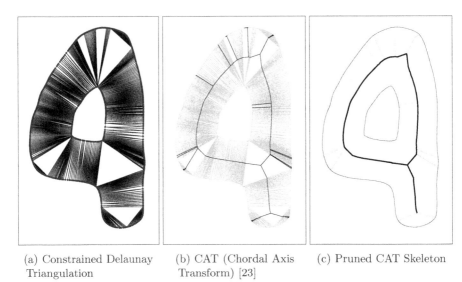

(a) Constrained Delaunay   (b) CAT (Chordal Axis   (c) Pruned CAT Skeleton
Triangulation   Transform) [23]

(d) *light red:* contours (after relevance filtering), *white:* pruned CAT skeletons

Fig. 7. GeoMaps representing contours, triangulations, and skeletons

Fig. 6 shows an example of such level-set contours: again, we can combine the advantage of high sub-pixel resolution with the advantage of common thresholding, which does not need any convolution filters and can thus be applied without implicit smoothing if the signal-to-noise ratio is high enough. In Fig. 6, this helps us in analyzing the cusps, which are important stroke characteristics.

4.4 Triangulation / Skeletonization

Our map is not only suitable for representing segmentation results, but it is also an adequate representation for triangulations or for skeletons (the latter requires the representation of bridges, see Sect. 3). Topological data structures have a long history in the computation of Delaunay triangulations and Voronoi diagrams (e.g. the quad-edge structure used in [24]).

The versatility of our GeoMap is illustrated in Fig. 7, which displays the result of the following example process:

1. First, we calculate sub-pixel watersheds of the original image from the spline-interpolated gradient magnitude.
2. (Simple relevance filtering) We iteratively merge regions until the difference between the average color of all adjacent regions is larger than a threshold (dark red contours in Fig. 7(d)).
3. Detect letters as hole regions which are darker than their parent face.
4. Apply a constrained Delaunay triangulation (CDT) to all letters, cf. Fig. 7(a).
5. (Chordal Axis Transform) Connect the mid-points of the inner chords to new edges, create a vertex for each inner ("join") triangle and connect vertices and edges to a CAT skeleton map, Fig. 7(b) (for details, see [23]).
6. (Simple pruning) Remove small branches: apply remove_bridge to edges shorter than two pixels with an end-vertex of degree 1 (Fig. 7(c), pruned parts in light gray).

Fig. 7(d) shows the contours from step 2 in red (note that the width of the original letter parts is less than two pixels, the whole region of interest is 64×19) and the pruned skeleton in white (the slight difference between the "5" and the "S" remains visible in the skeletons). The sampling of the "W" obviously violated Shannon's theorem and is hardly recognizable for a human, too.

This example is not meant to be a sophisticated, general feature extraction method, but it nicely illustrates the power of the GeoMap as a representation for planar graphs which offers convenient means to

- merge regions (step 2)
- manage statistical information on regions or edges, e.g. a regions' mean color (steps 2 and 3) or area (step 3), or the length of edges (step 6)
- inspect the geometry and decide upon inner / outer of regions (CDT, step 4)

5   Conclusion

Unified representations offering both topological and geometrical perspectives on a segmentation have been shown to be powerful as well as easy-to-use. In this paper, we extended the GeoMap formalism to achieve sub-pixel accuracy. We have shown that besides advanced sub-pixel segmentation techniques, triangulation and skeletonization can be performed equally well with our representation. Our experiments have shown that the advantages of the general planar map formalism still apply: our GeoMap framework allows for a significantly faster development of algorithms than without such a representation, and their formulations tend to become more concise due to the high level of abstraction. Algorithms with previously separate data structures can easily be compared and combined.

We are planning to release our implementation in the context of the VIGRA library. On the application side, we are currently working on the integration of learning methods and more sophisticated edge salience measures (e.g. based on boundary continuity or curvature) for relevance filtering.

References

1. Braquelaire, A.: Representing and segmenting 2d images by means of planar maps with discrete embeddings: From model to applications. In Brun, L., Vento, M., eds.: Graph-based Representations in Pattern Recognition, Springer (2005) 92–121
2. Meine, H., Köthe, U.: The GeoMap: A unified representation for topology and geometry. In Brun, L., Vento, M., eds.: Proc. Graph-Based Representations in Pattern Recognition, Springer (2005) 132–141
3. Brun, L., Kropatsch, W.: Construction of combinatorial pyramids. In Hancock, E., Vento, M., eds.: Graph-Based Repr. in Pattern Recognition, Springer (2003) 1–12
4. Mäntylä, M.: An Introduction to Solid Modeling. Computer Science Press (1988)
5. Brun, L., Domenger, J.P., Braquelaire, J.P.: Discrete maps: a framework for region segmentation algorithms. In: Graph-based Representations in Pattern Recognition, Springer (1998) 83–92
6. Meine, H.: XPMap-based irregular pyramids for image segmentation. Diploma thesis, Dept. of CS, University of Hamburg (2003)
7. Pavlidis, T.: Structural Pattern Recognition. Springer (1977)
8. Brun, L., Domenger, J.P.: Incremental modifications of segmented images. Technical Report RR112696, Université Bordeaux, LABRI (1996)
9. Brun, L., Domenger, J.P.: A new split and merge algorithm with topological maps and inter-pixel boundaries. In: Proc. WSCG'97. (1997)
10. Köthe, U.: XPMaps and topological segmentation - a unified approach to finite topologies in the plane. In Braquelaire, A., Lachaud, J.O., Vialard, A., eds.: Conf. on Discrete Geometry for Computer Imagery. Springer (2002) 22–33
11. Kropatsch, W.G.: Building irregulars pyramids by dual graph contraction. IEEE-Proc. Vision, Image and Signal Processing **142** (1995) 366–374
12. Canny, J.: A computational approach to edge detection. T-PAMI **8** (1986) 679–698
13. Kovalevsky, V.A.: Finite topology as applied to image analysis. Computer Vision, Graphics, and Image Processing **42** (1989) 141–161
14. Köthe, U.: Deriving topological representations from edge images. In Asano, T., Klette, R., Ronse, C., eds.: Geometry, Morphology, and Computational Imaging, WS on Theoretical Foundations of Computer Vision. Springer (2003) 320–334
15. Braquelaire, J.P., Brun, L.: Image segmentation with topological maps and inter-pixel representation. J. Visual Comm. and Image Representation **9** (1998) 62–79
16. Meine, H., Köthe, U.: Image segmentation with the exact watershed transform. In Villanucva, J., ed.: Proc. VIIP'05, ACTA Press (2005) 400–405
17. Unser, M., Aldroubi, A., Eden, M.: B-Spline signal processing: Part I and II. IEEE Trans. on Signal Processing **41** (1993) 821–848
18. Vincent, L., Soille, P.: Watersheds in digital spaces: an efficient algorithm based on immersion simulations. In: T-PAMI **13** (1991) 583–598
19. Steger, C.: Subpixel-precise extraction of watersheds. In: Proc. of 7th ICCV. Volume 2., IEEE Computer Society (1999) 884–890
20. Maxwell, J.C.: On hills and dales. Reprinted in W. D. Nivin (Ed.): The Scientific Papers of James Clerk Maxwell **2** (1952) 233–240. Dover Publications.
21. Marr, D., Hildreth, E.C.: Theory of edge detection. Proceedings of the Royal Society of London **B207** (1980) 187–217
22. Allgower, E.L., Georg, K.: Numerical path following. In: P.G. Ciarlet, J.L. Lions (Eds.), Handbook of Numerical Analysis **5** (1997) 3–207. North-Holland.
23. Prasad, L.: Morphological analysis of shapes. CNLS Newsletter **139** (1997)
24. Guibas, L.J., Stolfi, J.: Primitives for the manipulation of general subdivisions and the computation of voronoi diagrams. ACM Trans. on Graphics **4** (1985) 74–123

Feature Based Defuzzification at Increased Spatial Resolution

Joakim Lindblad[1] and Nataša Sladoje[2,*]

[1] Centre for Image Analysis, Swedish University of Agricultural Sciences,
Uppsala, Sweden
joakim@cb.uu.se
[2] Faculty of Engineering, University of Novi Sad,
Novi Sad, Serbia and Montenegro
sladoje@uns.ns.ac.yu

Abstract. Defuzzification of fuzzy spatial sets by feature distance minimization, recently proposed as an alternative to crisp segmentation, is studied further. Fully utilizing information available in a fuzzy (discrete) representation of a continuous shape, we present an improved defuzzification method, such that the crisp discrete representation of a fuzzy set is generated at an increased spatial resolution, compared to the resolution of the fuzzy set. The correspondence between a fuzzy and a crisp set is established through a distance between their representations based on selected features, where the different resolutions of the images to compare are taken into account. The performance of the method is tested on both synthetic and real images.

Keywords: fuzzy sets, defuzzification, multigrid resolution, distance measure, feature estimates.

1 Introduction

The advantages of representing objects in images as fuzzy spatial sets are numerous and have lead to increased interest for fuzzy approaches in image analysis [12]. Fuzziness is an intrinsic property of images. It is additionally introduced in digital image processing by discretization, and as a natural outcome of most imaging devices. Preservation of fuzziness implies preservation of important information about objects and images. Our previous results [2, 8, 11] show that an improved precision of shape description can be achieved if the description is based on fuzzy shape representation, where the fuzzy membership of a point reflects the level to which that point fulfils certain criteria to belong to the object. Among other shape descriptors, we have analysed perimeter, area, and moments of order up to two for shapes resulting from area coverage fuzzification. In this fuzzification approach, membership of a pixel is proportional to the part of its area covered by the observed object.

* Author is financially supported by the Ministry of Science of the Republic of Serbia through the Project ON144018 of the Mathematical Institute of the Serbian Academy of Science and Arts.

U. Eckardt et al. (Eds.): IWCIA 2006, LNCS 4040, pp. 131–143, 2006.

In spite of many advantages of utilizing fuzzy segmented images, a crisp representation of objects may still be needed. Reasons for that are, e.g., to facilitate easier visualization and interpretation. Even though it contains less information, a crisp representation is often easier to interpret and understand, especially if the spatial dimensionality of the image is higher than two. Moreover, analogues for many tools available for the analysis of binary images are still not developed for fuzzy images. This may force us to perform at least some steps in the analysis process by using a crisp representation of the objects.

In this paper, we are interested in generating a crisp representation of a fuzzy digital object. Such a process is known as defuzzification and we suggest to perform it by choosing the crisp representation that is closest to the given fuzzy set. The distance between two sets is expressed in terms of the difference between a number of selected quantitative features of the two sets. The novelty of the approach is that the crisp object is generated at higher spatial resolution, compared to the spatial resolution of the fuzzy object, by exploiting the additional information contained in the fuzzy representation. In this way, we propose a (crisp) segmentation technique that provides crisp objects represented at a higher spatial resolution than the given image resolution.

The paper is organized as follows: Section 2 gives an overview of the existing results related to defuzzification and lists the main definitions used in the paper. In Section 3 the main contribution of the paper, feature based defuzzification at increased spatial resolution, is presented. Section 4 contains examples of defuzzification method applied to one synthetic and two real images. Comments and concluding remarks are given in Section 5.

2   Background

We give a list of definitions and notions used in the paper and present existing results related to defuzzification.

2.1   Definitions

Definition 1. *A fuzzy set S on a reference set X is a set of ordered pairs $S = \{(x, \mu_S(x)) \mid x \in X\}$, where $\mu_S : X \to [0, 1]$ is the membership function of S in X.*

Being interested in applications in digital image analysis, we consider digital fuzzy sets, where $X \subset \mathbb{Z}^n$. In addition, when using digital approaches (computers) to represent, store, and analyse images, the (finite) number of grey-levels available is a natural limitation to the number of membership values that can be assigned to a digital point.

We denote by $\mathcal{F}(X)$ the set of fuzzy sets on a reference set X and by $\mathcal{P}(X)$ the set of crisp subsets of a set (the power set).

Definition 2. *An α-cut of a fuzzy set S, for $\alpha \in (0, 1]$, is the set*

$$S_\alpha = \{x \in X \mid \mu_S(x) \geq \alpha\}.$$

Definition 3. *The support of a fuzzy set S is the set*

$$Supp(S) = \{x \in X \mid \mu_S(x) > 0\}.$$

Definition 4. *The core of a fuzzy set S is the set*

$$Core(S) = \{x \in X \mid \mu_S(x) = 1\}.$$

Definition 5. *The moment $m_{p,q}(S)$ of a discrete fuzzy set S on a reference set $X \subset \mathbb{Z}^2$ is defined by*

$$m_{p,q}(S) = \sum_{(i,j) \in X} \mu_S(i,j)\, i^p j^q \ .$$

Definition 6. *The area (cardinality) of a discrete fuzzy set S on a reference set $X \subset \mathbb{Z}^2$ is*

$$A(S) = \sum_{(i,j) \in X} \mu_S(i,j).$$

Note that for a fuzzy set S, $A(S) = m_{0,0}(S)$, whereas the centroid of a fuzzy set S is defined as $C(S) = (C_x(S), C_y(S)) = \left(\dfrac{m_{1,0}(S)}{m_{0,0}(S)}, \dfrac{m_{0,1}(S)}{m_{0,0}(S)} \right)$.

Definition 7. *The perimeter of a fuzzy step subset S is*

$$P(S) = \sum_{\substack{i,j=1 \\ i<j}}^{m} \sum_{k=1}^{m_{ij}} |s_i - s_j| \cdot l(B_{ijk}),$$

where s_i, $i = 1, \ldots, m$, are m different membership values taken over the disjoint bounded constant-valued subsets of S, and $l(B_{ijk})$ is the length of the boundary between two neighbouring (constant-valued) fuzzy subsets having memberships s_i and s_j, determined as the overall length of the boundary between their supports, consisting of m_{ij} possibly disconnected parts.

Note 1: A fuzzy digital image can be understood as a fuzzy step set, i.e., as a disjoint union of a finite number of bounded subsets, each having a constant membership value. In a digital image, each pixel is seen as such constant-valued subset. A fuzzy step set is formally defined in, e.g., [1].

Note 2: We calculate $l(B_{ijk})$ and $P(S)$ of a discrete fuzzy set S as suggested in [11]; the perimeter of a fuzzy set is equal to the weighted sum of the perimeters of all the α-cuts of the fuzzy set. In the discrete case, a Marching Squares method is used to calculate the local contributions to the perimeter.

2.2   Related Work

Defuzzification is the process of replacing a fuzzy set with an appropriately cho-sen crisp set. It can be performed either as an inverse of fuzzification [6], with the intention to recover a fuzzified crisp original, or as a process independent of any fuzzification, but based on some pre-defined conditions that should be fulfilled for a crisp set to be the representation of a given fuzzy set [3, 7]. In image analysis the fuzzification function is rarely known, and practically never analytically defined; fuzzification of the image is a consequence of a combination of properties of the continuous original, discretization effects, and imaging conditions. Therefore, the inverse of a fuzzification function cannot, in general, be used for defuzzification in order to generate a good crisp discrete representation of the imaged object. Defuzzification is, instead, performed so that certain predefined criteria are respected in the process; the criteria are formulated with an intention to use the fuzzy representation as a source of valuable information about the geometric properties of the object that was fuzzified.

In our previous work [9, 10], we present a defuzzification method which generates a crisp object having area, perimeter, and centre of gravity as close as possible to the corresponding features of the fuzzy set, while keeping the similarity between the membership values of the points, as well as the gradient in each point, of the two sets as high as possible. The crisp object is generated at the same spatial resolution as the given fuzzy representation.

The results presented in [8, 11] show that the precision of estimates for perimeter, area, and higher order moments of a continuous shape, is significantly higher if a fuzzy discrete shape representation is used instead of a crisp discrete one. It is shown, either theoretically or through statistical studies, that a fuzzy approach can provide an alternative to increasing the spatial resolution of the image. This observation motivated the study presented in this paper: *Starting from a fuzzy shape representation, generate a crisp shape representation at an r times increased spatial resolution, while preserving features of a continuous original, estimated with a high precision from its fuzzy representation.*

2.3   Defuzzification by Feature Distance Minimization

Preservation of feature values is achieved by minimizing the distance between the given fuzzy set and the generated defuzzification, measured in an observed feature space. Details related to defuzzification by feature distance minimization, the distance measure definition and optimization, feature space construction, and search algorithms, are given in [10]. We recall here the notation and main definitions, used in the sequel.

Defuzzification. Given a fuzzy set $A \in \mathcal{F}(X)$, an optimal defuzzification $\mathcal{D}(A)$ of A, with respect to the distance measure d, is

$$\mathcal{D}(A) \in \{C \in \mathcal{P}(X) \mid d(A, C) = \min_{B \in \mathcal{P}(X)} [d(A, B)]\} . \tag{1}$$

Distance Measure. For an injective mapping Φ from $\mathcal{F}(X)$ into a metric space H, we define a metric on $\mathcal{F}(X)$ by requiring that Φ is an isometry.

We define the distance $d_p^\Phi(A, B)$ between fuzzy spatial sets A and B, on the same reference set X, as the Minkowski distance d_p between the representations of the sets A and B in the feature space $H \subset \mathbb{R}^n$:

$$d_p^\Phi(A, B) = d_p(\Phi(A), \Phi(B)). \qquad (2)$$

For $\mathbf{x}, \mathbf{y} \in \mathbb{R}^n$, the Minkowski distance is defined as

$$d_p(\mathbf{x}, \mathbf{y}) = \sqrt[p]{|x_1 - y_1|^p + |x_2 - y_2|^p + \cdots + |x_n - y_n|^p}.$$

By suitably designing the mapping Φ, the distance measure can be tuned to provide defuzzifications where both shape characteristics and membership values are taken into account. This enables defuzzification that fits the individual problem well, and provides a powerful family of defuzzification methods.

Optimization. In general, Equation (1) cannot be solved analytically. In addition, the search space $\mathcal{P}(X)$ is too big to be exhaustively traversed. As a consequence, we are forced to rely on heuristic search methods. In [10], two methods, floating search and simulated annealing, are used to find an approximate solution for Equation (1). Since the optimization task is a well separated problem, many other search methods can be used to approximatively solve Equation (1).

3   High Resolution Defuzzification

It has been shown in previous papers [2, 8, 11] that many shape features can be estimated with a higher precision when calculated from a fuzzy, instead of from a crisp representation of an object at a given image resolution. Utilizing this higher precision, it is natural to assume that a feature based defuzzification can provide a crisp representation of an object at a spatial resolution higher than the resolution of the given fuzzy image.

To perform defuzzification at increased spatial resolution, we need a distance measure that can relate fuzzy spatial sets represented at different resolutions. We observe an increase of the spatial resolution by an integer factor r; if r_F is the spatial resolution of the fuzzy set, and r_K is the spatial resolution of the crisp (defuzzified) set, then $r = \frac{r_K}{r_F}$. In that case, each pixel in the low resolution representation corresponds to a block of $r \times r$ pixels in the high resolution representation, as shown in Figure 1.

We note here that there are two approaches to perform multigrid studies: one is to observe the (r times) dilated object in the unchanged grid, whereas the other is to observe the unchanged object inscribed in the (r times) refined grid. These two approaches are dual. We use the first one, which implies that the size of the pixel is equal to 1 in all the observed grids, whereas the object features calculated in different grids are resolution-variant.

A main idea is to generalize membership similarity, which is a local feature, by interpreting it as a local area similarity. Instead of comparing pairs of corresponding pixels in the fuzzy and the crisp image, we relate the membership value

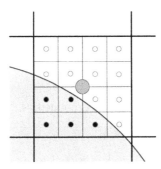

Fig. 1. One pixel in a low resolution (fuzzy) image, and the corresponding block of 4×4 pixels of a 4 times higher resolution (crisp) reconstruction

of a pixel in the fuzzy image (i.e., the pixel area) to the area of the corresponding block of $r \times r$ pixels (i.e., the number of (sub-)pixels with value 1) in the high resolution representation.

More formally, for a given membership $\mu(p_f) = m \in [0,1]$ of a pixel p_f to a spatial fuzzy set that represents the object, and a given resolution increase factor r, the area of the portion of the object within the corresponding block of $r \times r$ pixels can be estimated by mr^2. This correspondence is used as a basis for the high-resolution object reconstruction.

Equally important is to utilize the increased accuracy of estimates of global features – area, perimeter and moments of the first order – achieved when a fuzzy shape representation is used.

3.1   Feature-Based Representation and Distance Measure

In order to compare the corresponding features calculated in grids of different sizes, measures have to be rescaled with respect to the spatial resolution of the image and the dimensionality of the particular feature. It is taken into account that $P(S) = \mathcal{O}(r_s)$, $A(S) = \mathcal{O}(r_s^2)$, $m_{1,0}(S) = \mathcal{O}(r_s^3)$, $m_{0,1}(S) = \mathcal{O}(r_s^3)$ for a set S inscribed into a grid with spatial resolution r_S. To get resolution invariant global features, chosen for the task studied in this paper, we use

$$\tilde{P}(S) = \frac{P(S)}{P(X)}, \quad \tilde{A}(S) = \frac{A(S)}{A(X)}, \quad \tilde{C}_x(S) = \frac{C_x(S)}{C_x(X)}, \quad \tilde{C}_y(S) = \frac{C_y(S)}{C_y(X)},$$

where X is the reference set of resolution r_S. In this way, it is provided that $\tilde{P}(S) = \mathcal{O}(1)$, $\tilde{A}(S) = \mathcal{O}(1)$, $\tilde{C}_x(S) = \mathcal{O}(1)$, $\tilde{C}_y(S) = \mathcal{O}(1)$ for any grid resolution, which enables (meaningful) comparison of the feature values of fuzzy sets on different reference sets.

Note 3: The function $f(r) \geq 0$ is in the asymptotic complexity class $\mathcal{O}(g(r))$ (which is written as $f(r) = \mathcal{O}(g(r))$) iff there exists a constant $c \geq 0$ and a constant $r_0 \geq 1$ such that $f(r) \leq c \cdot g(r)$, for all $r \geq r_0$.

In addition to preserving the above global features, we also compare the (local) area of each block of $r \times r$ pixels of the crisp set K with the membership (area) of the one corresponding pixel of the fuzzy set F. When observed all together, global and local features should be of balanced relative size. Therefore, we scale down the local features in the representations for both fuzzy and crisp sets.

The distance between a fuzzy set

$$F = \{((x_i, y_i), \mu((x_i, y_i)) , \quad i = 1, \ldots, N\}$$

on a reference set of cardinality N, and a crisp set

$$K = \bigcup_{i=1,\ldots,N} K_i,$$

defined as a union of N blocks K_i, where each of the blocks is of size $r \times r$ and is of the form

$$K_i = \left\{ \left(rx_i - \frac{r-1}{2} + k, \ ry_i - \frac{r-1}{2} + l \right), \quad \text{for } k, l \text{ from } \{0, 1, \ldots, r-1\} \right\},$$

is determined as the Minkowski distance between the feature-based representations of the two sets, $\Phi(F)$ and $\Phi(K)$, in the observed feature space. According to the choice of features made for this study, we use the following representations:

$$\Phi(F) = \left(\frac{1}{\sqrt[p]{N}} \mu((x_1, y_1)), \ldots, \frac{1}{\sqrt[p]{N}} \mu((x_N, y_N)), \tilde{P}(F), \tilde{A}(F), \tilde{C}_x(F), \tilde{C}_y(F) \right),$$

$$\Phi(K) = \left(\frac{1}{\sqrt[p]{N}} \frac{1}{r^2} A(K_1), \ldots, \frac{1}{\sqrt[p]{N}} \frac{1}{r^2} A(K_N), \tilde{P}(K), \tilde{A}(K), \tilde{C}_x(K), \tilde{C}_y(K) \right).$$

The first N coordinates, corresponding to local features, are scaled down by the factor $\sqrt[p]{N}$, which preserves relative size of the contributions of the terms to the distance measure; it is provided that the N local features have the same impact on the overall Minkowski distance measure as one of the global features.

3.2   Defuzzification at Increased Resolution

Considering the scale-invariant feature representation introduced above, and Equations (1) and (2), a defuzzification K of a fuzzy set F at r times increased resolution is

$$\mathcal{D}(F) \in \{K \in \mathcal{P}(rX) \mid d(F, K) = \min_{B \in \mathcal{P}(rX)} [d(F, B)]\} . \tag{3}$$

Note 4: rX denotes the reference set X at r times increased resolution.

3.3   Search Algorithm

Simulated annealing [4] is a well known non-deterministic optimization algorithm. It is based on imitating the physical process of annealing, where the observed system, initially at high temperature and high energy, is slowly cooled and the energy of the system is gradually reduced towards a "frozen" ground state. Starting with an energy E and a temperature T, the initial configuration is perturbed and the resulting change in energy dE is computed. If the change in energy is negative, the new configuration is accepted. If the change in energy is positive, the new configuration is accepted with a probability which is dependent on the current temperature. This process is repeated sufficiently many times to give good sampling statistics for the current temperature. The temperature is then decreased and the entire process is repeated until a frozen state is achieved, at $T = 0$.

We apply this algorithm in order to find a crisp solution at a minimal distance to the given fuzzy set. The energy of the system is easily expressed in terms of the distance measure d_p^Φ, and a perturbation of the system as the addition or removal of one (sub-)pixel to the defuzzified set. To reduce the search space, pixels within the core of the fuzzy set are always included in the defuzzification and pixels outside the support of the fuzzy set are always excluded from the defuzzification. The starting configuration is taken to be a super-sampling of the α-cut at $\alpha = 0.5$ of the observed fuzzy set.

The speed of the cooling process allows a user-controlled trade-off between the speed and the quality of the optimization process. For the examples presented in the next section, the temperature is lowered in 3000 steps and at each temperature level, the number of perturbations tested is 20 times the number of pixels in the search space.

4   Examples

We present three examples illustrating the performance of the proposed defuzzification method. The defuzzification of a synthetic image is shown in Figure 2, whereas two real images are defuzzified and presented in Figures 4 and 5. Each figure displays five images: original, fuzzy segmented, and defuzzified object at 1, 4, and 8 times higher resolutions, respectively. Minkowski distance is calculated for $p = 2$. The core of the fuzzy set is indicated with a darker shade of grey in the defuzzified images. The centroid of the fuzzy set and of the defuzzified set are marked with "$\times$" and "$+$", respectively. When the two centroids coincide, the marks overlap and create "$*$".

For the synthetic image shown in Figure 2(a), the fuzzy segmented image (Figure 2(b)) is obtained by area coverage fuzzification of an 8 times downsampled image. Defuzzification is performed by minimizing the distance to the image in Figure 2(b) at increasing resolutions. The minimal obtained distance, corresponding to the defuzzification at each resolution, is given below the image. Figures 2(a) and (e) are at the same spatial resolution. Their visual comparison is enabled by Figure 3, which shows magnified the upper part of the crisp original object superimposed on the defuzzification. It is important to notice that the defuzzification (Figure 2(e)) is generated from an image (Figure 2(b)) which contains (approximately) 10 times less information than the original image (Figure 2(a)). (There are 8×8 more pixels in (a) each requiring one bit per pixel, while the 65 grey-levels in (b) require just above 6 bits per pixel.) In spite of that, the reconstruction shows very good visual similarity with the original.

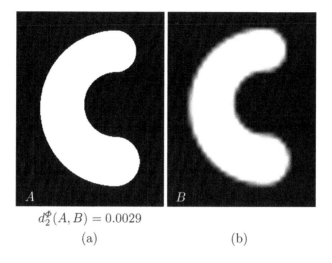

$$d_2^\Phi(A, B) = 0.0029$$

(a)   (b)

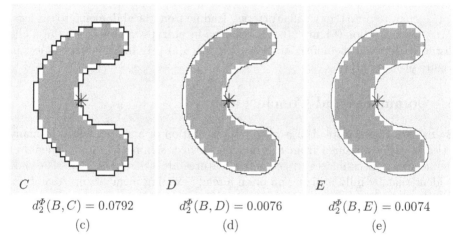

$d_2^\Phi(B, C) = 0.0792$ $d_2^\Phi(B, D) = 0.0076$ $d_2^\Phi(B, E) = 0.0074$

(c)   (d)   (e)

Fig. 2. Defuzzification of a synthetic object. (a) Synthetic crisp object A. (b) Fuzzification B of A at 8 times lower resolution. (c)-(e) Defuzzification C, D, and E of B at 1, 4, and 8 times increased resolution. Distances from the image in (b) are shown for each defuzzification and for the crisp original.

A second example is defuzzification of a part of a histological image of a bone implant (inserted in a leg of a rabbit), presented in Figure 4. The original is a colour image acquired by a light microscope. We tested our method on a part of the image (Figure 4(a)) containing a bone area (dark grey), surrounded by a non-bone area (light grey). Figure 4(b) shows a fuzzy object segmented by minimization of entropy as suggested in [5]. The distances between the defuzzifications (Figures 4(c)-(e)) and the fuzzy object in Figure 4(b) are given.

Our third example shows a slice of a magnetic resonance angiography (MRA) image of a human aorta (Figure 5(a)). The slice displays the region where the

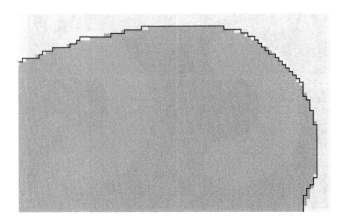

Fig. 3. The upper part of the reconstructed object E (Figure 2(e)) indicated with a black outline, superimposed on the original crisp object A (Figure 2(a)) shown in dark grey

aorta separates into the two iliac arteries, leading from the abdomen into the legs. Fuzzy segmentation (Figure 5(b)) is obtained by entropy minimization ([5]). The distances between the defuzzifications (Figures 5(c)-(e)) and the fuzzy object in Figure 5(b) are given.

5   Comments and Conclusions

By using a fuzzy, instead of a crisp, representation of a shape, which in many cases is easily obtained from the imaging device, significant improvements of the accuracy of estimates of geometric features are achievable. It has become evident that by fully utilizing an often already existing membership resolution, it is possible to overcome problems of insufficient available spatial resolution.

Defuzzification by feature distance minimization has shown to perform well as a crisp segmentation technique (in combination with a fuzzy segmentation), preserving important geometric properties of the object ([10]). In this paper, it is investigated how such a method can be applied to generate crisp representations of fuzzy objects at increased spatial resolutions.

Features to preserve are chosen to be area (zero-order moment of a shape), perimeter, and centroid of a shape (defined in terms of moments of order zero and one). It should be pointed out that the method for defuzzification presented in this paper is by no means limited to this set of features and our choice is just one of many available. The features to preserve should reflect the properties which are considered important for the object of study, and it should be possible to estimate these features with a high accuracy and precision from a fuzzy representation.

Some examples of the performance of the method are presented, showing promising results. The search space for the optimization grows quickly with increasing resolution. It is noticed that the simulated annealing optimization applied did not find the optimal solution for the 8 times increased resolution defuzzification of the synthetic test image in Figure 2; the original object in

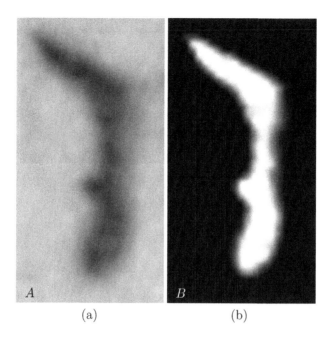

(a)   (b)

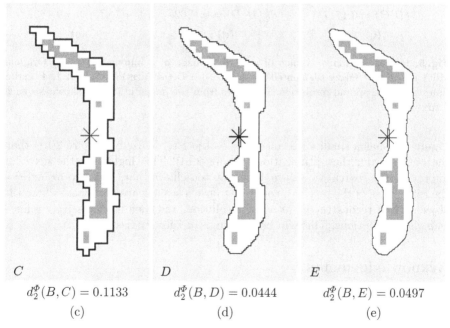

$d_2^\Phi(B,C) = 0.1133$ $d_2^\Phi(B,D) = 0.0444$ $d_2^\Phi(B,E) = 0.0497$

(c)   (d)   (e)

Fig. 4. Defuzzification of a selected part of a microscope image of a bone implant. (a) Selected part of an image. The dark grey area is bone, the light parts (light grey) are non-bone areas. (b) Fuzzy segmented bone area in (a). (c)-(e) Defuzzification of B at 1, 4, and 8 times increased spatial resolution. Distances from the image in (b) are shown for each defuzzification.

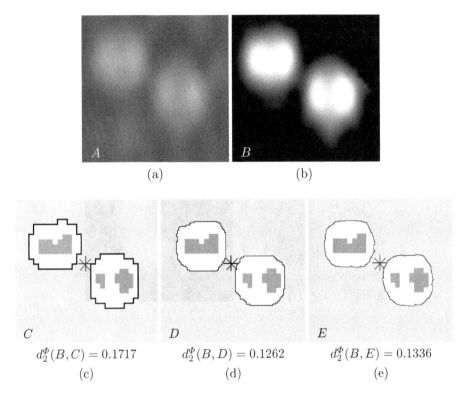

(a)   (b)

C   D   E

$d_2^\Phi(B,C) = 0.1717$ $d_2^\Phi(B,D) = 0.1262$ $d_2^\Phi(B,E) = 0.1336$

(c)   (d)   (e)

Fig. 5. Defuzzification of a slice of an MRA image of a human aorta. (a) Original MRA image. (b) Fuzzy segmented vessels. (c)-(e) Defuzzification of B at 1, 4, and 8 times increased spatial resolution. Distances from the image in (b) are shown for each defuzzification.

Figure 2(a) has a smaller distance to the object to defuzzify (Figure 2(b)) than the output from the optimization (Figure 2(e)). This highlights the need for improving the search procedure. One way to achieve that, and also to increase the robustness of the method, could be to use a scale space approach, where the object is first reconstructed at a low resolution, and then is successively refined at higher resolutions. This will be a topic of further studies.

Acknowledgements

Doc. Magnus Borga, Dept. of Biomedical Engineering, University Hospital, Linköping, Sweden, is gratefully acknowledged for initiating the study on high resolution defuzzification. Ministry of Science of the Republic of Serbia is acknowledged for financing the work of Nataša Sladoje on the Project ON144018. Dr. Lucia Ballerini and Dr. Xavier Tizon are acknowledged for providing the images in Figures 4(a) and 5(a), respectively.

References

1. A. Bogomolny. On the perimeter and area of fuzzy sets. *Fuzzy Sets and Systems*, 23:257–269, 1987.
2. J. Chanussot, I. Nyström, and N. Sladoje. Shape signatures of fuzzy sets based on distance from the centroid. *Pattern Recognition Letters*, 26(6):735–746, 2005.
3. W. V. Leekwijck and E. Kerre. Defuzzification: Criteria and classification. *Fuzzy Sets and Systems*, 108:159–178, 1999.
4. N. Metropolis, A. W. Rosenbluth, M. N. Rosenbluth, A. H. Teller, and E. Teller. Equation of state calculations by fast computing machines. *Journal of Chemical Physics*, 21(6):1087–1092, 1953.
5. S. Pal and A. Rosenfeld. Image enhancement and thresholding by optimization of fuzzy compactness. *Pattern Recognition Letters*, 7:77–86, 1988.
6. L. Rondeau, R. Ruelas, L. Levrat, and M. Lamotte. A defuzzification method respecting the fuzzification. *Fuzzy Sets and Systems*, 86:311–320, 1997.
7. E. Roventa and T. Spircu. Averaging procedures in defuzzification processes. *Fuzzy Sets and Systems*, 136:375–385, 2003.
8. N. Sladoje and J. Lindblad. Estimation of moments of digitized objects with fuzzy borders. In F. Roli and S. Vitulano, editors, *Proc. of 13th International Conference on Image Analysis and Processing (ICIAP 2005)*, volume 3617 of *LNCS*, pages 188–195, Cagliari, Italy, 2005. Springer-Verlag.
9. N. Sladoje, J. Lindblad, and I. Nyström. Defuzzification of discrete objects by optimizing area and perimeter similarity. In J. Kittler, M. Petrou, and M. Nixon, editors, *Proc. of 17th International Conference on Pattern Recognition (ICPR 2004)*, volume 3, pages 526–529, Cambridge, UK, 2004. IEEE Comp. Society.
10. N. Sladoje, J. Lindblad, and I. Nyström. Defuzzification of spatial fuzzy sets by feature distance minimization, 2005. Submitted.
11. N. Sladoje, I. Nyström, and P. Saha. Measurements of digitized objects with fuzzy borders in 2D and 3D. *Image and Vision Computing*, 23:123–132, 2005.
12. J. K. Udupa and G. J. Grevera. Go digital, go fuzzy. *Pattern Recognition Letters*, 23:743–754, 2002.

Extended Mumford-Shah Regularization in Bayesian Estimation for Blind Image Deconvolution and Segmentation

Hongwei Zheng and Olaf Hellwich

Computer Vision & Remote Sensing, Berlin University of Technology
Franklinstrasse 28/29, Office FR 3-1, D-10587 Berlin
{hzheng, hellwich}@cs.tu-berlin.de

Abstract. We present an extended Mumford-Shah regularization for blind image deconvolution and segmentation in the context of Bayesian estimation for blurred, noisy images or video sequences. The Mumford-Shah functional is extended to have cost terms for the estimation of blur kernels via a newly introduced prior solution space. This functional is minimized using Γ-convergence approximation in an embedded alternating minimization within Neumann conditions. Accurate blur identification is the basis of edge-preserving image restoration in the extended Mumford-Shah regularization. One output of the finite set of curves and object boundaries are grouped and partitioned via a graph theoretical approach for the segmentation of blurred objects. The chosen regularization parameters using the L-curve method is presented. Numerical experiments show that the proposed algorithm is efficiency and robust in that it can handle images that are formed in different environments with different types and amounts of blur and noise.

1 Introduction

Blur influences the automation, robustness and efficiency of many visual systems in many aspects. Blur identification, image restoration and recognition of blurred or unblurred regions or objects become more important, e.g., shown in Fig. 1. An ideal image f in the object plane is normally degraded by a linear space-invariant point spread function (PSF) h with an additive white Gaussian noise n using the lexicographic notation, $g = h * f + n$. The equation provides a good working model for image formation. The two-dimensional convolution is expressed as $h * f = Hf = Fh$, where H and F are block-Toeplitz matrices and can be approximated by block-circulant matrices.

Normally, the point spread function (PSF) of blur is neither known nor perfectly known. Such blur identification can be considered as blind image deconvolution. The challenge of blind image deconvolution (BID) is to uniquely define the optimized signals only from the observed images and is considered as an ill-posed problem in the sense of Hadamard [1]. However, knowledge of the direct model is not sufficient to determine an existent, unique and stable solution, and

U. Eckardt et al. (Eds.): IWCIA 2006, LNCS 4040, pp. 144–158, 2006.

(a)   (b)   (c)

Fig. 1. Blurred and unblurred regions in video data. (a) A video frame. (b) Blurred objects. (c) Unblurred object.

it is necessary to regularize the solution using some *a priori* knowledge. Mathematically, the *a priori* knowledge is often expressed through a regularization theory [2] which replaces an ill-posed problem by a well-posed problem with an acceptable approximation to the solution.

The regularization theory [2] presents numerous challenges as well as opportunities for further mathematical vision modeling to solve ill-posed problems. Compared to stochastic optimization [3], most blur identification and image restoration methods have been developed based on the deterministic regularization approach due to the efficiency of computation [4, 5, 6]. Different from the iterative Tikhonov regularization [4] and the total variational regularization [5], a general regularization method proposed by Mumford and Shah [6] has formulated image restoration, denoising and image segmentation in an energy minimization approach [7]. Currently, some Mumford-Shah (MS) based segmentation approaches combining with the level set method (LST) are intensively tested on the influences of noises or occlusions [8] and get successful results. However, curve-evolution based methods do not satisfactorily segment blurred regions or objects due to unstable and weak differences of gradients between blurred regions or objects and cluttered background.

Recently, variational regularization for image restoration [9] is investigated. Bar et. al. use a total variation method providing an initial value to a Mumford-Shah [6] functional for blur identification and edge preserving restoration. However, the initialization problem of regularization is still not progressively solved. This process needs more effective prior information and constraints to yield a unique solution. The Bayesian estimation framework provides a structured way to include prior knowledge concerning the quantities to be estimated. The Bayesian approach is, in fact, the framework in which most recent restoration methods have been introduced. Blake and Zisserman [10] proposed the use of a graduate non-convexity method, which can be extended to the blurring problem. Molina and Ripley [11] proposed the use of a log-scale for the image model in the Bayesian framework. Green [12] and Bouman et al. [13] used convex potentials in order to ensure uniqueness of the solution. Moreover, even if a unique solution exists, a proper initialization value is still intractable, e.g., when the cost function is non-convex, convergence to local minima often occurs without proper initialization. Molina et al. [14] have reported that the estimates for the PSF could vary significantly, depending on the initialization.

In this paper, we treat blur identification, image restoration and segmentation as a combinatorial optimization problem. Combination of blur identification, image restoration and segmentation in an extended MS functional is a reasonable strategy for such tasks due to the mutual support of edge-preserving image restoration and segmentation within a variational regularization. Firstly, Bayesian MAP estimation supports a good initial value for the optimization to the extended MS functional. Secondly, it is possible to get edge-preserving image restoration in the extended MS regularization via a Γ-convergence approximation [15, 9, 16]. Finally, an embedded alternate minimization method is introduced to achieve the outputs without scale problem between the estimates of the image and the PSF. One output of the finite sets of curves and object boundaries with different gradients can be considered as discrete analogues of graphs. It shows a theoretically and experimentally sound way of how a graph-theoretical approach is integrated to the extended MS functional for partitioning and grouping different gradient edges with blur information. The experimental results shows that the method yields explicit segmentation results as well as edge-preserving restoration under different kinds and amounts of blur.

The paper is organized as follows. In Sect. (2), the Γ-convergence MS approximation in the context of Bayesian estimation can be interpreted as an energy functional with respect to the estimation of the image and the PSF. The PSF learning is presented in Sect. (3). Optimization of three outputs in the newly designed embedded alternating minimization are presented in Sect. (4). Sect. (5) introduces a graph partitioning approach to group the detected edges with different gradients with blur constraints. These edges are directly computed from the extended MS functional. Experimental results are shown in Sect. (6). Conclusions are summarized in Sect. (7).

2   Extended MS Regularization in Bayesian Estimation

The Bayesian MAP estimation is utilized to get a *maximum a posteriori* (MAP) estimation using some prior knowledge. Following the Bayesian paradigm, the estimated image $\hat{f}$, the estimated PSF $\hat{h}$ and the observed image g is based on,

$$p(\hat{f}, \hat{h}|g) = \frac{p(g|\hat{f}, \hat{h})p(\hat{f}, \hat{h})}{p(g)} \propto p(g|\hat{f}, \hat{h})p(\hat{f}, \hat{h}) \tag{1}$$

Applying the Bayesian paradigm to the blind image deconvolution problem, we try to get convergence values from Eq. (1) with respect to the estimated image $\hat{f}$ and the estimated PSF $\hat{h}$. The MAP cost function E with respect to the estimated image $\hat{f}$ and the estimated PSF $\hat{h}$ from Eq. (1) are deducted according to the following,

$$E(\hat{f}|g, \hat{h}) \propto p(g|\hat{f}, \hat{h})p(\hat{f}), \ \ E(\hat{h}|g, \hat{f}) \propto p(g|\hat{f}, \hat{h})p(\hat{h}) \tag{2}$$

This Bayesian MAP approach can be computed in a regularization functional which optimizes two proposed cost functions in the image domain and the PSF domain. For the following description, we define the key symbols in Table. 1.

Table 1. List of key symbols

Symbol	Explanation
g, f, n, h	Degraded image, original image, additive noise, and blur.
$\hat{f}, \hat{h}, v$	Estimates to the original image, the blur and edge-curves.
$h_i(\theta)$	The i-th PSF parametric model with unknown parameters θ.
$E(\hat{f}\|g, \hat{h}), E(\hat{h}\|g, \hat{f})$	Image and blur-domain MAP cost function.
$E_\varepsilon(\hat{f}, \hat{h}, v), E_\varepsilon$	Approximated MS functional including all the estimates.
α, β, γ	Regularization parameters of image, edge and blur term.
$G = (V, E)$	A undirected weight graph with edges between vertices.

2.1   Prior Solution Space of Blur Kernels

Several forms of the prior distribution like Gibbs distribution [3], image smoothness or maximum entropy have been suggested by researchers from different disciplines but they are based on general knowledge about images. In reality, most real blurred images, whose power spectral densities vary considerably from low frequency domain in the uniform smoothing region to medium and high frequency domain in the discontinuity and texture regions. Also, most PSFs exist in the form of low-pass filters. Up to a certain degree, PSFs of numerous real blurred images satisfy parametric PSF models. Through these observations, we know that the performance of blur identification and image restoration should be according to their characteristics. The proposed prior solution space supports PSF prior in the Bayesian estimation. It attempts to address these asymmetries by integrating parametric blur knowledge into the scheme of the extended Mumford-Shah regularization.

We define a set Θ as a solution space of Bayesian estimation which consists of primary parametric PSF models as $\Theta = \{h_i(\theta), i = 1, 2, 3, ..., N\}$ in Fig. 2. $h_i(\theta)$ represents the ith parametric PSF with its own parameters θ, and N is the number of PSFs.

$$h_i(\theta) = \begin{cases} h_1(\theta) \propto h(x, y; L_i, L_j) = 1/K, & |i| \leq L_i \text{ and } |j| \leq L_j \\ h_2(\theta) \propto h(x, y) = K \exp(-\frac{x^2+y^2}{2\sigma^2}) & \\ h_3(\theta) \propto h(x, y, d, \phi) = 1/d, & \sqrt{x^2 + y^2} \leq D/2, \tan\phi = y/x \end{cases} \quad (3)$$

$h_1(\theta)$ is a pill-box blur kernel with a length of radius K. $h_2(\theta)$ is a Gaussian PSF and can be characterized by parameters with its variance σ^2 and a normalization constant K. $h_3(\theta)$ is a simple linear motion blur PSF with a camera direction motion d and a motion angle ϕ. The other blur structures like out-of-focus and uniform 2D blur [17], [18] have been also built in the solution space as *a priori* information.

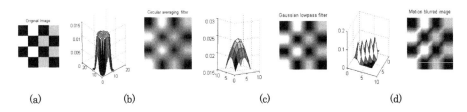

Fig. 2. (a) Original image. (b) Pillbox PSF. (c) Gaussian PSF. (d) Linear motion PSF.

2.2 Extended Γ-Convergence MS Regularization

After the discrete cartoon models from [3,10], the corresponding continuous model has been invented by Mumford and Shah [6]. The idea of the original MS functional is to subdivide an image into many meaningful regions (objects). It means to find a decomposition Ω_i of Ω and an optimal piecewise smoothing approximation f given a degraded image g. Thus, the estimated image f varies smoothly within each Ω_i, and discontinuously across the boundaries of Ω_i. The MS functional is formulated in an energy minimization equation,

$$E(f,C) = \frac{1}{2}\underbrace{\int_\Omega (g-f)^2 dxdy}_{\text{fidelityTerm}} + \alpha\underbrace{\int_{\Omega\setminus C}|\nabla f|^2 dxdy}_{\text{piecewiseSmoothing}} + \underbrace{\beta|C|}_{\text{totalEdgeLength}} \qquad (4)$$

Then we define

Ω is a connected, bounded and open subset R^2, $\Omega \subset R^2$,

f is the estimated image, $f \subset \Omega \setminus C$,

$C \subset \Omega$ is a finite set of segmenting curves and unit of object boundaries,

$|C|$ is the length of curve of C,

g is a bounded image-function with uniform feature intensity, $g : \Omega \to R$, and

$E(f,C)$ is energy function with respect to the estimates of image and curves.

It is hard to minimize this functional directly for achieving the set C numerically, keeping track of possible changes of its topology, and calculating its length. Likewise, the number of possible discontinuity sets is enormous even on a small grid. To solve such difficulties, Ambrosio et. al. [15] have introduced the Γ-convergence to the Mumford-Shah functional which means to replace the discontinuous C by a continuous variable v in the third term. An irregular functional $E(f,C)$ is then approximated by a sequence $E_\varepsilon(f)$ of regular functionals with a small constant ε, $\lim_{\varepsilon\to 0} E_\varepsilon(f) = E(f,C)$ and the minimization of E_ε approximates the minimization of E. The edge set is represented by a characteristic function $(1 - x_C)$ which is approximated by an auxiliary function $v(x)$ of the gradient edge integration map, i.e., $v(x) \approx 0$ of $x \in C$ for smoothing, and $v(x) \approx 1$ for getting edges. The equation is,

$$E_\varepsilon(f,v) \qquad (5)$$
$$= \frac{1}{2}\int_\Omega (g-f)^2 dxdy + \alpha\int_\Omega v^2|\nabla f|^2 dxdy + \beta\int_\Omega \left(\varepsilon|\nabla v|^2 + \frac{(v-1)^2}{4\varepsilon}\right)dxdy$$

Recently, Bar et.al. [9, 16] have combined this functional and the total variation functional for image restoration. Different from the work of [9, 16], we build a soft PSF learning term based on the PSF prior solution space for this functional. It improves the accuracy of the initial value of the PSF and also optimizes the PSF in a parametrical approach. The degradation model f is replaced by $h * f$ in the first fidelity term. The estimates to the original image and the blur kernel are denoted in $\hat{f}$ and $\hat{h}$ separately. The functional is formulated in the following,

$$E_\varepsilon(\hat{f}, \hat{h}, v) = \frac{1}{2} \int_\Omega (g - \hat{f} * \hat{h})^2 dxdy + \alpha \int_\Omega v^2 |\nabla \hat{f}|^2 dxdy \tag{6}$$

$$+ \beta \int_\Omega \left(\varepsilon |\nabla v|^2 + \frac{(v-1)^2}{4\varepsilon} \right) dxdy + \gamma \int_\Omega |\nabla \hat{h}|^2 dxdy + \delta |\hat{h} - \hat{h}_f|^2$$

where $\hat{h}_f$ is the final estimated PSF and $\hat{h}$ is the current estimated PSF. The fourth term $\gamma \int_\Omega |\nabla \hat{h}|^2 dxdy$ represents the regularization of the blur kernel. This term is necessary to reduce the ambiguity in the division of the apparent blur between the recovered image and the blur kernel. The flexibility of the last term $\delta |\hat{h} - \hat{h}_f|^2$ denotes the PSF decision learning error of the best-fit parametric model $\hat{h}_f$. The primary objective of this learning decision approach is to evaluate the relevance of parametric structure and integrate the information into the learning scheme accordingly. The effect of the PSF learning term is to pull the PSF MAP solution towards the PSF parametric model. It can adjust and incorporate the parametric model of the PSF throughout the process of blur identification and image restoration.

This functional can be interpreted as two cost functions with respect to the estimation of the image and the PSF, and one derived function for edge curves in the regularization functional. These functions are optimized alternatively in the embedded alternating minimization algorithm.

3   From Learnt PSF Statistics to PSF Estimation

In this paper, the estimation of PSF as an initial value is a starting point of the process for the image estimation. In the PSF domain, the PSF can be seen as the maximization of conditional probability. The cost function of the PSF from Eq. (2) is described using the extended Mumford-Shah functional,

$$E(\hat{h}|g, \hat{f}) = \arg \max_{\hat{h}} \left\{ p\left(g \middle| \hat{h}, \hat{f}\right) p_\Theta\left(\hat{h}\right) \right\} \tag{7}$$

$$= \frac{1}{2} \int_\Omega (g - \hat{h} * \hat{f})^2 dxdy + \gamma \int_\Omega \left|\nabla \hat{h}\right|^2 dxdy + \delta |\hat{h} - \hat{h}_f|^2$$

where $p(g|\hat{f}, \hat{h}) \propto \exp \left\{ -\frac{1}{2} \int_\Omega |g - \hat{h} * \hat{f}|^2 dxdy \right\}$, $p_\Theta(\hat{h}) \propto \gamma \int_\Omega \left|\nabla \hat{h}\right|^2 dxdy + \delta |\hat{h} - \hat{h}_f|^2$ is *a priori* knowledge including the PSF smoothing and PSF learning. $\hat{h}_f$ is a final selected and estimated parametric PSF model. The PSF estimation consists of the estimate of a sound PSF parametric model with supported size, and the coefficients of this PSF model. If the estimated PSF parametric model differs from the actual PSF, the blur coefficients and support size cannot be optimized and identified reliably during the following iterative optimization algorithm, and vice versa. Here, the PSF learning is used to find a suitable PSF parametric model using the prior PSF solution space which can largely decrease the searching space.

Since both the original and observed image represent intensity distributions that cannot take negative values, the PSF coefficients are always nonnegative, $h(x) \geq 0$. Furthermore, since image formation systems normally do not absorb or generate energy, the PSF should satisfy $\sum_{x \in \Omega} h(x) = 1.0$, $x \in \Omega$, $\Omega \subset R^2$. A MAP estimator is used to determine the best fit model $h_i(\theta^*)$ for the estimated PSF $\hat{h}$ in resembling the ith parametric model $h_i(\theta)$ in a multivariate Gaussian distribution, $h_i(\theta^*) \propto$

$$
\arg\max_{\theta} \log \left\{ \frac{1}{(2\pi)^{\frac{LB}{2}} |\sum dd|^{\frac{1}{2}}} \cdot \exp \left[-\frac{1}{2} \left(h_i(\theta) - \hat{h} \right)^T \sum_{dd}^{-1} \left(h_i(\theta) - \hat{h} \right) \right] \right\}
$$

where the first subscript i denotes the index of blur kernel. The modeling error $d = h_i(\theta) - \hat{h}$ is assumed to be a zero-mean homogeneous Gaussian distributed white noise process with covariance matrix $\sum_{dd} = \sigma_d^2 I$ independent of image. LB is an assumed support size of blur. Then the PSF learning likelihood is computed based on mahalanobis distance and corresponding model:

$$
l_i(\hat{h}) = \frac{1}{2} exp[(h_i(\theta) - \hat{h})^t \sum_{dd}^{-1} (h_i(\theta) - \hat{h})] \tag{8}
$$

In reality, most of blurs satisfy up to a certain degree of parametric structures. A best fit model $h_i(\theta)$ for $\hat{h}$ is selected according to the Gaussian distribution and a cluster filter. We use a K-NN rule to find the estimated output blur model $\hat{h}_f$ is obtained from the parametric blur models using

$$
\hat{h}_f = [l_0(\hat{h})\hat{h} + \sum_{i=1}^{C} l_i(\hat{h})h_i(\theta)]/[\sum_{i=1}^{C} l_i(\hat{h})] \tag{9}
$$

where $l_0(\hat{h}) = 1 - max(l_i(\hat{h}))$, $i = 1, ..., C$. The main objective is to assess the relevance of current estimated blur $\hat{h}$ with respect to parametric PSF models, and to integrate such knowledge progressively into the computation scheme. If the current blur $\hat{h}$ is close to the estimated PSF model $\hat{h}_f$, that means $\hat{h}$ belongs to a predefined parametric blur model. Otherwise, if $\hat{h}$ differs from $\hat{h}_f$ significantly, this means that current blur $\hat{h}$ may not belong to the predefined PSF priors.

4   Embedded Alternate Minimization

To achieve the results from Eq. (6), a scale problem arises between the minimization of the PSF and the image via steepest descent. The reason is that the $\partial E_\varepsilon / \partial \hat{h}$ is $\sum_{x \in \Omega} \hat{f}(x)$ times larger than $\partial E_\varepsilon / \partial \hat{f}$. Also, the dynamic range of the image $[0, 255]$ is larger than the dynamic range of the PSF $[0, 1]$. The scale factor changes dynamically with space coordinates (x, y). To avoid the scale problem, You and Kaveh [19] introduced an alternate minimization method following the idea of coordinate descent [20]. Later, Chan and Wong [21] demonstrated the efficiency of this method in TV $(L^1$ norm) based regularization for joint blur identification and image restoration.

To estimate the cost of E_ε, three outputs of the ideal image $\hat{f}$, the edge integration map v and the PSF $\hat{h}$ are computed for getting an optimized value from their partial differential equation of E_ε. The minimization of this equation with respect to v, $\hat{h}$ and $\hat{f}$ is carried out based on Euler-Lagrange equations. We can observe that Eq. (10) is a strictly convex and lower bounded with respect to the functions $\hat{f}$ and v if the other one and the blur PSFs are estimated and fixed. We have designed an embedded alternating minimization algorithm to get local minimum values simultaneously based on these three equations. These differentiations are

$$\frac{\partial E_\varepsilon(\hat{f}, v)}{\partial v} = 2\alpha v |\nabla \hat{f}|^2 + \beta(\frac{v-1}{2\varepsilon}) - 2\varepsilon\beta\nabla^2 v \tag{10}$$

$$\frac{\partial E(\hat{h}|g, \hat{f})}{\partial \hat{h}} = (\hat{f} * \hat{h} - g) * \hat{f}(-x, -y) - 2\gamma Div(\nabla \hat{h}) - 2\delta|\hat{h} - \hat{h}_f| \tag{11}$$

$$\frac{\partial E(\hat{f}|g, \hat{h})}{\partial \hat{f}} = (\hat{h} * \hat{f} - g) * \hat{h}(-x, -y) - 2\alpha Div(v^2 \nabla \hat{f}) \tag{12}$$

For solving these three equations, the Neumann conditions $\partial E_\varepsilon/\partial v = 0, \partial E_\varepsilon/\partial \hat{h} = 0$ and $\partial E_\varepsilon/\partial \hat{f} = 0$ correspond to the reflection of the image across the boundary with the advantages of not imposing any value on the boundary. ε is a small positive constants for discrete implementation. The small positive constant can help the estimation of image relatively stable in the minimization process. Based on an initial PSF value $h_0(x)$, the estimation of the ideal image $\hat{f}$ is initialized by the observed image g, edge parameter $v = 1$. The algorithm is described:

Initialization: $f_0(x) = g(x), v = 1, h_0(x)$ is random numbers
while $(nmse_1 > \varepsilon_1)$
    (1). nth it. $\hat{f}_n(x) = \arg\min(\hat{f}|\hat{h}_{n-1}, g)$, fix $\hat{h}(x)$
        while $(nmse_2 > \varepsilon_2)$
            (i). $v \propto argmin(v|\hat{f}) = \partial E_\varepsilon/\partial v$, fix $\hat{f}(x)$
            (ii). $f \propto argmin(\hat{f}|v) = \partial E_\varepsilon/\partial \hat{f}$, fix $v(x)$
        end
    (2). $(n+1)$th it. $\hat{h}_{n+1} = \arg\min(\hat{h}|\hat{f}_n, g)$, fix $\hat{f}(x)$
end

The global convergence can be reached given a small positive threshold ε_1 and ε_2 due to the nonnegativity of the image and the PSF. We use normalized mean square $(nmse)$ values of the PSF and the image to measure the minimization threshold respectively.

Since the convergence with respect to the PSF and the image are optimized alternately, the flexibility of this proposed algorithm allow us to use conjugate gradient algorithm for computing the convergence. Conjugate gradient method utilizes the conjugate direction instead of local gradient to search for the minima. Here, we use *gmres* method to optimize the cost functions. Therefore, it is faster and also requires less memory storage when compared with the other methods. If an image has $M \times N$ pixels, the above conjugate method will converge to the minimum of $E(\hat{f}|g, \hat{h})$ after $m \ll MN$ steps based on partial conjugate gradient method.

5   Graph Partitioning for Blurred and Unblurred Regions

Given a blur degraded image or video frame, we can observe that the gradient edge map v of foreground blurred objects are very weak and unstable comparing with the unblurred cluttered background, e.g., in Fig. 5 (b). We extend a spectral graph partitioning algorithm with a global criterion [22, 23] to the Mumford-Shah functional for segmenting the blurred regions or objects in video sequences. The combination of low level processing and mid or high level knowledge can be used to either confirm these groups or select some for further attention in repartitioning or grouping blurred and unblurred regions or objects in images.

To achieve the segmentation of such degraded images, we firstly consider a graph bisection problem. We can partition the vertices of a graph $G = (V, E)$ into two sets A and B to minimize the number of *cut edges*, i.e., edges with one endpoint in A and the other in B, where V are the vertices and E are the edges between these vertices. V can correspond to pixels in an image or set of connected pixels. The bisection problem can be formulated as the minimization of a quadratic objective function by means of the Laplacian matrix $L = L(G)$ of the graph G, $|\delta_{min}(A, B)| = min(x^T L x)$ with components $x_i = \pm 1$ and $\sum_{i=1}^{n} x_i = 0$, where $L = D - W$, $W = \{w_{ij}\}$ is the adjacency matrix of a graph, and D is the $n \times n$ diagonal matrix of the degrees of the vertices of G. Thus the bisection problem is equivalent to the problem of maximizing similarity of the objects within each cluster, or, find a *cut edge* through G with minimal weight in the form of $max(x^T W x) \Longleftrightarrow min(x^T L x)$.

The minimization problem is NP-complete. The approximation makes the optimization problem tractable by relaxing the constraints. To avoid unnatural bias for partitioning out small sets of points, and achieve the total dissimilarity between the different groups as well as the total similarity within the groups, Shi and Malik [23] proposed a new measure of the disassociation between two groups. Instead of looking at the value of total edge weight connecting the two partitions, the cut cost is computed as a fraction of the total edge connections to all the nodes in the graph. This disassociation measure is called the normalized cut (Ncut): $Ncut(A, B) = \frac{cur(A,B)}{asso(A,V)} + \frac{cut(A,B)}{asso(A,V)}$. A and B are two initial sets. The similar objects grouping algorithm is fully exploited by an eigensolver called the Lanczos method which speeds up the running time. The degree of dissimilarity between two pieces can be computed as total weight of the edges that have been removed. The two partition criteria in the grouping algorithm is to minimize the disassociation between the groups and maximize the association within the group.

The grouping algorithm is summarized as follows:

1. Given a set of features, set up an undirected weight graph $G = (V, E)$. Computing the weight on each edge, and summarize the information into W, and D.
2. Solve $(D - W)x = \lambda Dx$ for eigenvectors with the smallest eigenvalues.
3. Use the eigenvector with second smallest eigenvalue to bipartition the graph by finding the splitting point such that $Ncut$ is maximized. Note that Perona and Freeman [24] use the largest eigenvector.
4. Decide if the current partition should be subdivided by checking the stability of the cut, and make sure $Ncut$ is below pre-specified value.
5. Recursively repartition the segmented parts if necessary.

6   Numerical Experiments and Evaluation

Experiments on simulated data and real data are carried out to demonstrate the effectiveness of our algorithm.

Discrete Implementation. To solve the Γ-convergence to the MS functional, we use a discrete scheme called a cell-centered finite difference from [25, 9]. Following the way of discretization, Eq. (10) is written in a discrete form,

$$2\alpha v_{ij}[(\Delta_+^x \hat{f}_{ij})^2 + (\Delta_+^y \hat{f}_{ij})^2] + \beta \cdot \frac{v_{ij} - 1}{2\varepsilon} - 2\beta\varepsilon(\Delta_+^x \Delta_-^x v_{ij} + \Delta_+^y \Delta_-^y v_{ij}) = 0$$

where the forward and backward finite difference approximations of the derivatives $\partial \hat{f}(x, y)/\partial x$ and $\partial \hat{f}(x, y)/\partial y$ are denoted by $\Delta_\pm^x \hat{f}_{ij} = \pm(\hat{f}_{i\pm 1,j} - \hat{f}_{ij})$ and $\Delta_\pm^y \hat{f}_{ij} = \pm(\hat{f}_{i,j\pm 1} - \hat{f}_{ij})$. To minimize the column-stack ordering of $\{v_{ij}\}$, the system is of form $Mv = q$, where M is symmetric and sparse matrix and solve the minimization using the minimal residual algorithm. Let H denote the operator of convolution of different blur PSFs that are pre-estimated. Using the notation of [25], let $L(v)$ denote the differential operator $L(v)\hat{f} = -Div(v^2 \nabla \hat{f})$. Eq. (12) can be expressed as $\hat{H}^*(\hat{H}\hat{f} - g) + 2\alpha L(v)\hat{f} = 0$. Let $A(v)\hat{f} = \hat{H} * \hat{f} + 2\alpha L(v)\hat{f}$, we get $A(v)\hat{f} = \hat{H}^* \hat{g}$. $\hat{f}$ is iteratively determined. To obtain $\hat{f}^{n+1}$, a correction term $\hat{d}^n$ is added to the current value $\hat{f}^n$: $\hat{f}^{n+1} = \hat{f}^n + d^n \times d^n$ is estimated by $A(v)d^n = \hat{H}^* g - A(v)\hat{f}^n$ via the convergent descent method. Three outputs of gradient edges v, the restored image $\hat{f}$ and the estimated PSF can be achieved after the convergent optimization in the embedded alternate minimization.

Choosing Parameters for Regularization. The choice of regularization parameters is crucial due to the scale problem between the image and the PSF.

Fig. 3. (a)Original image. (b) Motion blur with 20dB Gaussian noise. (c)Restoration using iterative L^2 norm regularization. (d) Restoration using MS regularization.

Several papers have addressed the problem of estimating the optimal parameters [19, 21, 14]. We use L-curve [26] due to its robustness for correlated noise. It is a graphical tool for analysis of discrete ill-posed problems in a log-log plot for all valid parameters using the compromise between minimization of these quantities. The novelty is that no prior knowledge about the properties of the noise and the image (other than its "smoothness") is necessary, and required parameters are computed through this approach. There is a relatively general scale relation between α and γ with respect to the image and the PSF smoothing term. It is formulated as $\gamma/\alpha = \sum_{x \in \Omega} \hat{f}(x) \max_{x \in \Omega} \hat{f}(x)$. The order-of-magnitude of two parameters are given using the normalized local variance of image and PSF, $\alpha_i = 0.5/(1 + 10^3 \mathrm{var}(f(i)))$, $\gamma_i = 10^6/(1 + 10^3 \mathrm{var}(h(i))$ and $\delta_i = 10^6/(1 + 10^3 \mathrm{var}(d(i)))$, where $d = \hat{h} - \hat{h}_f$. However, our simulating tests have shown that blind restoration quality is usually not sensitive to regularization parameters. A meaningful measure called normalized mean square-error (NMSE) is used to evaluate the performance of the identified blur. $NMSE = (\sum_x \sum_y (h(x,y) - \hat{h}(x,y))^2)^{1/2}/(\sum_x \sum_y h(x,y))$.

Blur Identification and Image Restoration for Degraded Images. In the first experiment, we have compared the results of image restoration using the weighted L^2 norm regularization and the extended MS regularization. The parameters in the Mumford-Shah functional are tuned for the best performance

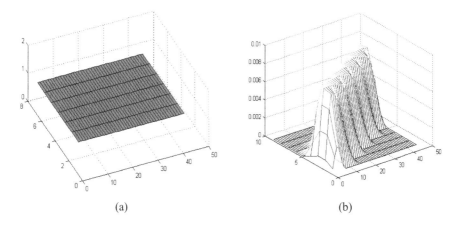

(a)   (b)

Fig. 4. Estimated PSFs based on the solution space. (a) A random initial PSF. (b) The estimated PSF with parameter adjustment during the iteration.

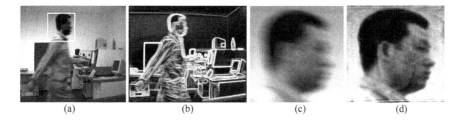

(a)   (b)   (c)   (d)

Fig. 5. (a) A blurred walking man. (b) the computed gradient edge map from MS. (c) Cropped blurred part. (d) Restored result.

$\alpha = 10^{-4}$, $\beta = 10^{-8}$, $\varepsilon = 10^{-3}$ with an estimated PSF from the regularized. From the results, we can observe that the results from Mumford-Shah functional is sharper and less ringing artifacts comparing with the weighted L^2 norm regularization using the same estimated PSF in Fig. 3. It highlights that the restoration of the extended MS regularization is towards edge-preserving restoration.

The second experiment has been tested on a real video sequence, shown in Fig. 5. During the embedded alternate minimization, the PSF for Fig. 5 (c) can be estimated using the suggested method shown in Fig. 4. The PSF in Fig. 4(a) is a random initial PSF value. The PSF in Fig. 4(b) is estimated based on prior PSF solution space and is adjusted using parametric structures of the selected PSF model. The noise is eliminated gradually using some low pass filters. The blur kernel is devised to model the computed blur with parametric optimization for the estimated PSF. The estimated PSF in Fig. 4(b) is a linear motion blur with certain support size. In Fig. 6, the blur identification and image restoration are illustrated. The PSF is optimized in the embedded alternating minimization. From the PSF profile, we can easily observe that some noise is still influencing the PSF in the pixel level. The degraded video frame is separated into RGB colour

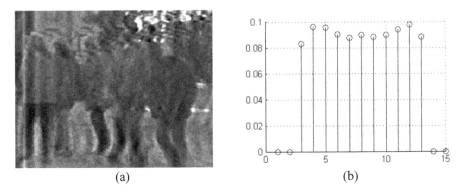

Fig. 6. Rstoration of Blurred video frame. (a) Restored result. (b) Estimated PSF.

channels and each channel is processed accordingly. Based on the estimated PSFs and regularization parameters, piecewise smooth and accurate PSF model helps to recover the blurred objects.

Segmentation of Partially-Blurred, Noisy Image Regions and Objects.
Segmentation of a blurred, noisy video sequence has good performance using the suggested method shown in Fig. 7. In Fig. 7(a), cluttered background objects

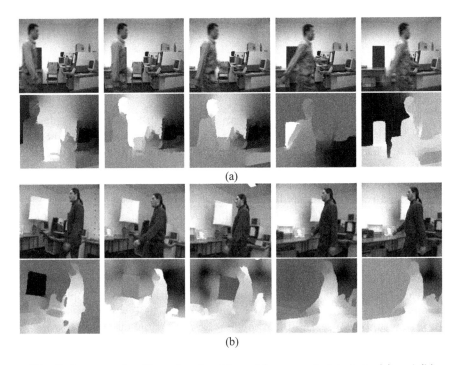

Fig. 7. Segmentation blurred and unblurred foreground objects in (a) and (b)

with stronger gradients do not influence the segmentation of blurred foreground objects with unstable and lower gradients. Fig. 7(b) shows unblurred foreground regions or objects segmented from blurred background. The MS functional can achieve accurate edge detection, v is initialized as 1, the edges are computed after a few iterations. These detected edges with different strengths of gradients are grouped via the extended graph-grouping and partitioning method with a global segmentation criterion (normalized cuts) into numerical groups. The segmentation result is labeled and color filled following the partitioned regions. However, the graph partitioning method for image segmentation needs more memory space and computation intensively.

7   Conclusions

This paper validates the hypothesis that the challenging tasks of blind image deconvolution and segmentation are implementable and demonstrated in the suggested approach. Blind image deconvolution is one kind of ill-posed inverse problem. Searching for the solution in the largest space is not a good strategy. *A priori* knowledge should be used from different viewpoints to improve the solution. Supply of accurate prior information directly to the computation is an excellent strategy since the approach improves the accuracy of initial value for the regularization. The Γ-convergence approximated MS regularization is extended to include cost terms for the estimation of blur kernels. The estimated PSF is not only based on the Bayesian MAP estimation but also optimized alternately in the extended MS regularization. Three outputs of the estimated image, the estimated PSF and edge curves are generated simultaneously from the extended MS functional. Furthermore, a graph spectral partitioning method is extended to group edges which is derived from the extended MS functional. These blurred and unblurred regions or objects can then be segmented accurately with a global segmentation criterion. It is clear that the proposed method is instrumental in image restoration and segmentation and can easily be extended in practical environments.

References

1. Hadamard, J.: Lectures on the Cauchy Problem in Linear Partial Differential Equations. Yale University Press (1923)
2. Tikhonov, A., Arsenin, V.: Solution of Ill-Posed Problems. Wiley, Winston (1977)
3. Geman, S., Geman, D.: Stochastic relaxation, gibbs distribution and the bayesian restoratio of images. IEEE Trans.Pattern Anal. Machine Intell. (1984) 721–741
4. Katsaggelos, A., Biemond, J., Schafer, R., Mersereau, R.: A regularized iterative image restoration algorithm. IEEE Tr. on Signal Processing **39** (1991) 914–929
5. Rudin, L., Osher, S., Fatemi, E.: Nonlinear total varition based noise removal algorithm. Physica D **60** (1992) 259–268
6. Mumford, D., Shah, J.: Optimal approximations by piecewise smooth functions and associated variational problems. Communications on Pure and Applied Mathematics **42** (1989) 577–684

7. Aubert, l., Kornprobst, P.: Mathematical problems in image processing: partical differential equations and the Calculus of Variations. Springer (2002)
8. Chan, T.F., Vese, L.A.: Active contours without edges. IEEE Trans. on Image Processing, **10** (2001) 266–277
9. Bar, L., Sochen, N., Kiryati, N.: Variational pairing of image segmentation and blind restoration. In Pajdla, T., Matas(Eds), J., eds.: ECCV 2004. Volume 3022 of LNCS., Berlin, Heidelberg, Springer (2004) 166–177
10. Blake, A., Zisserman, A.: Visual Reconstruction. MIT Press, Cambridge (1987)
11. Molina, R., Ripley, B.: Using spatial models as priors in astronomical image analysis. J. App. Stat. **16** (1989) 193–206
12. Green, P.: Bayesian reconstruction from emission tomography data using a modified em algorithm. IEEE Tr. Med. Imaging **9** (1990) 84–92
13. Bouman, C., Sauer, K.: A generalized gaussian image model for edge-preserving map estimation. IEEE Transactions of Image Processing **2** (1993) 296– 310
14. Molina, R., Katsaggelos, A., Mateos, J.: Bayesian and regularization methods for hyperparameters estimate in image restoration. IEEE Tr. on S.P. **8** (1999)
15. Ambrosio, L., Tortorelli, V.M.: Approximation of functionals depending on jumps by elliptic functionals via γ-convergence. Communications on Pure and Appled Mathematics **43** (1990) 999–1036
16. Bar, L., Sochen, N., Kiryati, N.: Image deblurring in the presence of salt-and-pepper noise. In Pajdla, T., Matas(Eds), J., eds.: Proc. International Conference on Scale-Space and PDE methods in Computer Vision. LNCS 3459, Hofgeismar, Germany, Springer (2004) 107–118
17. Banham, M., Katsaggelos, A.: Digital image restoration. IEEE S. P. **14** (1997) 24–41
18. Kundur, D., Hatzinakos, D.: Blind image deconvolution. IEEE Signal Process. Mag. **May** (1996) 43–64
19. You, Y., Kaveh, M.: A regularization approach to joint blur identification and image restoration. IEEE Tr. on Image Processing **5** (1996) 416–428
20. Luenberger, D.G.: Linear and nonlinear programming. Addison-Wesley Publishing Company (1984)
21. Chan, T., Wong, C.K.: Total variation blind deconvolution. IEEE Trans.on Image Processing **7** (1998) 370–375
22. Pothen, A., Simon, H.D.: Partitioning sparse matrices with eigenvectors of graphs. SIAM J. Matrix Analytical Applications **11** (1990) 430–452
23. Shi, J., Malik, J.: Normalized cuts and image segmentation. IEEE Transactions on Pattern Analysis and Machine Intelligence **Aug.** (2000) 888–905
24. Perona, P., Freeman, W.T.: A factorization approach to grouping. In Burkardt, H., Neumann(Eds.), B., eds.: Proc. ECCV. (1998) 655–670
25. Vogel, C.R., Oman, M.E.: Fast, robust total variation-based reconstruction of noisy, blurred images. IEEE Trans. on Image Processing **7** (1998) 813–824
26. Hansen, P., O'Leary, D.: The use of the l-curve in the regularization of discrete ill-posed problems. SIAM J. Sci. Comput. **14** (1993) 1487–1503

Polygonal Approximation of Point Sets

Longin Jan Latecki[1], Rolf Lakaemper[1], and Marc Sobel[2]

[1] CIS Dept., Temple University, Philadelphia, PA 19122, USA
{latecki, lakamper}@temple.edu
[2] Statistics Dept., Temple University, Philadelphia, PA 19122, USA
marc.sobel@temple.edu

Abstract. Our domain of interest is polygonal (and polyhedral) approximation of point sets. Neither the order of data points nor the number of needed line segments (surface patches) are known. In particular, point sets can be obtained by laser range scanner mounted on a moving robot or given as edge pixels/voxels in digital images. Polygonal approximation of edge pixels can also be interpreted as grouping of edge pixels to parts of object contours. The presented approach is described in the statistical framework of Expectation Maximization (EM) and in cognitively motivated geometric framework. We use local support estimation motivated by human visual perception to evaluate support in data points of EM components after each EM step. Consequently, we are able to recognize a locally optimal solution that is not globally optimal, and modify the number of model components and their parameters. We will show experimentally that the proposed approach has much stronger global convergence properties than the EM approach. In particular, the proposed approach is able to converge to a globally optimal solution independent of the initial number of model components and their initial parameters.

1 Introduction

Expectation Maximization (EM) is a very popular and powerful method that allows simultaneous estimation of model parameters and assignment of data points to components of the model. However, EM produces an optimal solution only if the number of model components is well estimated and the initial values of model parameters are close to the global optimum. If this is not the case EM is only guaranteed to produce a locally optimal solution. This is illustrated in Fig. 1, where (a) shows data points and the initial configuration of two straight line segments. The number of model components (2 line segments) is correctly initialized, but their position is not sufficiently close to the global optimum. Fig. 1(b) shows the final, locally optimal, result obtained by the classical EM algorithm. Fig. 1(c) shows the globally optimal approximation obtained by the proposed method on the same input.

Due to the local optimum problem, a correct estimation of the number of components and the initial parameters of a statistical model is crucial in all EM applications, and therefore, belongs to one of the most challenging problems in statistical reasoning. The proposed approach provides a solution to the problem

U. Eckardt et al. (Eds.): IWCIA 2006, LNCS 4040, pp. 159–173, 2006.

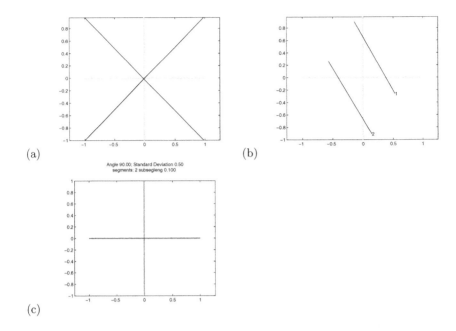

Fig. 1. (a) shows the data points and the initial position of model lines. (b) shows the optimal approximation of the data points obtained by EM. (c) shows the optimal approximation result obtained by the proposed method.

of local optimum in EM that is based on cognitively motivated local support evaluation of EM components. The example shown in Fig. 2 motivates the proposed approach. It is obvious to humans that the approximation in (c) of the underlying data points is significantly better then the approximation in (a). Observe the lack of local support in the data points of the middle part of the line in (a). This observation is the key argument for the proposed extension of EM. We will evaluate the support in data points of each EM component, and remove parts of components with insufficient support.

The existing approaches to determine the optimal number of EM components, of which AIC and BIC (Bayesian Information Criterion), which is equivalent to MDL (Minimum Description Length) [1], are most known, do not base their decision on the local support in data points of the model components. They assume only a fix cost per each model parameter. In particular, this means that a model component with high data support (i.e., positioned in a data region with high point density) costs the same as a component with low data support (i.e., positioned in a regions with low density of data points) although it is intuitively clear that a component with low data support is far less relevant than a component with large data support.

AIC, BIC, and MDL require separate EM runs until convergence for all possible number of model components, each run composed of several EM iterations, which may even be in the order of several thousands. For AIC, BIC, and MDL

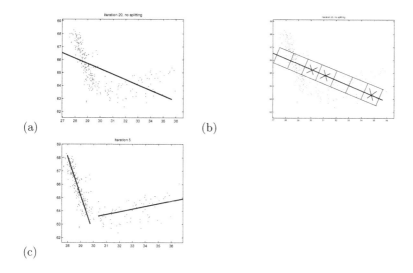

Fig. 2. It is obvious to us that the approximation in (c) of the underlying data points is significantly better then the approximation in (a). (a) shows the best possible approximation of the data points obtained by EM. (b) illustrates the line split (LS) based on subsegment removal. The removed subsegments are marked with crosses. (c) shows the final approximation result obtained by EM after the split.

to be successful, it is implicitly assumed that EM converges to global optimum in each run. However, as illustrated in Fig. 1(b) this is not always the case, since even with a correct number of components EM may get stuck in local minimum. Therefore, the correct number of two model components would not be selected by AIC, BIC, or MDL in our example. By locally evaluating the support in the data points of the two lines in Fig. 1(b), we can clearly determine that they form a bad approximation of the data points. By removing most of their parts, and retaining only small parts around the data points, we create a better input for the EM algorithm. This finally leads to a globally optimal approximation in Fig. 1(c).

The proposed approach provides a solution to the problem of local optimum in EM by adding two new steps that are well integrated with the standard E and M steps of EM. The two new steps are geometrically motivated and can be interpreted as split and merge steps in the context of line fitting. However, the proposed extension of EM is not restricted to any particular shape of model components. In the first new step, the split step, the model components obtained by a previous EM iteration are examined for support of the data points. The main idea (illustrated by the above example) is that higher point density around a model component (line segment in our application) indicates a presence of a linear structure in the data points around the segment. Parts of the segment that do not have sufficient support are removed. This may lead to segment removal but generally leads to a split of the segment into several subsegments. The second new step is merging similar model components. It prevents generating statistical models that overfit the data, i.e., fit noise in the data. This step requires a similarity measure of statistical model components. The merging step can be interpreted as perceptual grouping that dates back to the first results of Gestalt psychology in the beginning of 20th century [2].

We will show that integrating the split and merge operations in the EM framework leads to a globally optimal solution. In our experiments, we were able to obtain a globally optimal solution after just a few iterations (between 5 and 30). Two example applications of our approach are outlined in Fig. 3. (a) shows an original input toy image. (b) shows the edges obtained by Canny edge detector with a substantial amount of added noise, and the initial model for our algorithm. It consists of only two line segments. (c) shows an intermediate step of our algorithm. The final polygonal approximation obtained after 27 iterations is shown in (d). (e) shows an image obtained by sampling 3 ground truth segments (150 points) with a substantial amount of noise (2000 points). (f) shows the initial model segments for our algorithm. We present the results of our algorithm after 8 in (g) and 19 iterations in (h).

An overview of techniques for polygonal approximations of curves (when the order of data points is known), which have been studied at least since early seventies in computer vision, can be found in [3]. To some popular greed polygonal approximation methods in digital images belong [4] and [5, 6]. An overview of approaches to obtain polygonal maps from laser range data can be found in [7, 8].

We do not make any assumptions about the order of data points and extent of noise. The proposed method avoids the problem of a locally optimal solution and produces stable approximations not only to straight but also to curved lines. Moreover, the final number of fitted line segments depends on extent of noise. This means that the number of model components is adjusted to achieve the best possible approximation accuracy as the function of noise extent.

In order to show that the geometric and cognitively motivated split and merge steps can be incorporated into a statistical formalism, we introduce in Section 2 a new target function to be estimated in the EM framework, and reformulate the E and M steps in Section 3. Then we introduce statistical tests for the proposed split and merge steps in Section 4. Finally in Section 5, we describe the geometric parts of the split and merge steps.

2   Optimizing Kullback-Leibler Divergence

Our goal is to approximate the ground-truth density $q(x)$ with a member $p_\Theta(x)$ of a parametric family $\{p_\Theta(x) : \Theta \in S\}$ of densities. We use Kullback-Leibler divergence (KLD) to measure dissimilarity between the ground-truth and parametric family of densities. By definition, the KLD between the ground truth $q(x)$ and the density, $p_\Theta(x)$ is:

$$D(q(x)\|p_\Theta(x)) = \int \log \frac{q(x)}{p_\Theta(x)} q(x)dx$$

$$= \int \log q(x)q(x)dx - \int \log p_\Theta(x)q(x)dx \qquad (1)$$

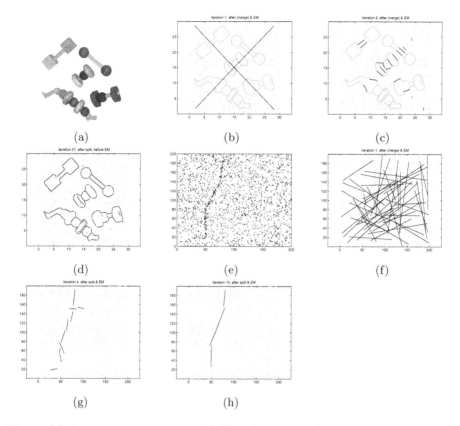

Fig. 3. (a) An original input image. (b) The edges obtained by Canny edge detector, and two initial line segments. (c) We see the polygonal approximation of the edge pixels obtained after (d) The final polygonal approximation obtained after 27 iterations is shown in (e)-(g) Illustrate our approach on simulated data generated by 3 ground truth segments with only 150 signal and 2000 noise points.

Observe that KLD is able to determine the optimal number of model components of p_Θ. This is due to the fact that KLD $D(q\|p_\Theta)$, viewed as a functional on the space $\{p_\Theta\}$ of Gaussian mixtures, is convex and hence has a unique minimum.

It can be easily derived that the parameters $\widehat{\Theta}$ minimizing (1) are given by

$$\widehat{\Theta} = \mathrm{argmax}_\Theta\Big\{\int \log p_\Theta(x)q(x)dx\Big\} \qquad (2)$$

We obtain the classical maximum likelihood estimator by applying the MC (Monte Carlo) integral estimator to (2) under the assumption that the observations $x_1, ..., x_n$ are i.i.d. (independently and identically distributed) sample points selected from the distribution $q(x)$.

$$\widehat{\Theta} = \mathrm{argmax}_\Theta \sum_i \log p_\Theta(x_i) \qquad (3)$$

However, as we derive below (equation (6)), if some proportion of the observations $x_1, ..., x_n$ are noisy, a more accurate estimator of Θ in (2) is given by:

$$\widehat{\Theta} = \operatorname{argmax}_\theta \sum_i \log p_\theta(x_i) sdd(x_i), \tag{4}$$

where *sdd* is called the **smoothed data density** and is defined in (7) below.

Equation (4) is the basis of the proposed approach. We demonstrate theoretically and experimentally that maximization in (4) yields substantially better results than the classical EM maximization in (3). To demonstrate the significance of (4), we consider the problem of estimating the optimal number of model components by minimizing the KLD $D(q(x)||p_\Theta(x))$ in Θ. The parametric family $\{p_\Theta(x)\}$, being a family of Gaussian mixture distributions, is convex. It follows that there is a unique member $p_{\widehat{\Theta}}(x)$ of the Gaussian mixture family with minimum KLD from q. This minimizing mixture must have the correct number of model components. However, it is well known that (3) cannot be used to estimate the correct number of model components, since (3) increases when the number of model components increases. In contrast, we are able to determine the correct number of model components when using (4) to estimate the KLD, $D(q(x)||p_\Theta(x))$. Thus, the modified EM algorithm that maximizes (4) is not only able to estimate model parameters but also the right number of model components.

One of the key steps in the derivation of (4) is the Monte Carlo (MC) estimate of the integral given by the right hand side of equation (1). Let $x_1, ..., x_n$ be i.i.d. sample points drown from the probability density function (pdf) $q(x)$. Then we can approximate the integral of a continuous function f by its MC estimate:

$$\int f(x)q(x)dx \approx \frac{1}{n} \sum_i f(x_i) \tag{5}$$

In the usual approach to inference, it is a commonly accepted assumption that sample data points $x_1, ..., x_n$ are distributed according to the (estimated) density $q(x)$. This assumption is the key to insuring that maximum likelihood estimators are appropriate for purposes of estimating parameters of interest. However, in all real applications, the sample data points are corrupted by a certain amount of noise. Usually the proportion of noisy points does not decrease when the number of sample points is increased. Due to the noise, the following equation provides a substantially better estimate

$$\int f(x)q(x)dx \approx \sum_i f(x_i) sdd(x_i). \tag{6}$$

Finally equation (4) clearly follows from (6) and (2).

The smoothed data density *sdd* is defined as

$$sdd(x) \propto \sum_{i=1}^{n} K\left(\frac{d(x, x_i)}{h}\right) = \frac{1}{nh} \sum_{i=1}^{n} G(d(x, x_i), 0, h), \tag{7}$$

where proportionality refers to the fact that $\sum sdd(x_i) = 1$, $d(x, y)$ is the Euclidean distance, and $G(d(x, y), 0, h)$ is a Gaussian with mean zero and the standard deviation (std) h. An intuitive motivation for sdd is as follows:

- If a given data point x_j is sampled from the true distribution $q(x)$, then x_j lies in a dense region of the observed sample points and consequently $sdd(x_j)$ is large.
- If a given data point x_j is sampled from the noise distribution, then x_j is likely to lie in a sparse region of the sample space, and consequently $sdd(x_j)$ is small.

To estimate the bandwidth parameter h, we can draw from a large literature on nonparametric density estimation [9, 10]. As we show in the presented experimental results, an accurate bandwidth estimation in not crucial in our approach.

3   E and M Steps

We introduce latent variables $z_1, ..., z_n$ which serve to properly label the respective data points $x_1, ..., x_n$. It is assumed that the pairs (x_i, z_i) for $i = 1, \ldots, n$ are i.i.d. with common (unknown) joint (ground truth) density, $q(x, z) = q(x)q(z|x)$; $q(x)$ is the marginal x-density and $q(z|x)$ is the conditional density of the label z given x. In this new framework, the KLD between the joint density $q(x, z)$ and a parametric counterpart density $p_\Theta(x, z)$ is

$$D(q(x, z)\|p_\Theta(x, z)) = D(q(x)q(z|x)\|p_\Theta(x)p_\Theta(z|x))$$

$$= \int_x \int_z \left\{ \log \left[\frac{q(x)}{p_\Theta(x)} \right] + \log \left[\frac{q(z|x)}{p_\Theta(z|x)} \right] \right\} q(x)q(z|x)dzdx$$

$$= \int_x \log \left[\frac{q(x)}{p_\Theta(x)} \right] q(x)dx + \int_x q(x) \int_z \log \left[\frac{q(z|x)}{p_\Theta(z|x)} \right] q(z|x)dz \tag{8}$$

We are now ready to introduce the expectation (E) and maximization (M) steps. Both steps aim at minimizing the same target function (8) in our framework. The expectation step yields the standard EM formula; considerations discussed above lead to a different solution for the maximization step.

Expectation Step: For a fixed set of parameters Θ, we want to find a conditional density $q(z|x)$ that minimizes $D(q(x, z)\|p_\Theta(x, z))$. Since KLD is always nonnegative, and the second summand in (8) is minimized for $q(z|x) = p_\Theta(z|x)$ (in which case it is equal to zero), we obtain from (8) that

$$q(z|x) = p_\Theta(z|x) \text{ minimizes } D(q(x, z)\|p_\Theta(x, z)).$$

In particular, for given sample points $x_1, \ldots, x_n$, we obtain

$$q(z_i = l|x_i) = p_\Theta(z_i = l|x_i) = p(z_i = l|x_i, \Theta) \tag{9}$$

$$= \frac{p(x_i|z_i = l, \Theta)p(z_i = l|\Theta)}{p(x_i|\Theta)} \tag{10}$$

$$= \frac{p(x_i|z_i = l, \Theta)p(z_i = l|\Theta)}{\sum_{j=1}^k p(x_i|z_i = j, \Theta)p(z_i = j|\Theta)} = \frac{p(x_i|z_i = l, \Theta)\pi_l}{\sum_{j=1}^k p(x_i|z_i = j, \Theta)\pi_j}, \tag{11}$$

where $\pi_l = p(z_i = l|\Theta)$ and $\pi_j = p(z_i = j|\Theta)$ are the prior probabilities of component labels l and j correspondingly.

Maximization Step: For the fixed marginal distribution $q(z|x) = p_\Theta(z|x)$, we want to find a set of parameters Θ that maximizes (8). Substituting $q(z|x) = p_\Theta(z|x)$ in (8), we obtain

$$D(q(x,z)||p_\Theta(x,z)) = \int \log(\frac{q(x)}{p_\Theta(x)})q(x)dx = D(q(x)||p_\Theta(x)) \qquad (12)$$

Thus, minimizing $D(q(x,z)||p_\Theta(x,z))$ in Θ is equivalent to minimizing $D(q(x)|| p_\Theta(x))$ in Θ. Using the estimate derived in equation (4), minimizing (12) in Θ is equivalent (in the MC setting discussed above) to maximizing the weighted marginal density

$$WM(\Theta) = \sum sdd(x_i) \log p_\Theta(x_i) = \sum sdd(x_i) \log p(x_i|\Theta)$$

$$= \sum_{i=1}^{n} sdd(x_i) \log[\sum_{l=1}^{k} p(x_i|z_i = l, \Theta)p(z_i = l|\Theta)]$$

$$= \sum_{i=1}^{n} sdd(x_i) \log[\sum_{l=1}^{k} p(x_i|z_i = l, \Theta)\pi_l] \qquad (13)$$

where $\pi_l = p(z_i = l|\Theta)$ are the prior probabilities of component labels $l = 1, \ldots, k$.

Now we explicitly use the incremental update steps of the EM framework. Using the prior probabilities of component labels $\pi_l^{(t)} = p(z_i = l|\Theta^{(t)})$ obtained at stage t for $l = 1, ..., k$, we obtain from (13) that an update of $WM(\Theta)$ is estimated by maximizing

$$WM(\Theta; \Theta^{(t)}) = \sum_{i=1}^{n} sdd(x_i) \log[\sum_{l=1}^{k} p(x_i|z_i = l, \Theta)\pi_l^{(t)}] \qquad (14)$$

in Θ with $\Theta^{(t)}$ denoting the value of Θ computed at stage t of the algorithm.

The crucial difference between this and the standard EM update is that our target function is weighted with terms $sdd(x_i)$. We note that the known convergence proofs for the EM algorithm apply in our framework, since adding the weights $sdd(x_i)$ in (14) does not influence the convergence.

4   Split and Merge

The proposed split and merge steps adjust the number of model components by performing compnent split and merge steps only if they increase the value of our target function (14). Our framework is very general in that it allows many possible selections of the candidate components for the split and merge steps. We present specific selection methods of the candidate components in Section 5.

They are based on a Maximum A Posteriori principle. In the following formulas, we assume that the candidate components are given.

Split: Assume that we are given two candidate model components l_1, l_2; we consider replacing the model component l with components l_1, l_2. Since our goal is maximizing $QM(\Theta; \Theta^{(t)})$ in formula (14), we simply need to check whether replacing l with l_1, l_2 increases WM, where $j \in \{1, \ldots, k\}$:

$$WM(\Theta; \Theta^{(t)}) = \sum_{i=1}^{n} sdd(x_i) \log[\sum_{j} p(x_i|z_i = j, \Theta)\pi_j^{(t)}]$$

$$< \sum_{i=1}^{n} sdd(x_i) \log[\sum_{j \neq l} p(x_i|z_i = l, \Theta)\pi_l^{(t)}$$

$$+ p(x_i|z_i = l_1, \Theta)\pi_{l_1}^{(t)} + p(x_i|z_i = l_2, \Theta)\pi_{l_2}^{(t)}] \tag{15}$$

We only need to perform 'local' computation to perform this test, i.e., we only need to compute the corresponding probabilities for the candidate components l_1, l_2, subject to the condition that $\pi_l^{(t)} = \pi_{l_1}^{(t)} + \pi_{l_2}^{(t)}$. The parameters are estimated following the sparse EM step in Neal and Hinton [11], (see equation (15)). In accordance with the results of [11] this local computation guarantees that the target function increases after each iteration (if (15) holds). Convergence is also guaranteed in this way.

Merge: Given a candidate component l, we merge two existing model components l_1, l_2 to l if for $j \in \{1, \ldots, k\}$

$$WM(\Theta; \Theta^{(t)}) = \sum_{i=1}^{n} sdd(x_i) \log[\sum_{j} p(x_i|z_i = j, \Theta)\pi_j^{(t)}]$$

$$> \sum_{i=1}^{n} sdd(x_i) \log[\sum_{j \neq l} p(x_i|z_i = l, \Theta)\pi_l^{(t)}$$

$$+ p(x_i|z_i = l_1, \Theta)\pi_{l_1}^{(t)} + p(x_i|z_i = l_2, \Theta)\pi_{l_2}^{(t)}] \tag{16}$$

Again we only need to perform 'local' computations to perform this test. For merge, we only need to compute the corresponding probabilities for the candidate component l, subject to the same constraint $\pi_l^{(t)} = \pi_{l_1}^{(t)} + \pi_{l_2}^{(t)}$. If (16) holds and we replace l_1, l_2 with l, the convergence of our algorithm follows from the results of [11].

We note that the proposed split and merge steps do not work in the classical EM framework. To see this, consider $sdd(x_i) = 1$ for all the data points $(i = 1, \ldots, n)$. The merge inequality (16) is not satisfied even if the ground truth model is assumed to be a single component, since multiple components can better fit the data, and consequently have a larger log likelihood value. Analogously, if the split inequality (15) holds for a reasonable selection of candidate component models, the classical EM framework incorrectly splits ground truth components. Thus, a mixture model of larger number of components is always prefered in the classical EM framework. In the proposed framework, *sdd* represents an estimated density of the data points. Consequently, in the proposed split and merge steps, the divergence of parametric components l, l_1, l_2 from the ground truth is evaluated with respect to this nonparametric density.

5   Line Segments as Components

We present specific details concerning our use of line segments as EM model components in the applications presented below. We stress that this section applies also to hyper planes in any dimensions, but the presentation is given in terms of line segments for purposes of simplification.

The proposed approach requires a minor extension of EM line fitting to work with line segments, which we will call Expectation Maximization Segment Fitting (EMSF). The difference between EMSF and EM line fitting is that our model components are line segments (rather than lines). The input, for our model, is a set of line segments and a set of data points. As with EM the proposed EMSF is composed of two steps:

(1) **E-step.** The EM probabilities are computed based on the distances of points to line segments instead of the distances of points to lines.
(2) **M-step.** Given the probabilities computed in the E-step, the new positions of the lines are computed by minimizing squared regression error weighted with these probabilities.

As in the case of EM line fitting, the output of the M-step is a new set of lines (not line segments). Since we need line segments as input to the E-step, we trim lines to line segment based on their support in the sample data. This is done by the split process described in Section 5.2.

Now we describe the specific details related to line segments for steps (1) and (2). In order to derive the solution of (14) for EM model components being line segments, we introduce so called EM weights. In the classical EM, the weight $w_{il}^{(t)} = p(z_i = l | x_i, \Theta^{(t)})$ represents the probability that x_i corresponds to segment s_l for $l = 1, \ldots, k$. We use the notation θ_l for the parameters of the line segment s_l itself. In our framework

$$w_{il}^{(t)} \propto sdd^{(t)}(x_i) \cdot p(z_i = l | x_i, \Theta^{(t)}), \tag{17}$$

and the weights are normalized so that $\sum_{l=1}^{k} w_{il}^{(t)} = 1$ for each i. After the E-step associated with the t'th iteration is accomplished, we obtain a new matrix $(w_{il}^{(t)})$. Intuitively, each row $i = 1, \ldots, n$ of this matrix corresponds to weighted probabilities that the data point x_i is associated with the corresponding line segments; each column $l = 1, \ldots, k$ can be viewed as weights representing the influence of each point on the computation of new line positions in the M-step. Below, we use the notation $x_i = (x_{ix}, x_{iy})$ with $(i = 1, \ldots, n)$ for the coordinates of the observed data points, and $(\bar{x}, \bar{y})$ for the coordinate averages. The line $\mathcal{L}_l$, constructed below, is constructed to go through the point $(\bar{x}, \bar{y})$. To obtain the solution of (14), we perform an orthogonal regression weighted with the matrix (w_{il}). The solution is given as the normal vector to line $\mathcal{L}_l$, which is the vector corresponding to the smallest eigenvalue of the matrix M_l defined as

$$\begin{bmatrix} \sum_{i=1}^{n} w_{il}(x_{ix} - \bar{x})^2 & \sum_{i=1}^{n} w_{il}(x_{ix} - \bar{x})(x_{iy} - \bar{y}) \\ \sum_{i=1}^{n} w_{il}(x_{ix} - \bar{x})(x_{iy} - \bar{y}) & \sum_{i=1}^{n} w_{il}(x_{iy} - \bar{y})^2 \end{bmatrix} \qquad (18)$$

Finally the parameters $\theta_l^{(t+1)}$ are given as parameters of the line segment $s_l^{(t+1)}$ obtained by trimming the line $\mathcal{L}_l$ to the data points.

We are now ready to introduce particular realization of split and merge for EM model components being line segments. The proposed split and merge EM segment fitting (SMEMSF) algorithm iterates the following three steps

(1) EMSF (2) Split (3) Merge

Split step is presented in detail in Section 5.2 while Merge step is described in Section 5.1. Split evaluates the support in the data points of lines obtained by EMSF and removes the parts that are weakly supported. Since we have a finite set of data points, this has the effect of trimming the lines to line segments. Finally the merge step merges similar line segments. Thus, split and merge steps adjust the number of model components to better fit the data.

5.1   Merging

If inequality (16) holds, we merge two model components represented by parameters l_1, l_2 into one model componet given by parameter l. While components l_1, l_2 are present at step t (they are line segments s_{l_1}, s_{l_2}), we did not yet specify hot to compute the candidate component l. Now we describe a particular method to generate a candidate component l in the particular case in which the model components are line segments. We stress that other methods are possible and that inequality (16) applies to them too.

A **support set** $S(s_j)$ for a given line segment s_j (model component l) is defined as set of points whose probability of supporting segment s_j is the largest, i.e.,

$$S(s_j) = \{x_i : w_{ij} = \max(w_{i1}, \dots, w_{ik})\}.$$

This maps each data point to a unique segment using the *Maximum A Posteriori* principle. Given two line segments s_{l_1}, s_{l_2}, the merged segment s_l is obtained by trimming the straight line obtained by regression on data points in $S(s_{l_1}) \cup S(s_{l_2})$. Trimming is performed by line split described in Section 5.2.

5.2   Line Split (LS)

A classical case of EM local optimum problem is illustrated in Fig. 2(a), where the line segment is in a locally optimal position. Clearly, the problem here is that we have a model consisting of one line only, while two line segments are needed. Fig. 2(b) illustrates a split operation described in this section. It is based on removal of subsegments that do not have sufficient support in the data points.

As the result we obtain two line segments. Finally, Fig. 2(c) shows the globally optimal approximation of the data points obtained by EM applied to the two segments.

The main idea is that higher point density along a segment indicates a presence of a linear structure in the data points around the segment. Each line or line segment is examined on having sufficient support in data points measured as point density around it. Only parts of segments that have sufficient support of the data points remain. This leads to split of existing lines or segments allowing us to adjust the number of the line segments (i.e., the number of EM model components) to better fit the input data points.

In [12] the number of points in discretely enumerated rectangular strips (e.g., neighborhoods of all possible segments with endpoints in some finite set) is counted. Then signal strips are selected based on the ratio of the number of points to the area. Then strips are grouped to polylines based on proximity of their endpoints and their angles. The main difference of our approach to the approach in [12] is that we do not select the signal segments, but evaluate the existing structures selected by EM. This makes our computation more efficient, since we do not need to numerate all possible strips, and more accurate, since the line segments are optimally fitted to the data points in our approach.

Line Split (LS) is composed of the following steps:

(2.1) Subsegment support computation.
(2.2) Removal of subsegments with insufficient support that satisfy inequality (15).

The input to LS are segments $s_1, \ldots, s_k$ obtained by clipping the lines $l_1, \ldots, l_k$ created in EMSF to the image rectangle. We divide each segment $s_j \in \{s_1, \ldots, s_k\}$ into subsegments of a predefined length $2r$, i.e., $s_j = I_1^j \cup \ldots \cup I_l^j$, so that two consecutive subsegments overlap in their common endpoint, where l is the number of subsegments. (For simplicity we assume that the length of s_j is exactly multiple of $2r$.) For each subsegment I_k^j, we define its support as the number of data points in the square $S(I_k^j)$ whose two sides are parallel to subsegment I_k^j and whose center is contained in I_k^j, i.e.,

$$support(I_k^j) = \#(\{x_i\} \cap S(I_k^j)).$$

A few such squares are illustrated in Fig. 2(b).

In each iteration a support threshold C is computed from the statistics of $support(I_k^j)$ values over all subsegments of all line segments. Finally subsegments I_k^j with $support(I_k^j) \leq C$ are removed. The subsegments to be removed are marked with crosses in Fig. 2(b). New segments are created as connected components of remaining subsegments of segment s_j. If inequality (15) holds, then the original input segment s_j is removed, and the newly created segments are added to the list of original segments for the next iteration of EMSF. If all its subsegments are removed, then a given segment is removed.

6   Applications

Two examples of approximations of point sets in digital images are illustrated in Fig. 3. An example application of our approach in robot mapping is outlined in Fig. 4. (a) shows an original data set of laser range scan points aligned with the algorithm presented in [13]. The original set is composed of 395 scans, each with 361 points. Thus, the original input map is composed of 142,595 points. We initialize our algorithm with 192 segments, the grid segments, as model components. (b) shows the output with 96 segments after the first iteration of our algorithm. The final polygonal map in (c), obtained after 6 iterations, is composed of 86 segments, i.e., of 172 points. Thus, the proposed approach yields the data compression ratio of 829:1. The mean distance of scan points to the closest line segments is 3.5cm. We selected this map, since it contains surfaces of curved objects. The obtained polylines in (c) illustrate that the proposed approach is

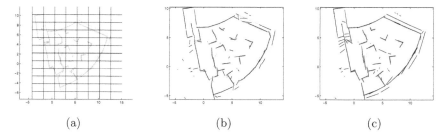

(a)   (b)   (c)

Fig. 4. (a) An original outdoor map is composed of 142,595 scan points obtained during the Rescue Robot Camp in Rome, 2004. We begin the approximation process with 192 line segments that form the grid. (b) shows the output after the first iteration of our algorithm with 96 segments. (c) The final polygonal map obtained after 6 iterations is composed of only 86 segments. The obtained compression rate is 829:1, and the approximation accuracy is 3.5cm.

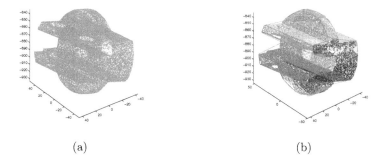

(a)   (b)

Fig. 5. (a) An input surface with 744,450 sample points. (b) Approximation with 27 planar patches.

well suited to approximate linear as well as curved surfaces. For more details on
the application of the proposed method in robot mapping see [14].

Examples illustrating fitting planar patches to 3D range data are given in [15].
Here we show only one example in Fig. 5. Fig. 5(a) shows a 3D projection of the
surface of some industrial part composed of 744,450 laser range scan points, obtained from `http://edge.cs.drexel.edu/Dmitriy/Scanned.tar.gz`. Fig. 5(b)
shows our approximation with 27 planar patches. The mean distance of each
point to the closest surface patch is 0.49 with the original object size of $100 \times 90 \times 100$.

7   Conclusions

The combination of Expectation Maximization Segment Fitting with alternating Segment Splitting and Merging was proven to be a powerful tool to gain a
polyline representation of edge points in digital images, leading to a geometrically higher representation and an excellent data compression rate. The newly
introduced, perceptual grouping based merging step balances the number of segments, created by partitioning and splitting, in a visually natural way and therefore allows for the number of starting segments for the EM step to be imprecise.
The extended EM algorithm is proven to yield globally optimal results.

Acknowledgment

This work was supported in part by the National Science Foundation under grant
NSF IIS-0534929. We also would like to acknowledge the support from the National Institute of Standards and Technology, award Nr. NIST 70NANB5H11119.
We thank Giorgio Grisetti (Univ. La Sapienza, Rome) for providing us the scan
data.

References

1. Beal, M.J., Ghahramani, Z.: The variational bayesian em algorithm for incomplete
   data. In: BAYESIAN STATISTICS 7. Oxford Univ. Press (2003)
2. Wertheimer, M.: Untersuchungen zur lehre von der gestalt ii. Psycologische
   Forschung **4** (1923) 301–350
3. Rosin, P.L.: Techniques for assessing polygonal approximations of curves. IEEE
   Trans. PAMI **19(3)** (1997) 659–666
4. Ramer, U.: An iterative procedure for the polygonal approximation of plane curves.
   Computer Graphics and Image Processing **1** (1972) 244–256
5. Latecki, L.J., Lakämper, R.: Shape similarity measure based on correspondence
   of visual parts. IEEE Trans. Pattern Analysis and Machine Intelligence **22(10)**
   (2000) 1185–1190
6. Latecki, L.J., Lakämper, R., Eckhardt, U.: Shape descriptors for non-rigid shapes
   with a single closed contour. In: Proc. of IEEE Conf. on Computer Vision and
   Pattern Recognition. (South Carolina, June 2000) 424–429

7. Sack, D., Burgard, W.: A comparison of methods for line extraction from range data. In: Proc. of the 5th IFAC Symposium on Intelligent Autonomous Vehicles (IAV). (2004)

8. Veeck, M., Burgard, W.: Learning polyline maps from range scan data acquired with mobile robots. In: Proc. of the IEEE/RSJ International Conference on Intelligent Robots and Systems (IROS). (2004)

9. Scott, D.W.: Multivariate Density Estimation: Theory, Practice, and Visualization. Wiley and Sons (1992)

10. Silverman, B.W.: Density Estimation for Statistics and Data Analysis. Chapman and Hall (1986)

11. Neal, R., Hinton, G.: A view of the em algorithm that justifies incremental, sparse, and other variants. In Jordan, M.I., ed.: Learning in Graphical Models, Kluwer (1998)

12. Arias-Castro, E., Donoho, D.L., Huo, X.: Adaptive multiscale detection of filamentary structures embedded in a background of uniform random points. Annals of Statistics (to appear)

13. Grisetti, G., Stachniss, C., Burgard, W.: Improving grid-based slam with rao-blackwellized particle filters by adaptive proposals and selective resampling. In: ICRA. (2005)

14. Latecki, L.J., Lakaemper, R.: Polygonal approximation of laser range data based on perceptual grouping and em. In: IEEE Int. Conf. on Robotics and Automation (ICRA). (2006)

15. Lakaemper, R., Latecki, L.J.: Decomposition of 3d laser range data using planar patches. In: IEEE Int. Conf. on Robotics and Automation (ICRA). (2006)

Linear Discrete Line Recognition and Reconstruction Based on a Generalized Preimage

Martine Dexet and Eric Andres

Laboratoire SIC, Université de Poitiers
Bât. SP2MI, bvd Marie et Pierre Curie, BP 30179
86962 Futuroscope Chasseneuil Cedex, France
{dexet, andres}@sic.univ-poitiers.fr

Abstract. A new efficient standard discrete line recognition method is presented. This algorithm incrementally computes in linear time all straight lines which cross a given set of pixels. Moreover, pixels can be considered in any order and do not need to be connected. A new invertible 2D discrete curve reconstruction algorithm based on the proposed recognition method completes this paper. This algorithm computes a polygonal line so that its standard digitization is equal to the discrete curve. These two methods are based on the definition of a new generalized preimage and the framework is the discrete analytical geometry.

1 Introduction

Several methods such as Marching Squares based methods (equivalent to the 3D Marching Cubes [1]) can be used to perform 2D discrete curve reconstruction. Another approach, called *discrete analytical reconstruction*, can also be considered. This method is composed of two steps: the discrete analytical line segment recognition and the curve polygonalization. The *recognition* step consists in determining if a set of pixels belongs to a same discrete line and the *polygonalization* one consists in replacing each recognized discrete segment by a Euclidean straight line segment in order to obtain a polygonal line. In this paper, we are interested in *invertible* 4-connected curve reconstruction methods (i.e. the digitization of the reconstructed object is equal to the original discrete curve).

Many discrete line recognition algorithms have been proposed in the last decades, mostly to recognize 8-connected discrete lines (see [2] for an overview on these methods). These algorithms can however be used to recognize 4-connected lines since 8-connected lines can be transformed into 4-connected ones and vice versa using shear transforms [3]. In [4], Debled-Rennesson et al. proposed a linear algorithm which provides at the end of the recognition process the analytical equation of only one corresponding discrete line. This being very restrictive, algorithms based on the notion of *preimage*, introduced by Dorst and Smeulders [5], have been proposed. These algorithms determine from a set of pixels the set of Euclidean straight lines the digitization of which contains the given pixels. These lines are deduced from the computation of the preimage of the

U. Eckardt et al. (Eds.): IWCIA 2006, LNCS 4040, pp. 174–188, 2006.

pixels which is the line set representation in a parameter space (see Sec. 2.2). A drawback of these algorithms is that they assume that pixels are located in the first octant. Several preimage computation algorithms have been proposed [6,7], with a minimal complexity of $\mathcal{O}(n\ log^2(n))$ for unconnected pixels, where n denotes the number of given pixels.

In this paper, we define the *generalized preimage* of a given set of pixels. The main difference between this definition and the preimage definition is that we make no assumption about the pixel locations. We then propose a new standard discrete line recognition algorithm based on this definition which computes the generalized preimage of a pixel set in linear time.

In the second part of this work, we present a new invertible 4-connected discrete curve reconstruction algorithm. In [8] the authors proposed a reconstruction method using a preimage based recognition algorithm [6]. The principle of this method is that for each recognized discrete segment, a straight line is chosen in the set of all possible solutions provided by the recognition algorithm (see Fig. 1a and 1b). However, the problem arising from this method is that the different lines can intersect outside the discrete line segments losing the reversibility property (see Fig. 1c). To avoid that, patches (small line segments) are added (see Fig. 1d).

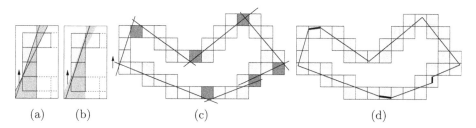

(a)   (b)   (c)   (d)

Fig. 1. Example of reconstruction 'with patches'. (a) Solution line set (in dark grey) computed for the recognized discrete segment. (b) Line choice. (c) Result of the recognition process. (d) Curve obtained after patches additions.

In [3] the authors proposed a way of reconstructing a discrete 4-connected curve without patches. The discrete segment recognition process is now constrained by a fixed point (see Fig. 2a and 2b) in order to force the first extremity of the reconstructed segment to lie inside the discrete curve, and then ensure the reversibility property. The advantage of this method is that no post-processing is needed, but the number of reconstructed segments is often greater than this obtained with the previous algorithm (see Fig. 2c).

The method that we propose in this article for reconstructing 2D 4-connected discrete curves is composed of a recognition and a polygonalization step. The principle of the recognition step is to constrain the discrete segment recognition by all points which are reached by the line set computed for the previous recognized segment. The recognition is thus less constrained than for the method without patches since the constraint of a fixed extremity for each straight line

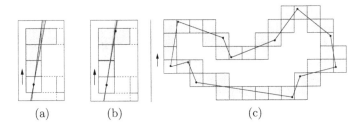

Fig. 2. Example of reconstruction 'without patches'. (a) All straight lines which cross the fixed point (in dark grey). (b) Solution line and next fixed point choices. (c) Result.

segment is relaxed. The polygonalization step works basically as follows: a point is fixed in the last recognized pixel, and a line is chosen in the solution set computed during the recognition step. Then, a straight line segment replaces the discrete one, and another point is chosen in the pixel extremity. And so on until the first pixel of the curve is reached (see Fig. 10).

In the next section, we give some useful recalls and definitions. Then, in Sections 3 and 4, we present our discrete line recognition algorithm and describe our discrete curve reconstruction method. We conclude in Section 5.

2   Definitions

In this section, we first give some recalls on the standard digitization model. Then, we present the parameter space we used in this work.

2.1   The Standard Analytical Model

The standard model [9] is a discrete analytical model, defined in any dimension, which allows in 2D the 4-connected digitization of any linear connected subset of $\mathbb{R}^2$. A standard line (see Fig. 3) is defined analytically as follows:

Definition 1 (2D Standard Line). *The standard line with parameters* $(a, b, c) \in \mathbb{R}^3$ *is the set of points* $(x, y) \in \mathbb{Z}^2$ *(also called discrete points) verifying* $-\omega \leq ax + by + c < \omega$ *where* $\omega = \frac{|a|+|b|}{2}$ *and* $a > 0$, *or* $a = 0$ *and* $b > 0$.

Remark 1. The standard digitization of a line also consists in all *pixels* (unit-size squares centered on discrete points) which are cut by the line, except when the line crosses a pixel vertex (see Fig. 3). In this case, only two or three pixels of the four adjacent ones belong to the digitization. This is due to the fact that one inequality in the standard line definition is strict.

Let O be a subset of $\mathbb{R}^2$ and $\mathbb{S}_t(O)$ be its standard digitization. Then, $\mathbb{S}_t(O_1 \cap O_2) \subseteq \mathbb{S}_t(O_1) \cap \mathbb{S}_t(O_2)$ and $\mathbb{S}_t(O_1 \cup O_2) = \mathbb{S}_t(O_1) \cup \mathbb{S}_t(O_2)$ (see [9] for more details).

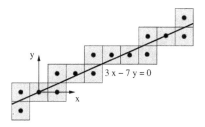

Fig. 3. The standard line with parameters $(3, -7, 0)$

2.2 Parameter Space

The notion of duality, and especially parameter spaces, is often used in image processing. For instance, the well known Hough transform (see [10] for a review on existing variations of this method) which is a very efficient tool used to recognize parametric shapes in an image, is performed in a parameter space.

Definition and Properties. In this work, we use the parameter space $(Oαβ) \subset \mathbb{R}^2$ where a point (a, b) stands for a straight line with equation $y = ax + b$ in the classical Euclidean space (Oxy). In the same way, a point (x, y) in the Euclidean space maps to the line with equation $β = -xα + y$ in the parameter space. In the following, if O is a subset of $\mathbb{R}^2$ in the Euclidean or parameter space, we denote by $Dual(O)$ the corresponding object (or *dual* object) in the other space. Let O_1 and O_2 be two subsets of $\mathbb{R}^2$. The following properties can be deduced from our definition of the duality: $Dual(O_1 \cap O_2) \subseteq Dual(O_1) \cap Dual(O_2)$ and $Dual(O_1 \cup O_2) = Dual(O_1) \cup Dual(O_2)$.

Remark 2. Let $p \in \mathbb{R}^2$ be a point. The dual of each point which lies in $Dual(p)$ is a line which crosses p (see Fig. 4a).

Dual of a Convex Polygon. Before we present our recognition algorithm, we need to define the dual of a convex polygon.

Definition 2 (Positive and Negative Extrusions). *Let $p = (x_p, y_p) \in \mathbb{R}^2$ be a point. The positive extrusion of p is defined by:*

$$p^+ = \{p' = (x_{p'}, y_{p'}) \in \mathbb{R}^2 | x_p = x_{p'} \ and \ y_p \le y_{p'}\}.$$

In the same way, the negative extrusion of p is defined by:

$$p^- = \{p' = (x_{p'}, y_{p'}) \in \mathbb{R}^2 | x_p = x_{p'} \ and \ y_p \ge y_{p'}\}.$$

Note that if O_1 and O_2 are two subsets of $\mathbb{R}^2$, then $(O_1 \cup O_2)^+ = O_1^+ \cup O_2^+$ and $(O_1 \cup O_2)^- = O_1^- \cup O_2^-$. Moreover, let $p \in \mathbb{R}^2$ be a point. Then, $Dual(p)^+ = Dual(p^+)$, and $Dual(p)^- = Dual(p^-)$. This property is illustrated in Fig. 4b.

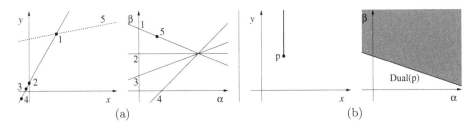

Fig. 4. Dual objects. (a) Some properties. (b) Positive extrusion of a point and its dual object: a half-space.

Let $P_c \subset \mathbb{R}^2$ be a convex polygon and $p_i = (x_{p_i}, y_{p_i}) \in \mathbb{R}^2, 1 \leq i \leq k$, its k vertices. We assume that p_1 is the vertex with the lowest abscissa and the lowest ordinate, and that the vertices are ordered on the border of P_c in the counter clockwise, i.e. the border of P_c is the closed polyline $(p_1, p_2, \ldots, p_k)$ (see Fig. 5a).

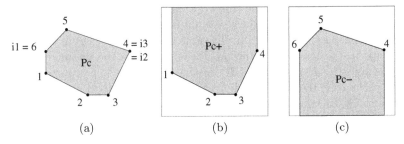

Fig. 5. Illustration of the positive and negative extrusions of a convex polygon. (a) A convex polygon P_c. (b) P_c positive extrusion. (c) P_c negative extrusion.

Let p_{i_1} be the vertex with the lowest abscissa and the highest ordinate. In the same way, let p_{i_2} (resp. p_{i_3}) be the vertex with the highest abscissa and the lowest (resp. highest) ordinate. Note that p_1 (resp. p_{i_2}) can be equal to p_{i_1} (resp. p_{i_3}). Finally, let $p_- = \{p_1, p_2, \ldots, p_{i_2}\}$ and $p_+ = \{p_{i_3}, p_{i_3+1}, \ldots, p_{i_1}\}$ be two sets of P_c vertices. The dual of a convex polygon can be defined as follows:

Theorem 1 (Dual of a Convex Polygon). *Let P_c be a convex polygon, p_+ and p_- the two point sets defined previously. Then:*

$$Dual(P_c) = \left[\bigcup_{p \in p_-} Dual(p)^+ \right] \cap \left[\bigcup_{p \in p_+} Dual(p)^- \right]$$

Proof. We can deduce from Definition 2 that $Dual(P_c) = Dual(P_c)^+ \cap Dual(P_c)^- = Dual(P_c^+) \cap Dual(P_c^-)$. Let us prove that $Dual(P_c^+) = \bigcup_{p \in p_-} Dual(p)^+$. The proof of $Dual(P_c^-) = \bigcup_{p \in p_+} Dual(p)^-$ can be obtained in the same way. Let P_l be the polyline $(p_1, p_2, \ldots, p_{i_2})$. Then, $P_c^+ = P_l^+ = (\bigcup_{j \in \{1, \ldots, i_2 - 1\}} [p_j, p_{j+1}])^+$ where $[p_m, p_n]$ denotes the line segment with endpoints p_m and p_n. We deduce that $P_c^+ = \bigcup_{j \in \{1, \ldots, i_2 - 1\}} [p_j, p_{j+1}]^+$ and then $Dual(P_c^+) = Dual(\bigcup_{j \in \{1, \ldots, i_2 - 1\}} [p_j, p_{j+1}]^+) = \bigcup_{j \in \{1, \ldots, i_2 - 1\}} Dual([p_j, p_{j+1}]^+)$.

Lemma 1. *Let $s = [p_1, p_2]$ be a line segment in $\mathbb{R}^2$. Then,*

$$Dual(s^+) = Dual(p_1{}^+) \cup Dual(p_2{}^+)$$

Proof. Let $p \in \mathbb{R}^2$ be a point in $Dual(s^+)$. $Dual(p)$ is a line which cuts s^+. Then, $Dual(p)$ cuts p_1^+ or p_2^+. Indeed, let us process by contradiction and assume that $Dual(p)$ does not cut neither p_1^+ nor p_2^+. Then, $Dual(p)$ passes under p_1 and under p_2 and then under s. We deduce that $Dual(p)$ does not cut s^+. The second inclusion $Dual(p_1{}^+) \cup Dual(p_2{}^+) \subset Dual(s^+)$ is obvious since $p_1{}^+ \subset s^+$ and $p_2{}^+ \subset s^+$.

By Lemma 1 it follows:

$$Dual(P_c^+) = \bigcup_{j \in \{1, \ldots, i_1 - 1\}} \left[Dual(p_j^+) \cup Dual(p_{j+1}^+) \right] = \bigcup_{p \in p_-} Dual(p^+)$$

and thus, $Dual(P_c^+) = \bigcup_{p \in p_-} Dual(p)^+$.

Theorem 1 allows us to simply compute the dual of a convex polygon from its vertices. It is a useful property that we will use in the following sections. An illustration of Theorem 1 is given in Fig. 6.

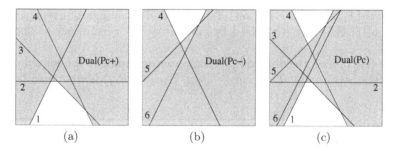

Fig. 6. Dual of the convex polygon P_c shown in Fig. 5a. (a) Dual of the P_c positive extrusion. (b) Dual of the P_c negative extrusion. (c) Dual of P_c.

3   Standard Line Recognition

In this section, we present our standard line recognition algorithm. The aim is to determine if a set of pixels belongs or not to a standard discrete line. The principle of our algorithm is to compute the set of straight lines (if it exists) which cross the given pixels. Each line in this set is the dual of a point in the parameter space which lies in a polygon that we call the *generalized preimage* of the pixels. Hence, if the generalized preimage is empty, the set of pixels does not belong to a standard discrete line, else, it belongs to it. In the following, we first give the definition of the generalized preimage. Then, we detail our recognition algorithm, and give some simplifications we can apply on it in order to improve its complexity.

3.1   The Notion of Generalized Preimage

As said previously, each point in the generalized preimage $\mathbb{G}_P$ of a pixel set $\mathcal{P}$ is the dual of a straight line which cuts all pixels in $\mathcal{P}$. We define the generalized preimage of a pixel set as follows:

Definition 3 (Generalized Preimage). *Let $\mathcal{P} = (P_1, \ldots, P_n)$ be a set of n pixels, and let $Dual(P_i)$ be the dual of P_i in the parameter space, $1 \leq i \leq n$. The generalized preimage $\mathbb{G}_P$ of $\mathcal{P}$ is defined by:*

$$\mathbb{G}_P(\mathcal{P}) = \bigcap_{i=1}^{n} Dual(P_i)$$

Remark 3. The standard digitization of many lines in the dual of the generalized preimage does not contain the given pixels (see Remark 1). However, we know that incorrect lines are located on the border of the generalized preimage since these lines cross pixel vertices. Thus, in order to obtain a correct line, it is sufficient to choose a point which is not on the generalized preimage border.

3.2   Recognition Algorithm

Let $\mathcal{P} = \{P_1, \ldots, P_n\}$ be a set of n pixels. The standard line recognition (see Algo. 1) is simply performed by computing the generalized preimage $\mathbb{G}_P$ of $\mathcal{P}$.

First, $\mathbb{G}_P(P_1)$, i.e. the dual of P_1, is computed according to the convex polygon dual definition given by Theorem 1. Then, $\mathbb{G}_P(\{P_1, P_2\})$ is computed by doing

---

Algorithm 1. Standard line recognition algorithm

Data: A set $\mathcal{P}$ of n pixels $P_1, \ldots, P_n$.
begin
 $\mathbb{G}_P \longleftarrow Dual(P_1)$;
 $i \longleftarrow 2$;
 while $\mathbb{G}_P \neq \emptyset$ **and** $i \leq n$ **do**
  $\mathbb{G}_P \longleftarrow \mathbb{G}_P \cap Dual(P_i)$;
  $i \longleftarrow i + 1$;
 if $\mathbb{G}_P \neq \emptyset$ **then**
  $\mathcal{P}$ belongs to a standard line.
 else
  $\mathcal{P}$ does not belong to a standard line.
end

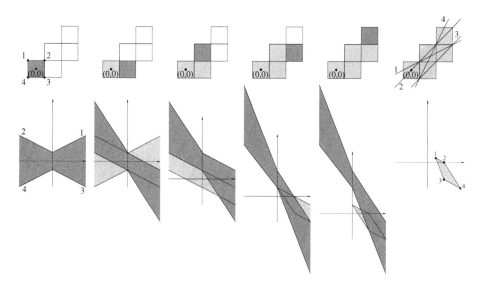

Fig. 7. Example of generalized preimage computation

the intersection between $\mathbb{G}_P(P_1)$ and $Dual(P_2)$. And so on until the computation of $\mathbb{G}_P(\{P_1,\ldots,P_n\})$ or until $\mathbb{G}_P$ becomes empty. Fig. 7 shows an illustration of the recognition process. Note that the pixels can be considered in any order, and do not need to be connected.

To perform the intersection operations, a first approach is to intersect directly the generalized preimage and the dual of each pixel. It is not an efficient method since the dual of a pixel is an open concave polygon. However, some improvements can be done. We present them in the next section.

3.3 Improvements

Several improvements can be applied to our algorithm. First, it is more interesting to separate the parameter space into two parts: one for $\alpha \geq 0$ and one for $\alpha \leq 0$. Indeed, computations are simplified since the dual of a pixel is then decomposed into two convex polygons. The second improvement comes from the fact that pixels can be connected. In this case, the generalized preimage is composed of at most two open or closed convex polygons, each one with at most four edges [5, 11]. Moreover, if the considered pixel is connected with a pixel the dual of which has already been intersected with the generalized preimage, it is not necessary to take into account the four dual lines of the current pixel vertices to perform the intersection. All different cases depending on the pixel connexity (4 or 8-connected) are shown in Fig. 8.

An illustration of these improvements applied on the example shown in Fig. 7 is given in Fig. 9. These improvements lead to a complexity for our algorithm of $\mathcal{O}(n)$ when applied on a 4-connected curve.

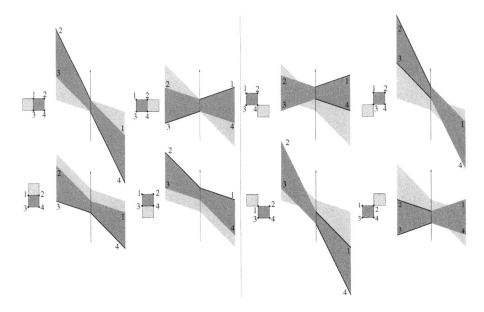

Fig. 8. Possible improvements (in dark grey: added pixel). Bold numbered lines do not need to be taken in account.

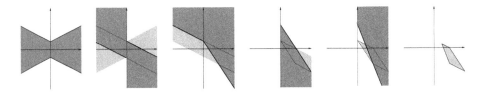

Fig. 9. Illustration of the improved recognition algorithm

4   Reconstruction of Standard Discrete Curves

In this section, we present a new invertible 4-connected discrete curve reconstruction algorithm based on the recognition algorithm presented in Section 3. We first briefly recall the principle of our algorithm, then detail the recognition and polygonalization steps. Finally, we compare results provided by our method and the methods with and without patches proposed in [8] and [3].

4.1   Principle

As stated previously, our algorithm works on 4-connected curves. Let $C = (P_1, \ldots, P_n)$ be such a curve and $P_1, \ldots, P_n$ its ordered pixels. The reconstruction of C according to our method is performed in two steps: the standard line segments recognition and the polygonalization.

During the recognition step, C is decomposed in several connected standard segments. For each one, a set of straight lines the standard digitization of which contains the segment is computed.

The recognition of the first segment S of C (see Fig. 10a) is performed as follows: first, S is composed of one pixel. Then, the set of lines which cut all pixels of S is computed. While this set is not empty, another pixel, adjacent to the previous one and chosen according to a recognition direction, is added to S. When the line set becomes empty, it means that the last considered pixel does not belong to S and a new recognition process starts with the last recognized pixel of S as starting pixel.

The method used to recognize the following segments (see Fig. 10b) is not the same. Indeed, let S_{prev} be the previous recognized segment, S_{new} the segment to be recognized, and P their common pixel. Throughout the recognition of S_{new}, we want to ensure that there exist straight lines which cut S_{new} and that these lines cross in P the lines computed for S_{prev}. In other words, let I be the point set in P which is crossed by the lines computed for S_{prev}. Then, as long as there exist lines which cut S_{new} and I, another pixel is added to S_{new}. The recognition of S_{new} ends when the previous conditions are no more verified.

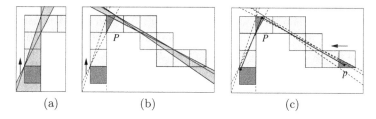

Fig. 10. Principle of our reconstruction method. (a) First segment recognition. The starting pixel is in dark grey and the direction indicated by the arrow. (b) Following segments recognition. (c) Polygonalization step. The polygonalization direction is indicated by the arrow.

The polygonalization step consists in replacing each recognized discrete segment by a Euclidean straight line segment in order to obtain a polyline so that its standard digitization is equal to C. The digitization of each reconstructed segment has to be equal to the corresponding discrete one. The polygonalization process starts with the last recognized discrete segment S_{last} and is performed in the reverse recognition direction (see Fig. 10c). A point p is fixed in the last recognized pixel and a straight line which passes through p is chosen in the computed line set of S_{last}. Then, a new point is chosen in the first recognized pixel P of S_{last} and another segment is computed from this new point since P is also the last recognized pixel of a segment. The polygonalization ends when the first recognized pixel of C is reached.

Starting Pixel and Recognition Direction. To ensure the unicity of the reconstruction, a starting pixel and a direction (white colored pixels and arrows in Fig. 1 and 2) are needed to perform the recognition. In this work, we use the two conventions proposed for open and closed curves by Sivignon et al. in [3]:

- For an open simple curve: the starting discrete point is the curve endpoint with lowest abscissa and lowest ordinate. The recognition direction is hence induced by the chosen starting point.
- For a closed simple curve: the starting point is the curve point with lowest abscissa and lowest ordinate. The recognition direction is the clockwise one.

Note that the starting discrete point choice is not rotation invariant. This is still an open question.

4.2   Straight Line Segment Recognition

The segment recognition is based on the recognition algorithm presented in Section 3. This algorithm computes in the parameter space defined in Section 2.2 the generalized preimage (see Definition 3) of a given set of pixels, i.e. the dual of all straight lines which cut the pixels, and thus determines if these pixels belong or not to a standard line.

In this work, the aim is to perform a constrained recognition, i.e. for each discrete segment S, we want to compute the set of lines which cut all pixels of S and cut a part, denoted I, of the first recognized pixel P of S (see Fig. 11a and 11b). If S is not the first segment of the curve to be recognized, this part is composed of the points which are reached by the line set L computed for the previous recognized segment, i.e. the intersection between L and P. Else, it is the whole pixel P. Moreover, I is a convex polygon (this property is simple to prove).

Hence, the recognition of a segment S is performed as follows: pixels are added one by one to S and the generalized preimage of S is computed at the same time. Note that improvements described in Section 3.3 can be applied to compute the generalized preimage since added pixels are 4-neighbours. The recognition ends when the starting (resp. last) pixel of the closed (resp. open) curve is reached or when the intersection between the dual of I and the generalized preimage of S is empty. Fig. 11 shows an example of recognized curve. The standard curve recognition method is detailed in Algorithm 2.

4.3   Curve Polygonalization

The polygonalization of the curve C (see Algo. 3) is performed starting from the last recognized pixel P of C in the reverse recognition direction (see Fig. 12). Note that if C is a closed curve then P is the recognition starting pixel. A first point p is chosen in the intersection I between P and the line set $L = Dual(\mathbb{G}_P(S))$ computed for the last recognized segment S. If C is closed then p is chosen in the intersection I' between I and the line set computed for the first recognized segment (if I' is not empty). Since I (resp. I') is a convex polygon, we can for instance choose the barycenter of I (resp. I') as p. Then, a

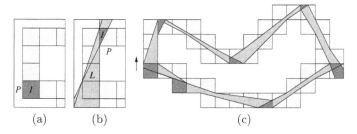

(a)   (b)   (c)

Fig. 11. Illustration of the recognition process. (a) First segment recognition constraint (I in the figure). Starting pixel (P). (b) Second segment recognition constraint. (c) Straight line sets resulting from the curve recognition process.

Algorithm 2. Straight line segments recognition

Data: A set $\mathcal{P}$ of n ordered pixels $P_1, \ldots, P_n$.
begin
  $i \longleftarrow 1$;
  $I \longleftarrow P_1$;
  while $i \leq n$ **do**
    $S \longleftarrow \{P_i\}$;
    $\mathbb{G}_P \longleftarrow Dual(I)$;
    while $\mathbb{G}_P \neq \emptyset$ **and** $i < n$ **do**
      $i \longleftarrow i + 1$;
      $S \longleftarrow S \cup \{P_i\}$;
      $\mathbb{G}_{P_{temp}} \longleftarrow \mathbb{G}_P$;
      $\mathbb{G}_P \longleftarrow \mathbb{G}_P \cap Dual(P_i)$;
    if $\mathbb{G}_P = \emptyset$ **then**
      $\mathbb{G}_P \longleftarrow \mathbb{G}_{P_{temp}}$;
      $S \longleftarrow S - \{P_i\}$;
      $i \longleftarrow i - 1$;
    $I \longleftarrow Dual(\mathbb{G}_P) \cap P_i$;
end

---

line l passing through p is chosen in L. This is done in the parameter space as follows: the intersection between $Dual(L) = \mathbb{G}_P(S)$ and $Dual(p)$ is computed. The dual of each point in this intersection is a line passing through p. A point is thus chosen in this intersection. Finally, the intersection between l and the first recognized pixel of S is performed, and the middle point of the obtained segment is chosen as new first point of the following segment reconstruction. The process is reiterated until the starting point of the curve is reached.

In case of closed curve, the reconstruction of the last straight line segment is double constrained, since the first fixed point lies into the starting pixel. Let p be this point and p_f be the first fixed point of the first recognized segment. If there exists a line passing through p and p_f, the last segment is reconstructed. Else, another point p_l is fixed in the starting pixel and a segment is added between p_l and p. Fig. 12 shows an example of reconstructed curve.

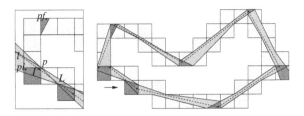

Fig. 12. Result of the polygonalization process (dotted line). The arrow indicates the reconstruction direction.

4.4 Results

Table 1 illustrates the performance of the three different reconstruction methods (with patches, without patches and our method) in terms of number of straight line segments obtained. The smaller this number, the better the polygonal description of the discrete object. We can see that our method provides a smaller number of line segments than equivalent methods with or without patches.

5 Conclusions and Future Work

In this paper, we have presented a new incremental standard discrete line recognition algorithm. Recognized pixels do not have to be connected and can be considered in any order. This algorithm is based on the notion of generalized preimage which provides the set of straight lines which cut a given set of pixels. The generalized preimage is obtained by intersecting the dual of the pixels in a parameter space and does not depend on their location in the space. Moreover, it is computed in linear time. Finally, we have proposed a new 4-connected curve reconstruction algorithm based on our recognition algorithm which improves the methods proposed in [8] and [3] (see Fig. 13). This algorithm has been

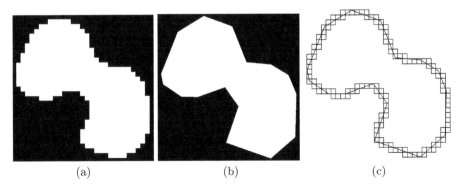

(a)   (b)   (c)

Fig. 13. Result of the reconstruction process. (a) Original digital image. (b) Reconstructed image. (c) Comparison between the discrete and reconstructed curves.

Algorithm 3. Curve polygonalization

Data: A set S of k connected recognized segments $S_1, \ldots, S_k$ with
$\qquad S_i = (P_{i_f}, \ldots, P_{i_l})$ and k solution line sets $L_1, \ldots, L_k$.

begin

$\quad$ | $\quad i \longleftarrow k$;

$\quad$ | Choose a real point p_{i_f} in P_{i_l};

$\quad$ | $p_f \longleftarrow p_{i_f}$;

$\quad$ | **while** $i > 1$ **do**

$\quad$ | $\quad$ | Choose a line l in L_i that passes through p_{i_f};

$\quad$ | $\quad$ | $s \longleftarrow l \cap P_{i_f}$;

$\quad$ | $\quad$ | Choose a new real point p_{i_l} on s;

$\quad$ | $\quad$ | Add a segment between p_{i_f} and p_{i_l};

$\quad$ | $\quad$ | $p_{i-1_f} \longleftarrow p_{i_l}$;

$\quad$ | $\quad$ | $i \longleftarrow i - 1$;

$\quad$ | Search a line l in L_1 that passes through p_{1_f} and p_f;

$\quad$ | **if** l *does not exist* **then**

$\quad$ | $\quad$ | Choose a line l in L_1 that passes through p_{1_f};

$\quad$ | $\quad$ | $s \longleftarrow l \cap P_{1_f}$;

$\quad$ | $\quad$ | Choose a new real point p_{1_l} on s;

$\quad$ | $\quad$ | Add a segment between p_{1_f} and p_{1_l} and between p_{1_l} and p_f;

$\quad$ | **else**

$\quad$ | $\quad$ | Add a segment between p_{1_f} and p_f;

end

Table 1. Comparison between the three reconstruction methods. The number of segments of the reconstructed curve is given.

Image	Curve pixels number	With patches	Without patches	Proposed method
	116	22	24	13
	356	50	62	39
	424	57	66	38
	574	71	104	57
	852	99	130	84

implemented in a discrete modeling system which combines continuous and discrete representations of a same object [12].

One of our future extensions will be to adapt our recognition algorithm to other digitization models and to irregular grids. Indeed, generalized preimage can be computed for any convex polygon. Another major advantage of the generalized preimage is that it can be easily extended to higher dimensions. The standard model being defined in arbitrary dimension, the next step is the extension of the recognition and reconstruction algorithms to dimension 3.

References

1. Lorensen, W.E., Cline, H.E.: Marching cubes: a high resolution 3d surface construction algorithm. In: SIGGRAPH. Volume 21 of Computer Graphics (ACM)., Anaheim, USA (1987) 163–169
2. Klette, R., Rosenfeld, A.: Digital straightness – a review. Discrete Applied Mathematics **139**(1–3) (2004) 197–230
3. Sivignon, I., Breton, R., Dupond, F., Andres, E.: Discrete analytical curve reconstruction without patches. Image and Vision Computing **23**(2) (2005) 191–202
4. Debled-Rennesson, I., Reveillès, J.: A linear algorithm for segmentation of digital curves. International Journal of Pattern Recognition and Artificial Intelligence **9**(6) (1995) 635–662
5. Dorst, L., Smeulders, A.W.M.: Discrete representation of straight lines. IEEE Transactions on Pattern Analysis and Machine Intelligence **6**(4) (1984) 450–463
6. Vittone, J., Chassery, J.M.: Recognition of digital naive planes and polyhedrization. In: DGCI. Volume 1953 of LNCS. (2000) 296–307
7. Cœurjolly, D.: Algorithmique et géométrie discrète pour la caractérisation des courbes et des surfaces. PhD thesis, Université Lumière Lyon 2, Lyon, France (2002)
8. Breton, R., Sivignon, I., Dupont, F., Andres, E.: Towards an invertible Euclidean reconstruction of a discrete object. In: DGCI. Number 2886 in LNCS, Naples, Italy (2003) 246–256
9. Andres, E.: Discrete linear objects in dimension n: the standard model. Graphical Models **65** (2003) 92–111
10. Maître, H.: Un panorama de la transformation de Hough – a review on Hough transform. Traitement du Signal **2**(4) (1985) 305–317
11. McIlroy, M.D.: A note on discrete representation of lines. AT&T Technical Journal **64**(2) (1985) 481–490
12. Dexet, M., Andres, E.: Hierarchical topological structure for the design of a discrete modeling tool. In: WSCG Full Papers Proceedings, Plzen, Czech Republic (2006) 1–8

Digital Line Recognition, Convex Hull, Thickness, a Unified and Logarithmic Technique

Lilian Buzer

Institut Gaspard Monge, A2SI Laboratory, ESIEE, 2 bd Blaise Pascal,
Cité Descartes, - BP 99, 93162 Noisy-Le-Grand Cedex, France
Buzerl@esiee.fr

Abstract. The recognition of discrete primitives as digital straight segments (DSS) is a deeply studied problem in digital geometry (see a review in [6]). One characterization of the DSS is purely geometrical: all the points must lie between two lines whose distance (relative to the infinite norm) is less than 1. A common approach used to solve this question is to compute the convex hull of the given points. Recent papers explain how to update the minimum distance when a point is inserted during an online (incremental) recognition in $O(\log n)$ time in the general case [2] or in $O(1)$ time with assumption [2, 4]. Nevertheless, for other cases like insertions mixed with deletions or the union of two DSS, we have no optimal method to compute the resulting width. Thus, we propose a unified, simple and optimal approach applicable for any configuration. Moreover, our function is called independently from the convex hull processing. This allows to reuse any existing library without any modification. Thereby, we offer an efficient tool that opens a new horizon for the applications.

Keywords: digital line, DSS, online, incremental, dynamic, union, recognition, convex hull, logarithmic complexity.

1 Introduction

1.1 Presentation

This paper focuses on the recognition of digital straight segments and, in a more general way, digital straight segments of fixed thickness (see Fig. 1).

A class of digital line recognition algorithms is based on the computation of the convex hull of the current set of points. Efficient algorithms [2, 4] process the recognition in linear time when the points are entered from left to right for example. For the online case, the known lower bound is $O(n \log n)$ [2, 3]. But, when the convex hull is known, we have no way to efficiently compute the vertical distance. The current optimal approach would be in $O(\log^2 n)$ time at least.

In this paper, we propose a completely new tool usable for this class of algorithms. It is a unified approach that leads to a logarithmic time complexity in any situation. Even with assumption (points entered from left to right), the optimal bound of $O(1)$ is preserved provided that we keep the result from the previous insertion. Our technique is based on a double combined binary search on the upper and lower borders of the convex hull.

U. Eckardt et al. (Eds.): IWCIA 2006, LNCS 4040, pp. 189–198, 2006.

In the first two sections we recall the problem of recognition, the role of the thickness in the digital line recognition problem and the importance of the convex hull. In section 4, we explain the previous approach used to compute the thickness of a given convex hull. In section 5, we present our new technique and sketch its proof. Throughout this paper, we will denote by $P.x$ the abscissa of a point P.

1.2   Applications

This technique does not distinguish itself by an important performance gain. Nevertheless it offers the possibility to deal with new problems. For example, if you use a hierarchical approach, you may want to determine if the union of two DSS is always a valid DSS. This time, you can use classical algorithms to compute the resulting convex hull of the union of the two known convex hulls and then apply our technique to determine the new thickness in $\log(n)$ time. Libraries of geometry can be reused without modification. It is also sufficient to later call our little routine. In the same way, we can reuse algorithms developed for dynamic convex hull computation. It means that points can be entered and deleted at any time during the recognition process. Our function is applied after each modification in $\log(n)$ time. Thus, we can now afford to tackle other questions that were a bit too complicate to deal with.

2   The α-Thick Digital Line

We hereafter recall the notion of α-thick digital line (see [5]). Its seminal definition was given by Reveillès in [8]. A digital line D in $\mathbb{Z}^2$ is described by a set of parameters: the *normal vector* $N = (a, b)$ in $\mathbb{Z}^2 \backslash \{0\}$ with $\gcd(a, b)$ that is equal to 1, the *inferior bound* γ in $\mathbb{Z}$ and the *arithmetic thickness* w in $\mathbb{Z}$. A point (x, y) of a digital line with parameters $(a, b), \gamma, w$ verifies the following diophantine inequality:

$$\gamma \leq a.x + b.y < \gamma + w \tag{1}$$

If we choose the arithmetic thickness to be equal to the infinity norm of the normal vector: $\|N\|_\infty = \sup\{|a|, |b|\}$, we obtain an 8-connected object called a *naive digital line* (see Fig. 1). Relative to this definition, we use another formulation, namely the α-*thick digital line*. The α value corresponds to a thickness

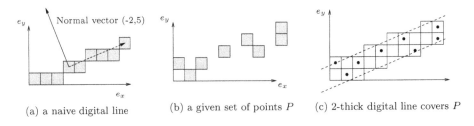

(a) a naive digital line   (b) a given set of points P (c) 2-thick digital line covers P

Fig. 1. The recognition problem using different thickness for the digital lines

ratio whose reference ($\alpha = 1$) is associated to the thickness of a naive digital line. A point (x, y) belongs to such a line if it verifies:

$$\gamma \leq a.x + b.y < \gamma + \alpha.\sup\{|a|, |b|\} \qquad (2)$$

3   Thickness Criterion

3.1   Introducing the Notion of Thickness

The definition of a digital line is intrinsically algebraic. We use an equivalent characterization which is more linked to the field of Euclidian geometry:

Lemma 1. *A set of points is a subset of an α-thick digital line if and only if these points can be covered by a strip of rational slope and of horizontal or vertical thickness strictly inferior to α.*

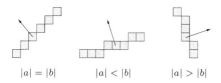

$|a| = |b|$ $|a| < |b|$ $|a| > |b|$

Fig. 2. Different types of digital line orientations

Fig. 3. Height and vertical thickness of a convex hull

3.2   Thickness and Convex Hull

In this subsection, we show that the notion of thickness (more precisely the thickness of the convex hull of the input points) plays an important role in the recognition problem.

Definition 1. *The height at abscissa β of a convex set C, denoted by $height(x)$, is defined to be the length of the segment resulting from the intersection of C with the vertical line $x = \beta$. The maximum reached by the function $height()$ is the vertical thickness of C (see Fig. 3). Width and horizontal thickness are defined analogously.*

Lemma 2. *A convex polygon N has a vertical thickness less than α iff there exists a strip of vertical thickness less than α that covers N.*

▶ Proof: let x denote the abscissa which corresponds to the maximum height of N. The upper and lower border of N at abscissa x can be linked to either a vertex or an edge. We can consider three different configurations:

1. edge-edge: this case only appears if both edges are parallel (see Fig. 4.b). If this were not the case, it would exist a greater value for the vertical thickness of N (see Fig. 4.a). As N is convex, it is included in the strip defined by these two edges. So N can be covered by a strip of correct thickness.

2. edge-vertex: as the maximum is achieved at this abscissa, the line passing through this vertex and parallel to this edge is tangent to N (see Fig. 4.c). We use this line and the previous edge to build a valid strip.

3. vertex-vertex: this case is a bit more complicated. Let e_{ul}, e_{ur}, e_{ll} and e_{lr} denote the emerging edges from these two vertices. As the polygon is convex, we have $slope_{e_{ur}} \leq slope_{e_{ul}}$ and $slope_{e_{lr}} \geq slope_{e_{ll}}$. The maximum height reached at this abscissa implies $slope_{e_{ul}} \geq slope_{e_{ll}}$ and $slope_{e_{ur}} \leq slope_{e_{lr}}$. If we choose e_{ll} as a border of our strip, e_{ul} and e_{lr} lie inside. Assume that $slope_{e_{ur}} \leq slope_{e_{ll}}$. In this case (see Fig. 4.d) N is included in a valid strip. In the opposite case where $slope_{e_{ur}} > slope_{e_{ll}}$ (see Fig. 4.e), this edge can not be chosen as a border for our strip. But, e_{ur} is a correct choice. Indeed, we have $slope_{e_{ur}} \leq slope_{e_{ul}}$ and $slope_{e_{ur}} \leq slope_{e_{lr}}$ by assumptions; as $slope_{e_{ur}} > slope_{e_{ll}}$ this finally implies that N is included in a strip of vertical thickness less than α. ◀

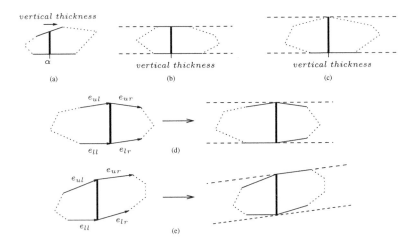

Fig. 4. The different configurations of the vertical thickness location

Using lemmas 1 and 2, we finally obtain the next property which links the thickness of a convex hull to the α-thick digital line recognition problem:

Property 1. A set of points is a subset of an α-thick digital line if and only if its convex hull has an horizontal or vertical thickness inferior to α.

3.3   The Importance of Convex Hull

We have shown that the knowledge of the convex hull thickness is intrinsically linked to the problem of digital line recognition. When points are inserted in a given direction (from left to right for example), simple and efficient algorithms exist [2, 4] and can be used in order to compute the convex hull thickness or to test wether the convex hull thickness is valid or not. Nevertheless, in other configurations, there is no specific algorithm. For example, when the convex hull has already been processed in a previous stage, we do not know how to efficiently compute its thickness. The next section introduces this method and explains why we were stuck to this lower bound.

4   Why This Problem Is Difficult?

Suppose we know the convex hull of a set of points. Now, the only thing we want to do is to efficiently compute the associated thickness. To introduce this problem, we first present a naive approach that has a linear time complexity and then we explain the $O(\log^2 n)$ time method. In the following, we denote by $U[]$ an array of size nU and $L[]$ an array of size nL that respectively represent the vertices of the upper and lower borders of the convex hull. This is obvious that $U[1] = L[1]$ and that $U[nU] = L[nL]$.

4.1   The Naive Method

We notice that the maximum of the function $height(x)$ is always associated to a vertex facing a segment (case *edge-edge* or case *edge-vertex*) or a vertex facing a vertex. Therefore, this function achieves its maximum at the abscissa defined by a vertex of the convex hull. Thus, to compute the thickness it is sufficient to compute the values of $height(x)$ at all the abscissae of the vertices of the convex hull. We can create a simple traversal of the upper and lower borders in order to easily check all the values of the function $height()$ at each vertex.

Let us imagine a sweep line moving from left to right. When the line passes trough two vertices at the same time, we process a case *vertex-vertex* and when only one vertex lies on the line, we obtain a case *edge-vertex* where the corresponding edge is the segment cut by the sweep line. To program this traversal, we only need to update the current rightmost vertex L on the lower hull and the current rightmost vertex U on the upper hull that are on the left of the sweep line. The location of the sweep line corresponds to the rightmost of these two vertices. When we want to shift the sweep line on the right, we must correctly update this couple of points. To do this, we pick the next vertex encountered by the sweep line. It can be computed in constant time by choosing the point between the successor of L and the the successor P that has the lowest abscissa. This vertex replaces its predecessor on the same border. As we always go forward, we finally obtain a complexity bounded by the number of vertices. Thus, this algorithm has a linear time complexity in the number of vertices. This is not efficient but when the number of vertices is small, it can be used. We hereafter give the source code of this method and show an example in figure 5.

4.2   Using Binary Searches

We do not use any mathematical consideration in the previous algorithm which leads to poor performance. Let $l(x)$ and $u(x)$ denote the two functions corresponding respectively to the lower and upper border. By definition, we notice

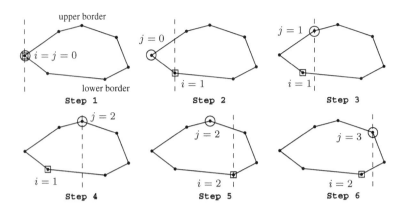

Fig. 5. Example of the naive approach

Function **COMPUTE_VERTICAL_THICKNESS** $(U[],nU,L[],nL)$
{
  $i = j = 0$
  $MH = 0$ // *maximal value of the function height*()
  while $((i \neq nU)$ *or* $(j \neq nL))$
  {
      if $(L[i].x == U[j].x)$ **CaseVertexVertex(MH,L[i],U[j])**
      else if $(L[i].x < U[j].x)$ **CaseEdgeVertex(MH,L[i]L[i+1],U[j])**
      else   **CaseEdgeVertex(MH,U[j]U[j+1],L[i])**

      // *shift the sweep line*
      if $((i < nU)$ *and* $(j < nL))$
         if $(L[i+1].x < U[j+1].x)$ $i++$
         else $j++$
      else if $(i < nL)$ $i++$
      else $j++$

  }
  return MH
}

that $height(x) = u(x) - l(x)$. As $u(x)$ is a concave function and $l(x)$ a convex function, then $height(x)$ is a concave function and so we can perform a binary search on it in order to find the abscissa of its maximum. Like in the previous algorithm we only consider abscissae where vertices are lying. Suppose we have nU vertices on the upper hull. We consider the middle point $MU = U[nU/2]$ and compute the value of $height(MU.x)$. Consider that an Oracle function tells us on which side of MU the maximum lies. Then, we can suppress one half of the points on the upper hull because we know that they will never be used again. Nevertheless no decimation has been done on the lower hull, so we perform the same approach on the other border one more time. Thus we are sure that one half of the vertices were rejected from the problem. So with these two steps, the size or our problem reduces by two at each iteration. Therefore, we can find in a logarithmic number of iterations the maximum value of the function $height()$.

At each step, we compute the value of $height(x)$ for the abscissa of a given vertex P and then call an Oracle function. To do this, we must find the facing edge or the facing vertex of P on the other border. This cannot be performed in constant time, we have to use another binary search on the abscissae of all the vertices lying on the other border. Thus, with a logarithmic cost, we can find the corresponding vertex $L[i]$ (resp. $U[j]$) of a point $U[j]$ (resp. $L[i]$).

We now describe the Oracle function. In fact, we can consider the derivative of $height(x)$ that is linked to the optimum location. This can be done in constant time when the indexes of the couple of vertices $L[i]$ and $U[j]$ are known. We simply subtract the slopes of the corresponding segments on the lower and upper borders to determine the value of the derivative. Notice that this function is not defined for the abscissae of the vertices because $l(x)$ and $u(x)$ are piecewise linear functions. But for our algorithm, this is not a problem. We only have to use the generalized derivative definition, a subgradient, in order to bypass this difficulty. Let u' denote the derivative of a function u. So the costly function is the binary search used to find the facing vertex. As this function is called a logarithmic number of times, our method has an $O(\log^2 n)$ time complexity where n denotes the number of vertices of the convex hull. An example is given in figure 6.

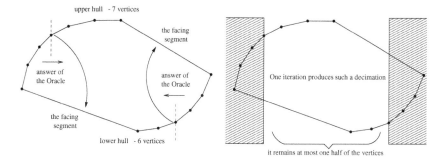

Fig. 6. Example of the double binary searches method

4.3   Conclusion

At this step, it seems impossible to beat the performance of $O(\log^2 n)$. Any attempt to evaluate the function $height(x)$ will force to set up a correspondence between the vertices of the lower hull and the upper hull. This cannot be achieved in constant time unless you use a preprocessing step. Nevertheless in an online (incremental) approach such preprocessing is not possible. As we must perform at least a binary search to compute the maximum of the concave function $height(x)$, we should conclude that this previous approach cannot achieve a logarithmic time performance.

5 The Logarithmic Algorithm

5.1 Introduction

We have previously seen that the evaluation of the function $height(x)$ was too expensive to achieve logarithmic performance. Thus, we have to bypass the computation of this function and try to preserve the idea of performing a binary search. Our new method leads to a simple implementation and elementary explanations are sufficient to describe its principle. Consider the function $height(x) = u(x) - l(x)$, we are looking for the abscissa where this function reaches its maximum. Thus, we seek the value of x where $height'(x)$ equals 0. Looked at another way, we can search for the abscissa where $u'(x) = l'(x)$. As $u(x)$ is a piecewise linear concave function, its derivative is a non-increasing function. In the same way, $l'(x)$ is a non-decreasing function.

5.2 Combined Binary Search

At this step, we consider the problem of finding the intersection between one piecewise constant non-decreasing function $l'(x)$ and one piecewise constant non-increasing function $u'(x)$. As in the previous algorithms, we only take care to the abscissae of the present vertices. Suppose we have nU vertices on the upper hull and nL vertices on the lower hull. Let $MU = U[nU/2]$ and $ML = L[nL/2]$ denote the two middle points on each border. We are not able to determine a value for the function $height(x)$ with this information. But, we can compute the generalized derivatives $u'(MU.x)$ and $l'(ML.x)$ in constant time using the points $U[nU/2 - 1]$, $U[nU/2 + 1]$, $L[nL/2 - 1]$ and $L[nL/2 + 1]$. For presentation convenience, we only consider the standard derivative of a function. With the knowledge of $MU.x$, $u'(MU.x)$, $ML.x$ and $l'(ML.x)$, we can determine the optimum location relative to the abscissa of $MU.x$ or relative to the abscissa of $ML.x$. In the following, we sketch the proof of our method.

Let us consider the case where $MU.x < ML.X$ and where $u'(MU.x) > l'(ML.x)$. See figure 7 for an example. The three other configurations are similar. We now create our Oracle function which determines the optimum location. As $u'()$ is a non-increasing function, we have for any $x < MU.X$, $u'(x) > u'(MU.x)$. In the same way, we notice that for any $x < ML.x$, we have $l'(x) < l'(ML.x)$. By assumption, $u'(MU.x) > l'(ML.x)$ and $MU.x < ML.x$, using these inequalities we obtain for any $x < MU.x$: $u'(x) > u'(MU.x) > l'(ML.x) > l'(MU.x) > l'(x)$. Thus, we can conclude that there exists no intersection between u' and l' on the left of MU. Therefore, we can delete one half of the vertices of the upper hull. This iteration only requires a finite number of comparisons and so it can be processed in constant time. All the vertices of the other border are preserved. In at most $\log_2(nU) + \log_2(nL)$ iterations, the number of vertices of the two borders decreases to a finite number. Thus, the resulting size of the problem is bounded and the last computations can be done in constant time. This approach achieves a logarithmic time complexity and avoids the computation of the values of $height(x)$.

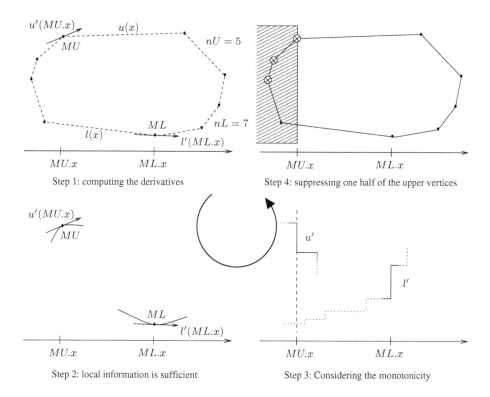

Fig. 7. The combined binary search

Geometrical interpretation. This method is equivalent to finding on which abscissa the tangents of the lower and upper hulls are parallel. We perform two combined binary search and their two cuts converge to the abscissa associated to the maximum value of $height(x)$.

6   Conclusion and Future Works

We focus on the digital line recognition problem. The class of the algorithms we study is based on the computation of the convex hull of the current set of points. Previous algorithms were optimal in the online case only. In other cases (deletion, union), awkward approaches were possible. We simplify and unify these techniques to propose a single, simple and efficient routine that is usable in any configuration and that preserves an optimal complexity of $O(\log n)$ time and of $O(1)$ time with assumption.

For example, no optimal algorithm was known to compute the vertical thickness of a given convex hull. We can now process such a case in logarithmic time. The proposed technique is new and it is based on a double and combined binary search on the tangents of the lower and upper convex hull borders. The implementation is simple and it leads to an efficient algorithm. To compute a dynamic convex hull (that supports insertion and deletion) or to unite two convex hulls, we can now reuse algorithms in the literature [1, 7] or existing implementations. At each step, we only have to call our small routine in $\log n$ time in order to determine the current thickness.

This method leads to an efficient implementation and if it is generalized to the three-dimensional space, it could offer an interesting and efficient algorithm for digital plane recognition.

Acknowledgement. The author thanks the reviewers for their helpful comments.

References

1. M. de Berg, M. van Kreveld, M. Overmars, and O. Schwarzkopf. *Computational Geometry: Algorithms and Applications (2nd ed.)*. Springer-Verlag, 2000.
2. L. Buzer. *An Elementary Algorithm for Digital Line Recognition in the General Case*, DGCI 2005, pp. 299-310, 2005.
3. L. Buzer. *Reconnaissance des plans discrets - Simplification polygonale*. Thèse, Université d'Auvergne, Clermont-Ferrand, 2002.
4. I. Debled-Rennesson, F. Feschet, J. Rouyer-Degli. *Optimal Blurred Segments Decomposition in Linear Time*. DGCI 2005, pp: 371-382, 2005.
5. I. Debled-Rennesson, R. Jean-Luc, J. Rouyer-Degli. *Segmentation of Discrete Curves into Fuzzy Segments*. 9th International Workshop on Combinatorial Image Analysis. Electronic Notes in Discrete Mathematics, Volume 12, 2003.
6. R. Klette and A. Rosenfeld. *Digital straightness - a review*. Discrete Applied Mathematics, 139, pp: 197-230, 2004.
7. F.P. Preparata, M.I. Shamos. *Computational Geometry*. Springer-Verlag, New-York, 1985.
8. J.P. Reveillès. *Géometrie discrète, calculs en nombre entiers et algorithmique*. Thèse d'état, Université Louis Pasteur, Strasbourg, 1991.

Incremental and Transitive Discrete Rotations

Bertrand Nouvel⋆ and Éric Rémila

UMR CNRS - ENS Lyon - UCB Lyon - INRIA 5668
Laboratoire de l'Informatique du Parallélisme
École Normale Supérieure de Lyon
46, Allée d'Italie 69364 LYON CEDEX 07 - France
{bertrand.nouvel, eric.remila}@ens-lyon.fr

Abstract. A discrete rotation algorithm can be apprehended as a parametric map f_α from $\mathbb{Z}[i]$ to $\mathbb{Z}[i]$, whose resulting permutation "looks like" the map induced by an Euclidean rotation. For this kind of algorithm, to be incremental means to compute successively all the intermediate rotated copies of an image for angles in-between 0 and a destination angle. The discretized rotation consists in the composition of an Euclidean rotation with a discretization; the aim of this article is to describe an algorithm which computes incrementally a discretized rotation. The suggested method uses only integer arithmetic and does not compute any sine nor any cosine. More precisely, its design relies on the analysis of the discretized rotation as a step function: the precise description of the discontinuities turns to be the key ingredient that makes the resulting procedure optimally fast and exact. A complete description of the incremental rotation process is provided, also this result may be useful in the specification of a consistent set of definitions for discrete geometry.

1 Introduction

The translation of the fundamental concepts of the Euclidean geometry into $\mathbb{Z}^n$ comprises the field of discrete geometry. As this theory of geometry is particularly suitable for combinatorial images and other data manipulated by computers [1], it would be interesting to provide a set of efficient algorithm for this theory that uses only integer-arithmetic; as this was suggested in [2].

Several attempts have been realized by various authors that wished to deliver back some properties of the Euclidean rotation to the discretized rotations widely used in computer graphics. A review of various resulting algorithms may be found in [3].

In this paper, we present an *incremental* algorithm: it successively computes all the rotated images according to the an increasing sequence of angles (starting from 0 to 2π). Since the set of rotated images is finite on a finite picture, this allows practically to compute *all* the intermediate rotated images. Moreover, the suggested procedure is *sound* and *accurate*: it returns exactly the same results as the discretized rotation. The procedure does not use any sine nor any cosine,

---

⋆ This author is supported by the french television channel TF1.

U. Eckardt et al. (Eds.): IWCIA 2006, LNCS 4040, pp. 199–213, 2006.

thus there is no precision problem due to the floating point arithmetic. Also, the algorithm is *fast*: to compute incremental rotations the algorithm computes only $\mathcal{O}(m^3 log(m))$ operations, instead of $\mathcal{O}(m^5)$ as needed by the naive algorithm. For the incremental rotations, the complexity of this algorithm if it uses pre-calculated tables, is $\mathcal{O}(m^3)$; it is optimal: The algorithm updates only the necessary pixels and only consider the necessary angles. Finally, since the algorithm uses the configurations that can be stored with very few states on the plane, we believe it is a good candidate for parallelization.

After a brief review of the motivations, and after the essential preliminary definitions, we proceed to a characterization of the discontinuities of the rotation process. Indeed, we will explain how to code the angles where the discontinuities happen. Also, with integer arithmetic only, we will specify how to perform the essential operations on the encoded angles. Naturally, a few technical lemmas are required to set up all this framework properly. Once this has been set, we will analyze the alterations that occur in the configuration at the discontinuities. Strengthened by previous results, we will then be ready to build the incremental discretized rotation procedure. The last section will be devoted to various extensions and miscellaneous details related to the theory that may lead to a better understanding of the discretized rotation process.

2   Groundwork

The first sections introduce the fundamental ideas, definitions and lemmas that matter to fully understand the algorithm. In this section, we review the motivations and we specify some vocabulary.

2.1   Motivations

The history of discrete geometry begins with the common will to give birth to an algorithmic theory of the geometry in the discrete spaces that would be consistent with the Euclidean geometry. We believe that a unified theory would provide a better understanding of both universe: continuous and discrete. Discrete rotations comprise the famous examples that have strengthened the idea that discrete and continuous spaces may be radically different. A review of the differences can be found in the prologue of [4].

More recently, discretized rotations have been an important issue in water-marking community. Water-marking algorithms that were robust under rotation were sought for by various teams the community; and it supports many discussions. More generally, the problem of finding algorithms for classification and recognition of patterns that are robust under discrete rotations is still a not-trivial issue in the conception of pattern related algorithms.

This paper is focused on an algorithm that uses similar principle as the one suggested by [5]; the basic idea is to compute a table of all the discontinuities of the discrete rotation process that occur in within a certain ball of radius m. Then this table will be used to drive the rotation process. Our paper provides a more accurate description of the discontinuities, as-well as a way to compute the rotation using integer arithmetic only.

The algorithm is designed to comprise the field of discrete geometry: our main problem is to provide a comprehensive and simple definition of algorithm for the discrete rotation, independantly of any specific usage, and that can be computed efficiently.

The usual discretized rotation algorithm is a bit chaotic when iterated: This is well illustrated by the following story: It has been implemented a physical simulation of the solar system which for specific reasons was coded with fixed precision. When the system has been tried, its inventors were surprised that its first conclusion was to predict a fatal collision in-between the earth and the sun within the next ten years; naturally there was a "bug". It was imputable to composition of rounding errors during the rotation process. More generally, decades of computer programmers have known by experience that they should avoid to compose rotations. The accumulation of the resulting errors may produce an unwelcome result. Some aspects of the dynamical system that is formed by the iterated action of a discretized rotation have already been studied; see [6] for iteration of $\pi/4$ rotations.

2.2   Conventions

We work in the *complex plane* $\mathbb{C}$, where $\mathbb{Z}[i]$ denotes the set of *Gaussian integers*, i.e. the set of complex numbers whose real and imaginary parts are both integer. Let m be a positive integer. We denote by $\mathbb{Z}[i]|_m$ the set of Gaussian integers whose modulus is at most m; $\mathbb{Z}[i]|_m = \{z \in \mathbb{Z}[i], |z| \le m\}$. Real and imaginary parts are denoted $\Re(z)$ and $\Im(z)$. Let x be a real number. We recall that the floor function $x \mapsto \lfloor x \rfloor$ is defined as the greatest integer less or equal to x. The *rounding function* is defined as : $[x] = \lfloor x + 0.5 \rfloor$; we also define the map $x \mapsto \{x\}$ by $\{x\} = x - [x]$. These maps are extended to complex numbers, by applying them independently on the real part and on the imaginary part. Let $\mathcal{H}$ be the set of complex number that have a semi-integer coordinate (in other terms $\mathcal{H} = (\mathbb{R} \times (\mathbb{Z} + \{\frac{1}{2}\})) \cup ((\mathbb{Z} + \{\frac{1}{2}\}) \times \mathbb{R})$). More generally $\mathcal{H}$ denotes the set of discontinuity points of the operator $x \mapsto [x]$.

Let α denote an angle in radians, i.e. an element of $\mathcal{A} = \mathbb{R}/(2\pi\mathbb{Z})$. The *Euclidean rotation* r_α is the bijective isometry of $\mathbb{C}$, $z \mapsto ze^{i\alpha}$. *The discretized rotation* $[r_\alpha]$ is precisely defined as the successive computation of the Euclidean rotation of angle α and of the discretization operator $z \mapsto [z]$. Thus, for each z of $\mathbb{C}$, $[r_\alpha(z)] = [ze^{i\alpha}]$. Remark that, for any $z \in \mathbb{C}$ we have $|[r_\alpha](z)| \le |z| + \frac{\sqrt{2}}{2}$, and that $|[r_\alpha](z) - r_\alpha(z)| \le \frac{\sqrt{2}}{2}$.

In this article, a *configuration* is a mapping from $\mathbb{Z}[i]$ to $\mathbb{Z}[i]$. Let m be a positive integer. A configuration C such that for all $z \in \mathbb{C}$, $\lim_{n \mapsto \infty} \frac{C([2^n z])}{2^n} - (e^{i\alpha}z) \to 0$ is called an α-*rotation map*. Thus, given a real α, the discrete rotation $[r_\alpha]$ is a configuration.

A partial configuration of radius m is a mapping from $\mathbb{Z}[i]|_m$ to $\mathbb{Z}[i]$. Each configuration induces a partial configuration. In this paper, we work on on partial

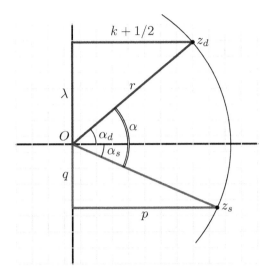

Fig. 1. An hinge angle and its generating pair for $\alpha = \alpha(9, -4, 7)$

configurations (denoted by $[r_\alpha]|_m$) induced by discretized rotations. Precisely, we study the mapping $\rho_m : \alpha \mapsto [r_\alpha]|_m$. Since $\mathbb{Z}[i]|_m$ is finite and $|[r_\alpha(z)]| \leq |z| + 1$, the set $\{\rho_m(\alpha), \alpha \in \mathbb{R}\}$ is finite.

Our aim is to produce an *exact incremental rotation algorithm* which, given an integer m successively produces all the values $\rho_m(\alpha)$ for $\alpha \in [0, 2\pi]$, in the order where they are reached (moving from 0 to 2π). Informally, it is a "video" algorithm which exhibits the successive configurations obtained along the rotation.

3   Hinge Angles

3.1   Definitions

The principal element that has influenced the design of the algorithm is a precise study the function ρ_m. We shall prove that ρ_m is a stair function (piecewise constant function). Thus, a precise study of its discontinuity steps gives the ability to recover the whole function. The discontinuities also correspond to the only updates of the configuration stored in the algorithm.

Definition 1. *An angle α is a* hinge angle *if there exists a source point z_s in $\mathbb{Z}[i]$ such that the destination point $z_d = z_s e^{i\alpha}$ has a (proper) semi-integer component (i.e. $z_d \in \mathcal{H}$) . For each hinge angle, the source point and the destination point form a generating pair.*

On a topological point of view, the hinge angles are the discontinuity points of the map $\alpha \mapsto [r_\alpha]$.

It is immediate from the definition that an angle is a hinge angle if and only if there exists integers p, q, k such that $2\,q\cos(\alpha) + 2\,p\sin(\alpha) = 2\,k + 1$.

The value $|z_s(\alpha)|^2$ is called the *order* of α. We can note that if (z_s, z_d) generates α then $(\bar{z}_s, \bar{z}_d)$ generates $-\alpha$. Note, also that if (z_s, z_d) generates α then $(i^Q z_s, i^Q z_d)$ generates the same angle, for any $Q \in \{0, 1, 2, 3\}$. We also define S_α the set of source points of α: $S_\alpha = \{z \in \mathbb{Z}[i] \,|\, e^{i\alpha} z \in \mathcal{H}\}$.

3.2 Fundamental Lemmas

The *Pythagorean angles*, as seen in [7] or [8], are such that $\alpha = \arctan(a/b)$ where a and b are issued from a Pythagorean triple $(a, b, c) \in \mathbb{N}^3$ (such that $a^2 + b^2 = c^2$). The positive integer c is the *radius* of the Pythagorean angle. An angle is Pythagorean if and only if its cosine and sine are rational.

Lemma 1. *Hinge angles and Pythagorean angles form disjoint sets.*

Proof: Assume that it exists an angle which is both Pythagorean and hinge. By definition, since it is a hinge angle, there exists a Gaussian integer z_s of $\mathbb{Z}[i]$ which is transformed by rotation in a point of $\mathcal{H}$. We can easily check that: $\{\Re(e^{i\alpha} z_s)\} = \{\Re(z_d)\} = \frac{1}{2}$. More exactly, $\Re(z_d) = \Re(z_s) \cos(\alpha) - \Im(z_s) \sin(\alpha)$. If α is Pythagorean, $\cos(\alpha) = \frac{a}{c}$ and $\sin(\alpha) = \frac{b}{c}$, where (a, b, c) is a primary Pythagorean triple (i.e. belongs to $\mathbb{Z}$ and $a^2 + b^2 = c^2$, $gcd(a, b, c) = 1$). $\{\Re(z_d)\}$ can be written as $\frac{n}{c}$ where n as an integer, while c is odd. But this is contradicting: $\{\Re(z_d)\} = \frac{1}{2}$

$\square$

A corollary of this lemma is that for any hinge angle, we have that $\cos(\alpha)$ or $\sin(\alpha)$ is an irrational quadratic number. They can be simultaneously irrational.

Let p, q, k be a triple of integers, such that $p^2 + q^2 > |k + \frac{1}{2}|$, we state $r^2 = p^2 + q^2$, and $\lambda = \sqrt{r^2 - (k + \frac{1}{2})^2}$. We define the angle $\alpha(p, q, k)$ by the equality $e^{i\alpha(p,q,k)}(p + qi) = k + \frac{1}{2} + \lambda i$; thus for this angle $((p + qi), (k + \frac{1}{2} + \lambda i))$ forms a generating pair.

Lemma 2 (Coding of the Hinge Angles). *Let (p, q, k) and (p', q', k') be two distinct generating triples such that $\alpha(p, q, k) = \alpha(p', q', k')$, then we have* $\det \begin{bmatrix} p & q \\ p' & q' \end{bmatrix} = 0.$

Proof: The proof is straightforward: Let $\alpha = \alpha(p, q, k)$, thus $2p \sin(\alpha) + 2q \cos(\alpha) = 2k + 1$, and similarly; $2p' \sin(\alpha) + 2q' \cos(\alpha) = 2k' + 1$. Consider these two equations as a linear system of $\cos(\alpha)$ and $\sin(\alpha)$. If $\det(\begin{bmatrix} p & q \\ p' & q' \end{bmatrix}) \neq 0$ then $\cos(\alpha)$ and $\sin(\alpha)$ have to be rational, and the angle have to be Pythagorean. But this would contradict the Lemma 1. Therefore (p, q) and (p', q') are colinear. $\square$

The lemma implies that $arg(z_s(\alpha))$ and $arg(z_d(\alpha))$ are uniquely defined for an angle α up to the choice of a quadrant. Thus, they are important characteristics of the hinge angle. An immediate corollary of Lemma 2 is that all generating pairs of $\alpha(z_s, z_d)$ are necessarily of the form $(\eta z_s, \eta z_d)$ or of the form $(i \eta z_s, i \eta z_d)$ with $\eta \in \mathbb{Q}$.

A triple (p, q, k) that generates an angle α, the triple (p, q, k) is called *primary* if p is positive and if it is minimal, i.e. it has the smallest $p^2 + q^2$ among the generating triples of α. Obviously, there exists for each hinge angle a unique primary generating triple.

Lemma 3 (Primary Generating Triple). *Let (p, q, k) be the primary generating triple of an angle α; the set of generating triples of α is $\{((2n+1)p, (2n+1)q, (2n+1)k+n), \text{ for } n \in \mathbb{Z}\}$.*

Proof: It is evident that if $z_s = (p+qi) \in S_\alpha$ is the source point issued from the primary generating triple (p, q, k) then $(2n+1)z_s \in S_\alpha$, since $\{\Re(2n+1)e^{i\alpha}z_s\} = \frac{1}{2}$ for any n in $\mathbb{Z}$. Also we can notice that for any $n \in \mathbb{Z}$, this is another generating triple $((2n+1)p, (2n+1)q, (2n+1)k+n)$ of the same angle. It is also evident that for any $n', n'' \in \mathbb{Z}$, $(2pn', 2qn', n'')$ cannot be a generating triple of α since $e^{i\alpha}(p+qi) \in \mathcal{H}$ implies that $e^{i\alpha}(2\,pn' + 2\,qn'i) \notin \mathcal{H}$. Now, we can notice $\beta p + \beta q i$ with $\beta \in \mathbb{R} \setminus \mathbb{Z}$ cannot be a source for arithmetic reasons: $\beta(p+qi) \notin \mathbb{Z}[i]$. Now, assume there exists $\frac{r}{s} \in \mathbb{Q} \setminus \mathbb{Z}$ and $k' \in \mathbb{Z}$ such that $(\frac{r}{s}p, \frac{r}{s}qi, k')$ is a generating pair of α, this would implie that $\frac{r}{s}(p+qi)$ is a Gaussian integer and thus $gcd(p, q)$ should be divisible by s (as r is not). Moreover $\Re(e^{i\alpha}(\frac{r}{s}p + \frac{r}{s}qi)) = \frac{2k'+1}{2}$. This can only happen if $2k'+1$ is also divisible by s. Now $p, q, 2k+1$ are all divisible by n, thus the generating pair is not primary, and this contradicts our hypothesis. Thus, $\beta(p+qi) \notin S_\alpha$ for $\beta \in \mathbb{R} \setminus \mathbb{Z}$. Thus, consequently to previous lemma and to these points, there is no other generating triples for the angle α than the listed ones.   □

3.3   Main Properties

From the previous lemmas, we can state the following properties that are useful for the rotation algorithm.

Proposition 1. *Any hinge angle can be uniquely described by its primary generating triple.*

This proposition is actually a corollary of Lemma 2 and 3.

Proposition 2. *The number of hinge angles of order at most m is lower than $8m^3$.*

This is easily proven by the fact that $|p| < m, |q| < m, |k + \frac{1}{2}| < m$

Upper bounds on the number of possible hinge angles can be found in [5]. This formula can be slightly refined by using $r_2(k)$ that represents the number of decomposition of an integer as the sum of two squares. The upper bound on the number of possible hinge angles is then: $\#(\mathcal{A}_\mathcal{H}|m) \leq \sum_{i=1}^{m} r_2(i)\lfloor \sqrt{i} - \frac{1}{2}\rfloor \leq (\sum_{i=1}^{m} r_2(i)) \lfloor \sqrt{m} - \frac{1}{2}\rfloor \leq (2m)^3$. Finally note that there is twice more rotation maps in $\rho_m(\alpha)$ than the number of hinge angles in $\mathcal{A}_\mathcal{H}|m$.

Proposition 3. *The elements of S_α forms 4 rays: $S_\alpha = \cup_{Q \in \{0,1,2,3\}}\{(2l + 1)i^Q(p + qi), l \in \mathbb{N}\}$*

This proposition is another consequence of the Lemma 3.

4   The Algorithm

The algorithm that we are going to explain takes as input an integer m that describes the side of the square $[-m, m]^2$, and it computes successively all α-rotation maps $[r_\alpha]|_m$ (for all the possible angles, in the trignometric order). The algorithm contains only one map that is updated progressively. Each time the map is in correct state the algorithm notifies this information, which can be used by another procedure for the analysis of the rotated configuration.

Schematically, the algorithm is structured in two main parts. During this first period of the algorithm, it starts with the enumeration of all the hinge angles whose order is smaller than the maximum order of the points of the picture. In the algorithm, the angles are not represented by floating point values but only via their associated triples (three small integer numbers). This encoding of the angles provides a convenient way to recover all the necessary informations required to transform one image into the next image during the incremental rotation process. This first part of the algorithm can be seen as a process similar to the generation of a sine table and it can be done once for all.

The second part of the process consists in a loop through the cycle of hinge angles. This loop applies successively the small transformations that are required to pass from $\rho_m(\alpha)$ to $\rho_m(\alpha')$ (where α' is the successive representative angle). Practically, the algorithm stores the map $\rho_m(\alpha)$, this allows to know directly the position of the image through the transformation $[r_\alpha]|_m$. However, for various usages, such rotations in cellular automata, the user should may prefer variant encoding of the transformation such as rotation configurations (maps from $\mathbb{Z}[i]$ to a finite set) [4]. The principle of this algorithm is translatable on these kind of configurations.

4.1   Enumerate and Sort Hinge Angles

To enumerate all hinge angles in $\mathcal{A}_\mathcal{H}|_m$, it is sufficient to enumerate all the Gaussian integers whose module is smaller than m (as source point), and to consider all possible semi-integers whose absolute value is smaller than m. This requires a time of $\mathcal{O}(m^3)$. Some angles will be enumerated twice, but the duplicates can be identified and removed during the sorting process.

The usual sorting algorithms handle n elements in $\mathcal{O}(n \log(n))$ operations, if we provide them a constant time comparison operator. The goal of this section is to explain how to compare two hinge angles via their generating triple coding. This comparison needs constant time. Using a quicksort on generated triples, the enumeration and the sorting of the hinge angles can therefore be computed in $\mathcal{O}(m^3 log(m))$.

Lemma 4 (Integer-Based Comparison). *Let $\alpha \in \mathcal{A}_\mathcal{H}$ which is associated to the triple (p, q, k) and $\alpha' \in \mathcal{A}_H$ which is associated to the triple (p', q', k'), it is possible to decide, in constant time, whether $\alpha < \alpha'$ using by knowing p, q, k, p', q', k'.*

Proof(sketch):   With the notations previously introduced we have: $\cos(\alpha) = (p(k + \frac{1}{2}) + q\lambda)/(p^2 + q^2)$ and $sin(\alpha) = (p\lambda - q(k + \frac{1}{2}))/(p^2 + q^2)$.

The key-argument is that the numbers used in the expressions of $\cos(\alpha)$ and $\sin(\alpha)$ are all integer, except λ, but 4λ is a square root of an integer, thus the other equations will simplify.

Thus the signs of $\cos(\alpha)$ and $\sin(\alpha)$ can be easily computed, reducing to the problem to the sign of an integer inequality that involves only usual operations. Hence, the "quadrant" of α can be computed with integer arithmetic-only.

If α and α' belongs to the same quadrant then we have to compare $\cos(\alpha)$ and $\cos(\alpha')$. This can also be reduced to determining the sign of an integer expression, that can be computed using only integer arithmetic.   □

4.2   Transforming $[r_{\alpha-}]$ into $[r_{\alpha+}]$ When $\alpha \in \mathcal{A}_{\mathcal{H}}$

Let $\alpha \in \mathcal{A}_{\mathcal{H}}$, we now know that there exists $z_s(\alpha) \in \mathbb{Z}[i]$ and $\Re(z_d(\alpha)) + \frac{1}{2} \in \mathbb{Z}$ and also $\Im(z_d(\alpha)) > 0$. Due to the trigonometric orientation, and to the chosen discretization operator, $\lim_{\alpha' \mapsto \alpha, \alpha' < \alpha} [r_{\alpha'}](z_s) = [r_\alpha](z_s)$ while $\lim_{\alpha' \mapsto \alpha, \alpha' > \alpha} [r_\alpha](z_s) = ([r_\alpha](z_s)-1)$. However, we can also notice that $\lim_{\alpha' \mapsto \alpha, \alpha' < \alpha} [r_{\alpha'}](-z_s) = ([r_\alpha](-z_s) + 1)$ while $\lim_{\alpha' \mapsto \alpha, \alpha' > \alpha} [r_{\alpha'}](-z_s) = [r_\alpha](-z_s)$. These results can be summarized in the following theorem:

Proposition 4. $\forall Q \in \{0, 1, 2, 3\}$, $\lim_{\epsilon \to 0, \epsilon > 0} [r_{\alpha+\epsilon}]((z_s i^Q) - [r_{\alpha-\epsilon}](z_s i^Q)) = i^{Q+2}$

Let ψ_{z_0} denote the map such that if z is in $(z_0 i^Q (2n+1))$, with $n, Q \in \mathbb{N}$, then $\psi_{z_0}(z, p) = p + i^{Q+1}$ else $\psi_{z_0}(z, p) = p$. Thus, it can now be stated that for any z in $\mathbb{Z}[i], [r_{\alpha+}](z) = \psi_{z_s}(z, [r_{\alpha-}](z))$.

4.3   The Sketch of the Algorithm

With the previous statements, we now have the necessary elements to understand the scheme of the algorithm. The algorithm may pass through *all* possible rotation maps. This version of the algorithm simply updates a discretized rotation map from on angle to another and it calls a function *doextproc* that is user-specified that is notified at each update. We will later see a more complex version which explains how to make rotate incrementally an image without storing any copy of the original. See Algorithm 1.

Most of the subtilty of the algorithm actually proceeds from the strategic sequence of the hinge angles (See Section 5.4). In the real implementations, various technical details shall be solved: Some additional code has generallly to be added to take in account the fact that most programs store images into rectangular buffers. Also, this version of algorithm actually jumps over the hinge angles, and it does not compute the $[r_\alpha]|_m$ that is associated to the hinge angle it self. This is easily fixable: Apply ψ on the two first quadrants, then call the procedure for notification; the algorithm has then to terminate the transformation associated to the hinge angle by computed the next two quadrants, and to call once more the notification procedure.

Algorithm 1. incremental_rotation_via_map_and_notifications(doextproc)

1: $\mathcal{A}_{\mathcal{H}}|_m \leftarrow list\_sort\_hinge\_angles(m)$
2: $\forall p \in \mathbb{Z}[i]|_m, R[p] \leftarrow p$
3: $i \leftarrow 0$
4: **while** true **do**
5:    $(z_s, z_d) \leftarrow \mathcal{A}_{\mathcal{H}}|_m[i]$
6:    $i \leftarrow (i+1) \bmod \#\mathcal{A}_{\mathcal{H}}|_m$
7:    **for** $Q \leftarrow 0$ to 3 **do**
8:       $k \leftarrow 0$
9:       **while** $|(2k+1)z_s| < m$ **do**
10:          $op \leftarrow R((2k+1)z_s i^Q)$
11:          $np \leftarrow op + i^Q$
12:          $R((2k+1)z_s i^Q) \leftarrow np$
13:          $k \leftarrow k+1$
14:       **end while**
15:    **end for**
16:    $doextproc(R, z_s)$
17: **end while**

---

r denotes the radius of the image, op means old position, np stands for new position, the function "doextproc" is a parameter function that is called each time the image has been set up in according to a configuration that corresponds to the image of a discretized rotation.

Description of the Algorithm 1: The lines 3 to 6, and 17 of the algorithm corresponds to the main loop of the algorithm that passes though all the hinge angles. The lines 7 to 15 are responsible of updating the configuration. In each quadrant Q, we will consider all the points where a change occurs (that stand within the map, i.e. "$|(2k+1)z_s| < m$") . R is the configuration that contains the map $[r_\alpha]|_m$. Line 10, we recover the code point whose image is alterated, line 11 we compute its new image, and line 12 the updated position of the image is stored in the configuration.

4.4   Application to Rotation of Images

The previous version of the algorithm computes an image of the rotation map, however the discretized rotation is not intrisically bijective on $\mathbb{Z}^2$ and it is not suitable to compute incrementally rotations of an image without any copy of the original. This version of the algorithm takes as input an interger m that describes the size of the of the α - rotation map to be computed, as well as an image (an application from $\mathbb{Z}[i]|_m$ to Q_C where Q_C is a the set of colors on which the image is defined.

It is well-known that a point of the discretized rotation has never more than two antecedents by discretized rotation (See [4]). Thus, to create lossless discretized rotation, one natural idea is to store both antecedents when the function is not injective. This requires one additional layer: hence, an image in our algorithm shall be an element of $Q_C^{(\mathbb{Z}[i] \times \{0,1\})}$, where Q_C is a set of colors.

Algorithm 2. *incremental_rotation_of_an_image(img$_0$, doextproc)*

1: $\mathcal{A}_{\mathcal{H}}|_m \leftarrow list\_sort\_hinge\_angles(m)$

2: $\forall p \in \mathbb{Z}[i]|_m, R[p] \leftarrow p, L[p] \leftarrow 0$
3: $i \leftarrow 0$
4: **while** true **do**
5:     $(z_s, z_d) \leftarrow ((\mathcal{A}_{\mathcal{H}}|_m)[i])$
6:     $i \leftarrow (i+1) \mod \#\mathcal{A}_{\mathcal{H}}|_m$
7:     **for** $Q \leftarrow 0$ to 3 **do**
8:         k=0
9:         **while** $|(2k+1)z_s| < m$ **do**
10:             $p \leftarrow (2k+1)z_s$
11:             $op \leftarrow R[p]$
12:             $ol \leftarrow L[p]$
13:             $np \leftarrow op + i^Q$
14:             $img_1(np) \leftarrow img_{ol}(op)$
15:             $R[p] \leftarrow np; L[p] \leftarrow 1;$
16:             $alone \leftarrow true;$
17:             **for** $d \leftarrow 0$ to 3 **do**
18:                 **if** $R[p+i^d] = op$ **then**
19:                     $alone \leftarrow false; p' \leftarrow p+i^d$
20:                 **end if**
21:             **end for**
22:             **if** $\neg alone$ and $L[p'] = 1$ **then**
23:                 $L[p'] \leftarrow 0$
24:                 $t \leftarrow img_1[op]$
25:                 $img_1[op] \leftarrow img_0[op]$
26:                 $img_0[op] \leftarrow t$
27:             **end if**
28:             $alone \leftarrow true;$
29:             **for** $d \leftarrow 0$ to 3 **do**
30:                 **if** $R[p+i^d] = np$ **then**
31:                     $alone \leftarrow false; p' \leftarrow p+i^d$
32:                 **end if**
33:             **end for**
34:             **if** $\neg$ alone **then**
35:                 **if** $p' < p$ **then**
36:                     $L[p'] \leftarrow 1; L[p] \leftarrow 1$
37:                 **else**
38:                     $L[p'] \leftarrow 2; L[p] \leftarrow 0$
39:                     $t \leftarrow img_1[np]$
40:                     $img_1[np] \leftarrow img_0[np]$
41:                     $img_0[np] \leftarrow t$
42:                 **end if**
43:             **else**
44:                 $L[p] \leftarrow 0$
45:                 $t \leftarrow img_1[np]$
46:                 $img_1[np] \leftarrow img_0[np]$
47:                 $img_0[np] \leftarrow t$
48:             **end if**
49:         **end while**
50:     **end for**
51:     $doextproc(R, z_s, img_0)$
52: **end while**

The notations are the similar to the one used for previous algorithm. Additionally, img$_0$ and img$_1$ denotes the two layers of the picture. L is a part of the configuration that is used to memorize the destination layer of a pixel.

Practically, the code is modified such that: the dataspace on which rotation are computed can support up to two "colors" for each position of $\mathbb{Z}[i]|_m$. We will use an arbitrary order (the lexical order or anything fast to compute) to decide which pixel will stand on the layer 1. The details that have been added allows to compute the 2-layers discretized rotation. See Algorithm 2.

Description of the Algorithm 2: The global structure of the algorithm is similar to the Algorithm 1: We have an initialization (lines 1 and 2) and line 3,4,5,6, and 52, one main loop that passes through all the possible hinge angles. It contains a loop that updates that picture quadrant by quadrant. In each quadrant, all the points where a change occurs within the configuration are considered. However the update process is here more complicated: lines 11, 13 and 15 are the same as in the previous process. At line 15 we also make the assumption that when a point moves to another then the cell is busy and the incoming datas has to wait on the second layer ($L = 2$). From line 17 to 21, the algorithm looks at if there was another point p' that was in the the the cell that was left. If there was one, then this point if it was on layer 1 has to fallback on the layer 0 and has to be displayed (lines 22 to 27). The code from lines 29 to 48 does similar process but for the destination position. We can see on line 35 that the algorithm takes in account a predefined order $<$ to decide which pixel shall stay on the layer 0.

4.5   Analysis of the Complexity

This algorithm has really a different complexity compared to the usual algorithm for rotations. This algorithm is slower than the usual one if you aim to do only one rotation, but it is faster when incremental rotations are needed. Note, it is generally assumed that each processor uses a sine/cosine table which allows to compute sine and cosine very efficiently.

Space Complexity: The second part of the algorithm uses the amount of memory required to store a 2 layer image and the rotation map. Thus, the algorithms uses about $3Km^2$, (where K is the memory cost to store one color, or one vector (assuming that these two data-types can be stored with the same number of bytes K)). The traditional algorithm uses about $2Km^2$ bytes of memory. Thus, for this part the space requirements are of the same order $\mathcal{O}(m^2)$ and similar in terms of multiplicative factors. The first part of the algorithm requires $\mathcal{O}(m^3 log(m))$ bytes of memory to construct a b-tree of hinge angles in $\mathcal{A}_{\mathcal{H}}|_m$. This table can be computed once for all.

Time complexity: In the first part, the list-and-sort procedure for hinge angles needs $\mathcal{O}(m^3 log(m))$ operations. The time of one iteration of the loop in the second part is intrisically linear in m (the complexity of the user contributed function, doextproc can of course decrease these performances). Although, the main loop is called $\mathcal{O}(m^3)$ times and contains another loop, the algorithm also requires only $\mathcal{O}(m^3)$ operations: we update only m^2 pixels and each pixel crosses at most $4m$ times the dual of the grid. To compute all $\rho(\alpha)|_m$, for $\alpha \in [0, 2\pi[$ with traditional rotation algorithms would have needed $O(m^5)$ operations.

Note: In the implementation, we have presented the incremental rotation with an endless loop. The algorithm can be easily modified to stop when it reaches a specified angle.

4.6   Open-Source Implementation

We have written an implementation of this algorithm. The C++ code of this program can be downloaded from the following URL: `http://perso.ens-lyon.fr/bertrand.nouvel/transitive-rotations/`. The implementation relies on similar ideas but it is actually slightly different: For historical reasons and other reasons that are related to the more general project in which we plan to use this code, it uses two types of configurations. The first kind is $[r_\alpha]|_m$ and is used to compute efficiently images by discrete rotations. While C_α (See [4]) is used for some additional local checking.

5   Transitive Discretized Rotations

An map f_α depending of a real parameter α is *transitive* on a set X, if for any element $x \in X$, and for any couple $\alpha_0, \alpha_1 \in \mathbb{R}$, we have $f_{\alpha_0} \circ f_{\alpha_1}(x) = f_{\alpha_0 + \alpha_1}(x)$.

Our aim in this section is to formalize the existence of a transitive discretized rotation operator $[r_\alpha]$ that acts a set X, where X is a cartesian product of the possible positions in the image, and of a configuration that represents the internal state of the algorithm. Thus there should exists an onto morphism h from X to $\mathbb{Z}[i]$. Moreover, the operator should verify:

- $\forall \alpha, \beta \in \mathcal{A}, \forall x \in X, [r_\alpha] \circ [r_\beta](x) = [r_{\alpha+\beta}](x)$ (*transitvity*)
- *For any $\alpha \in \mathcal{A}$, for any $x \in X$, we have $h([r_\alpha](x)) = [r_\alpha](h(x)) + z_\epsilon$, with $|z_\epsilon| \leq 3$. In other terms, the operator may use some additional informations, but it has to provide almost the same result as a discretized rotation, and it may correct the "mistakes" made by the discretized rotation when iterated.*

Since the goal of this operator is to formalize what does the incremental algorithm inside the the operator $[r_\alpha]$. We will consider that X is either:

$$X_m = \mathbb{Z}[i] \times \{0,1\} \times \{C \in \mathbb{Z}[i]|_{m+1}^{\mathbb{Z}[i]|_m} | \exists \alpha \in \mathcal{A}, C = [r_\alpha]|_m\}$$

$$X_\infty = \mathbb{Z}[i] \times \{0,1\} \times \{C \in \mathbb{Z}[i]^{\mathbb{Z}[i]} | \exists \alpha \in \mathcal{A}, C = [r_\alpha]\}$$

Since the set of real numbers is a dense and non countable set, transitive rotation operators implies that $Card(X) = Card(\mathbb{R})$. Thus, rotations of finite objects cannot lead to transitive rotations. However, we will establish a similar but weaker notion, ϵ-quasi-transitivity: An map f_α depending of real parameter α is ϵ-*quasi-transitive* on a set X if there exists a constant $\epsilon \in \mathbb{R}$, such that for any element $C \in X$, and for any couple $\alpha_0, \alpha_1 \in \mathbb{R}$, we have $f_{\alpha_0} \circ f_{\alpha_1}(C) = f_{\alpha_2}(C)$, with $|\alpha_2 - \alpha_1 - \alpha_0| < \epsilon$.

This section starts with some considerations on the angles that represent the equivalence classes of angles induce by the equality through $\rho_m(\mathcal{A})$. We will then discuss of the application of the rotation algorithm to infinite configurations. Everything will be set up properly for defining an ϵ-quasi-transitive rotation operator on finite configuration. This operator will turn to be transitive on infinite configurations. This operator is completely equivalent to the algorithm that has been explained. Hence, it is just a formalization of its application, it is an attempt to build a discrete framework where transitive discrete rotations would be possible.

5.1 Equivalence Classes of Angles for $\rho|_m(\alpha)$

We denote by $I_{\mathcal{H}}|_m$ the open intervals of angles of the form $]\alpha_0, \alpha_1[$ where α_0 and α_1 are two consecutive hinge angles of $\mathcal{A}_{\mathcal{H}}|_m$. For any angle $\alpha \in \mathcal{A}$ and any $m \in \mathbb{N}$ either $\alpha \in \mathcal{A}_{\mathcal{H}}|_m$ either there exists a unique $I \in I_{\mathcal{H}}|_m$ such that $\alpha \in I$. To each element of $I_{\mathcal{H}}|_m$ of the form $]\alpha_0, \alpha_1[$, we associate an angle $\Gamma(]\alpha_0, \alpha_1[)$ which is equal to $\frac{\alpha_0 + \alpha_1}{2}$.

Let $\hat{\mathcal{A}}_{\mathcal{H}}|_m = \Gamma(\mathcal{I}_{\mathcal{H}}|_m) \cup \mathcal{A}_{\mathcal{H}}|_m$. The set $\hat{\mathcal{A}}_{\mathcal{H}}|_m$ has the following properties:

- $\forall \alpha \in \mathcal{A}, \exists \alpha' \in \hat{\mathcal{A}}_{\mathcal{H}}|_m$, such that $\rho_m(\alpha') = \rho_m(\alpha)$
- $\forall \alpha_1, \alpha_2 \in \hat{\mathcal{A}}_{\mathcal{H}}|_m, (\alpha_1 \neq \alpha_2) \Leftrightarrow \rho_m(\alpha') \neq \rho_m(\alpha_2)$.

As ρ_m can be seen as a bijection from $\hat{\mathcal{A}}_{\mathcal{H}}|_m$ to $\rho_m(\mathcal{A})$, we define the function ϕ_m that associates to each configuration $\rho_m(\alpha)$ an the element α' of $\hat{\mathcal{A}}_{\mathcal{H}}|_m$ such that $\rho_m(\alpha) = \rho_m(\alpha')$.

5.2 Rotations of the Entire Grid $\mathbb{Z}^2$

If we consider infinite configurations, then $\{C \in \mathbb{Z}[i]^{\mathbb{Z}[i]} | \exists \alpha \in \mathcal{A} | C = [r_\alpha]\}$ is in bijection with $\mathcal{A}$. More precisely, for any $\alpha, \alpha' \in \mathcal{A}$, $[r_\alpha] = [r'_\alpha]$ if and only if and $\alpha = \alpha'$. This is true since $lim_{m \mapsto \infty}(\phi_m \circ \rho_m(\alpha)) - \alpha \to 0$. See also [4].

Moreover, there exists a convergent process that consists in computing gradually each $\rho_m(\alpha)$ having m that increments gradually from 1 to ∞.

5.3 Transitivity and ϵ-Quasi-transitivity

The following lemma is necessary to conclude:

Lemma 5. $\mathcal{A}_{\mathcal{H}}$ is dense subset of $\mathcal{A}$.

Proof. Let $\alpha_\pi = atan(a/b)$ be a Pythagorean angle with $a, b, c \in \mathbb{Z}$ such that $a^2 + b^2 = c^2$, and c odd. We have $c e^{i\alpha_\pi} = a + ib$. Thus, $((a + ib) z_s, c z_d)$ is again a generating pair since $(a + ib) z_s$ is again a Gaussian integer. Also the parity of c justifies that $c z_d$ belongs again to $\mathcal{H}$. Since the Pythagorean angles forms a dense subset of $\mathcal{A}$ and that hinge angle are closed under addition of a Pythagorean angle, we have the necessary elements to conclude.

We now have the necessary elements to prove the existence of a transitive discrete rotation operator on infinite configurations.

The Section 4.2 and the rotation algorithm specify which map ψ shall be applied to transform one rotation map $[r_{\alpha^-}]$ into another $[r_{\alpha^+}]$ (with respect to the transformations that have previously been applied).

Let $m \in \mathbb{N}$, we define $[\![r_\alpha]\!]$ as the application from X_m to X_m that does the same as the previous algorithm. More formally, assume $x \in X_m$ and $x = (z, l, R)$, where z is a point we consider; l is the current layer it relies on, and R is the initial rotation map "from which we will restart". There exists an angle $\alpha_{d_0} = \phi_m(R)$, such that $R = \rho_m(\alpha_{d_0})$. We put $[\![r_\alpha]\!](x) = (z', l', R')$. The application of the algorithm starting from R, and for a rotation of α radians, consists in applying all the maps ψ that are associated with the hinge angles that lie in-between α_{d_0} and α_{d_1}, with $\alpha_{d_1} = \alpha_{d_0} + \alpha$. Thus R' is equal to $(\psi_{z_{s,1}} \circ \psi_{z_{s,2}} \circ ...)(R)$, where $\{z_{s,i}\}$ is the sequence of the source points issued from the decreasing sequence of the hinge angles that stand in-between α_{d_0} and α_{d_1}. Also, we will have $z' = R'(y)$ where y is the unique Gaussian integer such that $R(y) = z$ and such that the destination layer of y is l. We define l' as the destination layer of z' (i.e. $[r_\alpha](y)$) according to the order that has been chosen in the algorithm. With this definition, $[r_\alpha]$ is ϵ - quasi-transitive on X_m, in the following meaning: $[r_\alpha] \circ [r_\beta] = [r_{\alpha+\beta+\epsilon(\alpha,\beta,m)}]$, where $\epsilon(\alpha, \beta, m)$ is the error that is made due to the fact the hinge angles of order at most m are a finite set. This error is bounded by the maximum distance between two consecutive hinge angles in $\mathcal{A}_{\mathcal{H}}|_m$.

From this discussion, we conclude that $[r_\alpha]$ can be interpreted as a transitive rotation operator on infinite configurations (on X_∞): the density of the hinge angles (Lemma 5) implies that $\epsilon(\alpha, \beta, m) = 0$ when m goes to infinity.

5.4   Additional Remarks

Center of the rotation. One important limitation of the operator $[r_\alpha]$ is that it can only been applied successively on rotations around the same center.

Remarks Dealing with Hinge Angles. Let's note that the hinge angles have many arithmetical properties which are beyond the scope of this paper. For instance, by the mean of arithmetical arguments, it can be stated that if α is a hinge angle then $k\alpha$ is a hinge angle for $k \neq 0 \bmod 3$ (by recurrence $k\alpha$ is a hinge angle then $(k + 3)\alpha$ is also a hinge angle) , if α is a $z_s(\alpha) \neq e^{ik\frac{\pi}{2}}$ then its sine and its cosine are rationally dependant, $\{\mathbb{Z}e^{i\alpha}\}$ forms span of $2gcd(x,y)$ lines of slopes $\frac{x}{y}$ in $(\mathbb{R}/\mathbb{Z})[i]$... Note also that the set of hinge angles that has been obtained and studied here is dependant of the discretization that has been used (and of the real center of the rotation).

Questions Related to the Computation of the Next Hinge Angle in $\mathcal{A}_{\mathcal{H}}|_m$**.** One of the remaining problem of our algorithm is that the only way to compute efficiently the successor angle is now, up to our current knowledge, to construct and use a table of the hinge angles.

The question of the structure of hinge angles seems to be linked with some famous arithmetic and number theory results, such as the Jacobi two-squares Theorem. Also, the hinge angles can be seen as the subset of the more general set of angles: $\alpha \in \mathbb{R}$ that verifies: $acos(\alpha) + b\sin(\alpha) = c$, with $a, b, c \in \{-m, ..., m\}$. Although this equation seems familiar due to its similarity to formulas for rotations, very few is known about its integer solutions.

6   Conclusions and Perspectives

This algorithm for discretized rotations has the numerous useful properties: Since it is an exact algorithm, that is valid on any size of data, it returns the same result as the direct discretized rotation. However, this algorithm proceeds incrementally, and it will be useful in any procedure that needs to do some checking along the rotation process. It is faster to compute than naive incremental rotations that would iterate the discretized rotation. More precisely, it leads to $\mathcal{O}(n^3)$ updates of pixel instead of $\mathcal{O}(n^5)$. As the discretized rotations, the process is very accurate in terms of spatial errors: The error made on the position of the image of point is bounded by $\frac{\sqrt{2}}{2}$ (which is optimal). The fact that it does not use any sine nor any cosine also contributes to the accuracy of the algorithm.

A similar study shall also be tractable for the 3-shears rotations. The main advantage that motivates this remark is that these rotations are natively bijective, thus apart the configuration, no additional layer would be required to implement incremental 3-shear rotations.

Finally, although the hinge angles are sortable using only integer arithmetic, the procedure is slowed down the necessity to use large integer numbers. To speed up the algorithm, as suggested in 5.4, a trend for ongoing research is to find an efficient procedure that computes the successor of an hinge angle in $\mathcal{A}_{\mathcal{H}}|_m$.

References

1. Klette, R., Rosenfeld, A.: Digital Geometry: Geometric Methods for Digital Picture Analysis. Morgan Kaufmann Publishers Inc. (2004)
2. Réveillès, J.P.: Géométrie discrète, calcul en nombres entiers, et algorithmique. Docent(Thèse d'État) (1991) Université Louis Pasteur.
3. Andrès, E.: Discrete Circles, and Discrete Rotations. PhD thesis, Université Louis Pasteur (1992)
4. Nouvel, B., Rémila, E.: Configurations induced by discrete rotations: Periodicity and quasiperiodicity properties. Discrete Applied Mathematics **127** (2005) 325–343
5. Amir, A., Butman, A., Crochemore, M., Landau, G.M., Schaps, M.: Two-dimensional pattern matching with rotations. Theoretical Computer Sciences **314** (2004) 173–187
6. Poggiaspalla, G.: Autosimilarité dans les Systèmes Isométriques par Morceaux. PhD thesis, Université d'Aix-Marseille II (Luminy) (2003)
7. Nouvel, B., Rémila, E.: Characterization of bijective discretized rotations. In: Proceedings of the International Workshop on Combinatorial Image Analysis (IWCIA). Number 3322 in LNCS (2004)
8. Voss, K.: Discrete Images, Objects and Functions in $\mathbb{Z}^n$. Springer, Berlin (1993)
9. Nouvel, B., Remila, E.: On colorations induced by discrete rotations. In: Proceedings of Discrete Geometry for Computer Imagery (DGCI). Number 2886 in LNCS (2003)

Discrete Homotopy of a Closed k-Surface

Sang-Eon Han

Department of Computer and Applied Mathematics
Honam University, Gwangju, 506-714, Korea
sehan@honam.ac.kr

Abstract. Let $SC_{k_i}^{n_i,l_i}$ be a simple closed k_i-curve in $\mathbf{Z}^{n_i}$ with l_i elements, $i \in \{1,2\}$. After doing a $(3^{n_1+n_2} - 1)$-homotopic thinning of $SC_{k_1}^{n_1,l_1} \times SC_{k_2}^{n_2,l_2}$ to obtain a closed $(3^{n_1+n_2} - 1)$-surface, we calculate the $(3^{n_1+n_2} - 1)$-fundamental group of $SC_{k_1}^{n_1,l_1} \times SC_{k_2}^{n_2,l_2}$ by the use of some properties of an $(8, 3^{n_1+n_2} - 1)$-covering.

Keywords: Digital image, Closed k-surface, Digital homotopy, (Local) (k_0, k_1)-homeomorphism, Digital (k_0, k_1)-covering, Digital k-fundamental group, k-homotopic thinning.

AMS Classification: 52XX, 52B05, 52Cxx, 57M05, 55P10, 57M10.

1 Introduction

Several approaches have been proposed for the study of topological properties of a binary digital image (X, k). Precisely, the digital(or discrete) topological approach, the connected order topological space(briefly $COTS$), the complex cell approach, and the Alexandroff topological approach were established[1, 2, 14]. In this paper we use the discrete topological approach to study a digital image with one of the general k-adjacency of $\mathbf{Z}^n$. For a subset $X \subset \mathbf{Z}^n$, the discrete topological subspace (X, D_X) is induced from $(\mathbf{Z}^n, D)$. Then consider a k_0-adjacency relation on $(X, D_X) \subset \mathbf{Z}^{n_0}$ and a k_1-adjacency relation on $(Y, D_Y) \subset \mathbf{Z}^{n_1}$.

The study of k-surfaces in $\mathbf{Z}^3$ has proceeded in order to find their discrete topological characterizations, such as the 3D Jordan theorem, a strong homotopy, local property of a strong 18- or 26- surface, a thinning algorithm within a digital Jordan surface, the digital k-topological number, and the digital k-linking number [1, 2, 15]. Moreover, a generalized simple closed k-surface was recently established in [6] with the restricted k-adjacency relations of $\mathbf{Z}^n$, $n \geq 3$, which is a generalization of a Malgouyres' 18-surface. Further, its digital topological properties were studied in relation with the digital connected sum in [6], the topological number, a minimal simple closed 18-surface and so forth [8]. Recently, the notion of (equivalent) (k_0, k_1)-covering was established in [7, 9, 10, 11], which is essentially used to calculate the digital fundamental group of some digital image.

In this paper we find some discrete topological property of some discrete k-surfaces in $\mathbf{Z}^n$ by the digital fundamental group, a k-homotopic thinning, and

U. Eckardt et al. (Eds.): IWCIA 2006, LNCS 4040, pp. 214–225, 2006.

a digital (k_0, k_1)-covering. Finally, each digital image (X, k) is assumed to be k-connected.

2 Preliminaries

For a positive integer m with $1 \leq m \leq n$, two distinct points $p = (p_1, p_2, \cdots, p_n)$, $q = (q_1, q_2, \cdots, q_n) \in \mathbf{Z}^n$, we say that p and q are adjacent according to m if

(1) there are at most m indices i such that $|p_i - q_i| = 1$; and
(2) for all other indices i such that $|p_i - q_i| \neq 1$, $p_i = q_i$.

In the following, the statement consisting of the two conditions (1) and (2) is called $(CON\star)$ [4, 5, 6, 7, 8, 9, 10, 11].

Precisely, as the generalization of the commonly used 4- and 8-adjacency of $\mathbf{Z}^2$; and 6-, 18- and 26-adjacency of $\mathbf{Z}^3$ in [1], given a natural number m in $(CON\star)$ with $1 \leq m \leq n$, m determines each of the general k-adjacency relations of $\mathbf{Z}^n$ in terms of $(CON\star)$ [4, 5, 6, 7, 8, 9, 10, 11] as follows.

$$k \in \{2n(n \geq 1), 3^n - 1(n \geq 2), 3^n - \sum_{t=0}^{r-2} C_t^n 2^{n-t} - 1(2 \leq r \leq n-1, n \geq 3)\}. \cdot (2-1)$$

For example, $(n, m, k) \in \{(2, 1, 4), (2, 2, 8); (3, 1, 6), (3, 2, 18), (3, 3, 26); (4, 1, 8),$ $(4, 2, 32), (4, 3, 64), (4, 4, 80); (5, 1, 10), (5, 2, 50), (5, 3, 130), (5, 4, 210), (5, 5, 242);$ $(6, 1, 12), (6, 2, 72), (6, 3, 232), (6, 4, 472), (6, 5, 664), (6, 6, 728)\}$ [4, 5, 6, 7, 8, 9, 10, 11].

In this paper the pair (X, k) is considered in a usual digital picture $(\mathbf{Z}^n, k, \bar{k}, X)$ in [1, 2, 3, 6], which is called a *digital image*, where $(k, \bar{k}) \in \{(k, 2n), (2n, 3^n - 1)\}$. Further, $k \neq \bar{k}$ except the case $(\mathbf{Z}, 2, 2, X)$ in $\mathbf{Z}$ owing to the *digital k-connectivity paradox*[13]. For $a, b \in \mathbf{Z}$ with $a \lneq b$, the set $[a, b]_{\mathbf{Z}} = \{n \in \mathbf{Z} | a \leq n \leq b\}$ is called a *digital interval*. A k-path from x to y in X is assumed to be a sequence $(x = x_0, x_1, x_2, \cdots, x_{m-1}, x_m = y)$ in X such that each point x_i is k-adjacent to x_{i+1} for $m \geq 1$ and $i \in [0, m-1]_{\mathbf{Z}}$. Then the number m is called the *length* of this path. If $x_0 = x_m$, then the k-path is said to be *closed*. For a digital image (X, k), two distinct points $x, y \in X$ are k-connected if there is a k-path from x to y in X, and if any two distinct points in X are k-connected, then X is called *k-connected*. For an adjacency relation k, a *simple k-path* with m elements in $\mathbf{Z}^n$ is assumed to be a sequence $(x_0, x_1, x_2, \cdots, x_{m-1}) \subset \mathbf{Z}^n$ such that x_i and x_j are k-adjacent if and only if either $j = i+1$ or $i = j+1$ [3, 6, 7, 8].

Now the following notion of digital continuity is efficiently used to study digital k-curves, k-surfaces in $\mathbf{Z}^n$[6, 8], and the digital (k_0, k_1)-covering in [5, 6, 7, 8, 9, 10, 11].

Proposition 1. *[4, 5, 6, 7, 8, 9, 10, 11] Let $(X, k_0) \subset \mathbf{Z}^{n_0}$ and $(Y, k_1) \subset \mathbf{Z}^{n_1}$ be digital images. A function $f : X \rightarrow Y$ is (k_0, k_1)-continuous if and only if for every $x_0 \in X, \varepsilon \in \mathbf{N}$, and $N_{k_1}(f(x_0), \varepsilon) \subset Y$, there is $\delta \in \mathbf{N}$ such that the corresponding $N_{k_0}(x_0, \delta) \subset X$ satisfies $f(N_{k_0}(x_0, \delta)) \subset N_{k_1}(f(x_0), \varepsilon)$, where $N_k(x_0, \varepsilon) := \{x \in X | \ l_k(x_0, x) \leq \varepsilon\} \cup \{x_0\}$, and $l_k(x_0, x)$ is the length of a shortest simple k-path from x_0 to x in X.*

Indeed, a *simple closed k-curve* with l elements in $X \subset \mathbf{Z}^n$ is the image of a $(2,k)$-continuous function $f : [0, l-1]_\mathbf{Z} \to X$ such that $f(i)$ and $f(j)$ are k-adjacent if and only if either $j = i + 1 (mod\, l)$ or $i = j + 1(mod\, l)$[3]. Now we use the notation $SC_k^{n,l}$ which can be assumed to be a sequence $(c_i)_{i \in [0, l-1]_\mathbf{Z}}$ with $f(i) = c_i$.

For a digital image (X, k) and its subset (A, k), we call (X, A) a *digital image pair* with k-adjacency. Furthermore, if A is a singleton set $\{x_0\}$, then (X, x_0) is called a *pointed digital image* [3]. A *relative digital homotopy* to a subset $A \subset X$, motivated by the *pointed digital homotopy* in [3], is efficiently used to study a digital k-surface in $\mathbf{Z}^n, n \geq 3$, and is essentially used to do a k-homotopic thinning.

Definition 1. *[6] Let (X, k_0) in $\mathbf{Z}^{n_0}$ and (Y, k_1) in $\mathbf{Z}^{n_1}$ be digital images and let (A, k_0) be a subset of (X, k_0). Let $f, g : X \to Y$ be (k_0, k_1)-continuous functions. Suppose there exist $m \in \mathbf{N}$ and a function $F : X \times [0, m]_\mathbf{Z} \to Y$ such that*

- *for all $x \in X, F(x, 0) = f(x)$ and $F(x, m) = g(x)$;*
- *for all $x \in X$, the induced function $F_x : [0, m]_\mathbf{Z} \to Y$ given by $F_x(t) = F(x, t)$ for all $t \in [0, m]_\mathbf{Z}$ is $(2, k_1)$-continuous;*
- *for all $t \in [0, m]_\mathbf{Z}$, the induced function $F_t : X \to Y$ given by $F_t(x) = F(x, t)$ for all $x \in X$ is (k_0, k_1)-continuous;*
- *for all $t \in [0, m]_\mathbf{Z}$, the induced map F_t on A is fixed, i.e., $F_t(x) = x$ for $x \in A$.*

Then we call F a (k_0, k_1)-homotopy relative to A between f and g, and we say f and g are (k_0, k_1)-homotopic rel A in Y.

Especially, if $A = \{x_0\} \subset X$ in Definition 1, then we say that F is a pointed (k_0, k_1)-homotopy at $\{x_0\}$[3]. When f and g are pointed (k_0, k_1)-homotopic in Y, we use the notation $f \simeq_{(k_0, k_1)} g$. Furthermore, we say that a digital image X is *k-contractible* if $1_X \simeq_k c_{\{x_0\}}$, where $c_{\{x_0\}}$ is a constant map for some point $x_0 \in X$[3]. We say that a (k_0, k_1)-continuous function $f : X \to Y$ is *k_1-nullhomotopic* in Y if f is k_1-homotopic in Y to a constant function $c_{\{y_0\}}$ for some $y_0 \in Y$[3]. Now we use the pointed (k_0, k_1)-homotopy for the digital fundamental group.

In order to understand the pointed digital k-homotopy in relation with the digital k-fundamental group, we need the notion of *trivial extension* in [3]: Precisely, if $m_f \leq m_{f'}$, we can obtain a trivial extension of a loop $f : [0, m_f]_\mathbf{Z} \to X$ to a loop $f' : [0, m_{f'}]_\mathbf{Z} \to X$ given by

$$f'(t) = \begin{cases} f(t) & \text{if } 0 \leq t \leq m_f; \\ f(m_f) & \text{if } m_f \leq t \leq m_{f'}. \end{cases}$$

Due to the pointed digital homotopy for a pointed digital image (X, x_0) and the notion of *trivial extension* in [3], the *(digital) k-fundamental group* $\pi_1^k(X, x_0)$ was originally established in [3, 12] with the Khalimsky operation in [12]. To be specific, let $F_1^k(X, x_0) = \{f \,|\, f$ is a $k - loop$ based at $x_0\}$. For members $f : [0, m_f]_\mathbf{Z} \to X$, $g : [0, m_g]_\mathbf{Z} \to X$ of $F_1^k(X, x_0)$, we obtain the map in [12] $f * g : [0, m_f + m_g]_\mathbf{Z} \to X$ defined by

$$f * g(t) = \begin{cases} f(t) & \text{if} \quad 0 \le t \le m_f; \\ g(t - m_f) & \text{if} \quad m_f \le t \le m_f + m_g. \end{cases}$$

If x_0 and x_1 belong to the same k-connected component of X, then $\pi_1^k(X, x_0)$ and $\pi_1^k(X, x_1)$ are proved isomorphic to each other[3]. Thus, for a k-connected digital image X, we need not fix a base point for the k-fundamental group.

In addition, for digital images (X, k_0) in $\mathbf{Z}^{n_0}$ and (Y, k_1) in $\mathbf{Z}^{n_1}$, a map $h : X \to Y$ is called a (k_0, k_1)-homeomorphism in [4, 5, 6, 7, 8, 9, 10] if h is (k_0, k_1)-continuous and bijective and further, $h^{-1} : Y \to X$ is (k_1, k_0)-continuous. Then we use the notation $X \approx_{(k_0, k_1)} Y$. If $k_0 = k_1$, we call it a k_0-homeomorphism [3, 4, 5, 6, 7, 8, 9, 10]. Besides, if X is pointed k-contractible, then $\pi_1^k(X, x_0)$ is proved to be trivial[3]. Further, a pointed k-connected digital image (X, x_0) is called *simply k-connected* if $\pi_1^k(X, x_0)$ is a trivial group[9].

By Definition 1 we have the following notion of strong k-deformation retract in [4, 11] which is useful for a k-homotopic thinning of a torus-like digital image $SC_{k_1}^{n_1,l_1} \times SC_{k_2}^{n_2,l_2}$ in Section 3.

Definition 2. *[4, 11] Suppose that (X, A) is a digital image pair with k-adjacency and $i : A \to X$ is the inclusion map, A is called a k-retract of X if and only if there is a k-continuous map $r : X \to A$ such that $r(a) = a$ for all $a \in A$. Then the map r is called a k-retraction of X onto A. For a digital image pair (X, A) with k-adjacency, we say that X is a strong k-deformation retract onto A if there is a k-homotopy relative to A $F : X \times [0, m]_{\mathbf{Z}} \to X$ such that $F(x, m)$ is a k-retract onto A for $x \in X$.*

Indeed, for a digital image (X, k), a *simple k-point* is one whose removal does not change the digital topological property of (X, k) up to a k-connectivity [1]. Then we usually say that a *k-thinning* is the processing of deleting some simple k-points from a digital image (X, k).

Definition 3. *[4, 11] For a digital image (X, k), we can delete some points from X in terms of a strong k-deformation retract. In other words, if (A, k) is a k-deformation retracted subimage of (X, k), then we say that A is a k-homotopically thinned digital image from X. Further, this processing is called a k-homotopic thinning.*

Indeed, there is a big difference between a usual k-thinning and the current k-homotopic thinning in [11] and Example 4(1).

Theorem 1. *[4, 11] If (A, x_0) is a strong k-deformation retract of (X, x_0), then $\pi_1^k(X, x_0)$ is isomorphic to $\pi_1^k(A, x_0)$.*

3   Closed $(3^{n_1+n_2} - 1)$-Surface Structure of the $(3^{n_1+n_2} - 1)$-Homotopically Thinned Digital Image from $SC_{k_1}^{n_1,l_1} \times SC_{k_2}^{n_2,l_2}$

We now need some terminologies in order to study a digital k-surface. A point $x \in X$ is called a *k-corner* if x is k-adjacent to two and only two points $y, z \in X$

such that y and z are k-adjacent each other[2]. Further, the k-corner x is called *simple* if y, z are not k-corners and if x is the only point k-adjacent to both y, z [1]. X is called a *generalized simple closed k-curve* if what is obtained by removing all simple k-corners of X is a simple closed k-curve in [2]. For a k-connected digital image (X, k) in $\mathbf{Z}^3$, we recall $|X|^x = N_{26}^*(x) \cap X$, $N_{26}^*(x) = \{x'|x \text{ and } x'$ are 26-adjacent$\}$[1, 2]. More generally, for a k-connected digital image (X, k) in $\mathbf{Z}^n, n \geq 3$, we can state $|X|^x = N_{3^n-1}^*(x) \cap X$, where $N_{3^n-1}^*(x) = \{x'|x \text{ and } x'$ are $(3^n - 1)$-adjacent$\}$. In other words, $|X|^x = N_{3^n-1}(x, 1) - \{x\}$ in $\mathbf{Z}^n$[6].

Indeed, the essential notions above allow us to have a generalized closed k-surface with one of the k-adjacency relations in (2-1) as the generalization of the digital k_1-surface in $\mathbf{Z}^3$[15, 17], where $k_1 \in \{6, 18, 26\}$. In this paper we will not consider the *orientability* of a closed k-surface in [17]. Indeed, a closed k-surface in $\mathbf{Z}^n$ was studied with the restricted $(3^n - 2^n - 1)$-adjacency in [6] and a simple closed k-surface in $\mathbf{Z}^n$ was introduced with one of the general k-adjacency relations in (2-1), where $k \neq 2n$. But it is too restrictive to define a closed $2n$-surface and a closed k-surface not simple, where $k \neq 3^n - 2^n - 1$.

Thus we now study the more generalized criterion of a closed k-surface in $\mathbf{Z}^n$, where the k-adjacency is taken from (2-1).

Definition 4. *[11] Let (X, k) be a digital image in $\mathbf{Z}^n, n \geq 3$, and $\bar{X} = \mathbf{Z}^n - X$. Then X is called a closed k-surface if it satisfies the following.*

(1) In case that $(k, \bar{k}) \in \{(k, 2n), (2n, 3^n - 1)\}$, where the k-adjacency is taken from (2-1) with $k \neq 3^n - 2^n - 1$, then
(a) for each point $x \in X$, $|X|^x$ has exactly one k-component k-adjacent to x;
(b) $|\bar{X}|^x$ has exactly two $\bar{k}$-components $\bar{k}$-adjacent to x; we denote by C^{xx} and D^{xx} these two components; and
(c) for any point $y \in N_k(x) \cap X$, $N_{\bar{k}}(y) \cap C^{xx} \neq \phi$ and $N_{\bar{k}}(y) \cap D^{xx} \neq \phi$.
Further, if a closed k-surface X does not have a simple k-point, then X is called simple.
(2) In case that $(k, \bar{k}) = (3^n - 2^n - 1, 2n)$, then
(a) X is k-connected,
(b) for each point $x \in X$, $|X|^x$ is a generalized simple closed k-curve.
Further, if the image $|X|^x$ is a simple closed k-curve, then the closed k-surface X is called simple.

The current cases (1) and (2) of Definition 4 are respectively generalizations of the closed k-surface in [17] and a closed 18-surface in [15], where $k \in \{6, 26\}$ with the pair $(k, \bar{k}) \in \{(6, 26), (26, 6)\}$. Obviously, we see that each closed 6-surface is simple.

Example 1. According to the criterion of a closed (simple) k-surface in Definition 4, examine the minimal simple closed 6-surface MSS_6 and two types of minimal simple closed 18-surfaces MSS_{18} and MSS_{18}' in [8].

Consider a simple closed k_i-curve with l_i elements in $\mathbf{Z}^{n_i}$, denoted by $SC_{k_i}^{n_i, l_i}$, $i \in \{1, 2\}$. Assume that $SC_{k_1}^{n_1, l_1} := (c_i)_{i \in [0, l_1 - 1]_{\mathbf{Z}}}$ and $SC_{k_2}^{n_2, l_2} := (d_j)_{j \in [0, l_2 - 1]_{\mathbf{Z}}}$

as sequences. Then we remind that the number m_i is always taken from the k_i-adjacency by $(CON\star)$, $i \in \{1,2\}$. Let us consider the following closed 4-, 8-curves in $\mathbf{Z}^2$ and closed 18-curves in $\mathbf{Z}^3$ used later in this paper.

$$
\left\{
\begin{array}{l}
SC_4^{2,8} := ((0,0),(0,1),(0,2),(1,2),(2,2),(2,1),(2,0),(1,0)),\\
SC_8^{2,4} := ((0,0),(1,1),(2,0),(1,-1)),\\
SC_8^{2,6} := ((0,0),(1,1),(1,2),(0,3),(-1,2),(-1,1)),\text{ and}\\
SC_{18}^{3,6} := ((0,0,0),(1,0,1),(1,1,2),(0,2,2),(-1,1,2),(-1,0,1)).
\end{array}
\right\} \cdots (3-1)
$$

Indeed, a torus-like digital image $SC_{k_1}^{n_1,l_1} \times SC_{k_2}^{n_2,l_2} \subset \mathbf{Z}^{n_1+n_2}$ is presented by an $(l_1 \times l_2)$-matrix with the component $t_{ij} := (c_i, d_j) \in SC_{k_1}^{n_1,l_1} \times SC_{k_2}^{n_2,l_2}$, which is indeed an $(n_1 + n_2)$-ordered pair in $\mathbf{Z}^{n_1+n_2}$ (see Example 2).

Example 2. A closed 50-surface $SC_8^{2,6} \times SC_{18}^{3,6} \subset \mathbf{Z}^5$ is simple but is not 50-contractible. Further, it is also a closed k-surface not 210-simple in $\mathbf{Z}^5$, where $k \in \{130, 210, 242\}$.

Proof: First, for $SC_8^{2,6}$ and $SC_{18}^{3,6}$ in (3-1), $SC_8^{2,6} \times SC_{18}^{3,6} := DT_{50} \subset \mathbf{Z}^5$ is presented in terms of the following row T_i, $i \in [0,5]_{\mathbf{Z}}$.

$$
\begin{pmatrix}
(0,0,0,0,0) & (0,0,1,0,1) & (0,0,1,1,2) & (0,0,0,2,2) & (0,0,-1,1,2) & (0,0,-1,0,1)\\
(1,1,0,0,0) & (1,1,1,0,1) & (1,1,1,1,2) & (1,1,0,2,2) & (1,1,-1,1,2) & (1,1,-1,0,1)\\
(1,2,0,0,0) & (1,2,1,0,1) & (1,2,1,1,2) & (1,2,0,2,2) & (1,2,-1,1,2) & (1,2,-1,0,1)\\
(0,3,0,0,0) & (0,3,1,0,1) & (0,3,1,1,2) & (0,3,0,2,2) & (0,3,-1,1,2) & (0,3,-1,0,1)\\
(-1,2,0,0,0) & (-1,2,1,0,1) & (-1,2,1,1,2) & (-1,2,0,2,2) & (-1,2,-1,1,2) & (-1,2,-1,0,1)\\
(-1,1,0,0,0) & (-1,1,1,0,1) & (-1,1,1,1,2) & (-1,1,0,2,2) & (-1,1,-1,1,2) & (-1,1,-1,0,1)
\end{pmatrix}
:=
\begin{pmatrix}
T_0\\ T_1\\ T_2\\ T_3\\ T_4\\ T_5
\end{pmatrix}
\cdot (3-2)
$$

Indeed, T_i and $T_{i+1(mod\,6)}$ are 50-adjacent and further, T^j and $T^{j+1(mod\,6)}$ are 50-adjacent, T^j stands for the j-th column of DT_{50}. Further, we see that DT_{50} satisfies the case (1) of Definition 4, which implies to be a closed 50-surface.

Similarly, we see that DT_{50} is also a closed k-surface, $k \in \{130, 242\}$.

Moreover, for each point $p \in DT_{50}$, since $|DT_{50}|^p = N_{242}(p,1) - \{p\}$ is a generalized simple closed 210-curve, DT_{50} is also proved a closed 210-surface, but DT_{50} is not 210-simple because for some point $x \in DT_{50}$, $|DT_{50}|^x$ is not a simple closed 210-curve.

Next, since neither of any 50-loops on $\{c_i\} \times SC_{18}^{3,6} \subset SC_8^{2,6} \times SC_{18}^{3,6}$ nor each 50-loop on $SC_8^{2,6} \times \{d_j\} \subset SC_8^{2,6} \times SC_{18}^{3,6}$ is 50-nullhomotopic in $SC_8^{2,6} \times SC_{18}^{3,6}$, $SC_8^{2,6} \times SC_{18}^{3,6}$ can not be 50-contractible. $\square$

In relation with Example 2 we obtain a closed k-surface from $SC_{k_1}^{n_1,l_1} \times SC_{k_2}^{n_2,l_2}$ by a strong k-deformation retract even if $m_1 \leq m_2$.

Theorem 2. $SC_{k_1}^{n_1,l_1} \times SC_{k_2}^{n_2,l_2}$ can be $(3^{n_1+n_2}-1)$-homotopically thinned to be a closed $(3^{n_1+n_2}-1)$-surface.

Proof: $SC_{k_1}^{n_1,l_1} \times SC_{k_2}^{n_2,l_2} := (c_i)_{i \in [0,l_1-1]_{\mathbf{Z}}} \times (d_j)_{j \in [0,l_2-1]_{\mathbf{Z}}}$ can be considered as an $(l_1 \times l_2)$-matrix with the component $t_{ij} := (c_i, d_j)$, $i \in [0, l_1-1]_{\mathbf{Z}}, j \in [0, l_2-1]_{\mathbf{Z}}$ so that it can be assumed to be the set $\cup_{i \in [0,l_1-1]_{\mathbf{Z}}} T_i = \cup_{j \in [0,l_2-1]_{\mathbf{Z}}} T^j$ in terms of the following row and column $T_i = \{c_i\} \times SC_{k_2}^{n_2,l_2}$ and $T^j = SC_{k_1}^{n_1,l_1} \times \{d_j\}$.

Now for $SC_{k_1}^{n_1,l_1} \times SC_{k_2}^{n_2,l_2}$, assume that $m_1 \leq m_2$. We say that $T_i = (t_{ij})$ $_{j\in[0,l_2-1]\mathbf{z}}$ and $T_{i+2(mod\,l_1)} = (t_{i+2(mod\,l_1)j})_{j\in[0,l_2-1]\mathbf{z}}$ are $(3^{n_1+n_2}-1)$-connected if t_{ij} and $t_{i+2(mod\,l_1)j}$ are $(3^{n_1+n_2}-1)$-connected for each $j \in [0,l_2-1]\mathbf{z}$.

If T_i and $T_{i+2(mod\,l_1)}$ are $(3^{n_1+n_2}-1)$-connected, then delete $T_{i+1(mod\,l_1)}$ from $\cup_{i\in[0,l_1-1]\mathbf{z}}T_i = SC_{k_1}^{n_1,l_1} \times SC_{k_2}^{n_2,l_2}$. For each $i \in [0,l_1-1]\mathbf{z}$, we do this processing consecutively. Thus we obtain the submatrix

$$\cup_{i\in[0,l_1'-1]\mathbf{z}}T_i \subset SC_{k_1}^{n_1,l_1} \times SC_{k_2}^{n_2,l_2}, \dots\dots\dots\dots\dots\dots\dots\dots\dots(3-3)$$

where $\{\{c_i\} \times SC_{k_2}^{n_2,l_2} := T_i | i \in [0,l_1'-1]\mathbf{z}\}$ is a subset of $\{\{c_i\} \times SC_{k_2}^{n_2,l_2} := T_i | i \in [0,l_1-1]\mathbf{z}\}$, $(c_i)_{i\in[0,l_1'-1]\mathbf{z}}$ is a subsequence of $(c_i)_{i\in[0,l_1-1]\mathbf{z}}$ and $l_1' = l_1-$ the cardinal number of the set $\{i+1(mod\,l_1)|$ T_i and $T_{i+2(mod\,l_1)}$ are $(3^{n_1+n_2}-1)-$ connected, $i \in [0,l_1-1]\mathbf{z}\}$.

Next, write the submatrix $\cup_{i\in[0,l_1'-1]\mathbf{z}}T_i$ in (3-3) by $\cup_{j\in[0,l_2-1]\mathbf{z}}T^j$, where $T^j := (c_i,d_j)_{i\in[0,l_1'-1]\mathbf{z}}$, $j \in [0,l_2-1]\mathbf{z}$. Then we say that $T^j = (c_i,d_j)_{i\in[0,l_1'-1]\mathbf{z}}$ and $T^{j+2(mod\,l_2)} = (c_i,d_{j+2(mod\,l_2)})_{i\in[0,l_1'-1]\mathbf{z}}$ are $(3^{n_1+n_2}-1)$-connected if (c_i,d_j) and $(c_i,d_{j+2(mod\,l_2)})$ are $(3^{n_1+n_2}-1)$-connected for all $i \in [0,l_1'-1]\mathbf{z}$. If T^j and $T^{j+2(mod\,l_2)}$ are $(3^{n_1+n_2}-1)$-connected, then delete $T^{j+1(mod\,l_2)}$ from the matrix $\cup_{j\in[0,l_2-1]\mathbf{z}}T^j$. For each $j \in [0,l_2-1]\mathbf{z}$, do this processing consecutively so that we obtain the submatrix

$$\cup_{j\in[0,l_2'-1]\mathbf{z}}T^j := \{(c_i,d_j)|i \in [0,l_1'-1]\mathbf{z}, j \in [0,l_2'-1]\mathbf{z}\}, \dots\dots(3-4)$$

where $l_2' = l_2-$ the cardinal number of the set $\{j+1(mod\,l_2)|$ T^j and $T^{j+2(mod\,l_2)}$ are $(3^{n_1+n_2}-1)$-connected, $j \in [0,l_2-1]\mathbf{z}\}$.

Then we denote by $DT_{3^{n_1+n_2}-1}$ the submatrix $\cup_{j\in[0,l_2'-1]\mathbf{z}}T^j$ in (3-4). Now we obviously see that $DT_{3^{n_1+n_2}-1}$ is taken from $SC_{k_1}^{n_1,l_1} \times SC_{k_2}^{n_2,l_2}$ in terms of such a kind of $(3^{n_1+n_2}-1)$-homotopic thinning and is a closed $(3^{n_1+n_2}-1)$-surface.□

Hereafter, by $DT_{3^{n_1+n_2}-1}$ we denote the $(3^{n_1+n_2}-1)$-homotopically thinned digital image in Theorem 2 and use it later in this paper.

Example 3. Consider the two digital images $X := SC_4^{2,10} := ((0,0),(1,0),(2,0),$ $(2,1),(2,2),(2,3),(1,3),(0,3),(0,2),(0,1)) := (c_i)_{i\in[0,9]\mathbf{z}}$ which is a simple closed 4-curve with ten elements in $\mathbf{Z}^2$ and $SC_{18}^{3,6} := (d_j)_{j\in[0,5]\mathbf{z}}$ in (3-1). Then the Cartesian product $X \times Y$ is assumed to be a (10×6)-matrix with the component $t_{ij} := (c_i,d_j) \in X \times SC_{18}^{3,6}$. Then, do 242-homotopic thinning on $X \times SC_{18}^{3,6}$ so that we obtain the 242-homotopically thinned digital image $X' \times SC_{18}^{3,6}$ from $X \times SC_{18}^{3,6}$ by Theorem 2, where $X' := ((1,0),(2,1),(2,2),(1,3),(0,2),(0,1)) := (c_i)_{i\in[0,5]\mathbf{z}}$. We see that $X' \times SC_8^{3,6} := DT_{242} \subset \mathbf{Z}^5$ is a closed 242-surface.

Indeed, Theorem 2 plays an essential role in calculating the $(3^{n_1+n_2}-1)$-fundamental group of $SC_{k_1}^{n_1,l_1} \times SC_{k_2}^{n_2,l_2}$ by the use of some properties of an $(8,3^{n_1+n_2}-1)$-covering map in Sections 4, 5(see Theorems 3, 4).

4 $(8, 3^{n_1+n_2} - 1)$-Covering, $(\mathbf{Z} \times \mathbf{Z}, p_1 \times p_2, DT_{3^{n_1+n_2}-1})$

Due to the digital (k_0, k_1)-covering theory in $[7, 9, 10, 11]$, the calculation of the digital fundamental group of some digital image was studied, e.g., $\pi_1^4(SC_4^{2,8})$ and $\pi_1^8(SC_8^{2,6})$ are, respectively, isomorphic to infinite cyclic groups, precisely, $(8\mathbf{Z}, +)$ and $(6\mathbf{Z}, +)$ $[9]$. Further, an equivalent presentation of the (k_0, k_1)-covering in $[9]$ is shown in $[10, 11]$ as follows.

Definition 5. $[10, 11]$ *Let (E, k_0) and (B, k_1) be digital images and let $p : E \to B$ be a (k_0, k_1)-continuous surjection. Suppose, for any $b \in B$, there exists $\varepsilon \in \mathbf{N}$ such that*

(1) for some index set M, $p^{-1}(N_{k_1}(b, \varepsilon)) = \cup_{i \in M} N_{k_0}(e_i, \varepsilon)$ with $e_i \in p^{-1}(b)$;
(2) if $i, j \in M$ and $i \neq j$, then $N_{k_0}(e_i, \varepsilon) \cap N_{k_0}(e_j, \varepsilon) = \phi$;
(3) the restriction map p on $N_{k_0}(e_i, \varepsilon)$ is a (k_0, k_1)-homeomorphism for all $i \in M$.

Then the map p is called an equivalent (k_0, k_1)-covering map and (E, p, B) is said to be an equivalent (k_0, k_1)-covering.

The k_1-neighborhood $N_{k_1}(b, \varepsilon)$ is called an *elementary k_1-neighborhood* of b with some radius ε and further, E is called a (k_0, k_1)-*covering space* of B. In the following, we use the equivalent (k_0, k_1)-covering when we state a (k_0, k_1)-covering. For example, let $p : \mathbf{Z} \to SC_k^{n,l} := (c_i)_{i \in [0,l-1]\mathbf{z}}$ be a map given by $p(r) = c_{r(mod l)}$, $r \in \mathbf{Z}$. Then the map p is obviously a $(2, k)$-covering map $[9, 10, 11]$.

Definition 6. $[5]$ *For digital images (X, k_0) in $\mathbf{Z}^{n_0}$ and (Y, k_1) in $\mathbf{Z}^{n_1}$, a (k_0, k_1)-continuous map $h : X \to Y$ is called a local (k_0, k_1)-homeomorphism if, for any $x \in X$, $N_{k_0}(x, 1) \subset X$ is (k_0, k_1)-homeomorphic to $N_{k_1}(h(x), 1) \subset Y$. If $n_0 = n_1$ and $k_0 = k_1$, then the map h is called a local k_0-homeomorphism.*

Definition 7. $[7, 10, 11]$ *A (k_0, k_1)-covering (E, p, B) is called a radius n-(k_0, k_1)-covering if $\varepsilon \geq n$ in Definition 5.*

Remark 1. If (E, p, B) is a (k_0, k_1)-covering, then the map p is a radius 1-(k_0, k_1)-covering map because a (k_0, k_1)-homeomorphism is a local (k_0, k_1)-homeomorphic bijection$[5]$. Thus we may take $\varepsilon = 1$ in Definition 5.

By the use of some properties of the $(2, k)$-covering $(\mathbf{Z}, p, SC_k^{n,l})$ we have the following.

Theorem 3. *In Theorem 2 consider the map $p_1 \times p_2 : \mathbf{Z} \times \mathbf{Z} \to DT_{3^{n_1+n_2}-1} := (c_i)_{i \in [0,l_1'-1]\mathbf{z}} \times (d_j)_{j \in [0,l_2'-1]\mathbf{z}}$ given by $p_1 \times p_2((r_1, r_2)) = (c_{r_1(mod\, l_1')}, d_{r_2(mod\, l_2')})$. Then the map $p_1 \times p_2$ is an $(8, 3^{n_1+n_2} - 1)$-covering map.*

Proof: Assume that $SC_{k_1}^{n_1,l_1} \times SC_{k_2}^{n_2,l_2} = \cup_{i \in [0,l_1-1]\mathbf{z}} T_i \subset \mathbf{Z}^{n_1+n_2}$, where each $T_i = \{c_i\} \times SC_{k_2}^{n_2,l_2}$ can be considered as the sequence $T_i := (t_{ij})_{j \in [0,l_2-1]\mathbf{z}}$, $t_{ij} = (c_i, d_j)$ and $i \in [0, l_1 - 1]\mathbf{z}$. Since

$$DT_{3^{n_1+n_2}-1} := (c_i)_{i \in [0,l_1'-1]\mathbf{z}} \times (d_j)_{j \in [0,l_2'-1]\mathbf{z}} \subset SC_{k_1}^{n_1,l_1} \times SC_{k_2}^{n_2,l_2}$$

is a closed $(3^{n_1+n_2} - 1)$-surface by Theorem 2, we now prove that $\mathbf{Z}^2$ is an $(8, 3^{n_1+n_2} - 1)$-covering space of $DT_{3^{n_1+n_2}-1}$.

First, the map $p_1 \times p_2 : \mathbf{Z} \times \mathbf{Z} \to DT_{3^{n_1+n_2}-1}$ given by

$$p_1 \times p_2((r_1, r_2)) = (c_{r_1(mod\, l_1')}, d_{r_2(mod\, l_2')})$$

is proved an $(8, 3^{n_1+n_2} - 1)$-continuous surjection. Precisely, for any element $t_{ij} := (c_i, d_j) \in DT_{3^{n_1+n_2}-1}$ and its $(3^{n_1+n_2} - 1)$-neighborhood with radius 1 in $DT_{3^{n_1+n_2}-1}$, there is the following 8-neighborhood with radius 1

$$N_8((r_1, r_2), 1) \subset \mathbf{Z} \times \mathbf{Z} \quad \text{such that}$$

$$p_1 \times p_2(N_8((r_1, r_2), 1)) = N_{3^{n_1+n_2}-1}(t_{ij}, 1),$$

where $r_1(mod\, l_1') = i$ and $r_2(mod\, l_2') = j$, which implies an $(8, 3^{n_1+n_2} - 1)$-continuous surjection of $p_1 \times p_2$. To be specific, $p_1 \times p_2(N_{3^{n_1+n_2}-1}(t_{00}, 1)) = \cup_{m,n\in\mathbf{Z}}N_8((ml_1', nl_2'), 1)$ in Fig.1.

Next, for any element $t_{ij} \in DT_{3^{n_1+n_2}-1}$, we prove that $N_{3^{n_1+n_2}-1}(t_{ij}, 1)$ is an elementary $(3^{n_1+n_2} - 1)$-neighborhood of the point t_{ij} with radius 1. Precisely,

$$(p_1 \times p_2)^{-1}(N_{3^{n_1+n_2}-1}(t_{ij}, 1)) = \cup_{(r_1, r_2)\in M_1 \times M_2 \subset \mathbf{Z}^2} N_8((r_1, r_2), 1), \quad \cdots (4\text{-}1)$$

where $r_1(mod\, l_1') = i$ and $r_2(mod\, l_2') = j$.

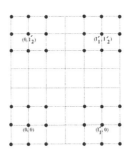

Fig. 1. $(p_1 \times p_2)^{-1}(N_{3^{n_1+n_2}-1}(t_{00}, 1))$ of the $(8, 3^{n_1+n_2} - 1)$-covering

In (4-1), for any $(r_1, r_2) \neq (r_1', r_2') \in M_1 \times M_2 \subset \mathbf{Z} \times \mathbf{Z}$, we see that

$$N_8((r_1, r_2), 1) \cap N_8((r_1', r_2'), 1) = \phi.$$

Moreover, the restriction map $p_1 \times p_2$ on $N_8((r_1, r_2), 1)$ is an $(8, 3^{n_1+n_2} - 1)$-homeomorphism for all $(r_1, r_2) \in M_1 \times M_2 \subset \mathbf{Z} \times \mathbf{Z}$.

Thus the proof of an $(8, 3^{n_1+n_2} - 1)$-covering $(\mathbf{Z} \times \mathbf{Z}, p_1 \times p_2, DT_{3^{n_1+n_2}-1})$ is completed, as required.   □

Example 4. (1) Consider $SC_4^{2,8} \times SC_8^{2,4} := (c_i)_{i\in[0,7]_\mathbf{z}} \times (d_j)_{j\in[0,3]_\mathbf{z}}$ with the following (8×4)-matrix presentation(see (4-2)).

$$
\begin{pmatrix}
(0,0,0,0) & (0,0,1,1) & (0,0,2,0) & (0,0,1,-1) \\
(0,1,0,0) & (0,1,1,1) & (0,1,2,0) & (0,1,1,-1) \\
(0,2,0,0) & (0,2,1,1) & (0,2,2,0) & (0,2,1,-1) \\
(1,2,0,0) & (1,2,1,1) & (1,2,1,2) & (1,2,0,3) \\
(2,2,0,0) & (2,2,1,1) & (2,2,1,2) & (2,2,0,3) \\
(2,1,0,0) & (2,1,1,1) & (2,1,1,2) & (2,1,0,3) \\
(2,0,0,0) & (2,0,1,1) & (2,0,1,2) & (2,0,0,3) \\
(1,0,0,0) & (1,0,1,1) & (1,0,1,2) & (1,0,0,3)
\end{pmatrix}
\rightarrow
\begin{pmatrix} T_1 \\ T_3 \\ T_5 \\ T_7 \end{pmatrix}
:= DT_{80} \cdots (4-2)
$$

Further, put $T^j := SC_4^{2,8} \times \{d_j\}$ and $T_i := \{c_i\} \times SC_8^{2,4}$, where $SC_4^{2,8} := (c_i)_{i \in [0,7]_{\mathbf{Z}}}$ and $SC_8^{2,4} := (d_j)_{j \in [0,3]_{\mathbf{Z}}}$. Thus we have the notation $SC_4^{2,8} \times SC_8^{2,4} := \cup_{i \in [0,7]_{\mathbf{Z}}} T_i := \cup_{j \in [0,3]_{\mathbf{Z}}} T^j$. Since T_i and $T_{i+2(mod\,8)}$ are 80-connected, $i \in \{1,3, 5,7\}$, we can delete the subset $\cup_{i \in \{0,2,4,6\}} T_i$ from the set $\cup_{i \in [0,7]_{\mathbf{Z}}} T_i$ in (4-2) in terms of the 80-homotopic thinning on $SC_4^{2,8} \times SC_8^{2,4}$. Consequently, we have a closed 80-surface 80-homotopically thinned.

Besides, we now consider an $(8, 80)$-covering map $p_1 \times p_2 : \mathbf{Z} \times \mathbf{Z} \to DT_{80}$ by the same method as Theorem 3.

(2) Similarly, by Example 2 we obtain that $(\mathbf{Z} \times \mathbf{Z}, p_1 \times p_2, SC_8^{2,6} \times SC_{18}^{3,6})$ is an $(8, 242)$-covering.   □

5 $(3^{n_1+n_2} - 1)$-Fundamental Group of $SC_{k_1}^{n_1,l_1} \times SC_{k_2}^{n_2,l_2}$

Even if $SC_{k_1}^{n_1,l_1} \times SC_{k_2}^{n_2,l_2}$ is not a closed k-surface, we can calculate the $(3^{n_1+n_2} - 1)$-fundamental group of $SC_{k_1}^{n_1,l_1} \times SC_{k_2}^{n_2,l_2}$ in terms of the $(3^{n_1+n_2}-1)$-homotopic thinning method in Theorem 2 and some properties of an $(8, 3^{n_1+n_2}-1)$-covering map in Theorem 3.

For three digital images $(E, k_0) \subset \mathbf{Z}^{n_0}$, $(B, k_1) \subset \mathbf{Z}^{n_1}$, and $(X, k_2) \subset \mathbf{Z}^{n_2}$, let $p : E \to B$ be a (k_0, k_1)-continuous map. For some (k_2, k_1)-continuous map f from X into B, as a digital analogue of a lifting in [16], we say that a *digital lifting* of f is a (k_2, k_0)-continuous map $\tilde{f} : X \to E$ such that $p \circ \tilde{f} = f$ [6,7,9,10,11]. Let $p : (E, e_0) \to (B, b_0)$ be a (k_0, k_1)-covering map which preserves the base point. Any k_1-path $f : [0, m]_{\mathbf{Z}} \to B$ beginning at b_0 has a *unique digital lifting* to a k_0-path $\tilde{f}$ in E beginning at e_0 [9,10,11]. Moreover, the following *digital homotopy lifting theorem* was originally introduced in [7].

Lemma 1. *[7] Let $((E, e_0), k_0)$ and $((B, b_0), k_1)$ be pointed digital images. Let $p : (E, e_0) \to (B, b_0)$ be a radius 2-(k_0, k_1)-covering map. For k_0-paths $g_0 : [0, m_0]_{\mathbf{Z}} \to (E, e_0)$ and $g_1 : [0, m_1]_{\mathbf{Z}} \to (E, e_0)$ that begin at $e_0 = g_0(0) = g_1(0)$, if $p \circ g_0$ and $p \circ g_1$ are k_1-homotopic rel $\{p(e_0), p(g_0(m_0)) = p(g_1(m_1))\}$ in B, then g_0 and g_1 are k_0-homotopic rel $\{e_0, g_0(m_0) = g_1(m_1)\}$ in E.*

In order to study the digital fundamental group of some digital closed k-surface, we need the following.

Definition 8. *[10, 11] For digital images* $((E, e_0), k_0)$, $((B, b_0), k_1)$, *let* $p :$ $(E, e_0) \rightarrow (B, b_0)$ *be a pointed* (k_0, k_1)-*covering map. If* $p_* \pi_1^{k_0}(E, e_0)$ *is a normal subgroup of* $\pi_1^{k_1}(B, b_0)$, *then* $((E, e_0), p, (B, b_0))$ *is called a regular* (k_0, k_1)-*covering.*

In Theorem 3 we obtain a base point preserving radius 2-$(8, 3^{n_1+n_2} - 1)$-covering $((\mathbf{Z} \times \mathbf{Z}, (0,0)), p_1 \times p_2, (DT_{3^{n_1+n_2}-1}, t_{00}))$. Then, by Lemma 1 we have the following which is the digital version of covering map property in [16]:

If $(p_1 \times p_2)^{-1}(t_{00})$ is a group, then we obtain the following: $(p_1 \times p_2)^{-1}(t_{00})$ is isomorphic to

$$\pi_1^{3^{n_1+n_2}-1}(DT_{3^{n_1+n_2}-1}, t_{00})/(p_1 \times p_2)_* \pi_1^8((\mathbf{Z} \times \mathbf{Z}, (0,0)) \cdots\cdots\cdots (5-1),$$

where $(p_1 \times p_2)_* : \pi_1^8((\mathbf{Z} \times \mathbf{Z}, (0,0)) \rightarrow \pi_1^{3^{n_1+n_2}-1}(DT_{3^{n_1+n_2}-1}, t_{00})$ is a group homomorphism induced from the covering map $p_1 \times p_2$.

Hereafter, for the product set $SC_{k_1}^{n_1,l_1} \times SC_{k_2}^{n_2,l_2} := \{(c_i, d_j) | i \in [0, l_1-1]_{\mathbf{z}}\}, j \in [0, l_2 - 1]_{\mathbf{z}}\}$, assume that

each of the subsets $\{c_i\} \times SC_{k_2}^{n_2,l_2}$ and $SC_{k_1}^{n_1,l_1} \times \{d_j\}$ in $SC_{k_1}^{n_1,l_1} \times SC_{k_2}^{n_2,l_2}$ is not $(3^{n_1+n_2} - 1)$-contractible. $\cdots\cdots\cdots\cdots\cdots\cdots\cdots\cdots\cdots\cdots\cdots\cdots (5-2)$

Indeed, for the digital image $SC_{k_1}^{n_1,l_1} \times SC_{k_2}^{n_2,l_2} := (c_i)_{i\in[0,l_1-1]_{\mathbf{z}}} \times (d_j)_{j\in[0,l_2-1]_{\mathbf{z}}}$, its $(3^{n_1+n_2}-1)$-fundamental group is stated by the fact that $\pi_1(SC_k^{n,l})$ is isomorphic to the infinite cyclic group, precisely, $(l\mathbf{Z}, +)[4, 5, 6, 7, 8, 9, 10]$.

Theorem 4. *With the hypothesis of (5-2)* $\pi_1^{3^{n_1+n_2}-1}(SC_{k_1}^{n_1,l_1} \times SC_{k_2}^{n_2,l_2})$ *is isomorphic to* $l_1'\mathbf{Z} \times l_2'\mathbf{Z}$, *where* $l_1' = l_1-$ *the cardinal number of the set of simple* $(3^{n_1+n_2} - 1)$-*points in* $SC_{k_1}^{n_1,l_1} \times \{d_j\} \subset SC_{k_1}^{n_1,l_1} \times SC_{k_2}^{n_2,l_2}$ *and* $l_2' = l_2-$ *the cardinal number of the set of simple* $(3^{n_1+n_2} - 1)$-*points in* $\{c_i\} \times SC_{k_2}^{n_2,l_2} \subset SC_{k_1}^{n_1,l_1} \times SC_{k_2}^{n_2,l_2}$.

Before proving Theorem 4, we need to show the hypothesis of (5-2) as follows: In case that $SC_{k_i}^{n_i,l_i} = SC_{2n_i}^{n_i,4}, i \in \{1,2\}$, we see that $\pi_1^{3^{n_i}-1}(SC_{2n_i}^{n_i,4})$ is trivial(see Example 2). Without the hypothesis of (5-2), Theorem 4 may not be true(for more details, see Remark 2).

Proof: By Theorem 2, after doing a $(3^{n_1+n_2}-1)$-homotopic thinning on $SC_{k_1}^{n_1,l_1} \times SC_{k_2}^{n_2,l_2}$, we have the $(3^{n_1+n_2} - 1)$-homotopically thinned digital image $DT_{3^{n_1+n_2}-1} \subset SC_{k_1}^{n_1,l_1} \times SC_{k_1}^{n_1,l_1}$ such that $(\mathbf{Z} \times \mathbf{Z}, p_1 \times p_2, DT_{3^{n_1+n_2}-1})$ is a radius $2 - (8, 3^{n_1+n_2} - 1)$-covering by Theorem 3. Since $\mathbf{Z} \times \mathbf{Z}$ is simply 8-connected, it is a radius $2 - (8, 3^{n_1+n_2} - 1)$-regular covering map. Thus, by (5-1) $(p_1 \times p_2)^{-1}(t_{00}) = l_1'\mathbf{Z} \times l_2'\mathbf{Z}$ is isomorphic to $\pi_1^{3^{n_1+n_2}-1}(DT_{3^{n_1+n_2}-1}, t_{00})$ which is isomorphic to $\pi_1^{3^{n_1+n_2}-1}(SC_{k_1}^{n_1,l_1} \times SC_{k_2}^{n_2,l_2}, t_{00})$ by Theorems 1 and 3. Thus the proof is completed, as required. $\square$

Example 5. By Theorem 3 we see that $\pi_1^{242}(Y \times SC_{18}^{3,6})$ is isomorphic to $8\mathbf{Z} \times 6\mathbf{Z}$, where $Y := SC_8^{2,8} := \{(0,0), (1,1), (1,2), (1,3), (0,4), (-1,3), (-1,2), (-1,1)\}$.

Proof: By the $(8, 242)$-covering $(\mathbf{Z} \times \mathbf{Z}, p_1 \times p_2, DT_{242} := Y \times SC_{18}^{3,6})$ in Theorem 3 and (5-1) the proof is completed. $\qquad\Box$

Remark 2. In view of Theorem 4 if either $\{c_i\} \times SC_{k_1}^{n_2,l_2}$ or $SC_{k_1}^{n_1,l_1} \times \{d_j\}$ is $(3^{n_1+n_2} - 1)$-contractible, then $\pi_1^{3^{n_1+n_2}-1}(SC_{k_1}^{n_1,l_1} \times SC_{k_2}^{n_2,l_2})$ need not be isomorphic to $l_1'\mathbf{Z} \times l_2'\mathbf{Z}$. Precisely, consider $SC_4^{2,8} \times SC_8^{2,6}$. Then, since $SC_4^{2,8}$ is 8-contractible [3, 6, 7, 8, 11], $SC_4^{2,8} \times \{d_j\}$ is 80-contractible. But $\{c_i\} \times SC_8^{2,6}$ is not 80-nullhomotopic in $SC_4^{2,8} \times SC_8^{2,6}$ because $SC_8^{2,6}$ is not 8-contractible [4, 5, 6, 7, 8, 9]. Thus $\pi_1^{80}(SC_4^{2,8} \times SC_8^{2,6}) \simeq 6\mathbf{Z}$ which is not isomorphic to $8\mathbf{Z} \times 6\mathbf{Z}$.

References

1. G. Bertrand, Simple points, topological numbers and geodesic neighborhoods in cubic grids, *Pattern Recognition Letters*, **15** (1994), 1003-1011.
2. G. Bertrand and R. Malgouyres, Some topological properties of discrete surfaces, *Jour. of Mathematical Imaging and Vision*, **20** (1999), 207-221.
3. L. Boxer, A classical construction for the digital fundamental group, *Jour. of Mathematical Imaging and Vision*, **10** (1999), 51-62.
4. S.E. Han, A generalized digital (k_0, k_1)-homeomorphism, *Note di Matematica*, **22(2)**(2004) 157-166.
5. S.E. Han, Algorithm for discriminating digital images w.r.t. a digital (k_0, k_1)-homoeomorphism, *Jour. of Applied Mathematics and Computing* **18**(1-2)(2005), 505-512.
6. S.E. Han, Connected sum of digital closed surfaces, *Information Sciences* **176**(3)(2006), 332-348.
7. S.E. Han, Digital coverings and their applications, *Jour. of Applied Mathematics and Computing* **18**(1-2)(2005), 487-495.
8. S.E. Han, Minimal simple closed 18-surfaces and a topological preservation of 3D surfaces, *Information Sciences* **176**(1)(2006), 120-134.
9. S.E. Han, Non-product property of the digital fundamental group, *Information Sciences* **171** (2005), 73-91.
10. S.E. Han, Equivalent (k_0, k_1)-covering and generalized digital lifting, *submitted, Information Sciences*.
11. S.E. Han, 32-Tori in $\mathbf{Z}^4$ and their comparisons by the 32-homotopy equivalence, *submitted, JKMS.*
12. E. Khalimsky, Motion, deformation, and homotopy in finite spaces, *Proceedings IEEE International Conferences on Systems, Man, and Cybernetics*, (1987), 227-234.
13. T. Y. Kong, A. Rosenfeld, Digital topology - A brief introduction and bibliography, Topological Algorithms for the Digital Image Processing, Elsevier Science, Amsterdam, 1996.
14. R. Kopperman, R. Meyer and R.G. Wilson, A Jordan surface theorem for three-dimensional digital spaces,*Discrete and computational Geometry*, **6** (1991), 155-161.
15. R. Malgouyres and G. Bertrand, A new local property of strong n-surfaces, *Pattern Recognition Letters*, **20** (1999), 417-428.
16. W.S. Massey, Algebraic Topology, Springer-Verlag, New York, 1977.
17. D.G. Morgenthaler and A. Rosenfeld, Surfaces in three -dimensional images, *Information and Control*, **51** (1981), 227-247.

Topology Preserving Digitization with FCC and BCC Grids

Peer Stelldinger[1] and Robin Strand[2]

[1] Cognitive Systems Group, University of Hamburg,
Vogt-Köln-Str. 30, D-22527 Hamburg, Germany
stelldinger@informatik.uni-hamburg.de
[2] Centre for Image Analysis, Uppsala University,
Lägerhyddsvägen 3, SE-75237 Uppsala, Sweden
robin@cb.uu.se

Abstract. In digitizing 3D objects one wants as much as possible object properties to be preserved in its digital reconstruction. One of the most fundamental properties is topology. Only recently a sampling theorem for cubic grids could be proved which guarantees topology preservation [1]. The drawback of this theorem is that it requires more complicated reconstruction methods than the direct representation with voxels. In this paper we show that face centered cubic (fcc) and body centered cubic (bcc) grids can be used as an alternative. The fcc and bcc voxel representations can directly be used for a topologically correct reconstruction. Moreover this is possible with coarser grid resolutions than in the case of a cubic grid. The new sampling theorems for fcc and bcc grids also give absolute bounds for the geometric error.

1 Introduction

In 3D image analysis one often has to deal with the huge amount of data of volumetric images. Using non-standard grids, like bcc and fcc grids for digitizing 3D objects is a very promising method in order to reduce the amount of data, since these grids have a very high packing density (i.e. the ratio between the volume of the largest ball completely enclosed in a voxel and the volume of the voxel itself) in comparison to cubic grids [2,3]. The advantages of the bcc and fcc grids has led to various applications in very different areas, e.g. in fuzzy segmentation [4] and in computer graphics and data visualization, [2,5,6]. Also, image processing algorithms on which many applications rely have been developed for these grids, e.g. weighted distance transforms [7] and multiscale representation of images [8].

The question addressed in this paper is, if these grids can be used for topology preserving digitization and if they are advantageous regarding this problem relatively to the common cubic grid.

While first solutions to topology preserving digitization in 2D have already been published in 1982 [9,10], the 3D generalization remained unproved for over 20 years. In fact it is proved that topological changes can never be avoided if one uses the digital reconstruction on a cubic grid, i.e. the union of the voxels

U. Eckardt et al. (Eds.): IWCIA 2006, LNCS 4040, pp. 226–240, 2006.

whose sampling points lie inside the object [11]. At least connectivity and the number of the objects components could be reliably detected with the digital reconstruction, as shown in [12]. But this is not all the topological information. Only recently one of the authors showed together with Siqueira and Latecki that one can digitize 3D r-regular objects with a sufficiently dense sampling grid, such that the whole topological information is unchanged, if one uses not the digital reconstruction based on the voxels, but certain other reconstruction methods [1].

As we will show in this paper, the use of a sufficiently dense bcc or fcc grid leads to a topologically correct digital reconstruction, i.e. one does not need to use more complicated reconstruction methods. Moreover the grid density which is necessary for proving the preservation of topology is for both fcc and bcc grids smaller than for cubic grids, i.e. one needs much less sampling points in order to guarantee the right topology.

2 Preliminaries

The *(Euclidean) distance* between two points x and y in $\mathbb{R}^n$ is denoted by $d(x,y)$, and the *Hausdorff distance* $d_H(\cdot,\cdot)$ between two subsets of $\mathbb{R}^n$ is the maximal distance between each point of one set and the nearest point of the other. Let $A \subset \mathbb{R}^n$ and $B \subset \mathbb{R}^m$ be sets. A function $f : A \to B$ is called *homeomorphism* if it is bijective and both it and its inverse are continuous. If f is a homeomorphism, we say that A and B are *homeomorphic*. Let A, B be two subsets of $\mathbb{R}^3$. Then A homeomorphism $f : \mathbb{R}^3 \to \mathbb{R}^3$ such that $f(A) = B$ and $d(x, f(x)) \le r$, for all $x \in \mathbb{R}^3$, is called an *r-homeomorphism* of A to B and we say that A and B are *r-homeomorphic*. A *Jordan curve* is a set $J \subset \mathbb{R}^n$ which is homeomorphic to a circle. Let A be any subset of $\mathbb{R}^3$. The *complement* of A is denoted by A^c. All points in A are *foreground* while the points in A^c are called *background*. The *open ball* in $\mathbb{R}^3$ of radius r and center c is the set $\mathcal{B}_r^0(c) = \{x \in \mathbb{R}^3 \mid d(x,c) < r\}$, and the *closed ball* in $\mathbb{R}^3$ of radius r and center c is the set $\overline{\mathcal{B}}_r(c) = \{x \in \mathbb{R}^3 \mid d(x,c) \le r\}$. Whenever $c = (0,0,0)$, we write $\mathcal{B}_r^0$ and $\overline{\mathcal{B}}_r$. We say that A is *open* if, for each $x \in A$, there exists a positive number r such that $\mathcal{B}_r^0(x) \subset A$. We say that A is *closed* if its complement, A^c, is open. The *boundary* of A, denoted ∂A, consists of all points $x \in \mathbb{R}^3$ with the property that if B is any open set of $\mathbb{R}^3$ such that $x \in B$, then $B \cap A \ne \emptyset$ and $B \cap A^c \ne \emptyset$. We define $A^0 = A \setminus \partial A$ and $\overline{A} = A \cup \partial A$. Note that A^0 is open and $\overline{A}$ is closed, for any $A \subset \mathbb{R}^3$. Note also that $\mathcal{B}_r^0(c) = (\overline{\mathcal{B}}_r(c))^0$ and $\overline{\mathcal{B}}_r(c) = \overline{\mathcal{B}_r^0(c)}$. The *$r$-dilation* $A \oplus \mathcal{B}_r^0$ of a set A is the union of all open r-balls with center in A, and the *r-erosion* $A \ominus \mathcal{B}_r^0$ is the union of all center points of open r-balls lying inside of A. We say that an open ball $\mathcal{B}_r^0(c)$ is *tangent* to ∂A at a point $x \in \partial A$ if $\partial A \cap \partial \mathcal{B}_r^0(c) = \{x\}$ and $\partial A \cap \mathcal{B}_r^0(c) = \emptyset$. We say that an open ball $\mathcal{B}_r^0(c)$ is an *osculating open ball of radius r to ∂A at point $x \in \partial A$* if $\mathcal{B}_r^0(c)$ is tangent to ∂A at x and either $\mathcal{B}_r^0(c) \subseteq A^0$ or $\mathcal{B}_r^0(c) \subseteq (A^c)^0$. Since all of the known topology preserving sampling theorems in 2D require the object to be r-regular [9,10,11], we will use the 3D generalization for our approach (refer to Fig. 1):

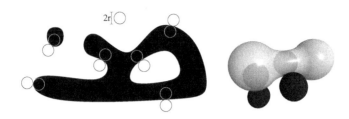

Fig. 1. For each boundary point of a 2D/3D r-regular set there exists an outside and an inside osculating open disc/ball of radius r

Definition 1. *A set $A \subset \mathbb{R}^3$ is called r-regular if, for each point $x \in \partial A$, there exist two osculating open balls of radius r to ∂A at x such that one lies entirely in A and the other lies entirely in A^c.*

Note, that the boundary of an r-regular set is a 2D manifold surface.

A countable set $S \subset \mathbb{R}^3$ of *sampling points* with $d_H(\mathbb{R}^3, S) \leq r'$ for some $r' \in \mathbb{R}_+$ such that $S \cap A$ is finite for each bounded set A, is called r'-*grid*. r' is called the covering radius. The *voxel* $\mathcal{V}_S(s)$ of a sampling point s is its Voronoi region, i.e. the set of all points lying at least as near to this point as to any other sampling point.

Given a translation vector t and a rotation matrix R in 3D, the *bcc* and *fcc* r'-*grids* are defined by $S := \{t + R \cdot \frac{2}{\sqrt{5}}(x_1, x_2, x_3) | x_1, x_2, x_3 \in \mathbb{Z}, x_1 \equiv x_2 \equiv x_3 (\mathrm{mod}2)\}$ and $S := \{t + R \cdot (x_1, x_2, x_3) | x_1, x_2, x_3 \in \mathbb{Z}, x_1 + x_2 + x_3 \equiv 0 (\mathrm{mod}2)\}$.

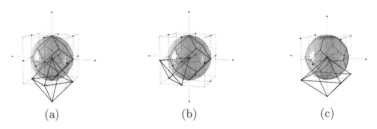

(a)   (b)   (c)

Fig. 2. Both fcc (a),(b) and bcc (c) grids can be embedded in cubic grids of higher resolution. The points are neighboring sampling points of the shown voxel of the fcc, respectively bcc grid, and the dashed lines show the cubic grid. The shown spheres have radius r'. The dual grid of an fcc r'-grid consists of octahedra (a) and tetrahedra (b). All line segments of them have length $\sqrt{2}r'$. The faces of both octahedra and tetrahedra are equilateral triangles. Based on the included cubic grid, a bcc r'-grid can be completely separated into octahedral configurations (c). Four of the line segments of the octahedra have length $\frac{4}{\sqrt{5}}r'$ and eight have length $2\sqrt{\frac{3}{5}}r'$. The faces of the octahedra are isosceles triangles with one angle of $2\arcsin\left(\frac{1}{\sqrt{3}}\right) \approx 70.53°$ and two angles of $\arccos\left(\frac{1}{\sqrt{3}}\right) \approx 54.74°$. Note that the fcc r'-grid is embedded in a cubic $\frac{\sqrt{3}}{2}$-grid, while the bcc r'-grid is embedded in a cubic $\sqrt{\frac{5}{3}}r'$-grid.

Due to the scaling factor these grids are r'-grids. Note that the Hausdorff distance between a 3D object and an r'-grid is at most r', thus different sampling grids with the same covering radius lead to results with a geometrically comparable accuracy (see Fig. 10(2) to (4)). Note, that both bcc and fcc grids can be embedded in a cubic grid, as is illustrated in Fig. 2.

The intersection of $A \subseteq \mathbb{R}^3$ with S is called the *digitization* of A with S. The *digital reconstruction* of A with S is the union of all voxels belonging to the sampling points of the digitization. Two voxels are *face-adjacent* or *adjacent* if they share a face. They are *vertex-adjacent* if they intersect in exactly one point (this is not possible for bcc grids).

3 Digital Reconstruction of r-Regular Sets

Let $A \subset \mathbb{R}^3$ be an r-regular object, let S be a bcc or a fcc r'-grid, and consider the digital reconstruction $\hat{A}$ of A with respect to S. Assume that no sampling point of S lies on ∂A. This assumption is not a restriction, as if some sampling point lies on ∂A, there always exists an $\varepsilon > 0$ such that the ε-dilation $A \oplus \overline{B}_\varepsilon$ is $(r - \varepsilon)$-regular with $r - \varepsilon > r'$, and $A \oplus \overline{B}_\varepsilon$ has the same digital reconstruction as A, thus by updating r to the value of $r - \varepsilon$, the assumption is true.

In this section we will show that certain configurations of neighboring voxels in an fcc or a bcc grid can not occur in a sufficiently dense digitization of an r-regular set and we will use this to show that the topology does not change during digitization. Therefore we need some definitions and lemmas about the local behavior of r-regular sets, which have already been introduced in [1]:

Definition 2. *Let x, y be two points in $\mathbb{R}^3$. Further let $s > d(x, y)$. Then, the intersection $P_s(x, y)$ of all closed s-balls containing x and y,*

$$P_s(x, y) = \bigcap \{\overline{B}_s(v) \mid x, y \in \overline{B}_s(v)\},$$

is called s-path region between x and y.

Now, let x, y, z be three points in $\mathbb{R}^3$, and assume that $s > \frac{1}{2} \max\{d(x, y), d(x, z), d(y, z)\}$. Then, the intersection $P_s(x, y, z)$ of all closed s-balls containing x, y and z,

$$P_s(x, y, z) = \bigcap \{\overline{B}_s(v) \mid x, y, z \in \overline{B}_s(v)\},$$

is called s-surface region between x, y and z.

A nonempty set B is called an r-simple-cut set if it is the intersection of (a maybe infinite number of) closed balls with radii smaller than r.

Lemma 1. *Let A be an r-regular set and x, y be two different points in A with $d(x, y) < 2r$. Further, let L be the straight line segment from x to y. Then, the function f mapping each point of L to the nearest point in A is well-defined, continuous and bijective, and the range of f is a simple path from x to y.*

Proof. Each point $L \cap A$ is its own nearest point in A. Since the intersection of surface normals of r-regular sets has a minimal distance of r to the surface [13], there exists for each point in $L \setminus A$ exactly one nearest point in ∂A because each of these points has a distance smaller than r to the boundary. Thus f must be a continuous function since if f would not be continuous at some point, this point would have more than one nearest point in ∂A. Note that any point of L lies on the normal vector of ∂A in its nearest boundary point. Now suppose one point p of ∂A would be the nearest point to at least two different points l_1 and l_2 of L. Then l_1 and l_2 both lie on the normal of ∂A in p. This implies that any point in L including x and y lies on this normal. Since the normal vectors of length r of an r-regular set do not intersect, the distance between x and y has to be at least $2r$ which contradicts $d(x, y) < 2r$. Thus f is bijective. Since every bijective continuous function of a compact metric space is continuous in both directions, f must be a homeomorphism. This implies that the range is a simple path from x to y. $\qquad\square$

Definition 3. *Let A be an r-regular set and x, y be two different points in A with $d(x, y) < 2r$. Further, let L be the straight line segment from x to y. Then, the range of the function f mapping each point of L to the nearest point in A is called the* direct path *from x to y regarding A.*

Lemma 2. *Let A be an r-regular set and x, y be two points both inside A or both outside A with $d(x, y) < 2r$. Then, $P_s(x, y)$ is a simple-cut set for any s with $\frac{1}{2} \cdot d(x, y) \leq s < r$, the direct path from x to y regarding A lies inside $A \cap P_s(x, y)$ and the direct path from x to y regarding $\overline{(A \ominus \overline{B}_\varepsilon)^c}$ lies inside $\overline{A^c} \cap P_s(x, y)$ for a sufficiently small $\varepsilon > 0$.*

Proof. First, let $x, y \in A$. Since $d(x, y) < 2r$, $P_s(x, y)$ is a simple cut set for any s with $\frac{1}{2} \cdot d(x, y) \leq s < r$. Now, suppose there exists a point p on the direct path lying outside of $P_s(x, y)$. Then the outside osculating open r-ball of A in p must cover either x or y which implies that they cannot lie on ∂A or inside A. Thus, the direct path has to be inside $P_s(x, y)$. If $x, y \in A^c$ the analog is true by looking at the $(r - \varepsilon)$-regular set $\overline{(A \oplus \overline{B}_\varepsilon)^c}$ for a sufficiently small $\varepsilon > 0$, since there always exists an ε such that x and y remain outside $A \oplus \overline{B}_\varepsilon$ and $s < (r - \varepsilon)$. $\qquad\square$

Lemma 3. *Let A be an r-regular set and let B be an s-simple-cut set with $s < r$. Further, let $B^0 \cap A^0 \neq \emptyset$ and $B \cap A^c \neq \emptyset$. Then, the intersection of the boundaries of A and B, $\partial A \cap \partial B$, is a Jordan curve.*

Proof. Let c_1 and c_2 be two arbitrary points in $B \cap A$ and let P be the direct path from c_1 to c_2. Then P lies inside of B due to lemma 2 and $P_s(c_1, c_2) \subset B$. This implies that $B \cap A$ must be one connected component.

Now, consider the two points c_1 and c_2 lying in $B \cap A^c$. Then the direct path does not necessarily lie in A^c since this set is open, but in $\overline{A^c}$. Thus for any open superset of the intersection of all r-balls containing c_1 and c_2 there exists a path from c_1 to c_2 inside this superset having a minimal distance to the direct path in $\overline{A^c}$. $(B \cap A^c)^0$ is such a superset, since $\partial(B \cap A^c)$ intersects the intersection only in c_1 and c_2.

Thus both $B \cap A$ and $B \cap A^c$ have to be one component and thus the intersection of the boundaries, $I = \partial A \cap \partial B$, must also be one component.

It remains to be shown that I is a Jordan curve. Since I separates ∂B in one part inside of A and one part outside of A, it is a Jordan curve if and only if there exists no point where B and A meet tangentially. Such a point would imply that either the inside or the outside osculating ball of A at this point covers B. Both cases are impossible since then $B^0 \cap A^0 = \emptyset$ or $B \cap A^c = \emptyset$. Thus, $\partial A \cap \partial B$ is a Jordan curve. □

Definition 4. *Let A be an r-regular set and let x, y, z be three arbitrary points inside of $A^0 \oplus \overline{B}_r$. Then, the inner surface patch $I_s(x, y, z)$ of x, y, z regarding A is the set defined by mapping each point of the triangle T spanned by points x, y, z to itself if it lies inside of A and mapping it to the nearest boundary point in ∂A otherwise.*

Now, let x, y, z be three arbitrary points inside of $A^c \oplus \overline{B}_r$. Then the outer surface patch $O_s(x, y, z)$ of x, y, z regarding A is the set defined by mapping each point of the triangle T between the points to itself if it lies inside of $(A \oplus \overline{B}_\varepsilon)^c$ and mapping them to the nearest boundary point $\partial \overline{(A \oplus \overline{B}_\varepsilon)^c}$ otherwise, with ε being half the minimal distance from the sampling points in A^c to ∂A.

Lemma 4. *Let A be an r-regular set and x, y, z be three points inside A with $\max\{d(x, y), d(x, z), d(y, z)\} < 2r$. Then $P_s(x, y, z)$ is a simple cut set for any s with $\frac{1}{2} \cdot d(x, y) \leq s < r$ and the inner surface patch is homeomorphic to a disc, lies inside $A \cap P_s(x, y)$ and is bounded by three paths, one going from x to y inside of $P_s(x, y, z) \cap P_s(x, y)$, another going from y to z inside of $P_s(x, y, z) \cap P_s(y, z)$ and the third going from z to x inside of $P_s(x, y, z) \cap P_s(z, x)$. The analog is true for x, y, z lying outside of A and the outer surface patch.*

Proof. The mapping used in definition 4 is a direct generalization of the mapping in definition 3 and it is a homeomorphism for the same reasons if x, y, z lie inside A and $\max\{d(x, y), d(x, z), d(y, z)\} < 2r$. Its boundaries are equal to the direct paths between each two of the three points. If x, y, z lie outside of A the proof is analog. □

In the next two subsections we will use these properties of r-regular objects for our proofs of the topology preservation. Therefore let s be an arbitrary but fixed number with $r' < s < r - \varepsilon$.

3.1   Digital Reconstruction on FCC Grids

Let us have a closer look at fcc grids. The Delaunay grid separates the space $\mathbb{R}^3$ into octahedra and tetrahedra due to the two types of voxel corners in an fcc grid with 6 respectively 4 neighboring voxels (see Fig. 2(a) and (b)). The digital reconstruction inside an octahedron respectively tetrahedron is totally determined by the sampling points at its vertexes. By our above assumption, each vertex of such an octahedron is either inside (i.e., a foreground point) or outside (i.e., a background point) A. So, there are at most $2^6 = 64$ distinct configurations for a octahedron and $2^4 = 16$ configurations for a tetrahedron with respect to the binary "status" of the vertices. However, up to rotational symmetry, reflectional symmetry, and complementarity (switching foreground and background points), these configurations are equivalent to the 6 canonical configurations for the octahedron and the 3 configurations for the tetrahedron shown in Fig. 3. In the following we will show that case 4 of the 6 octahedron

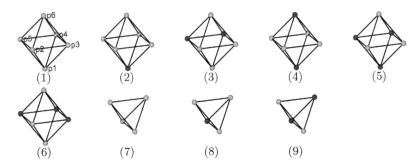

Fig. 3. There are $64+16$ distinct configurations in an fcc grid for neighboring sampling points that are either inside or outside a digitized set. However, up to rotational symmetry, reflectional symmetry, and complementarity (switching foreground and backgroud points), these 80 configurations are equivalent to the above 9 canonical configurations.

configurations can not occur if one digitizes an r-regular object with a sufficiently dense fcc grid. The problem of topology preserving digitization is that in case of configuration 4 a non-manifold surface is reconstructed, which can not be guaranteed to be topologically equivalent to the surface of the original object. This can be avoided by using a sufficiently dense sampling grid for digitizing an r-regular object, as shown by the following theorem:

Theorem 1. *Configuration 4 in Fig. 3 cannot occur in the digital reconstruction of an r-regular object with an fcc r'-grid with $\sqrt{2}r' < r$.*

Proof. In the following let the dark sampling points in Fig. 3 be in the foreground and the white sampling points in the background. Further, let the sampling points $p_1, p_2, \ldots p_6$ of an octahedron be numbered as shown in Fig. 3(a).

Suppose to the contrary, configuration 4 occurs in the digital reconstruction of an r-regular object A. Since the distance from p_1 to p_6 is $2r'$ and thus smaller than $2r$, there exists a foreground path between these points lying completely inside $P_s(p_1, p_6)$.

On the other side, the three background points p_2, p_3, p_4 have each a distance being smaller than $2r$. Thus, due to Lemma 4, there exists an outer surface patch between them. This patch lies inside $P_s(p_2, p_3, p_4)$ with its surface boundary lying inside the union of $P_s(p_2, p_3)$, $P_s(p_2, p_4)$ and $P_s(p_3, p_4)$. Analogously there exists a outer surface patch between the three background points p_2, p_4, p_5 with its boundary in $P_s(p_2, p_4)$, $P_s(p_2, p_5)$ and $P_s(p_4, p_5)$. Due their definition, the two outer surface patches have the boundary part inside $P_s(p_2, p_4)$ in common such that together they form a surface patch between the four points p_2, p_3, p_4, p_5. Fig. 4(b) and (c) show that $P_s(p_1, p_6)$ goes through this surface patch without intersecting the bounding r-path regions for $\sqrt{2}r' = r$ (Obviously this is also true for smaller r'). Since both p_1 and p_6 lie outside $P_s(p_2, p_3, p_4) \cup P_s(p_2, p_4, p_5)$, the path from p_1 to p_6 must go through the combined surface patch and thus there has to exist a point lying both in A and A^c. It follows that case 4 cannot occur in the digital reconstruction of an r-regular object if $\sqrt{2}r' < r$.   □

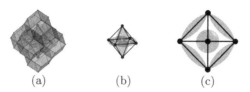

(a)   (b)   (c)

Fig. 4. Case 4 is impossible in dense digitizations (a) of r-regular objects with an fcc grid, since the path region crosses the outer surface region (b) and topview (c)

We have seen that case 4 is impossible if we use a sufficiently dense sampling grid. As one can see, in the remaining cases the intersection of one of the octahedra and the boundary of the digital reconstruction is always either empty (case 1) or homeomorphic to a disc, such that the octahedron is divided into an inner and an outer part, both homeomorphic to a ball (cases 2,3,5 and 6), see Fig. 5. This fact allows us to derive the following sampling theorem: The above theorem

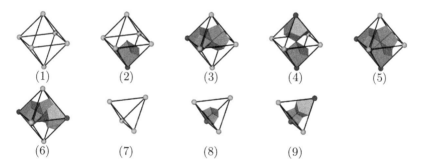

(1)   (2)   (3)   (4)   (5)

(6)   (7)   (8)   (9)

Fig. 5. Local reconstructions with the fcc grid inside the octahedra and tetrahedra. Note that for all cases except of cases 1, 4 and 7 the surface of the reconstruction inside the octahedra and tetrahedra is homeomorphic to a disc.

allows us to derive a sampling theorem for fcc grids, since it implies that the surface of the reconstruction is locally disc-shaped, which is also the case for the original r-regular object.

Theorem 2. *Let A be an r-regular object and S be an fcc r'-grid with $\sqrt{2}r' < r$. Then the digital reconstruction is $2r' + \varepsilon$-homeomorphic to A.*

Proof. Due to Theorem 1, the only cases which can occur in the digitization of an r regular object with an fcc r'-grid with $\sqrt{2}r' < r$ are cases 1 to 9 except of case 4.

Now consider a configuration of case 2,3,5 or 6 and let O denote the octahedron defined by the six sampling points. In these cases the intersection of ∂O and the boundary of the digital reconstruction is a Jordan curve (see Fig. 5). The same is true for cases 8 and 9 in the tetrahedron. Now each face F_i of the octahedron respectively tetrahedron is a triangle with its corner points having a distance of $\sqrt{2}r'$, i.e. smaller than $2r$. We define a new surface patch for such a triangle between three sampling points in the following way:

If all three sampling points p_1, p_2, p_3 lie inside of A, we take the inner surface patch. Analogously if all three sampling points lie outside of A, we take the outer surface patch.

(a)   (b)   (c)   (d)

Fig. 6. In order to construct a surface between three sampling points being not all in the foreground (a), we combine the inner (b) and the outer surface patch (c) such that the result (d) is cut by ∂A into exactly two parts

If only one sampling point p_1 lies inside of A, we use the mapping of the inner surface patch for each point lying inside the smaller triangle $\triangle(p_1, \frac{p_1+p_2}{2}, \frac{p_1+p_3}{2})$ and the mapping of the outer surface patch otherwise, see Fig. 6 for an illustration. In order to get a connected surface, we further add the straight line connections between the inner and the outer surface patch for any point lying on the straight line from $\frac{p_1+p_2}{2}$ to $\frac{p_1+p_3}{2}$. If one sampling point lies outside and the other two inside of A, we define the mapping analogously.

This leads to a surface patch between the three points which is always homeomorphic to a disc. Furthermore, since any of the added straight line connections follows a normal of ∂A and thus cuts ∂A exactly once, the intersection of the surface patch with ∂A is a simple curve.

By combining the surface patches of the octahedron respectively tetrahedron faces we get a surface homeomorphic to the octahedron respectively tetrahedron surface intersecting ∂A in a Jordan curve.

In order to guarantee that the new surface patches can be combined without topological errors, we have to show that they can only intersect in their boundaries. This is true if any two path regions can only intersect in common sampling points. Since any angle α of any of the equilateral octahedron or tetrahedron triangles is $60°$, we only have to ensure that the opening angle β of the cigar shaped path regions is smaller than $60°$. One can easily show that $\beta = 2 \arcsin\left(\frac{l}{2r}\right)$ with $l = \sqrt{2}r'$ being the distance between the two sampling points, and thus $\beta < 60°$ for $\sqrt{2}r' < r$.

If we have a octahedron of case 1 or a tetrahedron of case 7, we also can take the above surface patch construction, since it only consists of triangles lying completely inside respectively completely outside of A and thus the surface patches are well-defined. The resulting combined surface does not intersect ∂A at all.

Thus we have partitioned the whole space into deformed octahedral and tetrahedral regions separated by the new surface patches. The original object is homeomorphic to the result of the digital reconstruction inside each of the regarded regions. The combination of the local homeomorphisms (each being a $(2r' + \varepsilon)$-homeomorphism) leads to a global r-homeomorphism from A to the reconstructed set.   □

The above theorem states that an r-regular object can be reconstructed without any change in the topological information by using an fcc r'-grid with $\sqrt{2}r' < r$. This means that one only needs 4 sampling points per cube of sidelength $2r'$ and thus only a bit more than $\sqrt{2}$ sampling points per cube of sidelength r. This is much better than using a cubical sampling grid: The sampling theorem for cube grids derived in [1] needs $2r' < r$ and thus more than 8 sampling points per cube of sidelength r in order to guarantee topology preservation. This is more than 5.6 times the number of sampling points needed with an fcc-grid! See Fig. 10(5) to (7) for an example.

In addition to that topology preserving digitization on a cubic grid needs more complicated reconstruction methods than the digital reconstruction, as shown in [11] and [1] – whereas by using an fcc grid one can directly use the digital reconstruction.

There is another interesting implication of the above sampling theorem: Given an r-regular object we have to use a sampling grid with $2r' < r$ if we want to guarantee topology preservation and if we use cubic sampling grids. Now we make use of the fact that one can construct an fcc $\frac{2}{\sqrt{3}}r'$-grid by removing every second sampling point in a cubic r'-grid. Then we only need $\sqrt{2} \cdot \frac{2}{\sqrt{3}}r' < r$. Thus by throwing away half of the sampling points of the cubic grid and using the digital reconstruction on the resulting fcc grid, we can derive the correct topology at resolutions where this is not possible by reconstructing directly on the cubic grid, as can be seen in Fig. 10(8) to (13).

3.2   Digital Reconstruction on BCC Grids

In this subsection we will derive an analogous result to the ones of the last section for bcc grids. The voxels of a bcc grid are truncated octahedra having eight hexagonal and six square faces (see Fig. 2(c)). Thus two face-adjacent bcc voxels share either a hexagonal face or a square face.

Similarly to the fcc grid, we can partition the space $\mathbb{R}^3$ into octahedra based on the bcc grid. These octahedra are not directly given by the Delaunay grid. They are instead each a union of 4 tetrahedra of the Delaunay grid and they can be defined as follows: The bcc grid $S = \{\frac{2}{\sqrt{5}}(x_1, x_2, x_3)|x_1, x_2, x_3 \in \mathbb{Z}, x_1 \equiv x_2 \equiv x_3 (\mathrm{mod}2)\}$ contains the cubic grid $S' = \{\frac{4}{\sqrt{5}}(x_1, x_2, x_3)|x_1, x_2, x_3 \in \mathbb{Z}\}$, such that the remaining sampling points of $S \setminus S'$ are the center points of the dual cubes in S'. Now we can define for each face of a dual cube an octahedron by connecting its four corner points with the center points of the two adjacent dual cubes (see Fig. 2(c)). Each octahedron connects two bcc voxels sharing a square face and their common four neighbors (see Fig. 8(a)). The same partition into octahedra is used in [6] for building a polygonal surface reconstruction.

In contrast to the regular octahedra of the fcc grids, these octahedra are not regular, but square dipyramids with baselength $\frac{4}{\sqrt{5}}$ and pyramid height $\frac{2}{\sqrt{5}}$. Thus not 6 but 10 different canonical configurations have to be distinguished, see Fig. 7. Analogously to case 4 of the fcc grid we will now show that cases 4a and 4b can not occur if one digitizes an r-regular object with a sufficiently dense bcc grid.

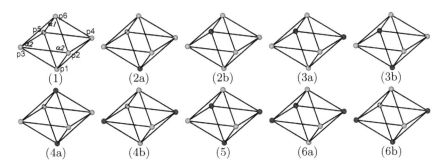

Fig. 7. There are 64 distinct configurations in a bcc grid for sampling points in an octahedron. However, up to rotational symmetry, reflectional symmetry, and complementarity (switching foreground and backgroud points), these 64 configurations are equivalent to the above 10 canonical configurations.

Theorem 3. *Configuration 4a and 4b in Fig. 7 cannot occur in the digital reconstruction of an r-regular object with a bcc r'-grid with $\sqrt{\frac{33}{10}}r' < r$.*

Proof. In the following let the dark sampling points in Fig. 7 be in the foreground and the white sampling points in the background. Further, let the sampling points $p_1, p_2, \ldots p_6$ be numbered as shown in Fig. 7(a).

Suppose configuration 4a occurs in the digital reconstruction of an r-regular object A. Since the distance from p_1 to p_6 is $\frac{4}{\sqrt{5}}r'$ and thus smaller than $2r$, there exists a foreground path between these points lying completely inside $P_s(p_1, p_6)$.

Analogously to the proof for the fcc grid, there exists a surface patch between the four points p_2, p_3, p_4, p_5, since their pairwise distance is at most $4\frac{\sqrt{2}}{\sqrt{5}}r' < 2r$. Fig. 8(b) and (c) show that $P_s(p_1, p_6)$ goes through this surface patch without intersecting the bounding r-path regions for $\sqrt{\frac{33}{10}}r' = r$ (Obviously this is also true for smaller r'). Since both p_1 and p_6 lie outside $P_s(p_2, p_3, p_4) \cup P_s(p_2, p_4, p_5)$, the path from p_1 to p_6 must go through the combined surface patch and thus there has to exist a point lying both in A and A^c.

Now suppose configuration 4b occurs in the digital reconstruction of an r-regular object A. Then the distance from p_2 to p_4 is $4\frac{\sqrt{2}}{\sqrt{5}}r'$ and thus smaller than $2r$, such that there exists a foreground path between these points lying completely inside $P_s(p_2, p_4)$.

In this case there can be constructed a surface patch between the four points p_1, p_3, p_5, p_6, since their pairwise distance is at most $4\frac{\sqrt{2}}{\sqrt{5}}r' < 2r$. Fig. 8(d) and (e) show that $P_s(p_2, p_4)$ goes through this surface patch without intersecting the bounding r-path regions for $\sqrt{\frac{33}{10}}r' = r$ (Obviously this is also true for smaller r'). Since both p_2 and p_4 lie outside $P_s(p_1, p_3, p_5) \cup P_s(p_3, p_5, p_6)$, the path from p_2 to p_4 must go through the combined surface patch and thus there has to exist a point lying both in A and A^c.

It follows that both cases 4a and 4b cannot occur in the digital reconstruction of an r-regular object if $\sqrt{\frac{33}{10}}r' < r$. $\qquad\qquad\square$

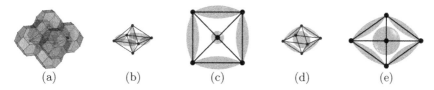

(a)   (b)   (c)   (d)   (e)

Fig. 8. Cases 4a (see (b) and (c)) and 4b (see (d) and (e)) are impossible in dense digitizations (a) of r-regular objects with a bcc grid

We have seen that cases 4a and 4b are impossible if we use a sufficiently dense sampling grid. Fig. 9 illustrates that again in any of the remaining cases the boundary of the digital reconstruction intersects the octahedron in a simple surface, being homeomorphic to a disc, such that we can derive the following sampling theorem for bcc grids:

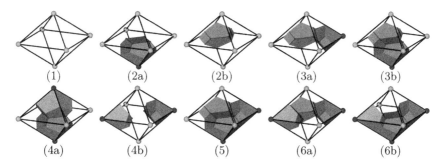

(1)   (2a)   (2b)   (3a)   (3b)

(4a)   (4b)   (5)   (6a)   (6b)

Fig. 9. Local reconstructions with the bcc grid inside the octahedra. Note that for all cases except of cases 1, 4a and 4b the surface of the reconstruction inside the octahedra is homeomorphic to a disc.

Theorem 4. *Let A be an r-regular object and S be a bcc r'-grid with $\sqrt{33/10}r'$ $< r$. Then the digital reconstruction is $4\sqrt{2/5}r' + \varepsilon$-homeomorphic to A.*

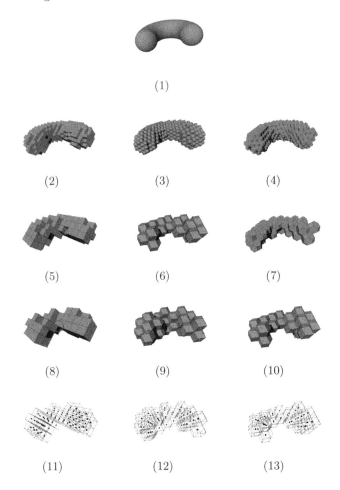

Fig. 10. Digitization of an r-regular object (1) with different r'-grids. Second row: Digital reconstruction with cubic (2), fcc (3) and bcc (4) r'-grid with $4r' = r$. As one can see, the quality of the reconstructions seems to be comparable for different grid types with the same r', while a different number of voxels is needed (972 voxels in (2), 774 voxels in (3) and 537 voxels in (4). Third row: Digital reconstruction with r' chosen near the critical values of the different grid types, i.e. $2r' = r - \varepsilon$ for the cubic grid (5), $\sqrt{2}r' = r - \varepsilon$ for the fcc grid (6) and $\sqrt{\frac{33}{10}}r' = r - \varepsilon$ for the bcc grid (7). For this example the topologically correct digital reconstruction needs 125 voxels in (5), 30 voxels in (6) and 43 voxels in (7). Fourth and fifth row: Digital reconstruction based on a cubic r'-grid with $1.5r' = r$. This resolution is not enough to guarantee topology preservation on cubic grids, as can be seen in the marked region in (8), but it allows topology preserving digitization on the two fcc subgrids (9) and (10). As can be seen in (11), (12) and (13), the sampling points of (8) are divided into two disjunct subsets of sampling points in fcc grid order.

Proof. Due to Theorem 3, cases 4a and 4b can not occur in the digitization of an r regular object with a bcc r'-grid with $\sqrt{33/10}r' < r$.

Now consider a configuration of one of the other cases except of case 1 and let O denote the octahedron defined by the six sampling points. In these cases the intersection of ∂O and the boundary of the digital reconstruction is a Jordan curve (see Fig. 9). Now each face F_i of the octahedron is a triangle with its corner points having a distance smaller than $2r$. We define a new surface patch for such a triangle between three sampling points in the same way as in Theorem 2.

Again, by combining the surface patches of the octahedron faces we get a surface homeomorphic to the octahedron surface intersecting ∂A in a Jordan curve.

In order to guarantee that the new surface patches can be combined without topological errors, we have to show that they can only intersect in their boundaries. This is true if any two path regions can only intersect in common sampling points. There exist two different angles $\alpha_1 = \arccos(1/3)$ and $\alpha_2 = 2\arcsin(1/\sqrt{3})$ in the corners of the octahedron triangles (see Fig. 7(a)). In both cases we have to ensure that these angles are bigger than the sum of the opening angles of the two corresponding path regions. While α_1 is between a triangle side of length $l_1 = 4/\sqrt{5} \cdot r'$ and a triangle side of length $l_2 = 2\sqrt{3/5}r'$, the angle α_2 is between two triangle sides of length l_2. Thus the two inequalities $\arcsin(l_1/2) + \arcsin(l_2/2) < \alpha_1$ and $2\arcsin(l_2/2) < \alpha_2$ have to be true, which is the case for $\sqrt{33/10}r' < r$.

Analogously to the proof of Theorem 2 it follows the existence of a homeomorphism from A to its digital reconstruction. This homeomorphism is a $4\sqrt{2/5}r' + \varepsilon$-homeomorphism, since $4\sqrt{2/5}r'$ is the maximal diameter of the octahedra. $\square$

Thus not only fcc grids but also bcc grids can directly be used to reconstruct an r-regular object without any change in the topological information. In case of bcc grids one only needs 2 sampling points per cube of sidelength $4/\sqrt{5} \cdot r'$ and thus only a bit more than $\frac{33}{64}\sqrt{33/2} \approx 2.09$ sampling points per cube of sidelength r. This is 3.8 times better than a cubic grid, although it is not as good as an fcc grid.

4   Conclusions

We have analysed the problems of topology preservation during digitization of r-regular objects in 3D with fcc and bcc grids. We showed that with a sufficient sampling density both fcc and bcc grids directly lead to a topology preserving reconstruction, which is not the case for cubic grids as shown in [1], where one needs more complicated reconstruction methods. Thus we derived the first sampling theorem for topology preserving digitization with non-standard grids in 3D.

Both fcc and bcc grids outperform cubic grids in the sense that (1) less sampling points are needed and (2) a bigger covering radius of the sampling grid is allowed if one wants to guarantee the correct topology of the digitized object. We got better results for the fcc grid than for the bcc one. We even showed that at certain resolutions, where a cubic grid does not guarantee topology preservation, one can remove half of the sampling points from the cubic grid such that one gets an fcc grid with a resolution which is enough to reconstruct the topology!

Acknowledgment

We thank Prof. Gunilla Borgefors for valuable comments and suggestions.

References

1. P. Stelldinger and L.J. Latecki. 3D Object Digitization with Topological and Geometric Guarantees. In *University Hamburg, Computer Science Department, Technical Report FBI-HH-M-334/05*, 2005.
2. Thomas Theußl, Torsten Möller, and Meister Eduard Gröller. Optimal regular volume sampling. In *VIS '01: Proceedings of the conference on Visualization '01*, pages 91–98, Washington, DC, USA, 2001. IEEE Computer Society.
3. Luis Ibanez, Chafiaa Hamitouche, and Christian Roux. Determination of discrete sampling grids with optimal topological and spectral properties. In *Proceedings of 6th Conference on Discrete Geometry for Computer Imagery, Lyon, France*, pages 181–192, 1996.
4. B. M. Carvalho, E. Gardu no, and G. T. Herman. Multiseeded fuzzy segmentation on the face centered cubic grid. In Springer-Verlag Ltd., editor, *Proceedings econd International Conference on Advances in Pattern Recognition ICAPR 2001*. ICAPR, Pattern Analysis and Applications journal.
5. Alois Dornhofer. A discrete fourier transform pair for arbitrary sampling geometries with applications to frequency domain volume rendering on the body-centered cubic lattice. Master's thesis, Vienna University of Technology, 2003.
6. Hamish Carr, Thomas Theußl, and Torsten Möller. Isosurfaces on optimal regular samples. In *VISSYM '03: Proceedings of the symposium on Data visualisation 2003*, pages 39–48. Eurographics Association, 2003.
7. Robin Strand and Gunilla Borgefors. Distance transforms for three-dimensional grids with non-cubic voxels. *Computer Vision and Image Understanding*, 100(3):294–311, dec 2005.
8. Robin Strand and Gunilla Borgefors. Resolution pyramids on the fcc and bcc grids. In Eric Andres, Guillaume Damiand, and Pascal Lienhardt, editors, *Discrete Geometry for Computer Imagery, 12th International Conference, DGCI 2005, Poitiers, France, April 13-15, 2005, Proceedings*, volume 3429 of *Lecture Notes in Computer Science*, pages 68–78. Springer, 2005.
9. T. Pavlidis. *Algorithms for Graphics and Image Processing*. Computer Science Press, 1982.
10. J. Serra. *Image Analysis and Mathematical Morphology*. Academic Press, 1982.
11. P. Stelldinger and U. Köthe. Towards a General Sampling Theory for Shape Preservation. *Image and Vision Computing Journal, Special Issue on Discrete Geometry for Computer Imagery*, 23(2):237–248, 2005.
12. P. Stelldinger. Digitization of Non-regular Shapes in Arbitrary Dimensions. *Submitted*.
13. L.J. Latecki, C. Conrad, and A. Gross. Preserving Topology by a Digitization Process. *Journal of Mathematical Imaging and Vision*, 8:131–159, 1998.

On the Notion of Dimension in Digital Spaces

Valentin E. Brimkov[1], Angelo Maimone[2], and Giorgio Nordo[2]

[1] Mathematics Department, SUNY Buffalo State College,
Buffalo, NY 14222, USA
`brimkove@buffalostate.edu`
[2] Dipartimento di Matematica, Università di Messina,
98166 Messina, Italy
`amaimone@dipmat.unime.it`, `giorgio.nordo@unime.it`

Abstract. Dimension is a fundamental concept in topology. Mylopoulos and Pavlidis [17] provided a definition for discrete spaces. In the present paper we propose an alternative one for the case of planar digital objects. It makes up certain shortcomings of the definition from [17] and implies dimensionality properties analogous to those familiar from classical topology. We also establish relations between dimension of digital objects and their Euler characteristic.

Keywords: digital topology, 2D binary object, dimension.

1   Introduction

Dimension is a fundamental concept in topology. It is a topological invariant [10] and plays an important role in defining and studying properties of basic geometric objects, such as curves and surfaces [11].

In digital topology the notion of dimension has attracted comparatively little attention, unlike some other topological notions (such as connectivity, tunnels, gaps, cavities, and others, see, e.g., [13, 12, 6]). In fact, already in 1971 Mylopoulos and Pavlidis [17] provided a definition of dimension for subsets of discrete spaces and, to our knowledge, it is the only one available in the literature. Recently Brimkov and Klette [5] applied that definition for defining digital curves and hypersurfaces. Overall, however, it has not been used very often in digital geometry.

In the present paper we first expose some shortcomings of the Mylopoulos-Pavlidis definition, for instance the fact that there may be 3-dimensional objects in a 2-dimensional digital space. Then we propose an alternative definition for the case of planar digital objects. Our definitions make up the "defects" of the one from [17] and imply dimensionality properties analogous to those familiar from classical topology. In particular, it makes possible to define a digital curve as a one-dimensional "digital" continuum (see Section 6), that parallels the classical definition of a curve proposed by Urysohn [19] and Menger [15]. We also provide characterization of dimension in terms of Euler characteristic, which is another basic topological invariant.

U. Eckardt et al. (Eds.): IWCIA 2006, LNCS 4040, pp. 241–252, 2006.

The paper is organized as follows. In Section 2 we introduce some notions and notations to be used in the sequel. In Section 3 we recall Mylopoulos-Pavlidis definition of dimension and discuss some undesirable phenomena within that framework. In Section 4 we present our main results. Characterization of dimension in terms of Euler characteristic is provided in Section 5. In Section 6 we use the proposed definition for defining digital curves. We conclude with some remarks in Section 7.

2   Preliminaries

Throughout we conform to the terminology used in [12] (see also [20, 13, 4]). All considerations take place in the *grid cell model* that consists of the grid cells of $\mathbb{Z}^2$, together with the related topology. In the grid cell model we represent pixels as squares, called *2-cells*. Their edges and vertices are called *1-cells* and *0-cells*, respectively. For every $i = 0, 1, 2$ the set of all *i*-cells is denoted by $\mathbb{C}_2^{(i)}$. Further, we define the space $\mathbb{C}_2 = \bigcup_{i=0}^2 \mathbb{C}_2^{(i)}$. We say that two 2-cells are 0-adjacent (1-adjacent) if $e \cap e' \in \mathbb{C}_2^{(0)}$ $(e \cap e' \in \mathbb{C}_2^{(1)})$. The relation of 0-adjacency (resp., 1-adjacency) is denoted by A_0^* (resp., A_1^*). Given a 2-cell p, by $A_0^*(p)$ and $A_1^*(p)$ we denote the A_0^* and A_1^* *adjacency neighborhood* of p, respectively, that are the sets of all 2-cells that are 0-adjacent (resp. 1-adjacent) to p. We also denote by $A_\alpha(p) = A_\alpha^*(p) \cup \{p\}$, $\alpha = 0$ or 1, the *incidence neighborhood* of p. The grid cell model can also be viewed as an *abstract cell complex* $(\mathbb{C}_2, <, dim)$ (see [14]). Here $<$ is a *bounding relation*, that is antisymmetric, irreflexive, and transitive, and such that for every $e, e' \in \mathbb{C}_2$, $e < e'$ if and only if eIe', where $I = A_0 \cup A_1$, and $dim(e) < dim(e')$. Hence $<$ is a strict partial order on $\mathbb{C}_2$ and the corresponding order topology $\tau(<)$ is called the *grid cell topology*. In this topology the *basic open sets* (i.e., the open sets of the base) are precisely the sets $U \subseteq \mathbb{C}_2$, such that, for every $u \in U$ and every $v \in \mathbb{C}_2$ with $u < v$, we have $v \in U$.

A *digital object* (or *digital picture*) D is any finite set of pixels in $\mathbb{C}_2^{(2)}$.

Next we recall some other notions of digital geometry. Let $\alpha = 0$ or 1. An α-*path* in a digital object D is a sequence of pixels from D such that every two consecutive pixels are α-adjacent. Two pixels are α-*joined* if there is an α-path between them. A digital object D is α-*connected* if there is an α-path joining any two pixels of D. Otherwise D is α-*disconnected*.

Let M be a subset of a binary picture S. If $S \setminus M$ is not k-connected, then the set M is said to be k-*separating* in S. Now let M be a finite set of pixels that is k-separating in $\mathbb{C}^2$ (note: $k = 0$ or $k = 1$). The infinite 1-component of $\mathbb{C}^2 \setminus M$ is called the *background component* of M, while the other (finite) 1-components of $S \setminus M$ are called *1-holes* of M (see Fig. 1).

Finally, we will call a *2-block* any 2×2 square of pixels, a *1-block* any 2×1 rectangle of pixels, and an *L-block* any 2-block with exactly one pixel missing (see Figure 2a,b).

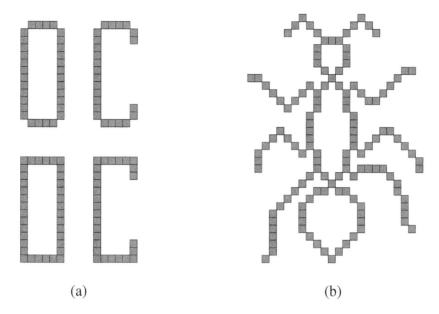

(a)   (b)

Fig. 1. a) Both '0's define one hole each, the former with respect to 0-adjacency and the latter to 1-adjacency. The 'C's a hole-free. b) A general digital 0-curve with three holes.

3   Review of Mylopoulos-Pavlidis Theory of Dimension

Mylopoulos and Pavlidis [17] proposed definition of dimension of a (finite or infinite) set of n-cells $D \subseteq \mathbb{C}_n$ with respect to an adjacency relation A_α (see [16] for more details; for its use see also [12]).

Let $\overline{A_\alpha}(c)$ be the union of $A_\alpha(c)$ with all n-cells c' for which there exist $c_1, c_2 \in A_\alpha^*(c)$ such that a shortest α-path from c_1 to c_2 not passing through c passes through c'. Note that for $n = 2$ we have $\overline{A_1}(c) = \overline{A_0}(c) = A_0(c)$.[1] We also denote $\overline{A_\alpha}^*(c) = \overline{A_\alpha}(c) \setminus \{c\}$.

A nonempty set $D \subseteq \mathbb{C}_n$ is called *totally α-disconnected* iff $A_\alpha^*(x) \cap D = \emptyset$ for each $x \in D$.

$D \subseteq \mathbb{C}_n$ is called *linearly α-connected* whenever $|A_\alpha^*(x) \cap D| \leq 2$ for all $x \in D$ and $|A_\alpha^*(x) \cap D| > 0$ for at least one $x \in D$.

Definition 1. *Let D be a digital object and A_α an adjacency relation on $\mathbb{C}_n$. The dimension $\dim_\alpha(D)$ is defined as follows:*

(1) $\dim_\alpha(D) = -1$ if and only if $D = \emptyset$,
(2) $\dim_\alpha(D) = 0$ if D is a totally α-disconnected nonempty set (i.e., there is no pair of cells $c, c' \in D$ such that $c \neq c'$ and $\{c, c'\}$ is α-connected),

---

[1] For higher dimensions these sets may not coincide; e.g., for $n = 3$, we have $\overline{A_2}(c) = \overline{A_1}(c) = A_1(c)$ and $\overline{A_0}(c) = A_0(c) \neq A_1(c)$, i.e., $\overline{A_2}(c) \neq \overline{A_0}(c)$.

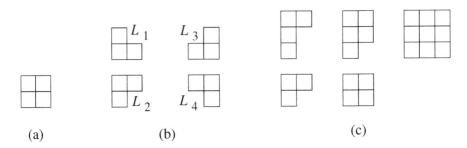

Fig. 2. a) 2-block. b) The four possible L-blocks. c) *Left:* 2-dimensional (top) and 1-dimensional (bottom) objects with respect to A_0; *Middle:* 2-dimensional (top) and 1-dimensional (bottom) objects w.r.t. A_1; *Right:* 3-dimensional object w.r.t. A_0.

    (3) $\dim_\alpha(D) = 1$ if D is linearly α-connected,
    (4) $\dim_\alpha(D) = \max\limits_{c \in D} \dim_\alpha(\overline{A_\alpha}^*(c) \cap D) + 1$ otherwise.

The following characterization of 2-dimensionality in $\mathbb{C}_2$ was given in [12]:

Proposition 1. $M \subseteq \mathbb{C}_2$ is two dimensional with respect to adjacency relation A_α if and only if:

    For $\alpha = 0$, M contains an L-block as a proper subset;
    For $\alpha = 1$, M contains a 2-block as a proper subset.

The above proposition suggests that a 2-dimensionality of a digital object is equivalent to existence of L- (resp. 2-) blocks in a digital object. Note however that, according to Definition 1, an L-block (resp. 2-block) itself, is one-dimensional with respect to A_0 (resp. A_1). To us, this is a shortcoming of this definition.

    Another "defect" of the definition, which seems to be even more serious to us, is that a digital object in the two-dimensional digital space $\mathbb{C}_2$ may have dimension three! This can be easily seen if we apply Definition 1 to an object that contains a (3×3)-block (see Figure 2c (right)).

    With the above examples in mind, in the next section we provide definitions of dimension, through which the above problems are resolved.

4   Main Results

Definition 2. Let D be a digital object and let the space $\mathbb{C}_2$ be equipped with an adjacency relation A_α, $\alpha \in \{0,1\}$. The dimension of D relative to A_α adjacency is denoted by $dim_\alpha(D)$ and defined as follows:

    1. $dim_\alpha(D) = -1$ if $D = \emptyset$,
    2. $dim_\alpha(D) = 0$ if D is totally α-disconnected,
    3. $dim_\alpha(D) = 1$ if: $\alpha = 0$ and D is not totally α-disconnected and does not contain any L-block; or $\alpha = 1$ and D is not totally α-disconnected and does not contain any 2-block,

4. $dim_\alpha(D) = 2$ otherwise (more precisely, if $\alpha = 0$ and D contains at least one L-block or $\alpha = 1$ and D contains at least one 2-block).

Remark 1. Note that in Definition 2, points (3) and (4) can be reformulated by using mathematical morphology [18, 9]. (Remember that in these cases D is not totally α-disconnected.) More precisely, we can define $dim(D) = 1$ iff $\alpha = 1$ and $\epsilon_B(D) = D \ominus B = \emptyset$ where the structuring element B is a 2-block, or $\alpha = 0$ and $\bigcup_{i=1}^4 \epsilon_{L_i}(D) = \bigcup_{i=1}^4 D \ominus L_i = \emptyset$ where L_i, $i = 1, \ldots, 4$, represents all possible L-blocks (see Figure 2b displaying the four possible L-blocks). Further, $dim(D) = 2$ iff $\alpha = 1$ and $\epsilon_B(D) \neq \emptyset$, or $\alpha = 0$ and $\bigcup_{i=1}^4 \epsilon_{L_i}(D) \neq \emptyset$.

In order to give a sort of "local" characterization of dimension, we now define dimension of a point of a digital object D.

Definition 3. *Let D be a nonempty digital object and $p \in D$. The local dimension of p within D with respect to A_0 is denoted $dim_0(p, D)$ and defined as follows:*

1. $dim_0(p, D) = 0$ if $A_0(p) \cap D = \emptyset$;
2. $dim_0(p, D) = 1$ if $A_0(p) \cap D$ is totally 0-disconnected;
3. $dim_0(p, D) = 2$ otherwise (i.e., if $A_0(p) \cap D$ is not totally 0-disconnected).

Lemma 1. *Let D be a nonempty digital object and $p \in D$. Then $dim_0(p, D) = 2$ iff p belongs to an L-block in D.*

Proof. Let $dim_0(\tilde{p}, D) = 2$, i.e, $A_0(p) \cap D$ is not totally 0-disconnected. Then there are at least two pixels $p_1, p_2 \in A_0(p) \cap D$ such that $(p_1, p_2) \in A_0$. Up to symmetries with respect to $\tilde{p}$, there are only two possible cases: $(p_1, p_2) \in A_1$ or $(p_1, p_2) \in A_0 \setminus A_1$. In both, the pixels $\tilde{p}, p_1, p_2$ form an L-block (see Figure 3a). Conversely, if p belongs to an L-block, then $A_0(p) \cap D$ is not totally 0-disconnected and, by Definition 3, $dim_0(p, D) = 2$. □

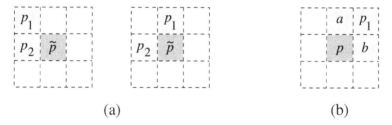

(a)   (b)

Fig. 3. a) Illustration to the proofs of Lemma 1. b) Illustration to the proofs of Proposition 4.

Remark 2. As a consequence of Lemma 1, we have that if in a digital object there is a pixel of dimension 2, then the same object must contain at least two other distinct pixels of the same dimension.

The next proposition characterizes 0-dimensionality in terms of local 0-dimensionality.

Proposition 2. *Let D be a nonempty digital object. Then $dim_0(D) = \max\{dim_0(p, D) : p \in D\}$.*

Proof. First let us suppose that $dim_0(D) = 0$, i.e., D is totally 0-disconnected. This is equivalent to saying that for every $p \in D$, $A_0(p) \cap D = \emptyset$, that is, $dim_0(p, D) = 0$. Hence $\max\{dim_0(p, D) : p \in D\} = 0$.

Now suppose that $dim_0(D) = 1$, i.e., there is no L-block in D. By Lemma 1, for every $p \in D$, $dim_0(p, D) \leq 1$ and at least one pixel has dimension one. Thus $\max\{dim_0(p, D) : p \in D\} = 1$. Conversely, suppose that $\max\{dim_0(p, D) : p \in D\} = 1$, i.e., for every $p \in D$, $dim_0(p, D) \leq 1$ and there exists some $\widetilde{p} \in D$ such that $dim_0(\widetilde{p}, D) = 1$. By Lemma 1 it follows that no pixel of D can belong to an L-block. So, by Definition 2, $dim_0(D) \leq 1$. Since $dim_0(\widetilde{p}, D) = 1$, it follows that $dim_0(D) \neq 0$. Hence, $dim_0(D) = 1$.

Finally, let $dim_0(D) = 2$. Then D contains at least one L-block L. By Lemma 1, every pixel $p \in L$ is such that $dim_0(p, D) = 2$. Thus $\max\{dim_0(p, D) : p \in D\} = 2$. Conversely, suppose that $\max\{dim_0(p, D) : p \in D\} = 2$. Then there exists at least one pixel $\widetilde{p} \in D$ such that $dim_0(\widetilde{p}, D) = 2$ and, by Lemma 1, $\widetilde{p}$ belongs to some L-block L. This implies that $dim_0(D) = 2$. □

We also have the following property.

Proposition 3. *Let D be a nonempty digital object and let $p \in E \subseteq D$. Then $dim_0(p, E) \leq dim_0(p, D)$.*

Proof. Suppose by contradiction that $dim_0(p, E) > dim_0(p, D)$. We have the following cases:

1. $dim_0(p, E) = 2$ and $dim_0(p, D) \leq 1$,
2. $dim_0(p, E) = 1$ and $dim_0(p, D) = 0$.

Consider first case 1. Since $dim_0(p, E) = 2$, we have that $A_0(p) \cap E$ is not totally 0-disconnected with respect to A_0. We have however that $A_0(p) \cap E \subseteq A_0(p) \cap D$. This means that $A_0 \cap (D)$ is not totally 0-disconnected with respect to A_0, too. Then $dim_0(p, D) = 2$, which is a contradiction.

Now consider case 2. Since $dim_0(p, E) = 1$, we have that $A_0(p) \cap E$ is a non-empty totally 0-disconnected set. However, $A_0(p) \cap E \subseteq A_0(p) \cap D$. So, $A_0(p) \cap D$ cannot be the empty set. This implies that $dim_0(p, D) \neq 0$, which is a contradiction. □

Propositions 4 and 3 immediately imply the following.

Corollary 1. *Let D be a nonempty digital object and let $E \subseteq D$. Then $dim_0(E) \leq dim_0(D)$.*

Definition 3 and the related results allow us to give the following definition of local dimension with respect to A_1 adjacency.

Definition 4. *Let D be a nonempty digital object and let $p \in D$. The local dimension of p within D with respect to A_1 is the nonnegative integer $dim_1(p, D) = dim_0(A_0(p) \cap D)$.*

Lemma 2. *Let D be a nonempty digital object and $p \in D$. Then $dim_1(p, D) = 2$ if and only if p belongs to a 2-block in D.*

Proof. Let $dim_1(p, D) = 2$. Then by Definition 4 we have that $dim_0(p, (A_0(p) \cap D)) = 2$. Therefore, by Definition 2, $A_0(p) \cap D$ contains at least one L-block L. This implies that $B = \{p\} \cup L$ is a 2-block of D and p a pixel of its. Now let p belong to a 2-block. Then we have that $A_0(p) \cap D$ contains an L-block. Therefore $dim_0(A_0(p) \cap D) = 2$. That is, by Definition 4, $dim_1(p, D) = 2$. □

We also have the following characterization of dimension with respect to A_1 adjacency.

Proposition 4. *Let D be a nonempty digital object. Then $dim_1(D) = \max \{dim_1(p, D) : p \in D\}$.*

Proof. Suppose first that $dim_1(D) = 0$. Then D is totally 1-disconnected. This is equivalent to saying that for every $p \in D$, $A_1(p) \cap D = \emptyset$. Then, for every $p \in D$, we have that $A_0(p) \cap D$ is totally 0-disconnected, i.e., $dim_0(A_0(p) \cap D) = 0$. In fact if, by contradiction, $A_0(p) \cap D$ was not totally 0-disconnected, then there should exist at least two 0-joined pixels $p_1, p_2 \in A_0 \cap D$. Wlog, consider pixel p_1. Clearly, it must belong to $A_0(p) \setminus A_1(p)$, since otherwise $A_1(p) \cap D \neq \emptyset$. However, p_2 is 0-adjacent to p_1. So, it must be in one of the positions a or b depicted in Figure 3c. Hence, $p_2 \in A_1(p) \cup D$, which is clearly a contradiction. All this implies that $\max\{dim_0(A_0(p) \cap D) = dim_1(p, D) : p \in D\} = 0$.

Now, let us suppose that $dim_1(D) = 1$, i.e., that there is no 2-block contained in D. By Lemma 2, for every $p \in D$, $dim_1(p, D) \leq 1$ and there is at least one pixel $\widetilde{p}$ such that $dim_1(\widetilde{p}, D) = 1$. Thus $\max\{dim_1(p, D) : p \in D\} = 1$. Conversely, suppose that $\max\{dim_1(p, D) : p \in D\} = 1$. Then for every $p \in D$, $dim_1(p, D) \leq 1$ and there exists some $\widetilde{p} \in D$ such that $dim_1(\widetilde{p}, D) = 1$. By Lemma 2, it follows that no pixel of D can belong to a 2-block. Therefore, by Definition 2, $dim_1(D) \leq 1$. Since $dim_1(\widetilde{p}, D) = 1$, $dim_1(D) \neq 0$. So, $dim_1(D) = 1$.

Finally, suppose that $dim_1(D) = 2$. Then D contains at least one 2-block B. By Lemma 2, every pixel $p \in B$ has dimension 2. Thus $\max\{dim_1(p, D) : p \in D\} = 2$. Conversely, suppose that $\max\{dim_1(p, D) : p \in D\} = 2$. Then there exists at least one pixel $\widetilde{p} \in D$ such that $dim_1(\widetilde{p}, D) = 2$ and, by Lemma 2, $\widetilde{p}$ belongs to some 2-block B. This implies that $dim_1(D) = 2$. □

Corollary 2. *Let D be a digital object. Then $dim_1(D) \leq dim_0(D)$.*

Proof. Let $p \in D$. Then $dim_0(p, (A_0(p) \cap D)) \leq dim_0(p, D) \leq \max\{dim_0(p, D) : p \in D\} = dim_0(D)$. Hence $dim_0(D)$ is greater than each element of the set $\{dim_0(p, A_0(p) \cap D) : p \in D\}$. So, $dim_1(D) = \max\{dim_1(p, D) : p \in D\} = \max\{dim_0(p, (A_0(p) \cap D)) : p \in D\} \leq dim_0(D)$. □

In a similar way one can prove the following properties that parallel well-known properties of dimension theory in $\mathbb{R}^n$.

Proposition 5. *Let D be a nonempty digital object and $p \in E \subseteq D$. Then $dim_1(p, E) \leq dim_1(p, D)$.*

Proposition 6. *Let D be a nonempty digital object and $E \subseteq D$. Then $dim_1(E) \leq dim_1(D)$.*

Proposition 7. *Let D_1 and D_2 be two mutually disjoint digital objects. Then $dim_\alpha(D_1 \cup D_2) = \max(dim_\alpha(D_1), dim_\alpha(D_2))$, where $\alpha \in \{0, 1\}$.*

5 Dimension and Euler Characteristic

In this section we establish relations between dimension of digital objects and their Euler characteristic. In combinatorial topology, Euler characteristic is a fundamental theoretic concept and basic topologic invariant. Recall that, given a subset D of the abstract cell complex $(\mathbb{C}_2, <, dim)$, its Euler characteristic is the number

$$\chi(D) = c_0 - c_1 + c_2, \tag{1}$$

where c_i is the number of the i-dimensional cells of D, $i = 0, 1, 2$. In describing our results, we will also use the notion of a *skeleton* of a digital object, which we introduce next.

Definition 5. *Let D be a non-empty digital object and let the space $\mathbb{C}_2$ be equipped with an adjacency relation A_α, $\alpha \in \{0, 1\}$. We call a skeleton of D the graph $S_\alpha(D) = (V, E)$ ($S(D)$, for short), whose set of vertices V are labeled by the elements of D (i.e., we may think that $V = D$), and, given two vertices p and q, $(p, q) \in E \iff p$ and q are α-adjacent.*

In what follows, we will characterize dimensionality in $\mathbb{C}_2$ with respect to A_1 adjacency, the characterization with respect to A_0 adjacency being similar. Because of Proposition 4, it is enough to consider the case of connected digital objects. We have the following theorem.

Theorem 1. *Let D be a 1-connected digital object with a skeleton $S(D) = (V, E)$. In terms of the denotations in equality 1, $|V| = |D| = c_2$. Let $|E| = m$. Then the following holds:*

1. *$dim_1(D) = -1$, if $c_2 = 0$*
2. *$dim_1(D) = 0$, if $c_2 \neq 0$ and $m = 0$*
3. *$dim_1(D) = 1$, if $c_2 > m > 0$*
4. *If $m = c_2 > 0$, then*
   (a) $dim_1(D) = 1$ if $\chi(D) = 0$
   (b) $dim_1(D) = 2$ if $\chi(D) > 0$
5. *If $m > c_2 > 0$,*
   (a) $dim_1(D) = 1$ if $\chi(D) < 0$
   (b) $dim_1(D) = 2$ if $\chi(D) \geq 0$

Before proving the above theorem, we recall some well-known elementary properties from graph theory (see, e.g., [3]).

Proposition 8. *Let G be a connected graph with n vertices and m edges. We have that:*

1. $n \leq m + 1$,
2. G is a tree iff $n = m + 1$, or, equivalently, if G has no cycle,
3. G has a unique cycle C_i of length $i \geq 3$ iff $m = n$.

We will also use the following simple lemma.

Lemma 3. *Let G be a connected graph with n vertices and m edges. Then G has at least two distinct cycles iff $m > n$.*

Proof. Suppose that G has at least two cycles. Then, by Part 2 of Proposition 8, G is not a tree, and $n < m + 1$. Since G has at least two cycles, by Part 3 of Proposition 8, we have that $n \neq m$. Hence, $m > n$.

Conversely, if $m > n$, by Part 2 of Proposition 8 we have that G is not a tree. Since G is connected, it has at least one cycle. If we assume by contradiction that there are no other cycles, by Part 3 of Proposition 8 we will have $m = n$ - a contradiction. □

Let us also list the following fact.

Lemma 4. *Let $D \subset \mathbb{C}^2$. Then*

$$c_0 - c_1 = c - h - c_2, \tag{2}$$

where c and h are the number of the (0-)connected components and (1-)holes of D, respectively.

Indeed, the following Euler-Poincare result is well-known in combinatorial topology:

$$\chi(D) = c_0 - c_1 + c_2 = \beta_0 - \beta_1 + \beta_2,$$

where β_0, β_1, and β_2 are the Betti numbers (see, e.g., [12]). These count respectively the number of connected components, tunnels, and cavities of a cell-complex. Since a plane digital object D is homotopic to a one-dimensional CW-complex [21], we clearly have $\beta_2 = 0$, from where we get the result stated.

We are now ready to prove Theorem 1.

Proof of Theorem 1. For Cases 1 and 2 the proof is trivial. We prove for the other three cases.

Case 3. Let $c_0 > m$. By Proposition 8 (Part 2), $S(D)$ is a tree. Suppose, by contradiction, that $dim_1 D \neq 1$. Since $n > m > 0$, it necessarily follows that $dim_1(D) = 2$, i.e., D admits at least one 2-block. Hence, $S(D)$ has at least one cycle C_4, which contradicts the fact that $S(D)$ is a tree.

Case 4. Let $c_0 = n$ and $\chi(D) > 0$. Suppose, by contradiction, that $dim_1(D) = 1$. Then D contains no 2-blocks or, equivalently, $S(D)$ contains no cycle C_4 as a subgraph. Then by Proposition 8 (Part 3), $S(D)$ must have at least one cycle C_i, where i is an even number greater than or equal to 8.

Since D is connected, $c = 1$ and the equality of Lemma 4 becomes $c_1 - c_0 = c_2 + h - 1$ $(c_0, c_1, c, h \neq 0)$. Since $c_1 - c_0 < c_2$, we have that $c_2 + h - 1 < c_2$. It follows that $h < 1$, that is, $h = 0$, a contradiction.

Now let $c_2 = m$ and $\chi(D) = 0$. Suppose, by contradiction, that $\dim_1(D) = 2$. So, D contains at least a 2-block and, consequently its skeleton has a C_4 subgraph.

As before, by Lemma 4, $c_1 - c_0 = c_2 + h - 1$. Since $\chi(D) = c_0 - c_1 + c_2 = 0$, we obtain $c_2 + h - 1 = c_2$, i.e., $h = 1$. By Lemma 3, this implies that $m > c_2$, a contradiction.

Case 5. Let $m > c_2 > 0$ and $c_0 - c_1 + c_2 \geq 0$. Suppose by contradiction that $\dim_1(D) = 1$. Then D has no 2-block and, consequently, $S(D)$ has no cycle C_4 as a subgraph. Since $m > c_2$, we have by Lemma 3 that $S(D)$ has at least two cycles different than C_4. Hence, $h > 1$. Since $\chi(D) = c_0 - c_1 + c_2 \geq 0$ and from Lemma 4, it follows that $h \leq 1$, a contradiction.   □

Remark 3. It is easy to see that Theorem 1 covers all possibilities. Thus a digital object D has dimension k, $-1 \leq k \leq 2$ only if D satisfies a corresponding condition of the theorem. Formally, the case "$c_2 = m$ and $\chi(D) < 0$" is not considered. However, it is easily verified to be non-admissible. In fact, by Proposition 8 (Part 3) we have that $S(D)$ has a unique cycle. Here we distinguish two subcases. Case (a): The unique cycle is C_4. We have $c = 1$ and $h = 0$. Case (b): The unique cycle is C_{2n}, $n > 3$. In this case $c = 1$ and $h = 1$. So, in both cases $1 = c \geq h$, i.e., $\chi(D) \geq 0$. However, $\chi(D) = c - h = 1 - h < 0$, which implies $1 < h$, a contradiction.

Lemma 4 and Theorem 1 imply the following corollary.

Corollary 3. *Let D be a 1-connected digital object whose skeleton $S(D)$ has c_2 vertices and m edges. Then the following implications hold:*

1. *If $c_2 > m$ then $\dim_1(D) = 1$,*
2. *If $c_2 = m$ and $h < 1$ then $\dim_1(D) = 2$,*
3. *If $c_2 = m$ and $h = 1$ then $\dim_1(D) = 1$,*
4. *If $m > c_2$ and $h \leq 1$ then $\dim_1(D) = 2$,*
5. *If $m > c_2$ and $h > 1$ then $\dim_1(D) = 1$.*

Characterization of dimension under 0-adjacency is similar to one under 1-adjacency. The points from 1 through 4 of Theorem 1 can be reformulated also for 0-adjacncy. Note that, since there may be 1-dimensional objects with $m > c_2 > 0$ and $\chi(D) < 0$, the last item of the theorem changes as follows:
    5′. If $m > c_2 > 0$ and $\chi(D) \geq 0$, then $\dim_0(D) = 2$.

6   Application to Digital Curves and Surfaces

Various definitions of digital curves and surfaces are available in the literature (see, e.g., [8]). A short survey of these is found in [5]. This last paper provided the first definitions of digital curves and hypersurfaces involving the notion of dimension as introduced in [17]. We give a more general definition of a digital curve next.

Recall that, since Urysohn [19] and Menger [15], a curve $\gamma \subset \mathbb{R}^2$ is known to be a one-dimensional continuum[2]. By analogy, we can define *digital continuum* as a nonempty, finite, and (α-)connected set of cells in a digital space. Then the classical Urysohn-Menger's definition would apply to the case of digital curves, as well:

Definition 6. *A digital curve $\gamma \subset \mathbb{C}_2$ (with respect to a certain adjacency relation) is a one-dimensional digital continuum.*

See Figure 4 for some illustration. The above definition straightforwardly generalizes for digital curves in a space of arbitrary dimension.

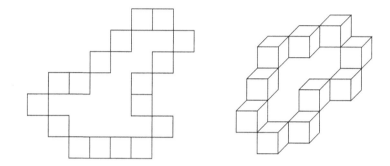

Fig. 4. Digital curves. *Left:* 0-curve in the two-dimensional digital space. *Right:* 1-curve in the three-dimensional digital space.

Similarly, one can define a digital surface and other topological objects. We believe this could be one more step towards developing a unified topological theory for both continuous and discrete spaces.

7   Concluding Remarks

In this paper we proposed definitions of dimension for planar digital objects for which the dimension is an essential characteristic. They serve as an alternative to the one proposed by Mylopoulos and Pavlidis [17], and make up some shortcomings of the latter. We believe that the notion of object dimension in digital spaces will play an increasing role in theoretical research, helping to make notions and results of digital topology compatible with those from classical topology. Ongoing research is focused on extending the presented definitions and results to higher dimensions and these will be included in the full-length journal version of the paper.

---

[2] *Continuum* in $\mathbb{R}^2$ is a nonempty subset of a topological space that is compact (closed and bounded) and topologically connected.

Acknowledgments

The authors are indebted to the three anonymous referees for their useful remarks and suggestions. The first author thanks Reinhard Klette for a number of useful discussions on the subject of this paper.

This research has been supported in part by a CNR (G.N.S.A.G.A.) and M.U.R.S.T. funds through "Gruppo Topologia e Geometria," Italy.

References

1. Andres, E., R. Acharya, and C. Sibata. Discrete analytical hyperplanes. *Graphical Models and Image Processing* **59** (1997) 302–309
2. Andres, E., Ph. Nehlig, and J. Françon, Tunnel-free supercover 3D polygons and polyhedra, In: D. Fellner and L. Szirmay-Kalos (Guest Eds.), *EUROGRAPHICS'97*, 1997, pp. C3-C13
3. Berge, C., *Graphs and Hypergraphs*, North-Holland, Amsterdam, 1976
4. Brimkov, V.E., E. Andres, and R.P. Barneva, Object discretizations in higher dimensions, *Pattern Recognition Letters* **23** (2002) 623–636
5. Brimkov, V.E., and R. Klette, Curves, hypersurfaces, and good pairs of adjacency relations, *LNCS 3322* (2004) 270–284
6. Brimkov, V.E., A. Maimone, G. Nordo, R. Barneva, and R. Klette, The number of gaps in binary pictures, In: G. Bebis et al. (Eds.), *International Symposium of Visual Computing, LNCS 3804* (2005) 35–42
7. Brimkov, V.E., A. Maimone, G. Nordo, Counting gaps in binary pictures, Proc. of *IWCIA'06*, LNCS (2005)
8. Chen, Li, *Discrete Surfaces and Manifolds*, Scientific & Practical Computing, 2004
9. Dougherty, E.R. and R. Lotufo, Hands on Morphological Image Processing, SPIE Press, Vol. TT59, Washington, 2003
10. Engelking, R., *General Topology*, Heldermann, Berlin, 1989
11. Hurewicz, W., and H. Wallman, *Dimension Theory*, Princeton U. Press, NJ, 1941
12. Klette, R., and A. Rosenfeld, *Digital Geometry - Geometric Methods for Digital Picture Analysis*, Morgan Kaufmann, San Francisco, 2004
13. Kong, T.Y., Digital topology, In: Davis, L.S., editor. *Foundations of Image Understanding*, Kluwer, Boston, Massachusetts, 2001, pp. 33–71
14. Kovalevsky, V.A., Finite topology as applied to image analysis, *Computer Vision, Graphics and Image Processing*, **46**(2) 141–161
15. Menger, K., *Kurventheorie*, Teubner, Leipzig, Germany, 1932
16. Mylopoulos, J.P., On the definition and recognition of patterns in discrete spaces, Ph.D. Thesis, Princeton University, Aug. 1970 [Tech. Rep. 84, Comput. Sci. Lab., Princeton University, Princeton, NJ]
17. Mylopoulos, J.P., and T. Pavlidis, On the topological properties of quantized spaces. I. The notion of dimension, *J. ACM* **18** (1971) 239–246
18. Serra, J., *Image Analysis and Mathematical Morphology*, Academic Press, London, 1982
19. Urysohn, P., Über die allgemeinen Cantorischen Kurven, Annual meeting, Deutsche Mathematiker Vereinigung, Marbourg, Germany, 1923
20. Voss, K., *Discrete Images, Objects, and Functions in* $\mathbf{Z}^n$, Springer Verlag, Berlin, 1993
21. Weisstein, W., CW-Complex, MathWorld – A Wolfram Web Resource, http://mathworld.wolfram.com/CW-Complex.html

Size and Shape Measure of Particles by Image Analysis

Weixing Wang

School of Electronic Engineering,
University of Electronic Science and Technology of China, Post code: 610054, China
wxwang@ee.uestc.edu.cn or znn525d@yahoo.com

Abstract. This paper presents an image analysis measurement algorithm - best-fit rectangle for particle size and shape. The best-fit rectangle approach is a combination of the Ferret method and the least 2nd moments minimization, only requiring calculation of three moments about the center of gravity, and maximum and minimum co-ordinates in a co-ordinate system oriented in the direction of the axis of least 2nd moments, and a simple area ratio. It is a simple rotation-invariance method, reflecting shape (Elongation and angularity). The algorithm is introduced theoretically in details, analyzed and compared to other widely used methods, and has been tested by a large number of solid particle samples in a laboratory. The test results show that by using this method, the results are very close to manual measurements.

1 Introduction

Image analysis techniques have been used for particle size measurement in the last twenty years [1-12]. As computers are widely used today, the cost of an image system is low, and particle size and shape analysis can be handled easily and fast. Within image analysis two dimensional images are normally analyzed. The image analysis techniques rely on first obtaining a digitized outline of each individual particle from a photograph or a video film, then measuring the size and shape parameters of the particle by computer and software. In image analysis, different ways of measuring particle size, such as chord size [1], equivalent circle diameter [2], maximum size [3], size of equivalent ellipse [4], and simple Ferret diameter [5] have been used. A reasonable representation of particle size should reflect the shape of particles. It can also be application dependent. Size determinations used in the previous studies [1-5], were often based on traditional measuring methods, not image analysis. Few researchers analyze if the determinations are reasonable in discrete image measuring procedures.

As investigated, if one uses a bounded rectangle to characterize the shape of a particle, the area ratio (R) between a particle and its bounded rectangle, may be a parameter to show the shape or rectangularity (angularity). It is convenient to classify the crushed solid particles into five basic shapes (see Fig. 1):

a) Triangle-like: a particle with three main sides, R will be about 50%;
b) Diamond-like: a particle with four main sides, but R will be about 50-60%;
c) Trapezoid-like: a particle with four main sides, R will be about 60-90%;
d) Polygon-like: a particle with more than four main sides, R can be about 60-90%;
e) Rectangle-like: a particle with four main sides, but R will be over 90%.

U. Eckardt et al. (Eds.): IWCIA 2006, LNCS 4040, pp. 253 – 262, 2006.

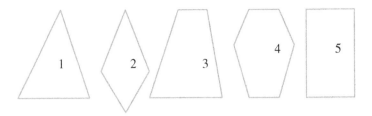

Fig. 1. Classification of crushed solid particles

The corners of categories (2)-(4) particles are somewhat rounded. Note also that we use the suffix "like" to stress that this is a crude shape description of the particle boundary. Figure 2, below, shows real particle shapes, where twelve solid particles of size 32-64 mm were randomly sampled from a rockpile in a quarry. They represent the five different shapes more or less. This is an underlying model in the discussion and evaluation of methods in this paper. The question is how to make the bounded rectangle – best-fit rectangle for each of the particles by image analysis, and to ensure the size and shape measurement rotation-invariant.

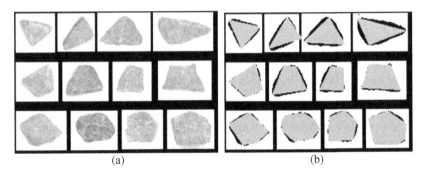

Fig. 2. Shapes of crushed solid particles. (a) The original image includes 12 particles. (b) Particles are constructed by triangles, diamonds, trapezoids and polygons.

A method of solid particle size measurement should consider the following aspects: (1) solid particles locate and orient randomly in an image, we cannot expect that orientations of particles are in a certain orientation in an image; (2) solid particles are irregular in shape; (3) in a quarry or mining industry, the boundaries of solid particles are *very rough*, owing to either their physical properties or technique of photograph or image digitizing and (4) even solid particles having the same size or elongation, they may have extremely different shapes.

Based on the above considerations, any method for aggregate size and shape measurements should meet the following requirements:

(1)   A method should have a size definition in which no matter how a particle is positioned in an image, the measured size should be unique, independent of image scanning direction or particle rotation, which is called rotational-invariance.

(2) The first shape parameter is Elongation. Flakiness (the ratio between thickness and length) or Elongation is a very important piece of information in aggregate industry in addition to particle size. Mora and Kwan, in their study [5-6], described several shape factors as the functions of Elongation.

(3) The second shape parameter is angularity and should be robust, in reference [5-6], the authors found that angularity definition varies among different researchers, and the traditional method or definition is un-reasonable. Here, we mean crude expression of angularity, not angularity description in detail. Two particles may have the same size and elongation, but their angularity is different. This parameter is also important information in aggregate industry.

(4) The measurement itself should be as simple as possible; Complicated algorithms or measurements are not necessarily suitable for multiple solid particle measurements (e.g. for an on-line system). As is well known, Fourier and Fractal measurement methods are very good for characterizing particle size and shape, but they are neither suitable for a single particle, nor for statistical analysis of multiple solid particles.

If a method of solid particle size measurement by image analysis meets the above four basic conditions, the measurement will be stable, and size and shape can be reproducible. Based on the four conditions, let us analyze and evaluate the existing and widely used methods.

Table 1 shows a comparison of six measurement methods. One can see that the Multiple Ferret method meets best the four conditions described above. The remaining question is to improve it, which is the topic of next section.

Table 1. Evaluation of the existing particle size and shape measurement methods

Methods	Rot. invariance	Elongation	Angularity	Simple
Chord sizing	No	No	No	Yes
Simple Ferret	No	No	No	Yes
Maximum diameter	Yes	No	No	No
Equivalent circle	Yes	No	No	Yes
Equivalent ellipse	Yes	Yes, not 100%	No	No
Multiple Ferret	Yes, not 100%	Yes	Yes, not 100%	No

2 New Method of Solid Particle Size and Shape Measurement

Empirically, as mentioned, the shapes of solid particles vary between triangular-like, diamond-like, trapezoidal-like, polygon-like and rectangular-like with intermediate shapes in between.

To make a measurement method rotationally invariant, reproducible, of low boundary roughness sensitivity and reflecting crude shape, a new measurement method, called the best-fit rectangle, has been developed. A summary of the measurement sequence is as follows:

(a) Obtain orientation of a particle by using least-second moment method (= rotationally invariant) which yields a simple closed formula for the orientation.

(b) Length and width can be obtained by using a Ferret box in the orientation of the least 2nd moment. Thus, rotation- invariant elongation is defined implicitly.
(c) The area ratio in the orientation of the axis of least 2nd moment yields approximate rectangularity. For crushed or blasted solid particles, the manufacturers want to know not only the elongation of a particle, but also angularity or rectangularity, so the area ratio obtained from the Ferret box in the direction of the least 2nd moment is of great practical importance. It is less practical to find the rectangularity by a discrete minimization procedure over possible Ferret boxes, since many directions need to be investigated in order to safeguard an exact minimum.

In order to resolve the problem of rotation dependence, some image analyzers measure a fixed number of Ferret diameters, usually 2, 4, 8 or more, and the user may be able to select how many of these are to be measured, then the maximum length can be obtained. In this method, only the length that is approximately close to maximum length can be obtained, the reproducibility and rotationally-invariant cannot be fully achieved. To meet the above conditions, we studied a new measurement method, which is a combination of multiple Ferret and the least second moments methods, as described as the follows.

2.1 Dot Product Method (Multiple Ferret)

Before the advent of automatic image analyzers, several quickly measurable parameters were defined, which would help particle size analysts to classify irregular-shaped particles by using a single linear measurement. One of the easiest to measure is the caliper diameter, the distance between two parallel tangents which are on opposite sides of the particle. This method was proposed by L.R..Ferret (1931) [7], and the measurement is often referred to by his name. In systems employing boundary-coding techniques, a single pass around the boundary of an object noting maximum and minimum x and y co-ordinates will yield vertical and horizontal Ferret diameters. So, Ferret diameter strongly depends on the direction of system scanning, which is not rotationally-invariant.

By co-ordinate transformation, it is possible to measure the Ferret diameter at any angle to the horizontal. The detailed description, so-called dot products, was mentioned by Fischler [8-9] can be summarized as

Let x_i, for $i=1, 2, ..., N$ be the sampled boundary points. Let u_j for $j=1, 2, ..., D$ be the so-called reference vectors. In one pass, traversing the points $i=1, 2, ..., N$ in any order, calculate:

$$x_i^T u_1, \ x_i^T u_2, \ ..., \ x_i^T u_D, \tag{1}$$

Save only: $MAX = \max\left(x_i^T u_j\right)$, and $MIN = \min\left(x_i^T u_j\right)$ $\qquad$ (2)

For each $j=1, 2, ...,D$, as well as those coordinates that give rise to the D maxima and D minima. The combined dot product method and convex hull method can be described as follows:

As before calculate for $i = 1, 2, ..., N$:

$x_i^T u_1$, $x_i^T u_2$, ..., $x_i^T u_D$,. Save only $MAX = \max(x_i^T u_j)$, together with all other values that are in the interval [MAX - T·(MAX - MIN), MAX], and $MIN = \min(x_i^T u_j)$ together with values in the interval [MIN + T·(MAX - MIN), MIN], $\forall j$, as well as those coordinates that give rise to these values. Call this new subset of points S. For all the points in S calculate the exact diameter using the convex hull method.

The multiple Ferret method, in the sense that $\max\{L_1, L_2, ..., L_D\}$ is chosen as the diameter, coincides with the dot product method, if new scanning directions are implemented using dot products with new coordinate axes. The only difference is that the directions chosen are multiples of 2, and multiple Ferret is often not implemented as a one-pass algorithm. Alternative, we have to find out the particle's orientation first by using the following method.

2.2 Orientation Definition

How precisely do we define the orientation for an object? The usual practice is to choose the axis of length. Least moment measurement is the only way to construct the minimal circumscribed rectangle for a particle. Let's review how to find the orientation of a particle by the use of least moment measurement with reference to Figure 4 [10]. Note that least moment measurements aim at determining a suitable orientation of an object, but combining the orientation with Ferret boxes has not been considered before.

Figure 3 is a two-dimensional equivalent of the axis of the least inertia. We search for the line for which the integral of the square of the distances between the points in the object and the line is a minimal:

$$E = \iint_I R^2 f(x, y)\,dxdy \tag{3}$$

where R is the perpendicular distance from the point (x, y) to the line sought after; $f(x, y)$ is a binary image.

From Figure 3, one has:

$$E = \frac{1}{2}(I_x + I_y) + \frac{1}{2}(I_x - I_y)\cos 2\theta - \frac{1}{2}I_{xy}\sin 2\theta$$
$$= I_x \sin^2 \theta - I_{xy}\sin\theta\cos\theta + I_y \cos^2 \theta \tag{4}$$

where $I_x = \iint_{I'}(x')^2 f(x, y)\,dx'dy'$, $I_y = \iint_{I'}(x'y')f(x, y)\,dx'dy'$, $I_y = \iint_{I'}(y')^2 f(x, y)\,dx'dy'$, $x' = x - \bar{x}$ and $y' = y - \bar{y}$; $(\bar{x}, \bar{y})$ is the centre of area of the object.

Minimization of E gives:

$$\sin 2\theta = \pm\frac{I_{xy}}{\sqrt{I_{xy}^2 + (I_x - I_y)^2}} \quad \text{and} \quad \cos 2\theta = \pm\frac{I_x - I_y}{\sqrt{I_{xy}^2 + (I_x - I_y)^2}} \tag{5}$$

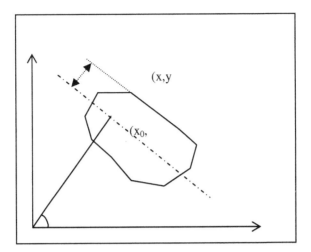

Fig. 3. Least second moment measurement for obtaining the orientation of a particle: The perpendicular distance from a point (x,y) to a line which pass through the center of an object

Of the two solutions, the ones with the plus signs give minimum for E, whereas ones with minus signs correspond to the maximum.

For a discrete binary image, we have

$$\tan^2 \theta + \frac{I(2,0)-I(0,2)}{I(1,1)}\tan \theta - 1 = 0 \qquad (6)$$

where

$$I(p,q) = \sum i^p j^q b(i,j), \; p,q = 0,1,2 \qquad (7)$$

with $b(i,j)$ the binary image, I and j are co-ordinates.

In conclusion, the best way to obtain the orientation of an object is by using least-second moment method, it is rotation-invariant. And the circumscribed rectangle will tend to be of minimal area.

To avoid small-scale fluctuations along boundaries of solid particles, polygonal approximation for every particle (object) is applied. In our case, the goal of a polygonal approximation is to smooth the boundary on a certain scale and obtain the significant boundary segment lines. The boundary can be approximated with varying precision. Here, we just use maximum eight segment lines to construct a polygon for each of the particles. The detailed information for the polygonal approximation we used can be found in [11]. The best-fit rectangles constructed by using our new method, is displayed in Figure 4(b). Particles' elongations and angularities are illustrated in Figure 5. The particles of triangle shape (No. 1-4 in Figure 5(a)), have angularities about 50% (or 0.5), but having varying elongations. The particles of trapezoid or polygon shape (No. 5-12 in Figure 5(a)) have a range of angularity between 0.58 to 0.70, and elongations varying from 0.62 to 0.98. This is reasonable situation: triangles (or diamond) and rectangles can be easily distinguished from other shapes by using

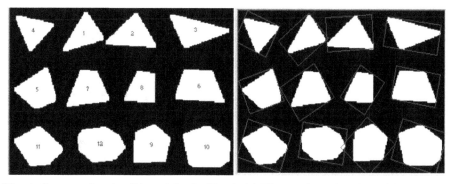

Fig. 4. Segmented real solid particles in Fig. 2. (a) The number of each particle has been marked; (b) The best-fit rectangle has been marked for each particle.

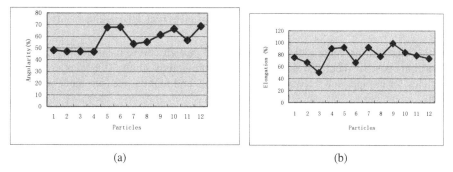

(a)   (b)

Fig. 5. Shape parameters for particles: (a) Angularity (rectangularity); (b) Elongation

our crude angularity definition, and elongation is independent on the angularity. Further detection of particle shapes is needed for more detailed analyses; see Conclusions and further studies.

Sieving size is just a size as measured by sieves; it cannot provide any other dimensional information. To compensate for this, we manually measured crudely three dimensions of every particle in our laboratory. As a general consideration, the manual measurement might be person-dependent. In order to eliminate subjective measurement errors, two engineers did the measurement together. In this manual measurement, we used two types of apparatus. One, as a standard apparatus, is for measuring length and an index of flakiness (the index only indicates if a particle has a flakiness over 1/3). The other one was made by author for measuring the thickness (T) and the width (W) of a particle.

For image analysis, solid particles were placed on a white plane to contrast with the color of the particles, and separated manually to ensure that every particle in a stable position. A camera was installed above the plane. Good illumination was controlled by light sources (both frontlighting and backlighting illuminations). The captured images are of a fairly good quality see Figure 6. The samples were analyzed by our image systems. Image segmentation programs are interactive programs which can be used by an operator to exactly delineate every particle in an image. The lengths,

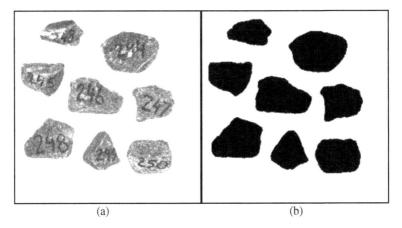

| (a) | (b) |

Fig. 6. Images taken under different illuminations. (a) Two bulbs around a camera, gives a diffuse frontlighting illumination. (b) One lighting box gives backlighting illumination.

widths, elongations, rectangularity and areas of particles were measured by using the algorithm- best-fit-rectangle described in the previous section. The number of solid particles for comparison between image analysis and manual measurement was up to 2000.

3   Comparison Between Sieving, Manual and Image Analyses

In the following weighted distribution curves, sieving and manual measurements are weighted by weight, and image measurements are weighted by area*λ (λ: experiment constant) [12]. In order to distinguish manual from image measurements, Lets assign a mark (m) for manual measurements and a mark (i) for image measurements.

Figure 7 shows the results, it can be seen that the width of the image analysis is compares well with that of manual measurements, but deviates from that of the

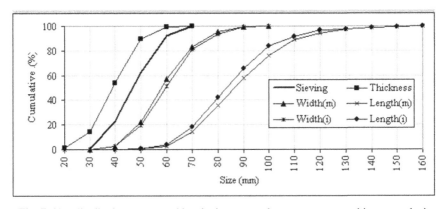

Fig. 7. Size distributions measured by sieving, manual measurements and image analysis

sieving analysis. The sieving analysis curve is located between in thickness and width (m) or width (i), they are parallel each other, which can be used for estimation of thickness and sieving analysis results. The comparison of the lengths shows that the image analysis curve has slightly shifted to the left compared with the curve of the manual measurements. The reason for this may be that it is difficult to obtain the exact lengths of particles by manual measurements. For a given width, a small length L will give a small elongation, which explains why the curve of elongation(i) is to the left of that of elongation (m), see Figure 8. In Figure 8, the flakiness curve has a much higher cumulative percentage than the other curves do. The comparison of the results also show that the particles in this sample have the property that W/L is close to T/W.

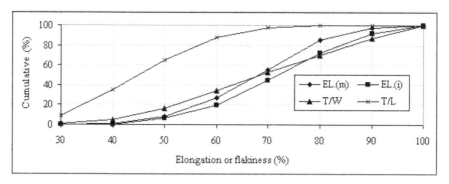

Fig. 8. Comparison between different elongation and flakiness

4 Conclusions and Further Studies

The best-fit rectangle method works very well for image analysis. This is particularly for solid particles such as crushed solid particles, which can be classified into five basic types of shape: Triangle-like, Diamond-like, Trapezoid-like, Polygon-like and Rectangle-like. The method is rotational-invariant, reflects well the shape (Elongation and angularity), and is simple. The test results show that by using this method, the measurement results agree well with manual measurements (for size, elongation and angularity). The accumulative width curve is parallel to the curves of sieving and thickness, implying that sieving analysis and thickness measurement can be easily estimated by the width.

The measurement method of the best-fit rectangle only gives general information about the shapes of particles. In order to obtain more detailed information on particle shape and boundary roughness, corner detection by polygonal approximation might be a possible way.

Acknowledgement

This contribution is part of the SCST (Science and Technology Bureau of Sichuan Province) project, Project No.05JY029-070.

References

1. Lin, C.L and Miller, J.D., The Development of a PC Image-Based On-line Particle Size Analyzer. J Minerals & Metallurgical Processing, No. 2, 1993, 29-35.
2. Maerz NH, Palangio TC, Franklin JA. WipFrag image based granulometry system. In: Franklin JA, Katsabanis T, editors. Measurement of Blast Fragmentation.Rotterdam: Balkema, 1996, p. 91-99.
3. Montoro JJ, Gonzalez E. New analytical techniques to evaluate fragmentation based on image analysis by computer methods. In: Proceedings of the 4th Int. Symposium Rock Fragmentation by Blasting, Vienna, Austria, 1993, p.309 -316.
4. Kemeny J. A practical technique for determining the size distribution of blasted benches, waste dumps, and heap-leach sites. J Mining Engineering, Vol. 46, No. 11, 1994, 1281-1284.
5. Mora CF, Kwan AKH. Sphericity, shape factor, and convexity measurement of coarse aggregate for concrete using digital image processing. Cement and Concrete Research 30, 2000, pp. 351-358.
6. Mora CF, Kwan AKH, Chan HC. Particle size distribution analysis of coarse aggregate using digital image processing. J Cement and Concrete Research, Vol. 28, No. 6, 1998, pp.921-932.
7. Wang WX, Fernlund, J. Shape Analysis of Aggregates. KTH-BALLAST Rapport no. 2, KTH, Stockholm, Sweden, 1994.
8. Arfken G.. Scalar or Dot Product, §1.3. In: Mathematical Methods for Physicists, 3rd ed. Orlando, FL: Academic Press, pp. 13-18, 1985.
9. Jeffreys H, Jeffreys BS. Scalar Product, §2.06. In: Methods of Mathematical Physics, 3rd ed. Cambridge, England: Cambridge University Press, pp. 65-67, 1988.
10. Horn BKP. Robot Vision. Londen, England, 1986, pp. 49 - 62.
11. Wang, W.X., 1998, Binary image segmentation of aggregates based on polygonal approximation and classification of concavities. Pattern Recognition, 31(10), 1503-1524.
12. Wang, W.X. and Fernlund, J., Shape Analysis of Aggregates, KTH-BALLAST Rapport no. 2, KTH, Stockholm, Sweden, 1994.

A Comparison of Shape Matching Methods for Contour Based Pose Estimation[*]

Bodo Rosenhahn[1], Thomas Brox[2], Daniel Cremers[2], and Hans-Peter Seidel[1]

[1] MPI for Informatics, Stuhlsatzenhausweg 85
66123 Saarbrücken, Germany
{rosenhahn, hpseidel}@mpi-inf.mpg.de

[2] CVPR Group, University of Bonn, Römerstr. 164, 53117 Bonn, Germany
{brox, dcremers}@cs.uni-bonn.de

Abstract. In this paper, we analyze two conceptionally different approaches for shape matching: the well-known iterated closest point (ICP) algorithm and variational shape registration via level sets. For the latter, we suggest to use a numerical scheme which was introduced in the context of optic flow estimation. For the comparison, we focus on the application of shape matching in the context of pose estimation of 3-D objects by means of their silhouettes in stereo camera views. It turns out that both methods have their specific shortcomings. With the possibility of the pose estimation framework to combine correspondences from two different methods, we show that such a combination improves the stability and convergence behavior of the pose estimation algorithm.

1 Introduction

Shape matching or shape registration is the basis for many computer vision techniques, such as image segmentation, pose estimation, and image retrieval, to name only few of them. As a consequence, a multitude of works on shape matching can be found in the literature, e.g., [4, 25, 11, 23, 14, 17]; see [24] for a survey.

Most of these approaches rely on classic explicit shape representations given by points that can be connected by lines or higher order curve segments to form a shape. A very popular shape matching method working on such representations is the iterated closest point (ICP) algorithm [2], at which we will take a closer look in Section 2.

An alternative to explicit shape models emerged in the form of implicit representations by means of level sets [10, 16]. Instead of representing a 2-D shape by the points on its contour, the contour is constituted implicitly by the zero-level line of a 2-D embedding function. Level set methods enjoy great popularity in the context of image segmentation with active contours. Recent methods in this field improve the results by integrating the knowledge of previously learned shapes [15], which

---

[*] We gratefully acknowledge funding by the DFG project CR250/1 and the Max-Planck Center for visual computing and communication.

U. Eckardt et al. (Eds.): IWCIA 2006, LNCS 4040, pp. 263–276, 2006.

involves matching the learned shape representation to the shape that is found in the image. Shape matching with level set representations has been suggested in this context by [15, 17, 9, 19]. In Section 3, we are concerned with such implicit shape representations and propose a numerical scheme from optic flow estimation for matching. In comparison to previous numerics in this field, this matching scheme ensures stability and provides a significant speedup.

Since the two mentioned classes of shape matchers are based on very different concepts, the question of superiority of the one or the other arises.[1] We have therefore compared both approaches in the case of one prominent application, namely silhouette based 2-D-3-D pose estimation. The relevance of shape matching in this context is briefly described in Section 4. The comparison in Section 5 shows that both matching concepts have their pros and cons. By a combination one can, at least in the context of silhouette based pose estimation, obtain the best of both approaches. The paper is concluded by a summary in Section 6.

2   Shape Matching with ICP

The goal of shape registration can be formulated as follows: Given two shapes and a distance measure, the task is to determine from a certain class of transformations one that leads to the minimum distance between the two shapes. The original ICP algorithm registers two point sets P and Q provided $TP \subseteq Q$ with the transformation T being a rigid transformation:

1. **Nearest point search**: for each point $p \in P$ find the closest point $q \in Q$.
2. **Compute registration**: determine the transformation T that minimizes the sum of squared distances between pairs of closest points (p, q).
3. **Transform**: apply the transformation T to all points in set P.
4. **Iterate**: repeat step 1 to 3 until the algorithm converges.

This algorithm converges to the next local minimum of the sum of squared distances between closest points. A good initial estimate is required to ensure convergence to the sought solution. Unwanted solutions may be found, if the sought transformation is too large, e.g. many shapes have a convergence radius in the area of 20° [7], or if the point sets do not provide sufficient information for a unique solution.

The original ICP algorithm has been modified in order to improve the rate of convergence and to register partially overlapping point sets. Zhang [25] uses a modified cost function based on robust statistics to limit the influence of outliers. The work also suggests to use a K-dimensions tree to partition the point set

---

[1] We want to note here that the ICP algorithm can also match surfaces and, hence, could as well be used to match implicit contours represented by level sets (which are surfaces, in fact). The focus of our comparison is on explicit and implicit contour representations. We take here the ICP algorithm as a representative for matching methods that work with explicit contours. Consequently, *ICP* has to be interpreted in this paper as *ICP with an explicit contour representation*.

and the author further reports on a significant speedup when registering large range image data sets. Bergevin et al. [1] extended the ICP algorithm to range images from multiple views. They ensure an even distribution of registration errors between overlapping views and report errors less than the range image measurement noise for multiple views of complex objects.

The accuracy of ICP depends on the *geometric information* (e.g. local curvatures) contained in the point sets. If insufficient shape information is available, inaccurate or incorrect registration may occur. Pennec et al. [18] developed a framework to characterize the uncertainty in point registration. Other approaches aim at the avoidance of local minima during registration subsuming the use of Fourier descriptors [21], color information [13], or curvature features [22].

The advantages of ICP algorithms are obvious: they are easy to implement and will provide good results, if the sought transformation is not too large. ICP algorithms have also been used for silhouette based 2D-3D pose estimation [20, 21]. In this context, additional problems arise due to ambiguities of transformations in direction of the projection rays. Sampling methods can be used to avoid some of these additional local optima, yet this is usually a very time consuming procedure.

3   Shape Matching with Level Sets and Optic Flow

3.1   Shape Representation with the Euclidean Distance Transform

In contrast to the point sets used for ICP algorithms, the method suggested in this section deals with shapes represented by an embedding function $\Phi : \Omega \subset \mathbb{R}^2 \to \mathbb{R}$. The contour can be obtained from such a representation as the zero-level line $C := \{\mathbf{x} \in \Omega | \Phi(\mathbf{x}) = 0\}$.

For a given contour, the representation by an embedding function is not unique. In general, one sets the values of Φ to the signed Euclidean distance of the next contour point

$$\Phi(\mathbf{x}) = \begin{cases} D(\mathbf{x}, C) & \mathbf{x} \text{ inside } C \\ -D(\mathbf{x}, C) & \mathbf{x} \text{ outside } C \\ 0 & \mathbf{x} \in C \end{cases} \tag{1}$$

where $D(\mathbf{x}, C)$ denotes the Euclidean distance of $x \in \Omega$ to the closest point $\tilde{\mathbf{x}}$ on the contour C. This choice of Φ has, among others, the nice property of being invariant under rotation and translation. It can be efficiently computed from a binary shape image by the algorithm given in [12]. The distance functions of two shapes are shown in Fig. 1.

Although the embedding of a shape in a higher dimensional space appears, on the first glance, to be less efficient than explicit representations, a closer look reveals many advantages. One such advantage is the flexibility of implicit shapes concerning their topology. While many explicit shape representations induce problems when a shape consists of several parts or contains enclosures, such cases are naturally handled in the level set framework.

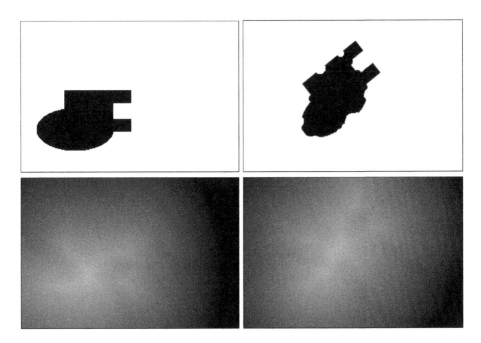

Fig. 1. Top: Source and target shapes that are to be matched. **Bottom:** Euclidean distance functions Φ of these shapes shown by gray value images in the range $[0, 255]$ where 128 marks the zero-level of Φ, i.e., dark areas show negative values of Φ, bright areas show positive values.

A straightforward distance measure for shapes being represented by embedding functions Φ_1 and Φ_2 is:

$$d^2(\Phi_1, \Phi_2) = \int_\Omega (\Phi_1(\mathbf{x}) - \Phi_2(\mathbf{x}))^2 \, \mathbf{dx}. \qquad (2)$$

Other distance measures for implicit shape representations as well as an analysis of their shortcomings can be found in [8]. The distance in (2) reveals a further important advantage of implicit shapes: one can not only measure a discrepancy for given points *on* the contour, but also for all points aside. Thus, matching two embedding functions not only takes the contour into account but also the area of the shapes. For instance the representation in (1) contains the skeleton [3] of the shape. Thus minimizing (2) also seeks to match the skeletons of two shapes.

Although the concept carries over to shapes of arbitrary dimension $D \geq 2$, in the following we will focus on the 2-D case.

3.2   Shape Matching with Optic Flow

Matching two shapes respective the distance measure in (2) can be formulated as the minimization over a group of transformations $\mathcal{T}$:

$$E(\mathcal{T}) = \int_\Omega (\Phi_1(\mathbf{x}) - \Phi_2(\mathcal{T}\mathbf{x}))^2 \, \mathbf{dx} \quad \rightarrow \quad \min. \tag{3}$$

The transformations may include, e.g., translation, rotation, and scaling, as in [9], or the group of perspective transformations, as in [19]. In [17] the transformation further comprises arbitrary deformations $\mathbf{w}(\mathbf{x}) := (u(\mathbf{x}), v(\mathbf{x}))^\top$ of the shape, i.e., $\mathcal{T}\mathbf{x} = \mathbf{x} + \mathbf{w}(\mathbf{x})$. As the minimization of (3) yields an ill-posed problem under these conditions, it was suggested to impose a regularization term to the deformation field:

$$E(u, v) = \int_\Omega (\Phi_1(\mathbf{x}) - \Phi_2(\mathbf{x} + \mathbf{w}))^2 + \alpha(|\nabla u|^2 + |\nabla v|^2) \, \mathbf{dx} \tag{4}$$

where $\alpha \geq 0$ is a regularization parameter that steers the influence of the regularization relative to the matching criterion.

In all existing works on shape matching, the minimization of such function(al)s is performed by means of gradient descent. However, this approach has its perfidies: for each optimization variable in (3), one has to choose a step size, and it is not sure, so far, how the step size has to be chosen to ensure convergence. Setting the step size too large can result in severe instabilities depending on the data. A gradient descent on (4), moreover, converges very slowly.

For an alternative numerical scheme, we suggest to make use of recent advances in optic flow estimation. Optic flow generally describes the 2-D motion field between images, and (4) is a well-known functional for computing optic flow. When regarding Φ_1 and Φ_2 as gray scale images, the estimation of the shape deformation field $\mathbf{w}$ yields an optic flow estimation problem. The first term in (4) contains the non-linearized optic flow constraint, which, in this case, implements the constraint that $\mathbf{w}(\mathbf{x})$ matches points with the same distance to the contour. The second term is a regularizer that penalizes variations in the flow field. This means, in particular, that there should be as few deformations as possible and the deformation field is sought to be smooth.

It has been shown in [5] that the minimization of such a nonlinear functional can be performed by solving a sequence $k = 0, ..., n$ of linear systems

$$\begin{aligned} (\Phi_x^k du^k + \Phi_y^k dv^k + \Phi_z^k) \, \Phi_x^k - \alpha \Delta(u^k + du^k) &= 0 \\ (\Phi_x^k du^k + \Phi_y^k dv^k + \Phi_z^k) \, \Phi_y^k - \alpha \Delta(v^k + dv^k) &= 0 \end{aligned} \tag{5}$$

with $\mathbf{w}^0 = 0$, $\mathbf{w}^{k+1} = \mathbf{w}^k + (du^k, dv^k)^\top$, the abbreviations $\Phi_x^k := \partial_x \Phi_2(\mathbf{x} + \mathbf{w}^k)$, $\Phi_y^k := \partial_y \Phi_2(\mathbf{x} + \mathbf{w}^k)$, $\Phi_z^k := \Phi_2(\mathbf{x} + \mathbf{w}^k) - \Phi_1(\mathbf{x})$, and $\Delta = \partial_{xx} + \partial_{yy}$ the Laplace operator. Note that in comparison to the more general functional in [5], the terms in (4) are both quadratic. Consequently, the inner fixed point iteration loop performed in [5] is not necessary. Quadratic terms for both the matching and the smoothness constraints are sufficient for the matching problem here, since there is basically no noise in Φ_1 and Φ_2 and discontinuities in the deformation field $\mathbf{w}$ are not desired. Robust non-quadratic regularizers or matching terms do not appear to be necessary but are conceivable.

This numerical scheme does not rely on a gradient descent. The linear systems in (5) are the outcome of a semi-implicit scheme that does not introduce a time step size. For solving the linear systems, one can employ iterative solvers such as Gauss-Seidel or SOR. These solvers always converge under the given conditions, independent from the data, and they converge much faster than a comparable gradient descent. A multi-resolution implementation as in [5] leads to an additional speedup and real-time performance (13 frames/sec with non-optimized C++ code on a 2GHz Laptop and 219×132 images). In the sequel, we will compare this optic flow based shape matcher with the ICP algorithm.

4 An Application: Silhouette Based 2-D-3-D Pose Estimation

We test the two shape matching methods in the context of contour based 2-D-3-D pose estimation. In this application, a known 3-D surface model (we use a tea pot here) is projected to the image plane to yield the object silhouette there. This silhouette is compared via shape matching to the contour extracted from the image by a segmentation method. This yields correspondences between points from the model silhouette to points from the contour, which are then used for a pose update. A summary of the algorithm is as follows:

1. **Surface projection**: project the surface with the initial pose to the image plane.
2. **Contour extraction**: segment the object region in the image.
3. **Shape matching**: register the two shapes by either ICP or optic flow.
4. **Pose update:** use the point correspondences from the matching for a pose update.
5. **Iterate:** repeat step 1 to 4 until convergence.

Point correspondences stemming from different camera views can be easily consolidated in step 4. For a detailed description of the method we refer to [6]. The critical issue, apart from the segmentation, is the shape matching. It is important that the matching can cope with noisy shapes due to segmentation errors as well as deformations due to 3-D rotations. In the following section, the performance of ICP as well as optic flow based point correspondences is evaluated. We also tested the simultaneous usage of correspondences from both matchers: Since both algorithms provide a set of 2D-3D correspondences, in step 4 they can be used together and solved simultaneously.

5 Experiments

We first analyzed the influence of the shape matching method on the accuracy and stability of the pose estimation when the images are disturbed by noise, partial occlusions, or changing lighting conditions. To this end, we used a stereo sequence consisting of 350 frames, to which we added Gaussian noise with a standard deviation of up to 80. Some sample frames without noise are shown in Fig. 2. During the whole sequence the object is not moving, thus, it is possible

Fig. 2. Some frames from a static stereo sequence (350 frames) with illumination changes and partial occlusions. **Top row:** left view. **Bottom row:** right view.

to regard the pose variance for a quantitative analysis. Note that the pose estimation method is able to capture moving objects, as well, as shown in a further experiment. The parameters were not tuned for the specific sequence.

The diagrams in Fig. 3 show the deviation from the mean pose when the method employed ICP, optic flow (OF), or their combination for matching (ICP+OF). Obviously, ICP provides a slightly more stable pose than the matching with optic flow. This can also be conjectured from Table 1 that lists the variances for the three matchers and different levels of noise. One can also see that the combination of point correspondences from both matching methods yields similar results as ICP alone. Figure 4 shows two example frames of the sequence. The pose is overlaid in the images.

While the first experiment evaluated the matchers in a situation where the shapes are already close to each other, in the second experiment, we tested the performance, when the model silhouette is far from the object contour in the images. For this purpose, we computed the contours as well as the pose in the first frame of the sequence as usual. We then disturbed the object's pose by a rotation in the area of $[-60\ldots60]$ degrees around the x, y, and z axes, or a translation in the area of $[-150\ldots150]$ mm along these axes. We generated 1000 samples in these intervals. Fig. 5 depicts for which rotations and translations the method was able (blue stars) or not able (red crosses) to converge back to the initially estimated pose. For an absolute translational deviation of less than 3mm, the pose was counted as converged, otherwise as failure. Obviously, the ICP matcher has more problems in case of large transformations than the optic flow based matcher. Combining both matchers yields a similar performance as for the optic flow matcher. Fig. 6 shows some exemplary rotations for which the method with the OF-ICP matcher converged, but the method using the plain ICP-algorithm did not. Table 2 summarizes the convergence rates in percent.

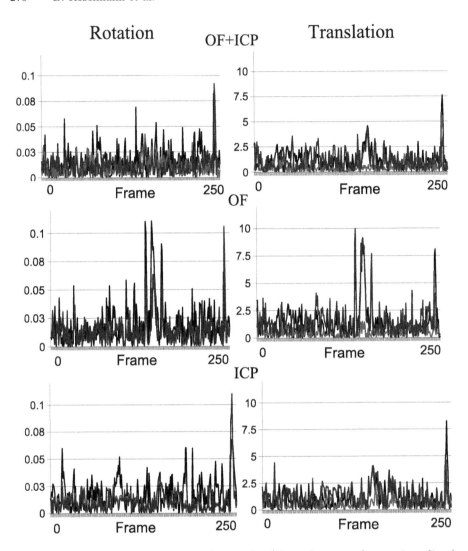

Fig. 3. Deviations from the mean pose for rotation (along the x, y and z-axes in radians) and translation (along the x, y and z-axes in mm) when the method uses the matching with optic flow, ICP, or the combination of both, respectively

According to the literature, as a rough rule for convergence, rotations must be below 20 degrees [7]. This is approximately the convergence radius we also obtained for ICP. Obviously, with the optic flow matcher, the convergence radius can be significantly larger (up to 40 degrees). A possible explanation for this outcome is the richer description of a shape by means of the signed distance function. This includes area based properties of a 2-D shape, which help in the usually ill-posed problem of fitting a 3-D surface to its projections in the image.

Table 1. Variances of the pose parameters (rotation, translation) for different matchers and noise levels (Gaussian noise with standard deviation 0, 40, and 80). The rotations are given in radians and the translations in millimeters.

Noise	Matcher	R_x	R_y	R_z	T_x	T_y	T_z
0	ICP+OF	0.0004	0.00009	0.00009	0.97	0.54	1.25
0	OF	0.0005	0.00017	0.00017	1.64	0.61	2.31
0	ICP	0.0007	0.0005	0.000045	1.39	0.6	1.84
40	ICP+OF	0.0005	0.00013	0.00013	1.79	0.31	1.76
40	OF	0.001	0.00034	0.000355	5.01	0.32	6.4
40	ICP	0.0003	0.0001	0.00009	0.96	0.48	1.49
80	ICP+OF	0.0004	0.0002	0.00019	2.29	0.53	2.16
80	OF	0.0069	0.00036	0.00036	4.58	0.57	5.08
80	ICP	0.00048	0.0002	0.0002	2.18	0.79	2.44

Fig. 4. Two exemplary pose results for the sequence with heavy noise

Finally, Fig. 7 shows pose results of a second stereo sequence, in which the tea pot is grabbed and moved around. We artificially distorted the images by overlaying rectangles of random size, position and color. The bottom row shows pose results in a virtual environment. Also for this sequence, we show a tracking diagram in Fig. 8. It compares the estimated x- y- and z-axis of the estimated

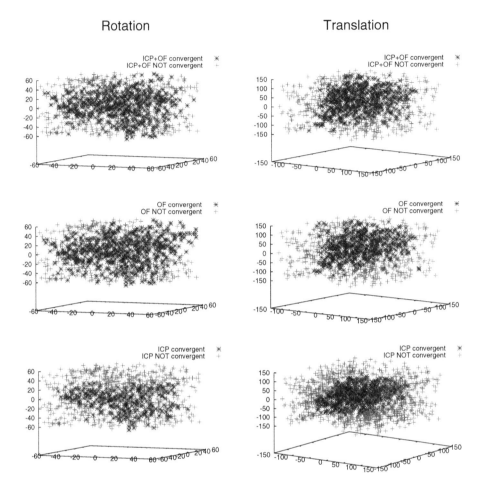

Fig. 5. Convergence of the three investigated methods for the same 1000 random disturbances in rotation (left) and translation (right). Blue stars show the disturbances for which the algorithm was able to converge back to the correct pose, red crosses show the cases of failure. The methods using the combination of ICP and optic flow (top) or solely the optic flow (middle) reveal a similar performance that is significantly better than the performance using the ICP algorithm alone (bottom).

Table 2. The convergence rate for the second experiment (see figure 5) in percent

Matcher	Rotation	Translation
ICP+OF	51,4 %	50,1 %
OF	55,7 %	46,7 %
ICP	27,8 %	32,9 %

Fig. 6. Example rotations, which still converged using the ICP+OF algorithm but failed with ICP alone

pose for different matching strategies (ICP+OF, OF, and ICP). Furthermore, we overlaid the result of the non-distorted sequence (which can be regarded as rough ground truth).

The distortions lead to errors in the pose estimation. With the ICP and ICP+OF matching these errors are mainly within a range of a few millimeter. Solely the results of the plain OF-approach show significantly larger errors in some parts of the sequence. One such part is indicated in the diagram. Obviously, the OF matcher is more sensitive to this kind of distortion than the ICP approach. This is because the distance function propagates the errors in the contour, whereas the ICP approach often ignores smaller occlusions by always taking the closest point for matching. Nonetheless, the combined ICP+OF approach still shows a stable tracking behavior.

Fig. 7. A second stereo sequence where the object is moved (345 frames). The handle of the tea pot temporarily vanishes behind the container and reappears. Finally the tea pot is moved around. **Top:** Pose results for different frames. **Bottom:** Synthetic visualization of the object and the estimated pose from different perspective views.

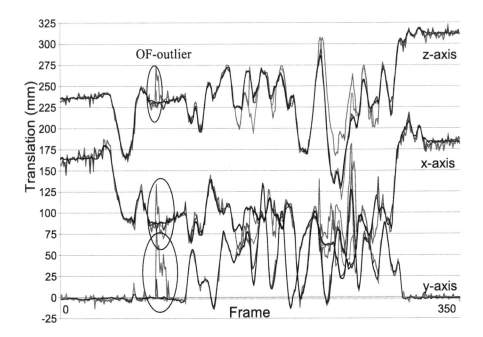

Fig. 8. Visualization of the x, y and z-axis while tracking the sequence in Fig. 7. The values show the results of the OF (red), ICP (gray), and ICP+OF (blue) matcher for the artificially distorted images. Furthermore, the result for the non-distorted images and the ICP+OF matcher is overlaid (black). All diagrams show a similar behavior, except for the OF matcher, which is more sensitive to this kind of distortion.

6   Conclusions

Two very different shape matching concepts have been investigated, one based on explicitly given contour points and another based on implicit shape representations via level sets. We have shown that matching shapes in the level set framework can be performed efficiently with a numerical scheme known from optic flow estimation. Moreover, we compared an ICP algorithm on an explicit contour representation and the level set based matching in the context of silhouette based 3-D pose estimation. It turned out that ICP on explicit contours yields estimates with less variations, whereas the optic flow matcher on the level set representation shows a clearly better convergence in case of large transformations. With the possibility to combine results from both matchers in the pose estimation framework, we demonstrated that one can obtain the best of both registration methods.

References

1. R. Bergevin, M. Soucy, H. Gagnon, and D. Laurendeau. Towards a general multi-view registration technique. *IEEE Transactions on Pattern Analysis and Machine Intelligence*, 18(8):540–547, 1996.

2. P. Besl and N. McKay. A method for registration of 3D shapes. *IEEE Transactions on Pattern Analysis and Machine Intelligence*, 12:239–256, 1992.

3. H. Blum. A transformation for extracting new descriptors of shape. In W. Wathen-Dunn, editor, *Models for the Perception of Speech and Visual Form*, pages 362–380. MIT Press, 1967.

4. R. W. Brockett and P. Maragos. Evolution equations for continuous-scale morphology. In *Proc. IEEE International Conference on Acoustics, Speech and Signal Processing*, volume 3, pages 125–128, San Francisco, CA, Mar. 1992.

5. T. Brox, A. Bruhn, N. Papenberg, and J. Weickert. High accuracy optical flow estimation based on a theory for warping. In T. Pajdla and J. Matas, editors, *Computer Vision - Proc. 8th European Conference on Computer Vision*, volume 3024 of *LNCS*, pages 25–36. Springer, May 2004.

6. T. Brox, B. Rosenhahn, and J. Weickert. Three-dimensional shape knowledge for joint image segmentation and pose estimation. In W. Kropatsch, R. Sablatnig, and A. Hanbury, editors, *Pattern Recognition*, volume 3663 of *LNCS*, pages 109–116. Springer, Aug. 2005.

7. D. Chetverikov, D. Stepanov, and P. Krsek. Robust Euclidean alignment of 3D point sets: The trimmed iterative closest point algorithm. *Image and Vision Computing*, 23(3):299–309, 2005.

8. D. Cremers and S. Soatto. A pseudo distance for shape priors in level set segmentation. In O. Faugeras and N. Paragios, editors, *Proc. 2nd IEEE Intl. Workshop on Variational, Geometric and Level Set Methods (VLSM)*, pages 169–176, 2003.

9. D. Cremers, N. Sochen, and C. Schnörr. Multiphase dynamic labeling for variational recognition-driven image segmentation. In T. Pajdla and J. Matas, editors, *Proc. 8th European Conference on Computer Vision*, volume 3024 of *LNCS*, pages 74–86. Springer, Berlin, May 2004.

10. A. Dervieux and F. Thomasset. A finite element method for the simulation of Rayleigh–Taylor instability. In R. Rautman, editor, *Approximation Methods for Navier–Stokes Problems*, volume 771 of *Lecture Notes in Mathematics*, pages 145–158. Springer, Berlin, 1979.

11. J. Feldmar and N. Ayache. Rigid, affine and locally affine registration of free-form surfaces. *International Journal of Computer Vision*, 18:99–119, 1996.

12. P. F. Felzenszwalb and D. P. Huttenlocher. Distance transforms of sampled functions. Technical Report TR2004-1963, Computer Science Department, Cornell University, Sept. 2004.

13. A. E. Johnson and S. B. Kang. Registration and integration of textured 3-D data. In *Proc. International Conference on Recent Advances in 3-D Digital Imaging and Modeling*, pages 234–241. IEEE Computer Society, May 1997.

14. L. J. Latecki and R. Lakämper. Shape similarity measure based on correspondence of visual parts. *IEEE Transactions on Pattern Analysis and Machine Intelligence*, 22(10):1185–1190, Oct. 2000.

15. M. E. Leventon, W. E. L. Grimson, and O. Faugeras. Statistical shape influence in geodesic active contours. In *Proc. 2000 IEEE Computer Society Conference on Computer Vision and Pattern Recognition (CVPR)*, volume 1, pages 316–323, Hilton Head, SC, June 2000.

16. S. Osher and J. A. Sethian. Fronts propagating with curvature-dependent speed: Algorithms based on Hamilton–Jacobi formulations. *Journal of Computational Physics*, 79:12–49, 1988.

17. N. Paragios, M. Rousson, and V. Ramesh. Distance transforms for non-rigid registration. *Computer Vision and Image Understanding*, 23:142–165, 2003.

18. X. Pennec and J. Thirion. A framework for uncertainty and validation of 3D registration methods based on points and frames. *International Journal of Computer Vision*, 25(3):203–229, 1997.

19. T. Riklin-Raviv, N. Kiryati, and N. Sochen. Unlevel-sets: geometry and prior-based segmentation. In T. Pajdla and J. Matas, editors, *Proc. 8th European Conference on Computer Vision*, volume 3024 of *LNCS*, pages 50–61. Springer, Berlin, May 2004.

20. B. Rosenhahn. *Pose Estimation Revisited*. PhD thesis, University of Kiel, Germany, Sept. 2003.

21. B. Rosenhahn and G. Sommer. Pose estimation of free-form objects. In T. Pajdla and J. Matas, editors, *Computer Vision - Proc. 8th European Conference on Computer Vision*, volume 3021 of *LNCS*, pages 414–427. Springer, May 2004.

22. S. Rusinkiewicz and M. Levoy. Efficient variants of the ICP algorithm. In *Proc. 3rd Intl. Conf. on 3-D Digital Imaging and Modeling*, pages 224–231, 2001.

23. K. Siddiqi, A. Shokoufandeh, S. Dickinson, and S. Zucker. Shock graphs and shape matching. *International Journal of Computer Vision*, 35:13–32, 1999.

24. R. Veltkamp and M. Hagedoorn. State-of-the-art in shape matching. Technical Report UU-CS-1999-27, Utrecht University, Sept. 1999.

25. Z. Zhang. Iterative points matching for registration of free form curves and surfaces. *International Journal of Computer Vision*, 13(2):119–152, 1994.

Relevance Criteria for Data Mining Using Error-Tolerant Graph Matching

Sidharta Gautama, Rik Bellens, Guy De Tré, and Johan D'Haeyer

TELIN, Ghent University, St.Pietersnieuwstraat 41, B-9000 Gent, Belgium
sidharta.gautama@ugent.be

Abstract. In this paper we present a graph based approach for mining geospatial data. The system uses error-tolerant graph matching to find correspondences between the detected image information and the geospatial vector data. Spatial relations between objects are used to find a reliable object-to-object mapping. Graph matching is used as a flexible query mechanism to answer the spatial query. A condition based on the expected graph error has been presented which allows to determine the bounds of error tolerance and in this way characterizes the relevancy of a query solution. We show that the number of null labels is an important measure to determine relevancy. To be able to correctly interpret the matching results in terms of relevancy the derived bounds of error tolerance are essential.

1   Introduction

Information Retrieval (IR) is a field of science concerned with searching for useful information in large, loosely structured or unstructured collections. The notion of relevance plays an extremely important part in Information Retrieval. The concept of relevance while seemingly intuitive, is nevertheless quite hard to define, and even harder to model in a formal fashion [1]. The core of published work on relevance is in the domain of document retrieval. Classic models of relevance are based on probabilistics models using the probabilistics ranking principle. These models explicitly attempt to model word occurrences in relevant and non-relevant classes of documents and use these models to classify the retrieved documents into the more likely class. While different models have been proposed [2], the absence of training data makes it difficult to estimate an efficient relevance model. Recent work using a language modeling approach focuses on viewing documents as models and queries as strings of text randomly sampled from these models [3].

For image retrieval, research on relevance has concentrated on integrating relevance feedback from the user into content-based image retrieval systems [4]. A straightforward way of getting the user into the retrieval loop is to ask the user to tune the system parameters during the retrieval process. This however proves too much a burden for the common. A more intuitive form of interaction is to ask the user to provide feedback regarding the relevance of the current retrieval results. The system then learns from these training examples to achieve better

U. Eckardt et al. (Eds.): IWCIA 2006, LNCS 4040, pp. 277–290, 2006.

Fig. 1. Example satellite images with the road data in overlay. Local inconsistencies between the two datasets are apparent.

performance in the next iteration. Interaction is given in the form of feedback for positive and negative examples as a binary decision or with a degree of (ir)relevance for each.

In our work, we focus on data mining in the context of assessment and control of the quality of the spatial data. The rapid growing number of sources of geospatial data, ranging from high-resolution satellite and airborne sensors, GPS pose severe problems for integrating data. Content providers face the problem of continuously ensuring that the information they produce is reliable, accurate and up-to-date (cfr.Fig. 1). In this field, automated detection of change and anomalies in the existing databases using image information can form an essential tool to support quality control and maintenance of spatial information. In such a system, digitized information extracted from images needs to be located within the spatial vector information in the GIS database. Fig. 2 illustrates the process. In this example, road junctions are extracted from the images. Each set of junctions is posed as a query to the GIS database and the spatial location of the set within the database is returned if overlap is found. In addition, a measure of correspondence should be reported for the overlapping region characterizing the spatial quality of the GIS data with respect to the image information. In some cases only partial or no spatial coordinates are available with the images (e.g. cost of accurate spatial registration) justifying the matching process. In our work, the query process, based on attributed graph matching, is driven by the spatial relations between the object features and takes into account different errors that can occur (e.g. spatial inaccuracy, data inconsistencies between image and vector data).

Paramount in this work is the reliability of the system. The relevance of the reporting should be high, meaning that the correct region should be found and if change between the image and vector region is reported, it should reflect the real life situation. This means we should be able to clearly define the meaning of relevant versus irrelevant results and the tolerance margins for inaccuracy. In this paper, we will discuss how error-tolerant graph matching is applied as

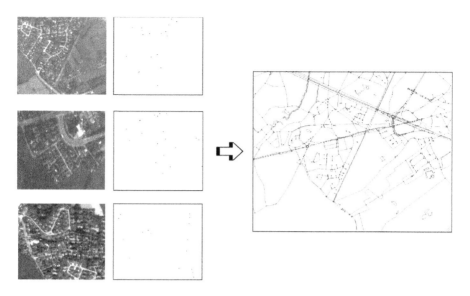

Fig. 2. Road junctions are detected in each image. The set of image junctions forms a query to find the corresponding set of junctions in the GIS database and a spatial quality measure should be reported.

flexible query mechanism to solve the spatial query. We derive an expression which characterizes the bounds where an image feature is identified as part of the object model or as a noise structure. This condition which maps a feature on the null label is a difficult constraint to model and has been traditionally set using heuristic rules-of-thumb. We show how the expected graph error of the object model can be used to determine this constraint.

The remainder of this paper is organized as follows. Section 2 introduces error-tolerant graph matching and derives the error bound to characterize acceptable over inacceptable inconsistencies. Section 3 gives experimental results on synthetic data which validates the derived bounds and applies the results to data mining.Section 4 concludes the paper.

2   Error-Tolerant Graph Matching for Data Mining

The problem can be represented as finding the correspondence between two sets of features: one set originating from the database and one set originating from the image. This can of course be generalized to situations other than image-to-GIS registration, like image-to-image or GIS-to-GIS. Given these features an abstract representation can be built as an attributed graph. The vertices of the graph represent image features and the vertex attributes can contain measurements on these features. The edges of the graph represent relations between features and the edge attributes can contain measurements on spatial relations. A similar graph can be built on the vector data, using data objects as vertices and relations between objects as edges. The problem of registration is represented as a graph matching problem, which seeks the correspondence of similar vertices between two attributed graphs.

In solving the correspondence problem, one should allow tolerance to imprecision and inconsistencies. Errors can occur on the location of the junction due to inaccurate detection, differences in spatial resolution of the data and data inconsistency. In addition false positives can be present in both datasets. In computer vision error tolerant graph matching techniques form an important class of techniques. These techniques seek a graph or subgraph morphism, which allows for distortions. A general distortion model defines the deletion and addition of graph vertices and edges, and replacement of attribute values. A similarity measure or distance function between two graphs is used that models the occurring distortions using heuristics. Early techniques were a generalization of string matching. More recent models are based on information theoretic principles or Bayesian modeling. Solving the correspondence problem is difficult and several optimal and approximate techniques have been proposed. These include, among others, search trees, dynamic programming, annealing and genetic algorithms. In this work, we examine relaxation labeling, a popular approximate technique which has low, polynomial time complexity [13, 8].

2.1   Graph Matching Defined as a Constraint Satisfaction Problem

The graph matching problem can be defined as a constraint satisfaction problem, which consists out of the following elements:

1. a set of objects $i \in \Omega_i$, corresponding to image features;
2. a set of labels $\lambda \in \Omega_\lambda$, corresponding to GIS features;
3. a neighbour relationship over the objects;
4. constraints on possible labels between pairs of neighbouring objects, $r_{ij}(\lambda, \lambda')$.

The constraints are defined as compatibility coefficients, $r_{ij}(\lambda, \lambda')$, for each pair of neighbouring objects i and j and for each pair of labels λ and λ'. These coefficients express the compatibility of assigning label λ to object i in combination with assigning label λ' to object j. Negative values express incompatibility, positive values compatibility. Different types of soft constraints can be checked in this way (spatial, topological, attribute relations). Several constraints can be applied simultaneously, where the total compatibility coefficient for a labeling is the sum over all constraints.

The optimal solution of the graph matching problem is then defined as the following quadratic problem. To each object i a probability distribution $\{p_i(\lambda)\}$ $\lambda \in \Omega_\lambda$ is associated that expresses that object i has label λ:

$$0 \leq p_i(\lambda) \leq 1, \quad \sum_{\lambda \in \Omega_\lambda} p_i(\lambda) = 1 \tag{1}$$

A labeling for the problem is specified by the match matrix $\bar{P} = \{p_i(\lambda)\}_{i \in \Omega_i, \lambda \in \Omega_\lambda}$. The matrix $\bar{P}$ is found such that the following objective function is maximized:

$$E(\bar{P}) = \sum_{i,j \in \Omega_i} \sum_{\lambda, \lambda' \in \Omega_\lambda} p_i(\lambda) p_j(\lambda') r_{ij}(\lambda, \lambda') \tag{2}$$

This quadratic function can be found in various forms in the literature. In graph theory, the problem is known as the maximum clique problem [7]. The optimal consistent correspondence is equivalent with the maximum clique in the association graph between two graphs. Based on a theory of Motzkin and Strauss, it has been shown how the global optima of a quadratic problem based on the adjacency matrix of a graph are equivalent to the maximum cliques [12]. The quadratic problem is similar to Eq. 2 where the compatibility matrix equals the adjacency matrix.

In computer vision, the graph matching problem has been studied using different techniques. In [8] a Bayesian model is developed which leads to a probabilistic relaxation scheme. In [10] it is shown how the updating equation of probabilistic relaxation can be written in a quadratic form similar to Eq. 2, with the compatibility coefficients related to conditional probabilities. In [11], a Markov random field approach is introduced to solve the matching problem using a relational description of the scene. The problem is modeled using normal distributions, which makes the modeling similar to the one used in [8]. The essential difference lies in the optimization technique which is applied. In the case of MRF, the optimal solution is found using annealing techniques. Probabilistic relaxation uses the relaxation equations, which are a form of gradient descent. After review of the literature, we find that the published techniques differ mostly in the optimization technique that is applied (i.e. search trees, annealing, genetic algorithms). We examine the modeling of the matching problem irrespective of the optimization technique used.

2.2   Defining the Correct Problem for Data Mining

In this work, we concentrate on the correct modeling of a problem. If a problem is badly modeled, it will always lead to the wrong solution, regardless of the optimization technique that is used. The problem lies in a good definition of the compatibility coefficients, $r_{ij}(\lambda, \lambda')$, since they determine the quadratic problem.

The value of the compatibility coefficient of a constraint is used in two ways. Firstly, it encodes the level of violation of the constraint, i.e. more negative values mean a higher violation. Secondly, it encodes the relative ordering of the constraints. Some constraints are more important than others, which is reflected in the value of their compatibility coefficient. Setting proper values to these coefficients is not straightforward and is often neglected in the literature.

To guarantee a good solution of the matching problem, the compatibility coefficients $r_{ij}(\lambda, \lambda')$ need to be determined correctly. In most applications, the value of these coefficients are determined using heuristics which basically impose a relative order on the constraints. Strong constraints receive a higher absolute value then weak constraints. The specific ratio between the constraints is usually determined through trial-and-error. For some constraints like the null assignment, it is however difficult to determine a correct value for the compatibility coefficient $r_{ij}(\lambda_\emptyset, \lambda')$. Since each object is a priori a possible null object, every assignment is consistent with the null assignment. The problem is to assess the relative importance of the null assignment with respect to the other constraints.

It should be avoided that the null solution is the most consistent solution of the system. On the other hand, false correspondences of spurious points should be less consistent than the null assignment.

To illustrate the importance of the coefficients, we take again the problem of data mining using point matching. Fig. 3 illustrates a general problem. On the left are images which are used as a query. These images contain a number of random points. The middle image is a reference image in the database. It is a noisy copy of the first image on the left (i.e. jitter on the spatial position of the points, missing and spurious points). Matching the top left image with the reference image gives the result illustrated in the top right image. This image shows the query and the reference image in overlay and reports the matched and null labels. For this example, the result is 100% correct: 3 points had been deleted in the reference image and are correctly assigned the null label; the other 5 points are correctly matched. When matching the bottom left image, which is not related to the reference image, the result is depicted in the bottom right image. The matching reports 2 matched points and 6 null matches. What we want to illustrate here is that the number of null matches is an important indicator whether or not the result of the query is relevant of not. A high number of null matches can indicate that two images are not related and it is therefor important to control the behaviour of the null condition with respect to other constraints.

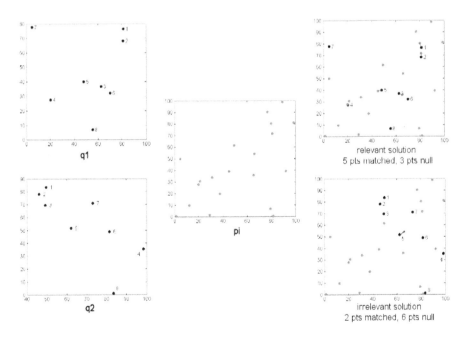

Fig. 3. Illustration of a query q_1 and q_2 with a pattern p_i in the database. The result is shown on the right with the matched points connected by a line and the number of null labels reported.

2.3   Consistency to Determine Correct Compatibility Coefficients

To determine the correct optimal solution of the quadratic problem, we rewrite Eq. 2 as a variational inequality, based on the mathematical concept of "consistency" [10]. The problem of finding consistent solutions of Eq. 2 is shown to be equivalent to solving a variational inequality. The concept of consistency then offers guidance in determining good compatibility coefficients.

The support of a label λ for the object i given by the correspondence $\bar{P}$ is defined as

$$s_i(\lambda) = s_i(\lambda, \bar{P}) = \sum_{j \in \Omega_i} \sum_{\lambda' \in \Omega_\lambda} r_{ij}(\lambda, \lambda') p_j(\lambda') \tag{3}$$

Given a non-ambiguous solution $\bar{P}$ (i.e. $p_i(\lambda) = 0$ or 1), with $\lambda_1, ..., \lambda_n$ the labels which are given to the resp. object $i, ...n$, then $\bar{p}$ is a consistent solution iff

$$s_i(\lambda_i, \bar{P}) \geq s_i(\lambda, \bar{P}), \quad \forall \lambda, \quad i = 1...n \tag{4}$$

For a non-ambiguous solution $\bar{P}$, this can be extended to the weighted sum of the support functions. $\bar{P}$ is a consistent solution iff

$$\sum_{\lambda \in \Omega_\lambda} p_i(\lambda) s_i(\lambda, \bar{P}) \geq \sum_{\lambda \in \Omega_\lambda} v_i(\lambda) s_i(\lambda, \bar{P}), \quad i = 1...n \tag{5}$$

for all labelings $\bar{V}$.

Eq. 5 defines the solution $\bar{P}$ through a system of n inequalities. Hummel and Zucker have shown that if the compatibility matrix $r_{ij}(\lambda, \lambda')$ is symmetric, the solution can be calculated as maximizing $E(\bar{P})$ given by Eq. 2. In the context of consistency, $E(\bar{P})$ depicts the average local consistency. The interpretation of the optimal solution in graph matching then becomes the solution $\bar{P}$ which maximizes the average local consistency.

The definition of consistency can be used to determine the correct values. The definition not only determines the optimal solution of the labeling problem, it also determines what values the compatibility coefficients should take for an "ideal" solution to become the optimal solution of the system. The ideal solution is the matching we wish to find given the noise properties of the detection. For a correct null assignment, we need to determine when the errors, which occur in the neighbour structure of a node, are acceptable and when the number of errors becomes too large so that the null label should be assigned. To analyse this, we should look at the support of the different assignments. In the case of the null assignment, the support can be written as:

$$s_i(\lambda_\emptyset, \bar{P}) = \sum_{j \in \Omega_i} \sum_{\lambda' \in \Omega_\lambda} r_{ij}(\lambda_\emptyset, \lambda') p_j(\lambda')$$

$$= w_\emptyset \sum_{j \in \Omega_i} \sum_{\lambda' \in \Omega_\lambda} p_j(\lambda') \tag{6}$$

$$= w_\emptyset d(i)$$

with $d(i)$ the degree of node i (i.e. the number of neighbours). We have simplified $r_{ij}(\lambda_\emptyset, \lambda') = w_\emptyset$ if $j \in \Omega_i$ (else $r_{ij}(\lambda_\emptyset, \lambda') = 0$). The constant factor $w_\emptyset$ is reasonable in the absence of prior knowledge of assignments.

The support for a non-null label can be split up into three classes Ω_i^+, Ω_i^- and Ω_i^0, namely positive coefficients which express compatibility, negative coefficients which express incompatibility and negative coefficients which control the null assignment. If we consider the first two classes of coefficients constant (resp. w_+ and w_-) within the neighbourhood of node i then the support for λ_i can be simplified to

$$s_i(\lambda_i) = \sum_{j \in \Omega_i^+} r_{ij}^+(\lambda_i, \lambda_j) + \sum_{j \in \Omega_i^-} r_{ij}^-(\lambda_i, \lambda_j) + \sum_{j \in \Omega_i^0} r_i^0$$
$$= w_+ n_+ + w_- n_- + w_\emptyset n_0 \tag{7}$$

Here n_+ is the number of compatible neighbours, n_- the number of incompatible neighbours and n_0 the number of null-neighbours, with $n_+ + n_- + n_0 = d(i)$. Eq.(6) and (7) give the following condition which holds in the optimal solution:

$$w_+ n_+ + w_- n_- + w_\emptyset n_0 > w_\emptyset d(i) \tag{8}$$

or equivalently

$$(1 - f^0)w_\emptyset < f^+ w_+ + f^- w_- \tag{9}$$

where f^+, f^- and $f^\emptyset$ are the fraction of compatible, incompatible and null assignments in the neighbourhood of object i for the ideal mapping.

Eq. 9 can be used to determine the weights for the compatibility matrix given the expected relational graph error. It allows to make a distinction between points showing small distortions, which should find a correspondent in the other dataset, and points showing severe distortions, which should be assigned the null label. As previous research usually relied on rules-of-thumb to determine these weights [13], the importance of this equation is that it allows precise definition of the weights of the graph matching problem with respect to the expected graph error of the system.

Figure 4 gives an example. The figure shows a mapping of an object i of dataset 1 which should be mapped on object λ of dataset 2. The immediate neighbourhood of each object is shown, where the neighbouring objects in the first dataset are mapped on their prime correspondent in the second set. A special case is the sixth object which is given the null label in the ideal mapping. The attributes of the graph edges are color coded in different shades of grey. The relational graph error in this case is expressed by a attribute error $f^- = 2/6 = 33.3\%$ (i.e. edges 2 and 4 change color when mapped) and a null error $f^\emptyset = 1/6 = 16.7\%$ (i.e. one null assignment in the neighbourhood of object i). Given these values of f^- and $f^\emptyset$, the following condition $w_\emptyset/w_- < 2/5$ should be taken into account when determining the weights. This ensures that objects which show more errors in their neighbourhood are mapped onto the null label.

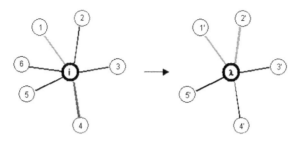

Fig. 4. Mapping of object i on λ and neighbouring objects mapped on their prime correspondent. Object 6 is given the null label. Graph edges having a different color have a different edge attribute. The mapping shows an attribute error $f^- = 2/6 = 33.3\%$ and a null error $f^\emptyset = 1/6 = 16.7\%$.

3   Experimental Results

A first set of experiments has been performed on images containing randomly scattered points. Each image is generated twice: one copy which serves as a reference and one copy which contains perturbations on the scattered points (e.g. noise on the position, added spurious points). The aim is to find the corresponding points between the two copies using graph matching while ignoring the spurious points in the data. The experiment is an abstraction of the correspondence problem between image and GIS data after features like road junctions have been detected in the image. To apply the technique to matching sets of points, we need to introduce the constraints which define similarity.

In our work, we use geometric invariants between subsets of corresponding points. The most simple constraints are binary relations like geometric relations (e.g. angle, distance) between a point and its neighbours to find correspondences. These are stable features, given the detection quality which can realistically be expected from road detection. In these experiments we rely only on the relative angle between pairs of points. In mapping a pair of points i and j on λ and λ' the relative angle between the lines ij and $\lambda\lambda'$ does not exceed a given $\triangle\alpha$. (e.g. $\pi/4$). If this constraint is violated, the compatibility coefficient $r_{ij}(\lambda, \lambda')$ is assigned a negative weight w_-.

The graph representation of the data is of course not restricted to angles and can be readily generalized to incorporate other measurements like connectivity, distance or other topological relations. In our case, the angle between junctions was chosen because it could be reliably measured in the image. Other measurements like connectivity between junctions are more difficult to measure in the image due to the degree of fragmentation in road detection. Compared to relative distance, we found angle measurements to be able to report more fine distortions [9]. Nevertheless, the graph matching technique is generic and applicable once image and GIS information are described in terms of attributed graphs.

3.1   Optimal Parameter Set

In the experiments, points are randomly scattered within an image of 512×512 pixels. The first set of points contains 30 points and the second set contains 100 points. Both sets have 20 points in common with a perturbation on their position using gaussian noise with a standard deviation between one and eight pixels. The matching result needs to make a distinction between points which are common between the two datasets (so called "real" points) and spurious points.

To determine the optimal weights of the graph matching process, Eq. 9 is used. In these experiments, the parameters of RL have been set at $\triangle \alpha = \pi/32$ and $w_- = -0.5$. Compatible matches are not awarded, meaning that $w_+ = 0$. The data contains a ratio of 10:30 outlier points so that $f^\emptyset = 1/3$. Eq. 9 can then be used to determine the weight $w_\emptyset$, which varies over the experiments since the graph label error f^- increases as more noise is added to the position. An added difficulty is that the label error f^- is a stochastic variable. To use Eq. 9, we need to determine the value of f^- which optimally makes the distinction between real distorted points and spurious points. This can be done by modeling the exhibited graph errors of real and spurious points as normal distributions with a certain mean and standard deviation, and taking the maximum likelihood estimate (MLE) as the optimal decision boundary f^-_{opt}. Label errors f^- below this threshold are then regarded as acceptable errors belonging to real points. Label errors f^- above this threshold are regarded as severe errors belonging to spurious points.

We measured the mean and standard deviation of the graph label error over a selection of 10 image pairs for a given amount of noise σ_{noise}. For real points, the ideal mapping is known and the graph label error for these points can be measured. Table 1 gives a summary of the label error statistics (m_1, σ_1) for the different amounts of noise. For spurious points, we selected the best matching corresponding point in the second dataset. Since this is a combinatorial problem, the match is approximated under the condition of a near ideal mapping, i.e. the real points are mapped on the correct correspondents, the other spurious points are mapped on the null label. Under these conditions, finding the best match for a point is a linear search. For this match, we measure the graph label error that would occur if a spurious point is mapped on his most likely candidate. Measured over the dataset, this gives an mean label error $m_n = 38.5\%$ with standard deviation $\sigma_n = 16\%$. Using MLE on these statistics, the threshold f^-_{opt} can be calculated and consequently $w_\emptyset^{opt}$ using Eq. 9. Table 1 gives the calculated $w_\emptyset^{opt}$. These calculated weights are compared to the measured optimal weights $w_\emptyset^{meas}$. The weights $w_\emptyset^{meas}$ have been determined by plotting the "receiver operating characteristic" (ROC) curve by varying $w_\emptyset$. For this curve, sensitivity and specificity are defined as follows:

$$sensitivity = \frac{TP}{TP + FN}$$
$$specificity = \frac{TN}{TN + FP}$$
(10)

where $\{TP, FP, TN, FN\}$ stands for true positive, false positive etc. If sensitivity is plotted along the X-axis and specificity along the Y-axis, the optimal performance is defined as the point on the ROC curve closest to the upper right corner $(1, 1)$. The weight associated with this sample is taken as the optimal measured weight $w_{\emptyset}^{meas}$. Table 1 shows a good correspondence between the calculated optimal weight $w_{\emptyset}^{opt}$ and the measured optimal weight $w_{\emptyset}^{meas}$. This illustrates the relevance of Eq. 9 to tune the graph matching process based on the expected graph error.

Table 1. Determining the null weight based on maximal likelihood with respect to the expected graph error. Statistics spurious points $m_n = 38.5\%$ and $\sigma_n = 16\%$.

stdev [pix]	m_1	σ_1	$w_{\emptyset}^{opt}$	$w_{\emptyset}^{meas}$
1	0.5%	1.1%	0.05	0.05
2	2.3%	2.8%	0.09	0.07
4	7.8%	6.2%	0.19	0.15
8	21.7%	10.4%	0.32	0.25
12	29.1%	12.2%	0.38	0.31

3.2 Data Mining Experiments

The previous experiments show how Eq. 9 can be used to determine the parameters for optimal graph matching performance. In these experiments, we show how the interpretation of the matching results can help in determining the relevancy of the query solution. Again we use datasets of random point patterns with the first set of points containing 30 points and the second set containing 100 points. The second set is a copy of the first point pattern with a perturbation on the position using gaussian noise and added random spurious points. When a reference point pattern is shown as a query, noisy copies of this pattern are regarded as relevant solutions while other point patterns are regarded as irrelevant solutions (cfr.Fig. 3). The dataset contains 10 unique point patterns. Each pattern is reproduced 10 times using gaussian noise with a standard deviation of 4, 8, 12, 16 and 20 pixels, which gives 50 relevant solutions for each pattern. The total dataset thus contains 500 patterns.

We perform matching using $w_0 = 0.3$ and $\triangle \alpha = \pi/32$. The plot in Fig. 5 shows the importance of the number of null labeling in determining relevancy. For each point pattern q which is posed as a query, the optimal correspondence $\bar{P}_i$ is sought with each pattern p_i in the database. Each correspondence $\bar{P}_i$ is characterized by a number of matched points and a number of null labels. The matched points can be used to measure the degree of correspondence between q and p_i by measuring the average graph distortion that is introduced under the mapping $\bar{P}_i$ of q on p_i. In our case, this is done by counting the number of label errors in a graph node neighbourhood, averaged over each graph node. This is similar to the classic graph edit distances used in the literature. Intuitively,

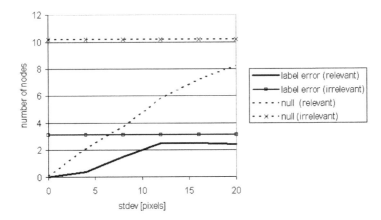

Fig. 5. Plot containing the average label error and average number of null labels for relevant and irrelevant query solutions for different noise levels

one can expect a higher degree of distortion to correspond with less relevant solutions. However looking at the results in Fig. 5, we see a saturation of the distortion measure for high standard deviation. This implies that the distortion measure cannot determine the correct order of relevancy above a noise level $\sigma_{noise} = 12$ pixels. The reason for this behaviour is that graph nodes which show a severe graph distortion in their neighbourhood are mapped on the null label and do not contribute to the average graph distortion.

In addition when we compare the average graph distortion of the relevant solutions with the distortion of the irrelevant solutions, we see that above noise level $\sigma = 12$ these values lie very close together.

Alternatively, we can use the number of null labels N_0 in a correspondence $\bar{P}_i$ as a measure of relevance. In Fig. 5, we see that this measure has a better behaviour. The number of null labels N_0 shows a linear increase with the noise level introduced in the point patterns. This means that N_0 can distinguish between mildly distorted and severely distorted relevant solutions. Additionally, the number of null labels N_0 of the relevant solutions is much lower than N_0 of the irrelevant solutions.

In Fig. 6, the statistics for correct relevant and irrelevant query solutions are summarized for each noise level σ_{noise}. For $w_0 = 0.30$, the threshold on N_0 is set to 9. This means that if $\bar{P}_i$ for the query solution p_i has less than 9 null labels, p_i is regarded as relevant, in the other case p_i is irrelevant. Fig. 6a shows a very good suppression of irrelevant point sets with only a few percent being returned as a relevant solution. Fig. 6b shows a high return of relevant point sets with a deterioration at higher noise levels as can be expected. At the highest level, some point patterns are regarded as irrelevant due to the degree of distortion. At this stage it is important to set a good value for the parameter w_0. Redoing

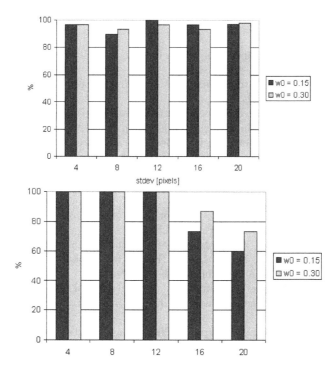

Fig. 6. Bar chart showing performance results of the mining process for different noise levels and different coefficients w_0: (above) correctly classified as irrelevant, (below) correctly classified as relevant

the experiment with $w_0 = 0.15$, we see the same performance for filtering the irrelevant point sets but a decrease in returning the correct solutions at high noise levels.

4   Conclusion

In this paper we presented a graph based approach for data mining geospatial information. The system uses error-tolerant graph matching to find correspondences between the detected image information and the road vector data. Spatial relations between objects are used to find a reliable object-to-object mapping. Graph matching is used as a flexible query mechanism to answer the spatial query. A condition based on the expected graph error has been presented which allows to determine the bounds of error tolerance and in this way characterizes the relevancy of a query solution. We show that the number of null labels is an important measure to determine relevancy. Determining when a null label is being assigned or not is dependent of the values of the compatibility coefficients which order the graph matching constraints. It is therefor important to correctly set these values to be able to control the behaviour of the graph matching process and to be able to interpret the matching results in terms of relevancy.

References

1. Mizzaro,S. 1997. Relevance: The whole history, Journal of the American Society for Information Science, 48(9), 810-832.
2. Lavrenko, V., 2002. Optimal Mixture Models in IR. Lecture Notes in Computer Science Vol.2291, 193-212.
3. Rosenfeld, R. 2000. Two decades of Statistical Language Modeling: Where Do We Go From Here? Proceedings of the IEEE, 88(8), 1270-1278.
4. Zhou, X., Huang, T. 2002. Relevance feedback in content-based image retrieval: some recent advances Source. Information SciencesApplications. Vol. 148(1-4), 129-137.
5. Tieu, K.,Viola P. 2004. Boosting image retrieval. International Journal of Computer Vision, 56(1), 17-36.
6. Besl, P., McKay, N. 1992. A method for registration of 3-D shapes. IEEE Trans Pat. Anal. and Mach. Intel. 14(2), 239-256.
7. Bomze, I.M., Budinich, M., Pardalos, P.M., Pelillo, M. (1999) The maximum clique problem. In: Du, D.Z., Pardalos, P.M. (Eds.), Handbook of Combinatorial Optimization, vol. 4. Kluwer Academic Publishers, Boston.
8. Christmas, W., Kittler, J., Petrou, M. 1995. Structural Matching in Computer Vision Using Probabilistic Relaxation. IEEE Trans Pat. Anal. and Mach. Intel. 17(8), 749-764.
9. Gautama, S., D'Haeyer,J., Philips, W. 2004. Image-based change detection of geographic information using spatial constraints. In: Flexible Querying and Reasoning in Spatio-Temporal Databases: Theories and Applications, R.De Caluwe, G.DeTre, G.Bordogna (ed.), Springer-Verlag, 351-368.
10. Hummel, R., Zucker, S. 1983. On the foundations of relaxation labeling processes. IEEE Trans Pat. Anal. and Mach. Intel. 5(3), 742-776.
11. Li, S. 1995. Markov Random Field Modeling in Computer Vision. Springer-Verlag, New-York.
12. Pelillo, M., Jagota, A. 1995. Feasible and infeasible maxima in a quadratic program for maximum clique. Journal of Artif. Neural Networks, 2(4), 411-420.
13. Wilson R. 1995. Inexact Graph Matching Using Symbolic Constraints. Ph. D. thesis, Department of Computer Science, University of York.

Computational Aspects of Digital Plane and Hyperplane Recognition

David Coeurjolly[1] and Valentin Brimkov[2]

[1] Laboratoire LIRIS - CNRS UMR 5205
Université Claude Bernard Lyon1
43 Bd du 11 novembre 1918
Villeurbanne, France
dcoeurjo@liris.cnrs.fr
[2] Mathematics Department
Buffalo State College
State University of New York
Buffalo, NY 14222, USA
brimkove@buffalostate.edu

Abstract. In these note we review some basic approaches and algorithms for discrete plane/hyperplane recognition. We present, analyze, and compare related theoretical and experimental results and discuss on the possibilities for creating algorithms with higher efficiency.

1 Introduction

In discrete geometry various definitions and properties of linear structures – such as straight lines, planes, or hyperplanes – have been proposed. On this basis, computational efficient analytic characterizations of these objects have been obtained. In many applications one considers a reverse problem: given a set of pixels or (hyper)voxels, decide if it is a portion of a discrete line or (hyper)plane. For this, a recognition algorithm is needed.

In dimension two, the arithmetic structure of discrete straight lines (DSL) has been exploited to design efficient algorithms, as both their asymptotic computational cost and practical efficiency have been studied (see [31] for a survey on the matter). In higher dimension, similar arithmetic structures still exist in digital planes and hyperplanes. However, to solve a recognition problem, one usually adapts algorithms from linear programming (LP) [13, 14, 19, 27] or computational geometry (CG) [23, 24, 35]. As a rule, there is a gap between the theoretical time complexity bounds obtained for linear programming or computational geometry problems and the practical efficiency of these algorithms when applied to discrete objects. Indeed, one can observe that the existing time complexity bounds are not tight when experimental analysis is performed.

In this presentation we review certain basic approaches for digital plane and hyperplane recognition and consider related computational aspects from both theoretical and practical point of view. In particular, we compare approaches and results related to computational geometry on one hand and integer linear programming on the other.

U. Eckardt et al. (Eds.): IWCIA 2006, LNCS 4040, pp. 291–306, 2006.

2   Approaches to Defining Digital Planarity

Throughout the paper we will refer to a set of integer points $S = \{p^1, p^2, \ldots, p^m\}$. In this section we review three basic approaches for defining digital planes and hyperplanes. Chronologically the first one is the following.

Consider a Euclidean hyperplane Γ defined by $\gamma_0 + \sum_{i=1}^n \gamma_i x_i = 0$ with $\{\gamma_i\} \in \mathbb{R}$, $|\gamma_i| \le |\gamma_n|$ for $1 \le i < n$ and $|\gamma_n| > 0$ (the axis x_n is called the major axis of the plane, see below). Let $p = (p_1, \ldots, p_n)$ be the intersection point of Γ and the straight line defined by $x_1 = r_1, x_2 = r_2, \ldots, x_{n-1} = r_{n-1}$ with $r_i \in \mathbb{Z}$, $1 \le i \le n$. The grid point $P = (r_1, r_2, \ldots, r_n) \in \mathbb{Z}^n$ with $p_n - \frac{1}{2} < r_n \le p_n + \frac{1}{2}$ is the *digital image* of p with respect to Γ. We have the following definition.

Definition 1 (Digital hyperplane [35]). *$S \subset \mathbb{Z}^n$ is a digital hyperplane iff there exists an Euclidean hyperplane H such that each grid point P of S is the digital image of a point $p \in H$.*

Another approach is based on the following definition.

Definition 2 (Digital flatness [36]). *Let $S \subset \mathbb{Z}^n$. S is called flat iff there exist $n + 1$ real numbers $\gamma_0, \ldots, \gamma_n$ such that:*

1. $\max\{|\gamma_0|, \ldots, |\gamma_n|\} = 1$ *;*
2. every point $P = (P_1, \ldots, P_n) \in S$ satisfies the condition

$$-\frac{1}{2} < \gamma_0 + \sum_{i=1}^n \gamma_i P_i \le \frac{1}{2}. \tag{1}$$

In [36] Veelaert proves that a discrete set satisfying Definition 1 is flat. Note that Veelaert's definition is more anisotropic since there is no constraints on the Euclidean hyperplane orientation (i.e., the conditions $|\gamma_i| \le |\gamma_n|$ for $1 \le i < n$ and $|\gamma_n| > 0$ in Definition 1).

Another way to define a digital hyperplane is the following.

Definition 3 (Discrete analytic hyperplane [3]). *A discrete analytic hyperplane P with coefficients $(a_1, a_2, \ldots, a_n, b) \in \mathbb{Z}^{n+1}$, $\gcd(a_1, \ldots, a_n) = 1$, and thickness $\omega \in \mathbb{N}^*$ is given by:*

$$P(a_1, a_2, \ldots, a_n, b) = \{(x_1, x_2, \ldots, x_n) \in \mathbb{Z}^n | 0 \le b + \sum_{i=1}^n a_i x_i < \omega\} \tag{2}$$

Thus a set $S \subset \mathbb{Z}^n$ is a subset of a discrete analytic hyperplane iff there exists a vector $(a_1, a_2, \ldots, a_n, b) \in \mathbb{Z}^{n+1}$ with $\gcd(a_1, \ldots, a_n) = 1$ and $\omega \in \mathbb{N}^*$, such that $S \subset P(a_1, a_2, \ldots, a_n, b)$.

This last definition is more general than the previous one since it allows to control the *thickness* of the set of grid point (see [3]). In what follows, we consider a class of discrete analytic hyperplanes, called *naive*, which have thickness $\omega = \max_{i=1..n}\{|a_i|\}$.

If the coefficients of the Euclidean plane in Definition 1 are rational numbers, then the digital hyperplane obtained by that definition is a discrete analytic hyperplane according to Definition 3 as well. Note that in plane recognition problems we usually have to consider finite subsets of grid-points. Hence, the case of irrational coefficients handled by Definitions 1 and 2 may not occur.

In the following, we will use the abbreviation DHP for a digital (or discrete) analytic hyperplane, and DHPS (Digital Hyperplane Segment) for a finite subset of grid points belonging to a DHP. In dimension three, we will denote digital planes (resp. digital plane segments) by DP (resp. DPS).

We conclude this section with one more technical notion to be used in the sequel.

Definition 4 (Major axis and DHP base). *Let S be either a digital hyperplane or a naive discrete analytic hyperplane. Then there exists a major axis x_j such that the projection $\bar{S}$ of S along x_j onto the plane $x_j = 0$ is a one-to-one and onto mapping. The $(n-1)$-dimensional set $\bar{S}$ is called the base of the DHP.*

It was proved in [10] that every digital plane has a major axis. For instance, in Definition 1 the major axis is x_n, in Definition 2 it is x_j provided that $|\gamma_j| = 1$, while in Definition 2 the major axis is x_j whenever $\omega = |a_j|$.

3   Survey on DHP Recognition Algorithms

In this section we review algorithmic solutions for the discrete (hyper)plane recognition problem that can be stated as follows: Decide if a given finite set $S \subset \mathbb{Z}^n$ is a DHPS. Furthermore, if the answer to this last question is positive, we would like to determine the DHP parameters.

We can distinguish two basic classes of recognition algorithms: ones based on computational geometry techniques and ones using linear programming.

3.1   Computational Geometry Algorithms

The first algorithms of this kind have been proposed in [23, 24] for DPS (*i.e.* for dimension $n = 3$). Let us define a *support* of a point set S to be a Euclidean plane such that all points from S lie on the same side of the plane. Then we have the following theorem.

Theorem 1. *[23] Let $S \subset \mathbb{Z}^3$. S is a DPS iff there is a support H of S, such that the distance between the points from S and H is less than 1.*

This theorem has been stated for $n = 3$ but it trivially extends to higher dimensions.

To find such a support plane, Kim considered the faces of the convex hull $CH(S)$ of S. He stated that S is a DPS iff there is a face of $CH(S)$ that induces a support plane satisfying the distance criterion of Theorem 1. In [23] and [24] Kim proposed several algorithms to discover such a face. The last statement, however, turned out to be wrong: as shown in [18, 12], the support of S can be defined by an edge of $CH(S)$. In such a case, S can be a DPS and there may be no face of its convex hull satisfying the distance criterion.

Another geometric approach to recognize DHPS is based on point set separability. We have the following theorem.

Theorem 2. *[35] Let $S \subset \mathbb{Z}^n$. S is a DHPS iff there exists an Euclidean hyperplane H that separates S from S', where S' is obtained by a translation of S at distance 1 along the major axis x_n of H.*

Thus the recognition problem is reduced to a separability test for two sets of gridpoints. In [35], two algorithms are detailed. One of them uses linear programming (see below). The other is based on computation of convex hull and polytopes intersection [29]. Specifically, S can be separated from S' by a plane iff $CH(S) \cap CH(S') = \emptyset$. Thus first $CH(S)$ and $CH(S')$ are found, then their intersection is computed. Convex hulls computation takes $O(m \cdot \log m + m^{\lfloor n/2 \rfloor})$ time, while the polytopes intersection can be found in $O(2^n \cdot m^{2^{n-2}n} \cdot \log m)$ (see [35]).

Finally, a class of algorithms are based on the notion of *thickness* of S (defined for $n = 3$). The thickness can be linked to the distance criterion proposed in Theorem 1. To this end, let us define the *chords set* of S as the set $\{P - P' | P, P' \in S\}$ [22]. Without loss of generality, suppose that the major axis of S is x_3. Then the geometric thickness of S is the x_3 coordinate of the intersection point of the convex hull of the chords set of S and the ray defined by x_3 and the origin O where $x_3 \geq 0$. The following theorem holds.

Theorem 3. *[22] Let $S \subset \mathbb{Z}^3$. S is a DPS iff its geometric thickness is less than 1.*

These definitions and results have been proposed for dimension $n = 3$, but admit easy generalizations. Some of the above-mentioned computational geometry algorithms are illustrated in Figure 1.

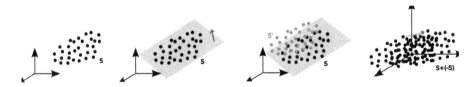

Fig. 1. Illustration of Computational Geometry recognition algorithms. *From left to right:* The input set of grid points S, recognition using a support, recognition using the separability test, and recognition based on the thickness of the chords set.

3.2   Linear Programming and Integer Linear Programming Algorithms

Since DHP definitions are based on inequalities, the aim of these recognition techniques is to directly solve a linear inequality system using tools from Linear Programming or Integer Linear Programming (ILP). Applying Definition 2 (resp.

Definition 3), we can associate to each grid-point P of S two inequalities with $n+1$ unknowns $\{\gamma_0, \ldots, \gamma_n\}$ in $\mathbb{R}$ (resp. $n+1$ unknowns $\{a_1, \ldots, a_n, b\}$ in $\mathbb{Z}$). For example using Definition 2, for each grid point $P \in S$ we have the inequalities $\gamma_0 - \frac{1}{2} + \sum_{i=1}^{n} \gamma_i P_i \leq 0$ and $\gamma_0 + \frac{1}{2} + \sum_{i=1}^{n} \gamma_i P_i > 0$. Using LP or ILP algorithms to these linear inequality systems, we can thus decide if the set S is a DHPS or not.

If we consider a specific orientation of the DHPS, *i.e.* if we restrict our attention to DHPS with a major axis x_n, the dimension can be reduced from $n + 1$ to n by considering the unknowns $\{\gamma_0/\gamma_n, \ldots, \gamma_{n-1}/\gamma_n, 1\}$ or $\{a_1/a_n, \ldots, a_{n-1}/a_n, 1, b/a_n\}$. Note that for the last system the unknowns are in $\mathbb{Q}$. Hence, for each grid point $P \in S$ we have, for example, the constraints:

$$\Gamma_0 - \frac{1}{2} + P_n + \sum_{i=1}^{n-1} \Gamma_i P_i \leq 0 \quad \text{and} \quad \Gamma_0 + \frac{1}{2} + P_n + \sum_{i=1}^{n-1} \Gamma_i P_i > 0. \qquad (3)$$

For solving such kind of linear programs one can take advantage of the rich arsenal of available linear programming algorithms (see, e.g., [32]). Optimal theoretical algorithms exist to decide if a set of grid points is a DHPS using Megiddo's theorem:

Theorem 4 ([27]). *Given a LP problem with m linear inequalities in $\mathbb{R}^n$ where n is fixed, an algorithm exists to solve the problem in $O(m)$ time.*

Unfortunately, the above bound $O(m)$ includes an implicit factor that is exponential in n (but is a constant when n is fixed). Considering the DHPS recognition problem based on ILP problems, complexity results can be founded in [13].

Algorithms Based on Preimage Computation. For the sake of clarity we will present separately the algorithms based on preimage computation, although these methods are deeply linked to those using linear programming.

Definition 5 (DHP preimage). *Given a DHP (resp. DHPS) S, its preimage is the set of Euclidean hyperplanes whose digitization coincides with/contains S.*

Note that for each DHP definition from Section 2 a digitization scheme can be specified. So in what follows, whenever a digitization scheme is considered, we will suppose that it is the one related to the particular DHP definition adopted.

Basically, the preimage is nothing but the feasible region of the LP inequality system associated with a set of grid points. If we suppose that the major axis is known (x_n, for example), the preimage is an n-dimensional polytope, possibly unbounded, whose vertices have rational coordinates. Obviously, if the preimage associated with a set S is empty, then S is *not* a DHPS. Note that if the major axis and the sign of each $\{\Gamma_i\}$ are known (*i.e.* if one knows the global orientation of the DHPS), the preimage polytope is initialized using the unit hypercube since $(\Gamma_0, \ldots, \Gamma_{n-1}) \in [0,1]^n$. A simple incremental DHPS recognition algorithm is given next. Considering $n-$dimensional DHPS, the step 6 in Algorithm 1 may have high computational cost due the combinatorial aspects of $n-$dimensional polytopes.

However, if $n = 3$, several efficient DPS recognition algorithms have been proposed [38, 15]. Indeed, if we denote by E the number of preimage vertices, the intersection of a Euclidean plane and a convex polyhedron can be computed in $O(E)$ time. It is not hard to realize that in this case the computational cost of the above algorithm is $O(m \cdot E)$. Moreover, the algorithm admits an online implementation which runs in time $O(E)$.

To have a tight computational cost bound, we need to bound E by the number of grid points m. Several approaches have been proposed to solve this problem for $n = 3$ (see [17]). It is not trivial, however. In dimension 3, we can easily see that E is bounded by the number of vertices of $CH(S)$. This can be proved using either the construction of the feasible region based on dual transformation proposed in [29] or simply by observing that the preimage corresponds to the set of Euclidean planes separating $CH(S)$ and $CH(S')$ in view of Theorem 2. In this case extremal Euclidean planes are defined by one or two grid points in $CH(S)$ and two or one grid point in $CH(S')$. Hence, the number of such extremal planes is bounded by $O(|CH(S)|)$. Finally, for $n = 3$, a bound on $|CH(S)|$ is also a bound on E.

In the experimental results presented in Section 5.3, E is always less than $|CH(S)|$ whatever the dimension. Further developments on that point is an important challenge since it is directly linked to the efficiency of Algorithm 1.

Algorithm 1. DHPS preimage based recognition

1: Let S be a set of grid points
2: Preimage initialization
3: **for each** grid point P of S **do**
4:     Let C_1 and C_2 be the two linear constraints of dimension $n - 1$ associated to P using Equation (3)
5:     **for each** constraint C in $\{C_1, C_2\}$ **do**
6:         Update the preimage cutting the polytope with the oriented hyperplane C
7:         **if** the preimage is empty **then**
8:             Stop the recognition, S is not a DHPS
9:         **end if**
10:    **end for**
11: **end for**

3.3   Other Algorithms

In this section we present some other algorithms that are not based on results from computational geometry or linear programming. The first one has been proposed by Veelaert [37]. It recognizes a DPS whose base is a *strip*, that is a 2-D set of the form $\{(x,y) \mid r \leq x \leq s\}$ with $r,\ s \in \mathbb{Z}$. The recognition process is based on the concept of *evenness* of a DHP [36].

Definition 6 ([36, 37]). *$S \subset \mathbb{Z}^n$ is called even iff its projection along the x_n axis is a one-to-one mapping, and for every quadruple (A, B, C, D) of points in S such that $A_{x_n=0} - B_{x_n=0} = C_{x_n=0} - D_{x_n=0}$, it holds $|(A_{x_n} - B_{x_n}) - (C_{x_n} - D_{x_n})| \leq 1$.*

This characterization is related to the Rosenfeld's DSL chord property [30] as well as to the Kim's DP chordal triangle property [23]. Veelaert showed that a strip of voxels S is a subset of a digital plane iff S is even [37]. (Note however that such an equivalence does not hold for arbitrary dimension.) This result implies an $O(m^2)$ algorithm for DPS recognition.

  Another approach is based on the parametrization of DP by least-squares fits. It is proved in [25] that, given a rectangular base DPS S, there is a one-to-one correspondence between least-squares plane fit parameters and the coefficient of the DPS S. Although these authors' goal was just to give a finite parametrization of DPS, the result can be used to obtain a simple recognition algorithm that consists of two stages:

1. Given an input set S, first the least-squares plane fit is computed, which provides the coefficients of the relevant Euclidean plane H. This stage takes $O(m)$ time since the fitting problem is of linear complexity.
2. Then it is verified if the digitization of the plane H obtained in Stage 1 coincides with S. If it is so, then S is a DPS that is precisely the digitization of H. Otherwise, S is not a DPS. It is shown that such a verification can be done in $O(m)$ time.

  It is easy to show that the following holds.

Theorem 5. *The least-squares plane fit algorithm solves correctly the rectangular base DPS recognition problem.*

Proof. The proof is straightforward: if the two sets coincide, then S is a DPS by Definition 1 or 2. Conversely, if S is a DPS, then because of the one-to-one correspondence between leas-squares fit parameters and DPS coefficients, there exists a unique least-squares plane fitting whose digitization coincides with S.□

Although the algorithm described above is restricted to DPS's with rectangular bases, generalizations to other base shapes as well as to higher dimension seem possible (see, e.g. [25] for a recent related results).

4   Efficiency of DHP Recognition Algorithms

Table 1 summarizes data about some basic DHPS recognition algorithms. One can see that optimal algorithms exist to recognize DHPS in any fixed dimension. However, these are only theoretical since no their implementation is available. In fact, only few algorithms have been implemented for $n = 3$. Among these are some algorithms based on point set thickness arguments [22], Fourier-Motzkin elimination [19, 20], direct LP [26], and preimage computation [38, 34].

Table 1. Survey of DPS and DHPS recognition techniques

Description	Sources	Dimension	Time	Online time	Remark
Convex hull width	[23, 24]	3	$O(m^2)$	-	
Convex hull separability	[35]	n fixed	$O(m \cdot \log m + m^{\lfloor n/2 \rfloor} + 2^n \cdot m^{2^{n-2}n} \cdot \log m)$	-	
Point set width	[22]	3	$O(m^7)$	yes	
Fourier-Motzkin Algorithm	[19]	3	n.a.	-	
Direct Linear Programming	[27]	n fixed	$O(m)$	$O(1)$ [14]	
Separability test based on LP	[35]	n fixed	$O(m)$	$O(1)$ [14]	
Integer Linear Programming	[13]	n fixed	$O(m \cdot \log D)^1$	-	
Arithmetic Preimage	[38]	3	$O(m^3 \cdot \log m)$	$O()$	
Arithmetic Preimage	[15]	3	$O(m \cdot \log^2 m)$	−	
Preimage		3	$O(m \cdot E)$	$O(E)$	
Evenness property	[37]	3	$O(m^2)$		strip base DPS
Least-squares fits	[25]	3	$O(m)$	no	rectangular base DPS
Arithmetic recognition	[28]	3		-	rectangular base DPS

Considering the asymptotic bounds on these algorithms' complexity, one can observe that there is a gap between those bounds and the estimations of algorithms' efficiency obtained through experiments. To illustrate this point, consider, for instance, the point set thickness algorithm from [22]. Its theoretic computational cost is $O(m^7)$ whereas in practice it features near linear time complexity when m increases.

Among the algorithms presented in Table 1, we will focus on the preimage based techniques. We have two serious arguments to do so. First, these are online algorithms and their incremental computational cost depends on the DPS specific characteristics (see below). Moreover, preimage computation provides a complete description of all Euclidean hyperplanes that satisfy the DP definitions. These two features appear to be important requirements in various applications, such as discrete surface segmentation into DPS and reversible polyhedrization of binary objects [20, 26, 33, 34, 16].

In the next section, we consider some results from number theory, theory of lattice polytopes, and integer linear programming in order to obtain time complexity bounds for DHPS recognition algorithms based on preimage computation. More precisely, we focus on bounds on the size of the convex hull of S that allows us to obtain bounds on the number of preimage vertices (see Section 3.2).

5   Towards Obtaining Tight Bounds on the Computational Cost of DP and DHP Recognition

5.1   Integer Programming and Associated Lattice Polytopes

Consider the integer linear programming problem (ILP) [32]:

$$\max cx \tag{4}$$
$$Ax \le b \tag{5}$$
$$x \in \mathbb{Z}^n \tag{6}$$

with $A = (a_{ij}) \in \mathbb{Z}^{m \times n}$, $b = (b_i) \in \mathbb{Z}^m$ and $c = (c_j) \in \mathbb{Z}^n$. The special case when $m = 1$ and all coefficients as well as the solution components are nonnegative is known as a *knapsack problem*.

It is well-known [21] that both ILP and KP are NP-complete, i.e., it is unlikely to have polynomial algorithms for their solution. Despite of this, many results have been obtained to characterize the set of grid-points in the feasible region of system (5) [32, 7, 39, 6].

Let us denote by $\mathcal{P}$ the convex polytope defined by Equation (5) and by P the convex hull of $\mathcal{P} \cap \mathbb{Z}^n$. Further, let $N(A, b)$ be the set of vertices of P and $|N(A, b)|$ its cardinality. Various upper and lower bounds on $|N(A, b)|$ have been found (see, for instance, [39] for a recent survey). In particular, we have the following theorem.

Theorem 6. *Let* $A = (a_{ij}) \in \mathbb{Z}^{m \times n}$, $b = (b_i) \in \mathbb{Z}^m$, $c = (c_j) \in \mathbb{Z}^n$ *and* $\alpha = \max\{|a_{ij}|, i = 1, \ldots, m, j = 1, \ldots, n\}$. *Then*

$$|N(A, b)| \le c_n m^{\lfloor n/2 \rfloor} \log^{n-1}(1 + \alpha) \tag{7}$$

where α *is an upper bound on the largest (by absolute value) coefficient in the ILP formulation and* c_n *is a quantity depending only on* n.

Moreover, upper bound (6) is tight, i.e., there is a class of matrices A and vectors b for which

$$|N(A, b)| \ge c'_n m^{\lfloor n/2 \rfloor} \log^{n-1}(\alpha)$$

where c'_n *is a constant depending only on* n.

Similar tight bounds hold for the number of the knapsack polytope vertices.

Now let P be a non-empty n-dimensional lattice polytope and $f_k(P)$ the number of its k-dimensional facets for $0 \le k < n$. In particular, f_0 is the number of vertices of P. Then the following theorem holds.

Theorem 7. [8]

$$f_k \le c_n (Vol\, P)^{\frac{n-1}{n+1}} \tag{8}$$

where c_n *is a quantity depending only on* n *and (Vol P) the volume of* P.

In dimension 2, the above result can be linked to the maximal number of edges $e(N)$ of a convex digital polygon included into an $N \times N$-grid [5, 2]:

$$e(N) = \frac{12}{(4\pi^2)^{1/3}} N^{2/3} + O(N^{1/3} log(N)).\tag{9}$$

Theorem 8. [8] *Let $K \in C(D)$ where $C(D)$ is the family of convex bodies with C^2 boundary and radius of curvature at every point and every direction between $1/D$ and D, $D \geq 1$. Let $\bar{K} = conv(K \cap \mathbb{Z}^n)$. If the diameter of K is enough large, then for every $n \geq 2$ there are constants $c_1(n)$ and $c_2(n)$ such that for all $k \in \{0, 1, \ldots, n-1\}$,*

$$c_n d^{n \frac{n-1}{n+1}} \leq f_k(\bar{K}) \leq c'_n d^{n \frac{n-1}{n+1}}.\tag{10}$$

In the following, we use these results to give bounds on the size of the convex hull and the preimage of S.

5.2   Application to DPS Recognition

In the following, we suppose that S is parametrized using the discrete analytical hyperplane definition (see Definition 3). Hence, let $\alpha = \max\{|a_1|, \ldots, |a_n|\}$. Obviously, diverse parametrizations exist for a DHPS. However, in a recognition process it is common to consider a parametrization that minimizes α and this parameter is usually bounded by the diameter of S (*i.e.* bounded by $O(N)$ in our framework).

Theorem 9.   *1. Let S be a DHPS with a hyper-rectangular base. Then*

$$|CH(S)| \leq c_n \log^{n-1}(1+\alpha);\tag{11}$$

2. Alternatively,

$$|CH(S)| \leq c_n N^{\frac{(n-1)^2}{n+1}};\tag{12}$$

where c, c_n are some quantities depending only on n.
3. If S is a DPS with digitally convex base containing the origin, then

$$|CH(S)| \leq cN^{\frac{2}{3}} \log^2(1+\max\{N,\alpha\}).\tag{13}$$

Proof. We first proof Equation (11). Given DHPS S with a hyper-rectangular base, we can construct $CH(S)$ using $2 \cdot n + 2$ linear constraints in dimension n. First, $2 \cdot n$ constraints are necessary to define the hyper-rectangular base (all these are given by $\{0 \leq x_j \leq b_j\}$ for $0 \leq j \leq n-1$ and $b_i \in \mathbb{N}$). Moreover, two additional constraints are needed to encode the two parallel Euclidean leaning hyperplanes associated with the DHPS. Finally, we can construct a matrix $A = (a_{ij}) \in \mathbb{Z}^{(2 \cdot n+2) \times n}$ with $\max\{|a_{ij}|, i = 1, \ldots, m, j = 1, \ldots, n\} = \alpha$ whose size depends only on n. Hence, using Theorem 6 with $m = 2 \cdot n + 2$, we obtain Equation (11).

To prove Equation (12), it suffices to observe that S is included in an N^n grid, therefore Vol $CH(S)$ is bounded by $O(N^{n-1})$. Indeed, the thickness of $CH(S)$ is necessarily lower than 1. Finally, using Theorem 7, we obtain the result stated.

Similarly, to prove Equation (13), we observe that if S is a DPS with a digitally convex base, then the base can be encoded by at most $O(N^{\frac{2}{3}})$ 2-D linear constraints (one constraint per edge of the 2-D convex hull of the base using Equation (9)). Furthermore, since S is included in an $N \times N \times N$ grid, the coefficients of the linear constraints are bounded by N. Finally, using the two constraints that define the 3-D discrete plane, $CH(S)$ can be represented by $O(N^{\frac{2}{3}})$ constraints overall. Then Equation (13) follows from Theorem 6.   □

Note that for dimension three, the bound for a DHPS with a hyper-rectangular base conforms to earlier results about rectangular base DPS presented in [17]. Combining geometric and number-theoretic approaches, the author also proves that, under some assumptions, the preimage has at most $O(\log N)$ facets.

Through some experiments presented in the next section, we quantify the differences between the size of the convex hull of S and its preimage.

5.3   Experimental Results

To evaluate the size of both the convex hull and the preimage of a DHPS, we have utilized a specific experimental framework. First, we need a DHPS random generator. Since no uniform random generator of DHPS is available, we created one that conforms to the following natural scheme:

Without loss of generality let us fix the major axis to be x_n. Then:

1. Construct the DHPS base in an N^{n-1}-grid;
2. Use a uniform normal vector generator to obtain the DHPS parameters;
3. *Raise* the base along the x_n axis using the parameters obtained in stage 2.

To generate the base, we consider two main classes of DHPS. The first is a hyper-rectangular one in which the lengths of the $(n-1)$-rectangle are given by independent random generators. DHPS's from the second class have digitally convex bases. In the latter case we generate a random set of cospherical Euclidean points that belong to the hyperparallelepiped $[-1, 1]^{n-1}$. Then the Euclidean convex hull of this set is computed and digitized for a given grid resolution in order to obtain digitally convex bases with *round* shapes. We choose this point distribution since it seems to be close to the worst-case regarding the convex hull size.

Figure 2 gives some examples of randomly generated DHPS's in dimension 3.

To test the theoretical results from Section 5.2, we use the well-known `qhull` program for n−dimensional convex hulls computation [1, 9]. To evaluate the number of vertices of n−dimension preimages, we first obtain a system of linear inequalities. For this, we apply Equation (3). Then we make use of the `lrs` software to convert an H-representation (half-space) of a polytope into a V-representation (vertex/ray) or vice versa, as exact arithmetic is used [4].

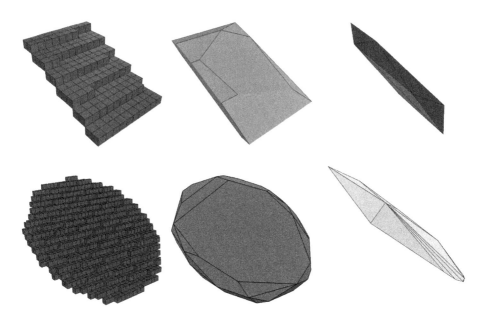

Fig. 2. *First row*: a rectangular base DPS, its convex hull and its preimage in the parameter space. *Second row*: a rounded base DPS, its convex hull and its preimage in the parameter space.

The above algorithm has been designed for problems of arbitrary dimension n, so it is not surprising that it is outperformed by the incremental Algorithm 1 from Section 3.2 on practical three-dimensional recognition problems.

We present results concerning the two classes of DHPS defined above with increasing N. In the graphs presented in Figure 3, 4 and 5, x-axis corresponds to the number of grid points given by the DHPS generators and the y-axis corresponds to either $|CH(S)|$ or $|Preimage(S)|$ with $n = \{3, 4, 5, 6\}$. Figure 3

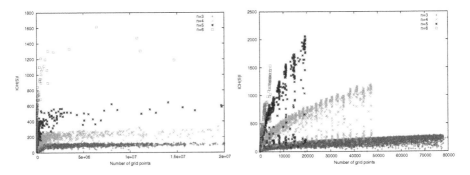

Fig. 3. Evaluation of $|CH(S)|$ on randomly generated DHPS. *Left:* the hyper-rectangular base class, and *Right:* the *cospherical* base DHPS class.

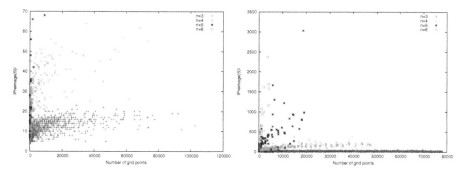

Fig. 4. Evaluation of $|Preimage(S)|$ on randomly generated DHPS. *Left:* the hyper-rectangular base class, and *Right:* the *cospherical* base DHPS class.

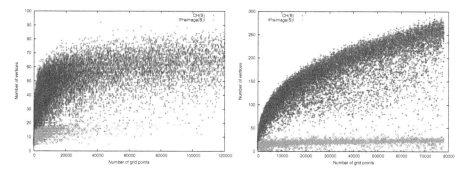

Fig. 5. Comparison between $|CH(S)|$ and $|Preimage(S)|$ for $n = 3$. *Left* the hyper-rectangular base class and *right*, the *cospherical* base DHPS class.

shows results on $|CH(S)|$ and Figure 4 the results on $|Preimage(S)|$. Finally, Figure 5 details a comparison between $|CH(S)|$ and $|Preimage(S)|$ for $n = 3$.

As expected in view of Theorem 9, the logarithmic behavior of both $|CH(S)|$ and $|Preimage(S)|$ for hyper-rectangular base DHPS clearly appears in the graphs. However, for the other class of DHPS with dimension greater than 3, the experimental results do not show a polynomial behavior of the curves. Further theoretical analysis is thus expected to lead to bounds closer to the experimental framework.

6   Conclusion and Future Works

In this article we have first reviewed algorithmic solutions to recognize DHPS whatever the dimension. We have theoretically optimal in time algorithm without efficient implementations and fast algorithms with quite high worst case computational costs. Our analysis suggests that specificities of DHPS recognition problems should be taken into account to obtain tights bounds. Based on results from integer programming and associated lattice polytopes, several theoretical results have been proposed to bound the number of vertices of the convex hull of a DHPS and the number of vertices of its preimage.

One may attempt to solve the recognition problem taking advantage of the theoretical results of Section 5.2. Our experimental results suggest the following conclusions. The direct use of convex hull computation may not be tractable since the design of incremental and output sensitive convex hull algorithms is a difficult task [11]. For dimension $n = 3$, our experiments on preimage computation showed that the quantity E defined in Section 3.2 is indeed quite low whatever the number of grid points. Thus Algorithm 1 appears to be a very simple and practically efficient incremental DPS recognition algorithm.

Several questions related to DHPS recognition are still open. Among these we list the following:

1. Are there tighter bounds on $|CH(S)|$ in higher dimensions?
2. Can arithmetical structures of DHPS speed up recognition algorithms?

Acknowledgments

The authors thank Isabelle Sivignon and Raphaëlle Chaine, Laboratoire LIRIS, Université Claude Bernard Lyon 1, for their fruitful comments on this work.

References

1. Qhull. http://www.qhull.org/.
2. D.M. Acketa and J.D. Žunić. On the maximal number of edges of convex digital polygons included into a $m \times m$-grid. *Journal of Combinatorial Theory*, Serie A(69):358–368, 1995.
3. E. Andres, R. Acharya, and C. Sibata. Discrete analytical hyperplanes. *Graphical Models and Image Processing*, 59(5):302–309, September 1997.
4. D. Avis. lrs: implementation as a callable library of the reverse search algorithm for vertex enumeration/convex hull problems. http://cgm.cs.mcgill.ca/~avis/C/lrs.html.
5. A. Balog and I. Bárány. On the convex hull of the integer points in a disc. In ACM-SIGACT ACM-SIGGRAPH, editor, *Proceedings of the 7th Annual Symposium on Computational Geometry (SCG '91)*, pages 162–165, North Conway, NH, USA, June 1991. ACM Press.
6. I. Barany. Random points, convex bodies, lattices. In *ICM: Proceedings of the International Congress of Mathematicians*, 2002.
7. I. Barany, Howe, and Lovasz. On integer points in polyhedra: A lower bound. *Combinatorica*, 12, 1992.
8. I. Bárány and D. G. Larman. The convex hull of the integer points in a large ball. *Math. Annalen*, 312:167–181, 1998.
9. C. B. Barber, D. P. Dobkin, and H. Huhdanpaa. The quickhull algorithm for convex hulls. *ACM Transactions on Mathematical Software*, 22(4):469–483, December 1996.
10. R. P. Barneva, V. E. Brimkov, and P. Nehlig. Thin discrete triangular meshes. *Theoretical Computer Science*, 246(1-2):73–105, 2000.

11. D. Bremner. Incremental convex hull algorithms are not output sensitive. *Discrete & Computational Geometry*, 21(1):57–68, 1999.

12. V. Brimkov, D. Coeurjolly, and R. Klette.   Digital planarity - a review. Technical report, Laboratoire LIRIS, Université Claude Bernard Lyon 1, 2004. `http://liris.cnrs.fr/publis/?id=1933`.

13. V. E. Brimkov and S. S. Dantchev. Complexity analysis for digital hyperplane recognition in arbitrary fixed dimension. In *International Conference on Discrete Geometry for Computer Imagery*, pages 287–298, 2005.

14. L. Buzer. A linear incremental algorithm for naive and standard digital lines and planes recognition. *Graphical Models*, 65(1-3):61–76, May 2003.

15. D. Coeurjolly. *Algorithmique et géométrie discrète pour la caractérisation des courbes et des surfaces*. PhD thesis, Université Lumière Lyon 2, Bron, Laboratoire ERIC, dec 2002.

16. D. Coeurjolly, Alexis Guillaume, and I. Sivignon. Reversible discrete volume polyhedrization using marching cubes simplification. In *SPIE Vision Geometry XII*, volume 5300, pages 1–11, San Jose, USA, 2004.

17. D. Coeurjolly, I. Sivignon, F. Dupont, F. Feschet, and J.-M. Chassery. On digital plane preimage structure. *Discrete Applied Mathematics*, 151(1–3):78–92, 2005.

18. I. Debled-Rennesson. *Etude et reconnaissance des droites et plans discrets*. PhD thesis, Université Louis Pasteur, 1995.

19. J. Françon, J.M. Schramm, and M. Tajine. Recognizing arithmetic straight lines and planes. In *6th Discrete Geometry for Computer Imagery*, volume 1176 of *LNCS*, pages 141–150. Springer-Verlag, 1996.

20. J. Françon and L. Papier. Polyhedrization of the boundary of a voxel object. In *8th International Conference in Discrete Geometry for Computer Imagery*, volume 1568 of *LNCS*, pages 425–434. Springer-Verlag, 1999.

21. M. R. Garey and D. S. Johnson. *Computers and intractability; a guide to the theory of NP-completeness*. W.H. Freeman, 1979.

22. Y. Gerard, I. Debled-Rennesson, and P. Zimmermann. An elementary digital plane recognition algorithm. *DAMATH: Discrete Applied Mathematics and Combinatorial Operations Research and Computer Science*, 151, 2005.

23. C. E. Kim. Three-dimensional digital planes. *IEEE Trans. on Pattern Analysis and Machine Intelligence*, 6:639–645, 1984.

24. C. E. Kim and I. Stojmenovic. On the recognition of digital planes in three-dimensional space. *Pattern Recognition Letters*, 12(11):665–669, 1991.

25. R. Klette, I. Stojmenovic, and J. D. Zunic. A parametrization of digital planes by least-squares fits and generalizations. *CVGIP: Graphical Model and Image Processing*, 58(3):295–300, 1996.

26. R. Klette and H. J. Sun. Digital planar segment based polyhedrization for surface area estimation. In *IWVF*, pages 356 366, 2001.

27. N. Megiddo. Linear programming in linear time when the dimension is fixed. *JACM: Journal of the ACM*, 31, 1984.

28. M. M. Mesmoudi. A simplified recognition algorithm of digital planes pieces. In *10th International Conference, Discrete Geometry for Computer Imagery*, volume 2301 of *LNCS*, pages 404–416, Bordeaux, France, april 2002. Springer Verlag.

29. F. P. Preparata and M. I. Shamos. *Computational Geometry : An Introduction*. Springer-Verlag, 1985.

30. A. Rosenfeld. Digital straight lines segments. *IEEE Transactions on Computers*, pages 1264–1369, 1974.

31. A. Rosenfeld and R. Klette. Digital straightness. In *Int. Workshop on Combinatorial Image Analysis*, volume 46 of *Electronic Notes in Theoretical Computer Science*. Elsevier Science Publishers, August 2001.

32. A. Schrijver. *Theory of Linear and Integer Programming*. Wiley and Sons, 1986.

33. I. Sivignon and D. Coeurjolly. From digital plane segmentation to polyhedral representation. In *Theoretical Foundations of Computer Vision "Geometry, Morphology, and Computational Imaging"*, number 2616 in LNCS, Springer-Verlag, pages 356–367, 2003.

34. I. Sivignon, F. Dupont, and J. M. Chassery. Decomposition of a three-dimensional discrete object surface into discrete plane pieces. *Algorithmica*, 38(1):25–43, 2003.

35. I. Stojmenović and R. Tosić. Digitization schemes and the recognition of digital straight lines, hyperplanes and flats in arbitrary dimensions. In *Vision Geometry, contemporary Mathematics Series*, volume 119, pages 197–212, American Mathematical Society, Providence, RI, 1991.

36. P. Veelaert. On the flatness of digital hyperplanes. *Journal of Mathematical Imaging and Vision*, 3:205–221, 1993.

37. P. Veelaert. Digital planarity of rectangular surface segments. *IEEE Pattern Analysis and Machine Intelligence*, 16(6):647–652, jun 1994.

38. J. Vittone and J.-M. Chassery. Recognition of digital naive planes and polyhedization. In *9th Discrete Geometry for Computer Imagery*, volume 1953 of *LNCS*, pages 296–307. Springer, 2000.

39. N. Y. Zolotykh. On the number of vertices in integer linear programming problems. Technical report, University of Nizhni Novgorod, 2000.

A Linear Algorithm for Polygonal Representations of Digital Sets

Helene Dörksen-Reiter[1] and Isabelle Debled-Rennesson[2]

[1] EMBL Hamburg
Building 25A, DESY
Notkestrasse 85
22603 Hamburg - Germany
doerksen@embl-hamburg.de
[2] LORIA Nancy – Campus Scientifique - BP 239
54506 Vandœuvre-lès-Nancy Cedex - France
debled@loria.fr

Abstract. Polygonal representations of digital sets with the same convexity properties allow a simple decomposition of digital boundaries into convex and concave parts.

Representations whose vertices are boundary points, i.e. are integer numbers, attract most attention. The existing linear Algorithm UpPolRep computes polygonal representations with some uncorresponding parts. However, the algorithm is unable to decide if a corresponding polygonal representation still exists and in the case of existence it is unable to compute the representation. Studying situations where uncorrespondences appear we extended the algorithm. The extention does not change the time complexity. If a digital set possesses a corresponding representation then it detects this representation. Otherwise, it recognizes that such representation does not exist.

Keywords: digital convexity, discrete lines and discrete curves, convex and concave parts of discrete curves, polygonal representation.

1 Introduction

Convex and concave parts of sets determine their visual components, i.e. such parts are meaningful for our perception. Decomposition boundaries into convex and concave parts has an important application. It is recognition objects by comparing with shapes from a database [10].

In the plane $\mathbb{R}^2$, the boundary of a polygonal set can be decomposed into convex and concave parts in an obvious way. In digital geometry it becomes a very difficult task (see e.g. [5]).

There are several techniques for the decomposition of digital sets into meaningful parts, e.g. proposed in [10] and [5]. Both methods have an approximative character. Another method for the decomposition of digital sets into convex and concave parts is proposed in [4]. It is exact. Here, the boundary of a digital set is decomposed into meaningful parts by means of Scherl's descriptors and fundamental segments. Descriptors introduced by Scherl [12] are points of local support with respect to a certain finite

U. Eckardt et al. (Eds.): IWCIA 2006, LNCS 4040, pp. 307–319, 2006.

number of directions. Fundamental segments are related to definition of arithmetical discrete lines [11, 1].

In [3] (see also [5]) following problem was studied: Can a digital set be represented by a polygonal set in the plane $\mathbb{R}^2$ such that it is a Jordan curve, it has only integer vertices and contains exactly the points of the set in its interior? One wants the representing polygon to have the same convexity properties as the digital set. A linear Algorithm UpPolRep for computing this kind of representations was proposed. It is based on the mentioned above exact method for decomposition digital sets into convex and concave parts [4] as well as the technique for segmentation digital curves into discrete line segments [1]. The origial algorithm computes representations with some uncorresponding parts. However, a corresponding polygonal representation with integer vertices may still exist [3]. Studying situations where uncorrespondences appear we extended the algorithm. If a digital set possesses a corresponding representation then our algorithm detects this representation. Otherwise, it recognizes that such representation does not exist.

Section 2 defines theoretical preliminaries. These contain the definition for digital $(0,1)$-curves and notations about discrete lines. In Section 3 we sketch the method proposed in [4] for the decomposition of $(0,1)$-curves into convex and concave parts by means of fundamental segments. Section 4 shows how to decompose the boundary of a digital set into convex and concave parts. It is based on the method for $(0,1)$-curves.

In Section 5 we recall notations and main results about polygonal representations from [3]. Then, in Section 6 critical situations where uncorrespondences appear are studied. The extention of Algorithm UpPolRep is presented. At the end of this section we shortly discuss about another method designed for the same purpose.

2   Preliminaries

We focus on 8-neighborhood structure for sets of $\mathbb{Z}^2$ and 4-neighborhood structure for complements of sets.

Definition 1. *Given an 8-connected digital set $\mathcal{K} \subseteq \mathbb{Z}^2$. $\mathcal{K}$ is called a* digital 8-curve *whenever each point $x \in \mathcal{K}$ has exactly two 8-neighbors in $\mathcal{K}$ with the possible exception of at most two points, the so-called* end points *of the curve, having exactly one neighbor in $\mathcal{K}$.*

A curve without end points is named a closed *curve.*

Each digital 8-curve can be ordered (or oriented) in a natural manner by means of a simple compact ordered data structure. It contains the coordinates of one element of $\mathcal{K}$ and a sequence of code numbers $\{0, 1, \cdots, 7\}$. It indicates for each point of $\mathcal{K}$ which of its neighbors will be the next point on the curve. This data structure was proposed by Freeman [7] and is known as the *chain code*.

Definition 2. *Given an ordered digital 8-curve $\mathcal{K} = (\kappa_1, \cdots, \kappa_n)$. For a code number $k \in \{0, \cdots, 7\}$, the curve $\mathcal{K}$ is called a $(k, k+1 (\mathrm{mod}\ 8))$-curve whenever the chain code representation of $\mathcal{K}$ consists exclusively of at most two chain codes k and $k+1 (\mathrm{mod}\ 8)$.*

For a code number $v \in \{0, \cdots, 7\}$, *a* level *of* $\mathcal{K}$ *is a maximal subset of the curve whose chain code representation consists only of the code number* v. *The number of successive elements of a level is called the* length *of the level.*

Since each $(k, k+1 \pmod 8)$-curve is rotation of some $(0,1)$-curve, in the later sections we may focus, without loss of generality, on $(0,1)$-curves.

If the convex hull of a finite $(0,1)$-curve possesses only elements of the curve then it is a segment of a *discrete line*. The arithmetical definition of discrete lines was introduced by J.-P. Reveillès [11].

Definition 3. *A discrete line* $\mathcal{D}(a,b,\mu,\omega)$ *with slope* a/b, $b \neq 0$ *and pgcd* $(a,b) = 1$, *lower bound* μ, *arithmetical thickness* ω *(all parameters are integer numbers) is the set of grid points which satisfies the double diophantine inequality*

$$\mu \leq ax - by < \mu + \omega.$$

We note the preceding discrete line $\mathcal{D}(a,b,\mu,\omega)$. We are mostly interested in *naïve* lines which verify $\omega = \sup(|a|, |b|)$, we shall note them $\mathcal{D}(a,b,\mu)$. Without loss of generality we may consider discrete lines under restrictions $a, b > 0$ and $a < b$, therefore $\omega = \max(a,b) = b$.

The real straight lines $ax - by = \mu$ and $ax - by = \mu + b - 1$ are called *upper leaning line* and *lower leaning line* of $\mathcal{D}(a,b,\mu)$, respectively. The grid points satisfying the leaning line equalities are called *upper* and *lower leaning points*.

Let $\mathcal{K}$ be a segment of a discrete line $\mathcal{D}(a,b,\mu)$. The problem to determine the convex hull of the elements of $\mathcal{K}$ is solved in [2]. The convex hull of $\mathcal{K}$ is a closed polygonal curve which can be subdivided into two polygonal curves joining its first and last points : the *lower frontier* and *upper frontier* of the convex hull. How to detect all points which belong to the lower and upper frontier is shown in [2, Proposition 3, p.120]. Since the curve $\mathcal{K}$ is a segment of a discrete line, the intersection of $\mathcal{K}$ and its convex hull consists only of elements of $\mathcal{K}$.

Definition 4. *A digital curve* $\mathcal{K}$ *is said to be* lower digitally convex *if there is no grid point between* $\mathcal{K}$ *and the lower frontier of the convex hull of* $\mathcal{K}$.

3   Convex and Concave Parts of $(0,1)$-Curves

In [4] we introduced a method how to decompose a digital curve into meaningful parts. This technique is based on the concept of *fundamental segments*, which are segments of discrete lines having maximal possible lengths.

Definition 5. *Let* $\mathcal{K} = (\kappa_1, \cdots, \kappa_n)$ *be a* $(0,1)$-*curve. Parameters* a *and* b *in discrete line segments considered below are assumed to be minimal. A part* $(\kappa_i, \cdots, \kappa_j)$ *is called a* fundamental segment *of* $\mathcal{K}$ *whenever one of the following conditions is true:*

- $i = 1$, $j = n$ *and* $(\kappa_1, \cdots, \kappa_n)$ *is a segment of* $\mathcal{D}(a,b,\mu)$. *Then* $\mathcal{K}$ *consists of one single fundamental segment.*

- $i = 1$, $j < n$ and $(\kappa_1, \cdots, \kappa_j)$ is a segment of $\mathcal{D}(a, b, \mu)$ such that $(\kappa_1, \cdots, \kappa_{j+1})$ is not a segment of any discrete line. Here, $(\kappa_1, \cdots, \kappa_j)$ is the first fundamental segment of $\mathcal{K}$.
- $i > 1$, $j = n$ and $(\kappa_i, \cdots, \kappa_n)$ is a segment of $\mathcal{D}(a, b, \mu)$ such that $(\kappa_{i-1}, \cdots, \kappa_n)$ is not a segment of any discrete line. Here, $(\kappa_i, \cdots, \kappa_n)$ is the last fundamental segment of $\mathcal{K}$.
- $i > 1$, $j < n$ and $(\kappa_i, \cdots, \kappa_j)$ is a segment of $\mathcal{D}(a, b, \mu)$ such that $(\kappa_{i-1}, \cdots, \kappa_j)$ and $(\kappa_i, \cdots, \kappa_{j+1})$ are not segments of any discrete line.

The fundamental segment $(\kappa_i, \cdots, \kappa_j)$ *will be denoted by* $\mathcal{F}(a, b, \mu)$.

All fundamental segments can be ordered in the sense of the oriented curve, we mark these $\mathcal{F}_i(a_i, b_i, \mu_i)$, $i = 1, \cdots, m$.

The problem to find decomposition of a $(0, 1)$-curve into fundamental segments is equivalent to the problem to determine subsets of the curve having constant tangents. A linear algorithm is proposed in [6]. An example of a $(0, 1)$-curve with its fundamental segments is demonstrated in Figure 1.

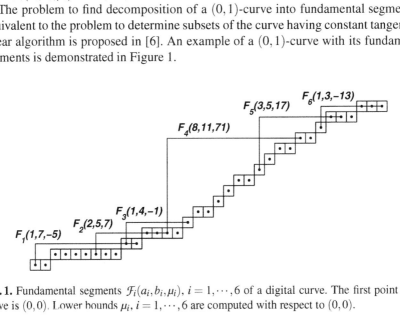

Fig. 1. Fundamental segments $\mathcal{F}_i(a_i, b_i, \mu_i)$, $i = 1, \cdots, 6$ of a digital curve. The first point of the curve is $(0, 0)$. Lower bounds μ_i, $i = 1, \cdots, 6$ are computed with respect to $(0, 0)$.

In [4] we proved that a digital curve is lower digitally convex if and only if the sequence of the slopes of its fundamental segments is increasing. Then, parts of the curve, whose fundamental segments possess increasing slopes, were defined as convex, such possessing decreasing slopes were defined as concave. In this manner, we introduced *maximal convex* and *maximal concave parts* of curves.

Definition 6. *Let* $\mathcal{K}$ *be a finite* $(0, 1)$-*curve and* $\mathcal{F}_i(a_i, b_i, \mu_i)$, $i = 1, \cdots, m$ *be fundamental segments of* $\mathcal{K}$. *A part consisting of successive fundamental segments* $\mathcal{F}_u(a_u, b_u, \mu_u)$, $\cdots, \mathcal{F}_v(a_v, b_v, \mu_v)$, $1 \leq u \leq v \leq m$ *is called a* maximal convex part *of* $\mathcal{K}$ *whenever one of the following conditions is true:*

- $u = 1$, $v = m$ and $\frac{a_j}{b_j} < \frac{a_{j+1}}{b_{j+1}}$, $1 \leq j \leq m - 1$.
- $u \neq 1$, $v \neq m$, $\frac{a_{u-1}}{b_{u-1}} > \frac{a_u}{b_u}$, $\frac{a_v}{b_v} > \frac{a_{v+1}}{b_{v+1}}$ *and* $\frac{a_j}{b_j} < \frac{a_{j+1}}{b_{j+1}}$ *for all* $u \leq j \leq v - 1$.

- $u = 1, v \neq m, \frac{a_v}{b_v} > \frac{a_{v+1}}{b_{v+1}}$ and $\frac{a_j}{b_j} < \frac{a_{j+1}}{b_{j+1}}$ for all $1 \leq j \leq v - 1$.
- $u \neq 1, v = m, \frac{a_{u-1}}{b_{u-1}} > \frac{a_u}{b_u}$ and $\frac{a_j}{b_j} < \frac{a_{j+1}}{b_{j+1}}$ for all $u \leq j \leq m - 1$.

A maximal concave part *of $\mathcal{K}$ is defined in the same manner by replacing the signs '<'
and '>' in the above definition.*

The curve from Figure 1 has four maximal parts: two maximal convex parts which
are $(\mathcal{F}_1, \mathcal{F}_2)$ and $(\mathcal{F}_3, \mathcal{F}_4)$, two maximal concave parts which are $(\mathcal{F}_2, \mathcal{F}_3)$ and $(\mathcal{F}_4,
\mathcal{F}_5, \mathcal{F}_6)$.

Polygonal curves on $\mathbb{R}^2$, whose edges are leaning lines of fundamental segments, are
called *fundamental polygonal representations* (see [4]). Obviously, fundamental polygonal representations possess the same convexity properties as the curve.

4   Segmentation Boundaries of Digital Sets into $(0, 1)$-Curves

We use the term *(oriented) boundary* of digital sets. For a comprehensive view of this
concept, we refer to books and articles on digital geometry and topology (e.g. [9, 14, 8]).

Boundary of a digital set can be decomposed into $(0, 1)$-curves by means of *Scherl's
descriptors* (see [3, Chapter 2]). Descriptors introduced by Scherl [12] are boundary
points of a digital set belonging to local extrema of linear functions with following
main directions: $0°$, $45°$, $90°$, $135°$, $180°$, $225°$, $270°$ and $315°$. Thus, descriptors are
segments of horizontal, vertical and diagonal grid lines and belong to locally convex or
concave parts of the set. These are respectively named *T*- or *S-descriptors*. In Figure 2
Scherl's descriptors of a digital set are shown.

The succession of descriptor points on the oriented boundary of a digital set is not
arbitrary [5, 12]. It can be shown that the boundary of an 8-connected digital set can
be decomposed into $(k, k + 1 (\mathrm{mod}\ 8))$-curves such that the part on the boundary between two such successive curves consists only of descriptor points [5]. Since *T*- and
S-descriptors belong to locally convex and concave parts, respectively, technique for

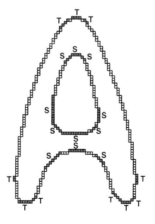

Fig. 2. Scherl's descriptors of a digital set. T- and S-descriptors are indicated.

partition curves mentioned in Section 3 can be extended to decomposition boundaries of digital sets into convex and concave parts (see also [4]).

An alternative definition for Scherl's descriptors was given in [3] and a linear algorithm for computing descriptor points was proposed.

5 Polygonal Representations of Digital Sets

In this section we recall notations and main results about polygonal representations of digital sets from [3].

Definition 7. *Given a digital set $S_\triangle \subseteq \mathbb{Z}^2$. A polygonal representation of $S_\triangle$ is a polygonal set $\Pi \subseteq \mathbb{R}^2$ with vertices V such that: $x \in S_\triangle \iff x \in \Pi \cap \mathbb{Z}^2$.*
Π is called:

– **discrete** *if all vertices of Π are in $\mathbb{Z}^2$;*
– **faithful** *if the succession of convex and concave parts of the boundary of $S_\triangle$ corresponds to the succession of convex and concave parts of the boundary of Π.*

The chain code representation of a digital set contains its all boundary points. It is discrete, but not faithful. The fundamental polygonal representation mentioned in Section 3 is a faithful one. This representation is in general not discrete. Here, polygonal representations which are both discrete and faithful attract most attention.

The previous sections have shown that the calculation of polygonal representations of digital sets can be restricted to the calculation of representations of $(0,1)$-curves. The polygonal representation of a $(0,1)$-curve $\mathcal{K} = (\kappa_1, \cdots, \kappa_n)$ can be subdivided into two polygonal curves between κ_1 and κ_n lying above and below $\mathcal{K}$. They are called *upper* and *lower polygonal representation*, respectively. Obviously, the lower polygonal representation of $\mathcal{K}$ is an upper polygonal representation of the curve $\tilde{\mathcal{K}} = (\kappa_n \cdot A, \cdots, \kappa_1 \cdot A)$, where $A = \begin{pmatrix} -1 & 0 \\ 0 & -1 \end{pmatrix}$.

In the literature it is more used to consider lower polygonal representations. Here, upper polygonal representations are analysed since they are more convenient for computation [3].

5.1 Upper Representations of Concave Curves

Given a concave $(0,1)$-curve $\mathcal{K}$ with fundamental segments $\mathcal{F}_i(a_i, b_i, \mu_i)$, $i = 1, \cdots, m$. We note U_F the upper leaning point of a discrete line segment whose x-coordinate is minimal, U_L is the leaning point if it is maximal. Upper leaning points of fundamental segments are successive elements on the concave curve (dual case of [4, Proposition 4.1]). Further, vertices of the upper polygonal representation between U_{F_1} and U_{L_m} are: $\{U_{F_1}, U_{L_1}, U_{F_2}, U_{L_2}, \cdots, U_{F_m}, U_{L_m}\}$ (dual case of [4, Theorem 4.1]). Note that for the numerical implementation it is not recommendable to determine fundamental segments. In spite of linearity, this technique is applicable only to concave curves.

Algorithm UpPolRep [3] detects a discrete faithful polygonal representation. It is constructed on the basis of Algorithm AddPoint [2]. Algorithm AddPoint determines the segment $(\kappa_1, \cdots, \kappa_j)$ of a discrete line belonging to $\mathcal{K} = (\kappa_1, \cdots, \kappa_n)$ such that $j = n$

or $(\kappa_1,\cdots,\kappa_{j+1})$ is not a segment of a discrete line. Further, if the second case appears then it begins to determine the next segment of a discrete line from the last calculated leaning point U_L.

The next proposition shows the theoretical background behind this procedure if the curve is concave. It is the dual case of [3, Proposition 4.2]. Here, we give an improved proof for it.

Proposition 1. *Let $\mathcal{K}$ be a concave $(0,1)$-curve and $\mathcal{F}_i(a_i,b_i,\mu_i)$, $i = 1,\cdots,m$ are its fundamental segments. Then $U_{L_j} \in \mathcal{F}_{j+1}(a_{j+1},b_{j+1},\mu_{j+1})$ for all $1 \leq j \leq m-1$.*

Proof. It holds $x_{U_{L_j}} \leq x_{U_{F_{j+1}}}$ for all $1 \leq j \leq m-1$ (dual case of [4, Proposition 4.1]). The case $x_{U_{L_j}} = x_{U_{F_{j+1}}}$ is trivial. We concentrate on $x_{U_{L_j}} < x_{U_{F_{j+1}}}$. Since $\mathcal{K}$ is concave then by duality of [4, Proposition 4.2] there is no another grid point between U_{L_j} and $U_{F_{j+1}}$ and the real line through U_{L_j} and $U_{F_{j+1}}$. Thus, the segment between U_{L_j} and $U_{F_{j+1}}$ belongs to a discrete line such that first leaning point of it is U_{L_j}, last leaning point is $U_{F_{j+1}}$. Assume $U_{L_j} \notin \mathcal{F}_{j+1}(a_{j+1},b_{j+1},\mu_{j+1})$. Then it holds $U_{F_{j+1}} \notin \mathcal{F}_j(a_j,b_j,\mu_j)$. Further, assume that first and last elements of fundamental segments $\mathcal{F}_j(a_j,b_j,\mu_j)$ and $\mathcal{F}_{j+1}(a_{j+1},b_{j+1},\mu_{j+1})$ are κ_p, κ_{p+q} and κ_s, κ_{s+t}, respectively. Elements on $\mathcal{K}$ between κ_p, $U_{F_{j+1}}$ and U_{L_j}, κ_{s+t} are not discrete line segments. We deduce that elements between U_{F_j}, $U_{F_{j+1}}$ and U_{L_j}, $U_{L_{j+1}}$ are not discrete line segments, too. Hence, we can find another fundamental segment between $\mathcal{F}_j(a_j,b_j,\mu_j)$ and $\mathcal{F}_{j+1}(a_{j+1},b_{j+1},\mu_{j+1})$ that leads to a contradiction of their succession.   □

Algorithm AddPoint detects not all leaning points of fundamental segments. It is demonstrated in Figure 3.

The solution of this problem is given in [2]. For a segment $(\kappa_1,\cdots,\kappa_n)$ of a discrete line $\mathcal{D}(a,b,\mu)$ and $r(x,y) = ax - by$ holds: vertices of the upper frontier between κ_1 and U_F are given by the maximal sequence of points $\{P_{i=1,\dots,k}\}$ such that $r(\kappa_1) > r(P_1) > \cdots > r(P_k) > r(U_F)$, vertices between U_L and κ_n are given by the maximal sequence such that $r(\kappa_n) > r(P_1) > \cdots > r(P_k) > r(U_L)$.

UpPolRep. Upper polygonal representation of a digital curve

$V \leftarrow \emptyset$; /* vertices of the upper polygonal representation */
$START \leftarrow \kappa_1$;
repeat
    Determine the segment $(START,\cdots,\kappa_j)$ of a discrete line such that $j = n$ or
    $(START,\cdots,\kappa_{j+1})$ is not a segment of any discrete line;
    $V \leftarrow V\cup$ maximal sequence $\{P_{i=1,\dots,k}\}$ between $START$ and U_F;
    $V \leftarrow V\cup U_F \cup U_L$;
    if $(START,\cdots,\kappa_{j+1})$ *is not a segment of any discrete line* **then**
      | $START \leftarrow U_L$;
    else
      | $STOP$;
    end
until $STOP$;
$V \leftarrow V\cup$ maximal sequence $\{P_{i=1,\dots,k}\}$ between U_L and κ_n;

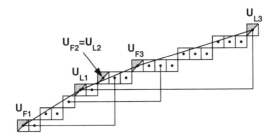

Fig. 3. The vertex $U_{F_2} = U_{L_2}$ of the upper discrete faithful polygonal representation of this concave curve will be not detected

5.2 Upper Representations of Convex Curves

Assume now that the curve $\mathcal{K} = (\kappa_1, \cdots, \kappa_n)$ is convex. In difference to concave curves, $\mathcal{K}$ may have no discrete faithful representation, or may have more than one. Curves which do not have discrete and faithful polygonal representations are not exceptional. The sequences of points of the upper frontier between κ_1, U_{F_1} and U_{L_m}, κ_n, if not empty, belong to each upper discrete polygonal representation and they are concave vertices. Here, there exists no discrete faithful representation. Another difference to concave case is the fact that leaning points of fundamental segments of a non concave curve can be successive elements or not as demonstrates Figure 4.

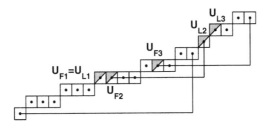

Fig. 4. For upper leaning points of fundamental segments of this convex curve holds $x_{U_{L_1}} < x_{U_{F_2}}$, however, $x_{U_{L_2}} > x_{U_{F_3}}$

Studying of these problems led to the results which are represented in the following two propositions and the lemma. Their proofs can be found in [3].

Proposition 2. *Given a convex $(0,1)$-curve $\mathcal{K}$ with 2 fundamental segments. If $x_{U_{L_1}} < x_{U_{F_2}}$, then there exists no upper faithful polygonal representation of $\mathcal{K}$ which is discrete. Moreover, there exists at least one point between U_{L_1} and U_{F_2} which is a concave vertex of each discrete polygonal representation.*

Examples for Proposition 2 are given in Figure 5.

Proposition 3. *Given a convex $(0,1)$-curve $\mathcal{K}$ with 2 fundamental segments. If one of the following conditions is true:*

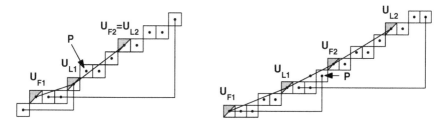

Fig. 5. Examples for convex curves having no upper discrete faithful representations. On the left curve, Algorithm UpPolRep computes the vertex P which is concave. On the right curve, it computes the convex vertex P such that U_{L_1} and U_{F_2} are concave.

1. $x_{U_{F_2}} = x_{U_{L_1}}$,
2. $x_{U_{F_2}} < x_{U_{L_1}}$ and there is no element $\kappa \in \mathcal{K}$ such that $x_{U_{L_1}} < x_\kappa < x_{U_{L_2}}$ and κ is lying above the real line through U_{L_1} and U_{L_2},
3. $x_{U_{F_2}} < x_{U_{L_1}}$ and there is no element $\kappa \in \mathcal{K}$ such that $x_{U_{F_1}} < x_\kappa < x_{U_{F_2}}$ and κ is lying above the real line through U_{F_1} and U_{F_2},

then there exists an upper discrete and faithful polygonal representation between U_{F_1} and U_{L_2}.

An example for Proposition 3 is demonstrated in Figure 6.

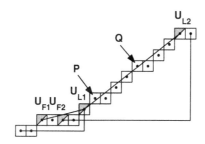

Fig. 6. Points $P, Q \in \mathcal{K}$ of this convex curve are lying above real line segment between U_{L_1} and U_{L_2}. Condition 1 is violated. Algorithm UpPolRep computes the concave vertex P. Condition 2 is true, i.e. U_{F_1}, U_{F_2} and U_{L_2} are vertices of a discrete faithful representation (between U_{F_1} and U_{L_2}). If the orientation is reversed then Algorithm UpPolRep computes this representation.

Lemma 1. *Let $\mathcal{K}$ be a convex $(0,1)$-curve with $m \geq 2$ fundamental segments. For each fundamental segment $\mathcal{F}_j(a_j, b_j, \mu_j)$, $j = 1, \cdots, m-1$ the sequence of first leaning points $\{U_{F_{j+i}}\}$, $i \geq 1$, $j+i \leq m$ such that $x_{U_{F_{j+i}}} \leq x_{U_{L_j}}$ is not empty. Assume for the sequence $\{U_{F_{j+i}}\}$ the index i is maximal and the following condition is true:*

there is no element $\kappa \in \mathcal{K}$ such that $x_{U_{L_j}} < x_\kappa < x_{U_{L_{j+i}}}$ and κ is lying above the real line through U_{L_j} and $U_{L_{j+i}}$,

then there exists an upper discrete and faithful polygonal representation of the segment of $\mathcal{K}$ between U_{F_1} and U_{L_m}.

Figure 7 represents an example which satisfies this lemma.

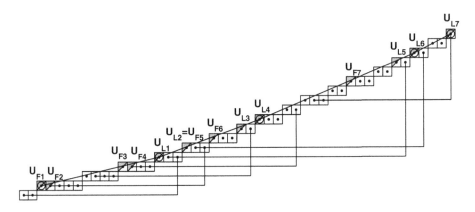

Fig. 7. Convex curve with fundamental segments $\mathcal{F}_i(a_i, b_i, \mu_i)$, $i = 1, \cdots, 7$ satisfying Lemma 1 is demonstrated. Vertices $\{U_{F_1}, U_{L_1}, U_{L_4}, U_{L_6}, U_{L_7}\}$ of an upper discrete and faithful polygonal representation between U_{F_1} and U_{L_7} will be calculated by Algorithm UpPolRep. The slopes of the representation are 0.2308, 0.3636, 0.4118, 0.5000.

6   Extention of Algorithm UpPolRep

In this section we will study situations where uncorrespondences between digital sets and their polygonal representations appear and we will extend Algorithm UpPolRep.

Assume Algorithm UpPolRep is applied to an arbitrary curve. Then it calculates a discrete representation. Examples in Figure 5 and Figure 6 present the only critical situations where the calculated representations may be not faithful. We will study these situations deeply.

Assume Algorithm UpPolRep is applied to the curve from Figure 5, right. Here, the situation is characterised by the fact that Algorithm UpPolRep stops twice at the same point. We define it as a double-step. In this case we immediately know that we are on a convex part. On a concave part, Algorithm UpPolRep cannot have double-steps.

Now the curve is extended on the left and right by one fundamental segment at each side. We have four fundamental segments. The elements of the original curve are in the second and third fundamental segments of the extended curve, this part is convex. If the part consisting of first and second fundamental segments is concave and the part consisting of third and fourth ones is concave as well, then Algorithm UpPolRep calculates a discrete faithful representation. Otherwise, there exists no discrete and faithful representation. For numerical implementations we have to check if successive calculated slopes increase. We must also take into account that Algorithm UpPolRep can have double-steps (at most in each loop). The slope of a double-step will not be considered.

The next situation: Algorithm UpPolRep is applied to curve from Figure 5, left. This curve does not possess a discrete faithful representation. Algorithm UpPolRep needs two steps. The computed leaning lines have increasing slopes. It indicates that we are on a convex curve. Further, the property, that in the corresponding successive steps U_L (previous) and U_F (next) are not identical, characterises this situation.

Now the curve is extended on the right side and has three fundamental segments. If successive calculated slopes increase, then this curve does not have a discrete and faithful representation. Also here slopes of double-steps will not be considered.

Assume Algorithm UpPolRep is applied to the curve from Figure 6. The curve possesses a discrete faithful representation, however, Algorithm UpPolRep will not compute it. This situation is characterised by increasing slopes and by the property that U_L (previous) and U_F (next) are not identical. A faithful representation will be found if we apply Algorithm UpPolRep to the curve with reversed orientation.

Considering the described critical situations, we construct the extention for Algorithm UpPolRep. Here we decide about the existence of a faithful representation comparing successive slopes. We replace vertex if a similar situation to the one from Figure 6 appears. The slope of a double-step is not considered and at the beginning it will be assumed that the representation is discrete and faithful.

In Figure 8 extended Algorithm UpPolRep is applied to the curve from Figure 1.

Extention of Algorithm UpPolRep

/* representation is discrete and faithful */;
if *double-step* **then**
  | **if** *previous slopes increase* **or** *next slopes increase* **then**
  | | /* there exists no discrete and faithful representation */;
  | **end**
end
if *slopes increase* **and** *corresponding U_L (previous) and U_F (next) are not identical* **then**
  | **if** *next slopes increase* **then**
  | | /* there exists no discrete and faithful representation */;
  | **end**
end
if *slopes increase* **and** *corresponding U_L (previous) and U_F (next) are not identical* **and** *for reversed curve U_L (previous) and U_F (next) are identical* **then**
  | /* replace vertex */;
end

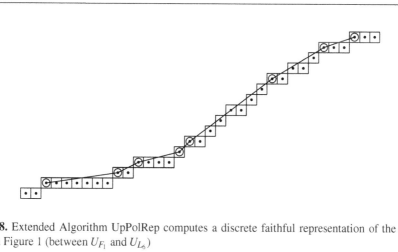

Fig. 8. Extended Algorithm UpPolRep computes a discrete faithful representation of the curve from Figure 1 (between U_{F_1} and U_{L_6})

There is another technique for computing polygonal representations suggested in [5]. It uses the concept of so-called *convex* and *concave exposed points*. Exposed points define vertices of a polygonal representation. There is a difficulty by calculating exposed points since it relies on the *Hübler-transform* which is initially not known. Moreover, the algorithm based on this technique is unable to decide if calculated representation is faithful and if such a representation actually exists.

7 Conclusions

The precise examination fundamental segments of a discrete curve makes us enable to settle if a discrete faithful polygonal representation exists or not. Leaning points and leaning lines gives the most improtant characteristics.

We developed an extention for linear Algorithm UpPolRep [3]. In its original version Algorithm UpPolRep computes polygonal representations with some uncorresponding parts. Unfortunately, Algorithm UpPolRep is unable to decide if a discrete faithful representation still exists and in the case of existence it is unable to compute this representation. Now if a digital set possesses a discrete faithful representation then the extention detects it. Digital sets for which discrete faithful polygonal representations do not exist will be recognized, too. Furthermore, extended Algorithm UpPolRep has linear time complexity.

References

1. Debled-Rennesson, I. and Reveillès, J.-P.: A linear algorithm for segmentation of digital curves. *International Journal of Pattern Recognition and Artificial Intelligence* **9**:635–662, 1995.
2. Debled-Rennesson, I., Rémy, J.-L. and Rouyer-Degli, J.: Detection of discrete convexity of polyominoes. *Discrete Applied Mathematics* **125**:115–133, 2003.
3. Dörksen, H.: *Shape Representations of Digital Sets based on Convexity Properties*. Dissertation. Fachbereich Mathematik. Universität Hamburg, 2004.
   http://www.sub.uni-hamburg.de/opus/volltexte/2005/2332/
4. Dörksen-Reiter, H. and Debled-Rennesson, I.: Convex and Concave Parts of Digital Curves. In Klette, R., et al, editors: *Geometric Properties from Incomplete Data*, volume 31 of Computational Imaging and Vision. Springer Verlag, 2005.
5. Eckhardt, U. and Reiter, H.: Polygonal Representations of Digital Sets. *Algorithmica* **38**(1):5–23, 2004.
6. Feschet, F. and Tougne, L.: Optimal time computation of the tangent of a discrete curve: application to the curvature. *Lecture Notes in Computer Science* **1568**:31–40, 1999.
7. Freeman, H.: On the encoding of arbitrary geometry configurations. IRE Trans. EC **10**:260–268, 1961.
8. Klette, R. and Rosenfeld, R.: *Digital Geometry*. Morgan Kaufmann publishers, Elsevier, 2004.
9. Hübler, A.: *Diskrete Geometrie für die Digitale Bildverarbeitung*. Dissertation B. Friedrich-Schiller-Universität Jena, 1989.
10. Latecki, L. and Lakämper, R.: Convexity rule for shape decomposition based on discrete contour evolution. *Computer Vision and Image Understanding* **73**(3):441–454, 1999.

11. Reveillès, J.-P.: *Géométrie discrète, calcul en nombres entiers et algorithmique*. Thése d'État, Strasbourg, 1991.
12. Scherl, W.: *Bildanalyse allgemeiner Dokumente*. (Informatik-Fachberichte, Band 131), Springer-Verlag, Berlin, Heidelberg, New York, London, Paris, Tokyo, 1987.
13. Valentine, F. A.: *Convex Sets*. McGraw-Hill Series in Higher Mathematics. McGraw-Hill Book Company, New York, San Francisco, Toronto, London, 1964.
14. Voss, K.: *Discrete Images, Objects and Functions in $\mathbb{Z}^n$*. Algorithm and Combinatorics, vol. 11, Springer-Verlag, Berlin, Heidelberg, New-York, 1993.

The Monogenic Curvature Scale-Space[*]

Di Zang and Gerald Sommer

Cognitive Systems Group
Institute of Computer Science and Applied Mathematics
Christian Albrechts University of Kiel, 24118 Kiel, Germany
{zd, gs}@ks.informatik.uni-kiel.de

Abstract. In this paper, we address the topic of monogenic curvature scale-space. Combining methods of tensor algebra, monogenic signal and quadrature filter, the monogenic curvature signal, as a novel model for intrinsically two-dimensional (i2D) structures, is derived in an algebraically extended framework. It is unified with a scale concept by employing damped spherical harmonics as basis functions. This results in a monogenic curvature scale-space. Local amplitude, phase and orientation, as independent local features, are extracted. In contrast to the Gaussian curvature scale-space, our approach has the advantage of simultaneous estimation of local phase and orientation. The main contribution is the rotationally invariant phase estimation in the scale-space, which delivers access to various phase-based applications in computer vision.

1 Introduction

It is well know that corners and junctions play an important role in many computer vision tasks such as object recognition, motion estimation, image retrieval, see [1, 2, 3, 4]. Consequently, signal modeling for such structures is of high significance. There are bulk of researches for intensity-based modeling, e.g. [5, 6, 7]. However, those approaches are not stable when the illumination varies. Phase information carries most essential structure information of the original signal [8]. It has the advantage of being invariant with respect to the illumination change. Hence, we intend to design a model for local structures with phase information contained. For 2D images, there are three types of structures, which can be associated with the term intrinsic dimension [7]. As a local property of multi-dimensional signals, it expresses the number of degrees of freedom necessary to describe local structures. The intrinsically zero dimensional (i0D) signals are constant signals. Intrinsically one dimensional (i1D) signals represent lines and edges. Corners, junctions, line ends, etc. are all intrinsically two dimensional (i2D) structures which have certain degrees of curvatures. There are lots of related work for 2D structures modeling. The structure tensor [5] estimates the main orientation and the energy of 2D structures. However, phase information is neglected. The Gaussian curvature scale-space [9, 10] enables the extraction

---

[*] This work was supported by German Research Association (DFG) Graduiertenkolleg No. 357 (DZ) and Grant So-320/2-3 (GS).

U. Eckardt et al. (Eds.): IWCIA 2006, LNCS 4040, pp. 320–332, 2006.
© Springer-Verlag Berlin Heidelberg 2006

of signal curvature in a multi-scale way with no phase contained. A nonlinear curvature scale-space was proposed in [11] for shape representation and recognition, but it is impossible to extract phase information in this framework. Bülow and Sommer [12] proposed the quaternionic analytic signal, which enables the evaluation of the i2D signal phase. But it has the drawback of being not rotationally invariant. The monogenic signal [13] is a novel model for i1D signals. It is a generalization of the analytic signal in 2D and higher dimensions. However, the monogenic signal captures no information of the i2D part. A phase model is proposed in [14], where the i2D signal is split into two i1D signals and the corresponding two phases are evaluated. Unfortunately, steering is needed and only i2D patterns superimposed by two perpendicular i1D signals can be correctly handled.

In this paper, we present a novel approach to model i2D structures in a multi-scale way. Combining methods of tensor algebra, monogenic signal and quadrature filter, the monogenic curvature signal, as a novel model for i2D structures, is derived in an algebraically extended framework. It is unified with a scale concept by employing damped spherical harmonics as basis functions, which results in a monogenic curvature scale-space. Local amplitude, phase and orientation, as independent local features, are extracted. In contrast to the Gaussian curvature scale-space, our approach has the advantage of simultaneous estimation of local phase and orientation, which enables many phase-based applications in computer vision tasks.

2   Geometric Algebra Fundamentals

Geometric algebras [15, 16, 17] constitute a rich family of algebras as generalization of vector algebra. Compared with the classical framework of vector algebra, the geometric algebra enables a tremendous extension of modeling capabilities. By embedding our problem into a certain geometric algebra, more degrees of freedom can be obtained, which makes it possible to extract multiple features of i2D structures. For the problem we concentrate on, 2D image data is embedded into the Euclidean 3D space. Therefore, an overview of geometric algebra over Euclidean 3D space is given. The Euclidean space $\mathbb{R}^3$ is spanned by the orthonormal basis vectors $\{\mathbf{e}_1, \mathbf{e}_2, \mathbf{e}_3\}$. The geometric algebra $\mathbb{R}_3$ of the 3D Euclidean space consists of $2^3 = 8$ elements,

$$\mathbb{R}_3 = span\{1, \mathbf{e}_1, \mathbf{e}_2, \mathbf{e}_3, \mathbf{e}_{23}, \mathbf{e}_{31}, \mathbf{e}_{12}, \mathbf{e}_{123} = I_3\} \tag{1}$$

Here $\mathbf{e}_{23}$, $\mathbf{e}_{31}$ and $\mathbf{e}_{12}$ are the unit bivectors and the element $\mathbf{e}_{123}$ is a trivector or unit pseudoscalar. In this geometric algebra, vectors square to one, bivectors and trivector all square to minus one. A general combination of these elements is called a multivector

$$M = a + b\mathbf{e}_1 + c\mathbf{e}_2 + d\mathbf{e}_3 + e\mathbf{e}_{23} + f\mathbf{e}_{31} + g\mathbf{e}_{12} + hI_3 \tag{2}$$

The geometric product of two multivectors M_1 and M_2 is indicated by juxtaposition of M_1 and M_2, i.e. $M_1 M_2$. The multiplication results of the basis elements are shown in table 1. The geometric product of two vectors $\mathbf{x} = x_1\mathbf{e}_1 + x_2\mathbf{e}_2$ and $\mathbf{y} = y_1\mathbf{e}_1 + y_2\mathbf{e}_2$ can be decomposed into their inner product ($\cdot$) and outer product ($\wedge$)

$$\mathbf{xy} = \mathbf{x} \cdot \mathbf{y} + \mathbf{x} \wedge \mathbf{y} \qquad (3)$$

where the inner product of $\mathbf{x}$ and $\mathbf{y}$ is $\mathbf{x} \cdot \mathbf{y} = x_1 y_1 + x_2 y_2$ and the outer product is $\mathbf{x} \wedge \mathbf{y} = (x_1 y_2 - x_2 y_1)\mathbf{e}_{12}$.

Due to the orthogonality of basis vectors, their outer products are equivalent to their geometric products.

$$\mathbf{e}_1 \wedge \mathbf{e}_2 = \mathbf{e}_1\mathbf{e}_2 = \mathbf{e}_{12} \qquad (4)$$

$$\mathbf{e}_2 \wedge \mathbf{e}_3 = \mathbf{e}_2\mathbf{e}_3 = \mathbf{e}_{23} \qquad (5)$$

$$\mathbf{e}_3 \wedge \mathbf{e}_1 = \mathbf{e}_3\mathbf{e}_1 = \mathbf{e}_{31} \qquad (6)$$

The k-grade part of a multivector is obtained from the grade operator $\langle M \rangle_k$. A blade of grade k, i.e. a k-blade B_k, is the outer product ($\wedge$) of k independent vectors $\mathbf{x}_1, ..., \mathbf{x}_k \in \mathbb{R}^3$

$$B_k = \mathbf{x}_1 \wedge ... \wedge \mathbf{x}_k = \langle \mathbf{x}_1...\mathbf{x}_k \rangle_k \qquad (7)$$

Hence, $\langle M \rangle_0$ is the scalar part of M, $\langle M \rangle_1$ represents the vector part, $\langle M \rangle_2$ indicates the bivector part and $\langle M \rangle_3$ is the trivector part, which commutes with every element of $\mathbb{R}_3$.

Table 1. The geometric product of basis elements

	1	$\mathbf{e}_1$	$\mathbf{e}_2$	$\mathbf{e}_3$	$\mathbf{e}_{23}$	$\mathbf{e}_{31}$	$\mathbf{e}_{12}$	I_3
1	1	$\mathbf{e}_1$	$\mathbf{e}_2$	$\mathbf{e}_3$	$\mathbf{e}_{23}$	$\mathbf{e}_{31}$	$\mathbf{e}_{12}$	I_3
$\mathbf{e}_1$	$\mathbf{e}_1$	1	$\mathbf{e}_{12}$	$-\mathbf{e}_{31}$	I_3	$-\mathbf{e}_3$	$\mathbf{e}_2$	$\mathbf{e}_{23}$
$\mathbf{e}_2$	$\mathbf{e}_2$	$-\mathbf{e}_{12}$	1	$\mathbf{e}_{23}$	$\mathbf{e}_3$	I_3	$-\mathbf{e}_1$	$\mathbf{e}_{31}$
$\mathbf{e}_3$	$\mathbf{e}_3$	$\mathbf{e}_{31}$	$-\mathbf{e}_{23}$	1	$-\mathbf{e}_2$	$\mathbf{e}_1$	I_3	$\mathbf{e}_{12}$
$\mathbf{e}_{23}$	$\mathbf{e}_{23}$	I_3	$-\mathbf{e}_3$	$\mathbf{e}_2$	-1	$-\mathbf{e}_{12}$	$\mathbf{e}_{31}$	$-\mathbf{e}_1$
$\mathbf{e}_{31}$	$\mathbf{e}_{31}$	$\mathbf{e}_3$	I_3	$-\mathbf{e}_1$	$\mathbf{e}_{12}$	-1	$-\mathbf{e}_{23}$	$-\mathbf{e}_2$
$\mathbf{e}_{12}$	$\mathbf{e}_{12}$	$-\mathbf{e}_2$	$\mathbf{e}_1$	I_3	$-\mathbf{e}_{31}$	$\mathbf{e}_{23}$	-1	$-\mathbf{e}_3$
I_3	I_3	$\mathbf{e}_{23}$	$\mathbf{e}_{31}$	$\mathbf{e}_{12}$	$-\mathbf{e}_1$	$-\mathbf{e}_2$	$-\mathbf{e}_3$	-1

The dual of a multivector M is defined to be the product of M with the inverse of the unit pseudoscalar I_3

$$M^* = MI_3^{-1} = -MI_3 \qquad (8)$$

The modulus of a multivector is obtained by $|M| = \sqrt{\langle M\widetilde{M} \rangle_0}$, where $\widetilde{M}$ is the reverse of a multivector defined as $\widetilde{M} = \langle M \rangle_0 + \langle M \rangle_1 - \langle M \rangle_2 - \langle M \rangle_3$.

If only the scalar and the bivectors are involved, the combined result is called a spinor

$$S = a + e\mathbf{e}_{23} + f\mathbf{e}_{31} + g\mathbf{e}_{12} \qquad (9)$$

All spinors form a proper subalgebra of $\mathbb{R}_3$, that is the even subalgebra $\mathbb{R}_3^+$. A spinor represents a scaling-rotation, i.e. $S = r\exp(\theta B)$, where B is a bivector indicating the rotation plane, θ is the rotation angle within that plane and r refers to the scaling factor. It is shown in table 1 that the square of the bivector or trivector equals -1. Therefore, the imaginary unit i of the complex numbers can be substituted by a bivector or a trivector, yielding an algebra isomorphism. A vector-valued signal $\mathbf{f}$ in $\mathbb{R}_3$ can be considered as the result of a spinor acting on the $\mathbf{e}_3$ basis vector, i.e. $\mathbf{f} = b\mathbf{e}_1 + c\mathbf{e}_2 + d\mathbf{e}_3 = \mathbf{e}_3 S$. The transformation performed under the action of the spinor delivers access to both the amplitude and phase information of the vector-valued signal $\mathbf{f}$ [18]. From the logarithm of the spinor representation, two parts can be obtained. They are the scaling which corresponds to the local amplitude and the rotation which corresponds to the local phase representation. The $\mathbb{R}_3$-logarithm of a spinor $S \in \mathbb{R}_3^+$ takes the following form

$$\log(S) = \langle\log(S)\rangle_0 + \langle\log(S)\rangle_2 = \log(|S|) + \frac{\langle S\rangle_2}{|\langle S\rangle_2|}\mathrm{atan}\left(\frac{|\langle S\rangle_2|}{\langle S\rangle_0}\right) \tag{10}$$

where atan is the arc tangent mapping for the interval $[0, \pi)$. The scalar part $\langle\log(S)\rangle_0 = \log(|S|)$ illustrates the logarithm of the local amplitude, hence, local amplitude is obtained as the exponential of it

$$|S| = \exp(\log|S|) = \exp(\langle\log(S)\rangle_0) \tag{11}$$

The bivector part of $\log(S)$ indicates the local phase representation

$$\langle\log(S)\rangle_2 = \frac{\langle S\rangle_2}{|\langle S\rangle_2|}\mathrm{atan}\left(\frac{|\langle S\rangle_2|}{\langle S\rangle_0}\right) \tag{12}$$

3 Damped Spherical Harmonics

In the light of the proposal in [14], 2D damped spherical harmonics are employed as basis functions. Since we are more interested in the angular portions, the polar representation of damped spherical harmonics is used instead of the Cartesian form. Assume the angular behavior of a signal is band limited, therefore, only damped spherical harmonics from order zero to three are applied, otherwise, aliasing would occur. Damped spherical harmonics in the spectral domain have much simpler forms than that in the spatial domain. An nth order damped spherical harmonic in the Fourier domain reads

$$H_n = \exp(n\alpha\mathbf{e}_{12})\exp(-2\pi\rho s) = [\cos(n\alpha) + \sin(n\alpha)\mathbf{e}_{12}]\exp(-2\pi\rho s) \tag{13}$$

where n indicates the order of the damped spherical harmonic, ρ and α represent the polar coordinates, s is the scale parameter. The 2D damped spherical harmonics can be alternatively regarded as 2D spherical harmonics $\exp(n\alpha\mathbf{e}_{12})$ combined with a Poisson kernel $\exp(-2\pi\rho s)$ [19]. The Poisson kernel is a low-pass filter which, like the Gaussian kernel, will result in a linear scale-space, called Poisson scale-space. As a result, local signal analysis can be realized in a multi-scale approach. Except for the zero order, every damped spherical harmonic consists of two orthogonal components. In the spatial domain, damped spherical harmonics from order 0 to 3 are illustrated in Figure 1. The first or-

Fig. 1. From left to right are damped spherical harmonics from order 0 to 3 in the spatial domain (white:+1, black:-1). Except for zero order, every damped spherical harmonic consists of two orthogonal components.

der damped spherical harmonic H_1 is basically identical to the conjugate Poisson kernel [14]. When the scale parameter is set to zero, the conjugate Poisson kernel equals the Riesz transform [13].

4   The Monogenic Curvature Scale-Space

It is a well-known fact that 1D analytic functions correspond directly to 2D harmonic fields. In mathematics, these functions are also called holomorphic. Such functions are characterized by having a local power series expansion about each point [20]. This generalizes to 2D such that monogenic functions correspond to 3D harmonic fields. In Clifford analysis, the term monogenic is used to express the multidimensional character of the functions. Since in this paper, we present a novel approach which is to some degree a generalization of the analytic signal to the i2D case, it thus is called the monogenic extension of a curvature tensor. The monogenic curvature signal, as a novel model for i2D structures, can be derived from it. The monogenic scale-space, shown in Figure2, is formed by the monogenic curvature signal at all scales.

4.1   Monogenic Extension of the Curvature Tensor

Motivated from the differential geometry, the curvature tensor can be constructed. Two dimensional intensity data can be represented as surfaces in 3D Euclidean space. Such surfaces in geometrical terms are Monge patches of the form

$$\mathbf{f} = \{x\mathbf{e}_1, y\mathbf{e}_2, f(x,y)\mathbf{e}_3\} \tag{14}$$

This representation makes it easy to use differential geometry to study the properties of the surface. The primary first-order differential quantity for an image is the gradient, which is defined as

$$\nabla \mathbf{f} = \sum_{i=1}^{2} g_i \mathbf{e}_i \frac{\partial \mathbf{f}}{\partial x_i} \tag{15}$$

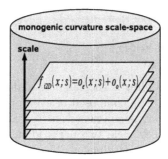

Fig. 2. The monogenic curvature scale-space

where g_i indicates the following basis

$$g_1 = \begin{bmatrix} 1 \\ 0 \end{bmatrix} \quad g_2 = \begin{bmatrix} 0 \\ 1 \end{bmatrix} \tag{16}$$

Thereby, the gradient is reformulated as

$$\nabla \mathbf{f} = \begin{bmatrix} \mathbf{e}_1 \frac{\partial}{\partial x} f(x,y) \mathbf{e}_3 \\ \mathbf{e}_2 \frac{\partial}{\partial y} f(x,y) \mathbf{e}_3 \end{bmatrix} = \begin{bmatrix} f_x \mathbf{e}_{13} \\ f_y \mathbf{e}_{23} \end{bmatrix} \tag{17}$$

Analogously, as second-order differential quantity, the Hessian matrix H is given by

$$H = \begin{bmatrix} \mathbf{e}_1 \frac{\partial}{\partial x} f_x \mathbf{e}_{13} & \mathbf{e}_2 \frac{\partial}{\partial y} f_x \mathbf{e}_{13} \\ \mathbf{e}_1 \frac{\partial}{\partial x} f_y \mathbf{e}_{23} & \mathbf{e}_2 \frac{\partial}{\partial y} f_y \mathbf{e}_{23} \end{bmatrix} = \begin{bmatrix} f_{xx} \mathbf{e}_3 & -f_{xy} \mathbf{e}_{123} \\ f_{xy} \mathbf{e}_{123} & f_{yy} \mathbf{e}_3 \end{bmatrix} \tag{18}$$

This representation belongs to a hybrid matrix geometric algebra $M(2, \mathbb{R}_3)$, which is the geometric algebra of a 2×2 matrix with elements in $\mathbb{R}_3$, see [21].

According to the derivative theorem of Fourier theory [22], the Hessian matrix in the spectral domain reads

$$\mathcal{F}\{H\} = \begin{bmatrix} (-4\pi^2 \rho^2 \frac{1+\cos(2\alpha)}{2} \mathbf{F}) & (4\pi^2 \rho^2 \frac{\sin(2\alpha)}{2} \mathbf{F}) \mathbf{e}_{12} \\ (-4\pi^2 \rho^2 \frac{\sin(2\alpha)}{2} \mathbf{F}) \mathbf{e}_{12} & (-4\pi^2 \rho^2 \frac{1-\cos(2\alpha)}{2} \mathbf{F}) \end{bmatrix} \tag{19}$$

where $\mathcal{F}$ indicates the Fourier transform and $\mathbf{F}$ is the Fourier transform of the original signal $\mathbf{f}$. The angular parts of the derivatives are related to spherical harmonics of even order 0 and 2. It is well known that the Hessian matrix contains curvature information. Based on it, i0D, i1D and i2D signals can be separated by computing the trace and determinant. Therefore, we are motivated to construct a curvature tensor T_e, which is related to the Hessian matrix. The curvature tensor can be obtained from a tensor-valued filter, i.e. $T_e = \mathcal{F}^{-1}\{\mathbf{F}H_e\}$, where $\mathcal{F}^{-1}$ means the inverse Fourier transform and H_e indicates a tensor-valued filter in the frequency domain with the following form

$$H_e = \frac{1}{2}\begin{bmatrix} H_0 + \langle H_2 \rangle_0 & -\langle H_2 \rangle_2 \\ \langle H_2 \rangle_2 & H_0 - \langle H_2 \rangle_0 \end{bmatrix} = \frac{1}{2}\begin{bmatrix} 1 + \cos(2\alpha) & -\sin(2\alpha)\mathbf{e}_{12} \\ \sin(2\alpha)\mathbf{e}_{12} & 1 - \cos(2\alpha) \end{bmatrix} \exp(-2\pi\rho s)$$

$$= \begin{bmatrix} \cos^2(\alpha) & -\frac{1}{2}\sin(2\alpha)\mathbf{e}_{12} \\ \frac{1}{2}\sin(2\alpha)\mathbf{e}_{12} & \sin^2(\alpha) \end{bmatrix} \exp(-2\pi\rho s) \tag{20}$$

The angular portion of this filter is the same as that of the Hessian, according to equations (13) and (19), it is composed of even order 2D spherical harmonics H_0 and H_2. In this filter, components $\cos^2(\alpha)$ and $\sin^2(\alpha)$ are two angular windowing functions. They yield two perpendicular i1D components of the 2D image along the $\mathbf{e}_1$ and $\mathbf{e}_2$ coordinates. The other components of the filter are also the combination of two angular windowing functions, i.e. $\frac{1}{2}\sin(2\alpha) = \frac{1}{2}(\cos^2(\alpha - \frac{\pi}{4}) - \sin^2(\alpha - \frac{\pi}{4}))$. These two angular windowing functions result in two i1D components of the 2D image, which are oriented along the diagonals of the plane spanned by $\mathbf{e}_1$ and $\mathbf{e}_2$. All of the angular windowing functions are shown in Figure 3. They make sure that i1D components along different orientations are extracted. Consequently, this even filter enables the extraction of differently oriented even i1D components of the 2D image. Since the conjugate

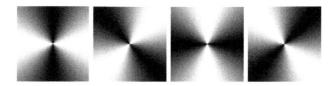

Fig. 3. From left to right are the angular windowing functions of $\cos^2(\alpha), \sin^2(\alpha - \frac{\pi}{4}), \sin^2(\alpha)$ and $\cos^2(\alpha - \frac{\pi}{4})$ with white:+1 and black:0

Poisson kernel H_1 [14] is able to evaluate the corresponding odd information of the i1D signal, the odd representation of the curvature tensor is obtained by employing H_1 to its elements. Besides, the odd representation of the curvature tensor, denoted as T_o, can also result from a tensor-valued odd filter H_o, i.e. $T_o = h_1 * T_e = \mathcal{F}^{-1}\{H_o\mathbf{F}\}$ with h_1 referring to the spatial representation of the conjugate Poisson kernel. Thereby, the odd filter H_o can be obtained from the even filter by employing the conjugate Poisson kernel, i.e. $H_o = H_1 H_e$. In the spectral domain, the odd filter thus takes the following form

$$H_o = \frac{1}{2}\begin{bmatrix} H_1(H_0 + \langle H_2 \rangle_0) & H_1(-\langle H_2 \rangle_2) \\ H_1(\langle H_2 \rangle_2) & H_1(H_0 - \langle H_2 \rangle_0) \end{bmatrix} \tag{21}$$

Combing the curvature tensor and its odd representation forms a general 2D image representation, i.e. $T = T_e + T_o$. This algebraically extended representation can also be regarded as the monogenic extension of the curvature tensor.

4.2 The Monogenic Curvature Signal

Analogous with the differential geometry approach, 2D structures can be classified by computing the determinants and traces of the tensor pair T_e and T_o.

Since the non-zero determinant indicates the existence of i2D structures, the even and odd parts of i2D structures are obtained from the determinants of the curvature tensor and its odd representation, respectively. The even part of i2D structures reads

$$o_e(\mathbf{x}; s) = \det(T_e)\mathbf{e}_3 = A\mathbf{e}_3 \tag{22}$$

The determinant of the curvature tensor is scalar valued. Therefore, same as the monogenic signal, the even part of i2D structures is embedded as the $\mathbf{e}_3$ component in the 3D Euclidean space. The odd part of i2D structures is

$$o_o(\mathbf{x}; s) = \det(T_o)\mathbf{e}_2 = B\mathbf{e}_1 + C\mathbf{e}_2 \tag{23}$$

Because $\det(T_o)$ is spinor valued, by multiplying the $\mathbf{e}_2$ basis from the right, $o_o(\mathbf{x}; s)$ takes a vector valued representation. Hence, a local representation for i2D structures is obtained by combining the even and odd parts of i2D structures. This local representation for i2D structures is called the monogenic curvature signal and it takes the following form

$$\mathbf{f}_{i2D}(\mathbf{x}; s) = o_e(\mathbf{x}; s) + o_o(\mathbf{x}; s) = A\mathbf{e}_3 + B\mathbf{e}_1 + C\mathbf{e}_2 \tag{24}$$

The original scalar signal $f(\mathbf{x}), \mathbf{x} \in \mathbb{R}^2$ is thus mapped to a vector-valued signal $\mathbf{f}_{i2D}(\mathbf{x}; s)$ in $\mathbb{R}_3$ as a local representation of i2D signals.

4.3   Local Features and Geometric Model

According to the introduction in Section 2, local features of the monogenic curvature signal can be defined using the logarithm of $\mathbb{R}_3^+$. The spinor field which maps the $\mathbf{e}_3$ basis vector to the monogenic curvature signal $\mathbf{f}_{i2D}(\mathbf{x}; s)$ is given by $\mathbf{f}_{i2D}(\mathbf{x}; s)\mathbf{e}_3$. The local amplitude and local phase representation are obtained as

$$|\mathbf{f}_{i2D}(\mathbf{x}; s)| = \exp(\langle \log(\mathbf{f}_{i2D}(\mathbf{x}; s)\mathbf{e}_3)\rangle_0) = \exp(\log(|\mathbf{f}_{i2D}(\mathbf{x}; s)\mathbf{e}_3|)) \tag{25}$$

$$\arg(\mathbf{f}_{i2D}(\mathbf{x}; s)) = \frac{\langle \mathbf{f}_{i2D}(\mathbf{x}; s)\mathbf{e}_3\rangle_2}{|\langle \mathbf{f}_{i2D}(\mathbf{x}; s)\mathbf{e}_3\rangle_2|} \mathrm{atan}\left(\frac{|\langle \mathbf{f}_{i2D}(\mathbf{x}; s)\mathbf{e}_3\rangle_2|}{\langle \mathbf{f}_{i2D}(\mathbf{x}; s)\mathbf{e}_3\rangle_0}\right) \tag{26}$$

where $\arctan(\cdot) \in [0, \pi)$ and $\arg(\cdot)$ denotes the argument of the expression. As the bivector part of the logarithm of the spinor field $\mathbf{f}_{i2D}(\mathbf{x}; s)\mathbf{e}_3$, this local phase representation describes a rotation from the $\mathbf{e}_3$ axis by a phase angle φ in the oriented complex plane spanned by $\mathbf{f}_{i2D}(\mathbf{x}; s)$ and $\mathbf{e}_3$, i.e. $\mathbf{f}_{i2D}(\mathbf{x}; s) \wedge \mathbf{e}_3$. The orientation of this complex plane indicates the local main orientation. Therefore, the local phase representation combines local phase and local orientation of i2D structures. The dual of the complex plane $\mathbf{f}_{i2D}(\mathbf{x}; s) \wedge \mathbf{e}_3$ is a rotation vector

$$\mathbf{r}(\mathbf{x}; s) = (\arg(\mathbf{f}_{i2D}(\mathbf{x}; s)))^* = \langle \log(\mathbf{f}_{i2D}(\mathbf{x}; s)\mathbf{e}_3)\rangle_2^* \tag{27}$$

The rotation vector $\mathbf{r}(\mathbf{x}; s)$ is orthogonal to the local orientation and its absolute value represents the phase angle of the i2D structure. With the algebraic embedding, a geometric model for the monogenic curvature signal can be visualized as

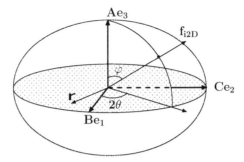

Fig. 4. The geometric model for the monogenic curvature signal. Here, φ is the phase, 2θ denotes the main orientation in terms of double angle representation, $\mathbf{r}$ indicates the rotation vector.

is shown in Figure 4. The geometric model is an ellipsoid, which looks very similar to that of the monogenic signal. However, each axis encodes totally different meaning. The even part of the i2D structure is encoded within the $\mathbf{e}_3$ axis, and the odd information is encoded within the plane spanned by $\mathbf{e}_1$ and $\mathbf{e}_2$ axes. The angle φ represents the phase and 2θ is the main orientation in a double angle representation form. The rotation vector $\mathbf{r}$ lies in the plane orthogonal to $\mathbf{e}_3$ since it is dual to the bivector $\mathbf{f}_{i2D} \wedge \mathbf{e}_3$. Combining the local amplitude and local phase representation, the monogenic curvature signal for i2D structures, can be reconstructed as

$$\mathbf{f}_{i2D} = |\mathbf{f}_{i2D}| \exp\left(\arg\left(\mathbf{f}_{i2D}\right)\right) \tag{28}$$

Having a definition for the i2D local features, we recognize that local amplitude, phase and orientation are scale dependent. However, they are independent of each other at each scale.

Gaussian curvature scale-space [9, 10] and the morphological curvature scale-space [11] are suitable for recovering invariant geometric features of a signal at multiple scales. However, the definition of curvatures and the scale generating operator are totally different from our approach. Besides, no phase information is contained in those frameworks. In contrast to these methods, our approach enables the simultaneous estimation of local amplitude, local phase and orientation information in a common scale-space concept. Consequently, the monogenic curvature scale-space has a unique advantage if a quadrature relationship concept is required.

5   Experimental Results

In this section, we show some experimental results in the framework of the monogenic curvature scale-space. A synthetic image superimposed by an angular and a radial modulation is adopted as the test image. The blobs in this image are regarded as i2D structures. The monogenic curvature signal at a certain scale

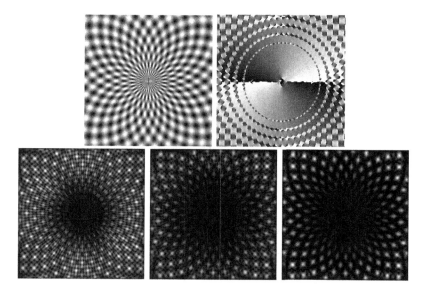

Fig. 5. Top row: from left to right are the test image and the orientation estimation at a certain scale of the monogenic curvature scale-space. Bottom row: local energy outputs at three different scales.

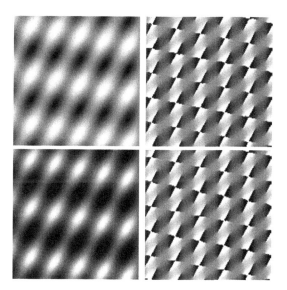

Fig. 6. Top row: from left to right are the test image and its phase estimation. Bottom row: the illumination changed test image and its phase evaluation.

can be obtained to characterize i2D structures. The test image and extracted local features are illustrated in Figure 5.

Fig. 7. Top row: test image, the energy of the monogenic signal and its phase. Bottom row: Energy and phase of the monogenic curvature signal.

The estimated orientation denotes the continuous main orientation at a certain scale. Because the evaluated orientation has a value between 0 and π, it is wrapped along the horizontal axis. The local energy output, i.e. square of the local amplitude, indicates the existence of i2D structures. Besides, it also demonstrates the rotation-invariant property of the monogenic curvature signal. Local energy outputs of the test image at three different scales are shown.

Another test image is composed of two cosine signals with different frequencies, amplitudes and orientations. The test image and the estimated phase information are shown in Figure 6. Because the local amplitude and local phase are independent of each other, when the illumination of the original image varies, the estimated local phase is still stable. This delivers access to many phase-based processing techniques in computer vision.

The monogenic curvature signal is a novel model for i2D structures, which handles the type of structure that the monogenic signal cannot correctly deal with. The third experiment aims to show the difference of these two models. Figure 7 demonstrates local energies and local phases extracted from the monogenic signal and the monogenic curvature signal. It is obvious that the monogenic signal responds to i1D structures and the monogenic curvature signal extracts features from i2D structures.

6   Conclusions

We present the monogenic curvature scale-space in this paper. Coupling methods of tensor algebra, monogenic signal and quadrature filter, the monogenic curvature signal, which characterizes i2D structures is obtained. Employing damped spherical harmonics as basis functions unifies a scale concept with the monogenic curvature signal. The monogenic curvatures scale-space is thus formed by the monogenic curvature signals at all scales. Local amplitude, local phase and local orientation of i2D structures, as independent features, can be extracted. Compared with the Gaussian curvature scale-space and the morphological curvature scale-space, our approach has remarkable advantage of simultaneous estimation of local phase and local orientation, which delivers access to various applications in the computer vision.

References

1. Costabile, M.F., Guerra, C., Pieroni, G.G.: Matching shapes: a case study in time varying images. Computer Vision, Graphics and Image Processing **29** (1985) 296–310
2. Han, M.H., Jang, D.: The use of maximum curvature points for the recognition of partially occluded objects. Pattern Recognition **23** (1990) 21–33
3. Liu, H.C., Srinath, M.D.: Partial classification using contour matching in distance transformation. IEEE Transactions on Pattern Analysis and Matchine Intelligence **12** (1990) 1072–1079
4. Wang, H., Brady, M.: Real-time corner detection algorithm for motion estimation. Image and Vision Computing **13** (1995) 695–703
5. Förstner, W., Gülch, E.: A fast operator for detection and precise location of distinct points, corners and centers of circular features. In: Proc. ISPRS Intercommission Conference on Fast Processing of Photogrammetric Data, Interlaken, Switzerland (1987) 281–305
6. Köthe, U.: Integrated edge and junction detection with the boundary tensor. In: Proceeding of 9th Intl. Conf. on Computer Vision. Volume 1. (2003) 424–431
7. Krieger, G., Zetzsche, C.: Nonlinear image operators for the evaluation of local intrinsic dimensionality. IEEE Transactions on Image Processing **5** (1996)
8. Oppenheim, A.V., Lim, J.S.: The importance of phase in signals. IEEE Proceedings **69** (1981) 529–541
9. Mokhtarian, F., Suomela, R.: Curvature scale space for robust image corner detection. In: Proc. 14th International Conference on Pattern Recognition (ICPR'98). Volume 2. (1998) 1819–1821
10. Mokhtarian, F., Bober, M.: Curvature scale space representation: theory, applications, and MPEG-7 standardization. Kluwer Academic Publishers (2003)
11. Jalba, A.C., Wilkinson, M.H.F., Roerdink, J.B.T.M.: Shape representation and recognition through morphological curvature scale spaces. IEEE Trans. Image Processing **15** (2006) 331–341
12. Bülow, T., Sommer, G.: Hypercomplex signals - a novel extension of the analytic signal to the multidimensional case. IEEE Transactions on Signal Processing **49** (2001) 2844–2852
13. Felsberg, M., Sommer, G.: The monogenic signal. IEEE Transactions on Signal Processing **49** (2001) 3136–3144
14. Felsberg, M.: Low-level image processing with the structure multivector. Technical Report 2016, Christian-Albrechts-Universität zu Kiel, Institut für Informatik und Praktische Mathematik (2002)

15. Lounesto, P.: Clifford Algebras and Spinors. Cambridge University Press (1997)
16. Ablamowicz, R.: Clifford Algebras with Numeric and Symbolic Computations. Birkhäuser, Boston (1996)
17. Hestenes, D., Li, H., Rockwood, A.: Geometric computing with clifford algebras. In Sommer, G., ed.: New Algebraic Tools for Classical Geometry. Springer-Verlag (2001) 3–23
18. Sommer, G., Zang, D.: Parity symmetry in multi-dimensional signals. In: Proc. of the 4th International Conference on Wavelet Analysis and its Applications, Macao (2005)
19. Felsberg, M., Sommer, G.: The monogenic scale-space: A unifying approach to phase-based image processing in scale-space. Journal of Mathematical Imaging and Vision **21** (2004) 5–26
20. Stein, E., Weiss, G.: Introduction to Fourier Analysis on Euclidean Spaces. Princeton University Press, New Jersey (1971)
21. Sobczyk, G., Erlebacher, G.: Hybrid matrix geometric algebra. In Li, H., Olver, P.J., Sommer, G., eds.: Computer Algebra and Geometric Algebra with Applications. Volume 3519 of LNCS., Springer-Verlag, Berlin Heidelberg (2005) 191–206
22. Bracewell, R.: Fourier Analysis and Imaging. Kluwer Academic / Plenum Publishers, New York (2003)

Combinatorial Properties of Scale Space Singular Points

Atsushi Imiya and Tomoya Sakai

Institute of Media and Information Technology, Chiba University
Yayoi-cho 1-33, Inage-ku, 263-8522, Chiba, Japan
{imiya, tsakai}@faculty.chiba-u.jp

Abstract. Singular points in the linear scale space provide fundamental features for the extraction of dominant parts of an image. Employing the geometrical configuration of singular points, it is possible to construct a tree in scale space. This tree expresses a hierarchical structure of dominant parts. In this paper, we clarify the graphical grammar for the construction of this tree in the linear scale space and morphological scale space. Furthermore, we show a combinatorial structure of singular points in the linear scale space and morphological scale space using conformal mapping from Euclidean space to the spherical surface.

1 Introduction

In this paper, we analyse deep structure properties of linear scale space and morphological scale space. In the linear scale space general functions are generated by the convolution of the input function and Gauss kernel. The morphological scale space analysis is the set theoretical version of scale space analysis.

The singular-point configuration in the linear scale space yielded by Gaussian blurring of function is called deep structure in the linear scale space. Morphological scale space is introduced from view point of set theoretical image analysis. Deep structure such as medial axis and skeleton first introduced in digital image processing [2, 1] and later described using morphological operations [3]. We extend the deep structure analysis to the binary morphological scale space of binary signals and images. We derive an algorithm for the construction of stationary tree for as a deep structure of binary signals and images.

Recently, Kuijper [17, 18] introduced a method for image hierarchical analysis based on the trajectory of singular point in the Gaussian scale space. The Gaussian scale-space analysis [4, 5, 7, 6, 8, 9, 10] is an established image analysis tool which provides multi-resolution analysis and expression of steel images and sequence of images[11, 19]. Hereafter, we use DSSS for the abbreviation of deep structure in the linear scale space. DSSS describes hidden topological nature of the original functions dealing with gray values of a n-variable function in the scale space as a $(n + 1)$-dimensional topographical maps [12, 13, 14, 15, 16, 17, 18, 19]. The configuration of singular points in the scale space depends on the scale. If the scale changes the configuration of singular points transformed. This transition of the singular-point configuration defines a tree for a function [12, 13, 14, 16].

U. Eckardt et al. (Eds.): IWCIA 2006, LNCS 4040, pp. 333–346, 2006.
© Springer-Verlag Berlin Heidelberg 2006

In this paper, we clarify the graphical grammar for the construction of tree from singular-point configurations in the linear scale space and in the morphological scale space. Furthermore, we show a relation between this grammar and the deformation of polytopes which are defined by the configuration of singular points in each scale.

Kuijper et al [17, 18] dealt with the singular points with zero second-derivatives as a DSSS feature. This class of singular point in the linear scale space is called the top points or the critical points. A top-point is a singular point in the scale space on which both first and second derivatives are zero. Since in the higher dimensional space, the second order local properties of surface are described by Hessian, the sign of second derivative is expressed by the signs of the eigenvalues of the Hessian matrix.

Iijima [4] defined the singular points with the first derivative is zero. Iijima called this class of singular points in the scale space the stationary points. Furthermore, Iijima showed that the stationary points define the centres of view fields which extract a dominate portions of an image for a fixed scale [4]. Zhao and Iijima proposed [12, 13] a tree construction strategy in the linear scale space using the configuration of their stationary points.

In sections 3, and 4, we show the symbolic structure of DSSS for 2D images in the linear scale space and morphological scale space , respectively. Furthermore, we introduce a grammatical structure which describes the transition of the singular-points configuration when the scale parameter increases.

2   Mathematical Preliminary

2.1   Linear Scale Space

In the n-dimensional Euclidean space $\mathbf{R}^2$, for an orthogonal coordinate system x-y defined in $\mathbf{R}^2$, a vector in $\mathbf{R}^2$ is expressed by $\boldsymbol{x} = (x, y)^\top$ where $\cdot^\top$ is the transpose of a vector. Setting $|\boldsymbol{x}|$ to be the length of $\boldsymbol{x}$, the linear scale-space transform for function $f(\boldsymbol{x})$, such that

$$f(\boldsymbol{x}, \tau) = \frac{1}{4\pi\tau} \int_{-\infty}^{\infty} \int_{-\infty}^{\infty} f(\boldsymbol{y}) \exp(-\frac{|\boldsymbol{x} - \boldsymbol{y}|^2}{4\tau}) d\boldsymbol{y}, \tag{1}$$

defines the general image of function $f(\boldsymbol{x})$. Therefore, function $f(\boldsymbol{x}, \tau)$ is defined in $\mathbf{R}^2 \times \mathbf{R}_+$ [4]. The function $f(\boldsymbol{x}, \tau)$ is the solution of the linear diffusion equation

$$\frac{\partial f(\boldsymbol{x}, \tau)}{\partial \tau} = \varDelta f(\boldsymbol{x}, \tau), \ \tau > 0, \ f(\boldsymbol{x}, 0) = f(\boldsymbol{x}). \tag{2}$$

The solution of eq. (2) is formally expressed

$$f(\boldsymbol{x}, \tau) = \exp(\varDelta\tau) f(\boldsymbol{x}) \tag{3}$$

using the theory of Lie group [9].

Stationary points for the topographical maps in the scale space [4, 12, 14] are defined as the solutions of the equation $\nabla f(\boldsymbol{x}, \tau) = 0$. The stationary-curves in the scale space are the collections of the stationary points. We denote the trajectories of the stationary points as $\boldsymbol{x}(\tau)$. Setting $\boldsymbol{H}$ to be the Hessian matrix of $f(\boldsymbol{x}, \tau)$, Zhao and Iijima [12] showed that the stationary-curves for a n-dimensional image are the solution of,

$$\boldsymbol{H}\frac{d\boldsymbol{x}(\tau)}{d\tau} = -\nabla\Delta f(\boldsymbol{x}(\tau), \tau) \tag{4}$$

and clarified topological properties of the stationary-curves for two-dimensional patterns Since the Hessian matrix is always singular for singular points, this equation is valid for nonsingular points. The definitions are formally valid to functions defined in $\mathbf{R}^n$ for $n \geq 3$. Using the second derivations of $f(\boldsymbol{x}, \tau)$, we classify the topological properties of the stationary points on the topographical maps. In the neighbourhood of the point $\boldsymbol{x}$ which satisfies the relation $\nabla f(\boldsymbol{x}, \tau) = 0$, we have the equation

$$\frac{d^2 f}{d\, \boldsymbol{n}^2} = \boldsymbol{n} \cdot \nabla(\boldsymbol{n} \cdot \nabla f) = \boldsymbol{n}^\top \boldsymbol{H} \boldsymbol{n} \tag{5}$$

Equation (5) means that the eigenvectors of Hessian matrix of $f(\boldsymbol{x}, \tau)$ gives the extremal of D^2 and that the extremal are achieved by the eigenvalues of the Hessian of $f(\boldsymbol{x}, \tau)$, since $\alpha_1 \geq \boldsymbol{n}^\top \boldsymbol{H} \boldsymbol{n} \geq \alpha_n$ for $|\boldsymbol{n}| = 1$. Furthermore, the rank of the Hessian matrix in the higher-dimensional space classifies the properties of the singular points.

Definition 1. *For 2D functions, a point is the singular point, if the rank of the Hessian matrix at the point is one.*

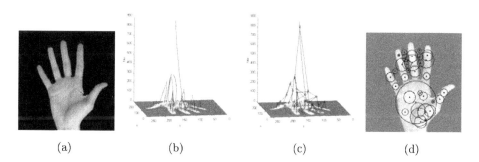

(a) (b) (c) (d)

Fig. 1. Scale space singularity of an image. (a) An image. (b) Stationary curve of the image (a). (c) Tree extracted from (c). (d) Field of view with respect to scales extracted by the tree (c).

Figure 1 shows scale space singularity of an image and extraction of hierachical structure of the image in linear scale space. (a) is an original image. (b) shows stationary curve of the image (a). (c) is the tree extracted by detecting stationary points on the curve. (d) shows fields of view with respect to scales extracted by the tree (c).

2.2 Morphological Scale Space

Mathematical Preliminary. Setting $\mathbf{A}$ to be a finite closed set in the n-dimensional Euclidean space $\mathbf{R}^2$, the Minkowski addition and subtraction of sets are defined as

$$\mathbf{A} \oplus \mathbf{B} = \bigcup_{x \in A, y \in B} (x + y), \quad \mathbf{A} \ominus \mathbf{B} = \overline{\overline{\mathbf{A}} \oplus \overline{\mathbf{B}}}. \tag{6}$$

The inner and outer boundary of point set $\mathbf{A}$ with respect to radius λ are defined as

$$\Delta_\delta^+ \mathbf{A} = (\mathbf{A} \oplus \delta \mathbf{D}) \setminus \mathbf{A}, \ \Delta_\delta^- \mathbf{A} = \mathbf{A} \setminus (\mathbf{A} \ominus \delta \mathbf{D}) \tag{7}$$

for the unit disc such that $\mathbf{D} = \{x \mid x \leq 1\}$, where $\lambda \mathbf{A} = \{\lambda x \mid x \in \mathbf{A}\}$ for $\lambda > 0$. We call

$$\mathbf{A}_\delta = \Delta_\delta^+ \mathbf{A} \bigcup \Delta_\delta^- \mathbf{A} \tag{8}$$

the boundary belt of $\mathbf{A}$ with respect to λ. Geometrically, we have the relation

$$\lim_{\delta \to +0} \mathbf{A}_\delta = \partial \mathbf{A}, \tag{9}$$

where $\partial \mathbf{A}$ is the boundary curve of set $\mathbf{A}$.

Set Theory Definition. Let τ_{ij} such that

$$r_i + r_j + \tau_{ij} = |c_i - c_j| \tag{10}$$

is the minimum of distances among $\mathbf{F}_i$
For a small positive constant ε, we define

$$\tau_+ = \tau_{ij} + \varepsilon, \quad \tau_- = \tau_{ij} - \varepsilon. \tag{11}$$

If $\tau = \tau_-$, $\tau = \tau_{ij}$, and $\tau = \tau_+$, $\mathbf{F}_i$ and $\mathbf{F}_j$ are disjunct, touching and overlapping, respectively. Therefore, we set

$$\mathbf{F}_{\tau_1} = \left(\bigcup_{k \neq i,j} \mathbf{F}_k \right) \bigcup \mathbf{F}_{ij} \tag{12}$$

where

$$\mathbf{F}_{ij} = r_{ij} \mathbf{D} \oplus \{c_{ij}\}, \tag{13}$$

for

$$c_{ij} = \frac{r_j + \tau_1}{r_i + r_j + 2\tau_1} c_i + \frac{r_i + \tau_1}{r_i + r_j + 2\tau_1} c_j, \tag{14}$$

and

$$r_{ij}^2 = (r_i + \tau_1)^2 + (r_j + \tau_1)^2. \tag{15}$$

We set

$$\mathbf{F}_{\tau_+} = \bigcup_{k=1}^{n(\tau_+)} \mathbf{F}_k \tag{16}$$

by reordering the suffices of elements for $n(\tau_+) = n(\tau_-) - k$ for $k \geq 1$, where k is the number of merged disks. These relations lead to the conclusion that for a collection of disjoint sets the minimum of distances among sets defines the life time of merging of elements in the collection.

Figure 2 shows evolution of discs. Two discs in (a) are merged to a disc in (c) through the configuration in (b).

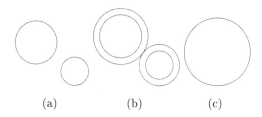

(a)   (b)   (c)

Fig. 2. Evolution of discs. Two discs in (a) are merged to a disc in (c) through the configuration in (b).

Functional Definition. Setting

$$f(\boldsymbol{x}) = \begin{cases} 1, \text{ if } \boldsymbol{x} \in \boldsymbol{F}, \\ 0, \text{ otherwise}, \end{cases} \tag{17}$$

we can have a binary function from set of points in a Euclidean space.

$$f(\boldsymbol{x}, \tau) = \begin{cases} 1, \text{ if } \boldsymbol{x} \in \boldsymbol{F}_\tau, \\ 0, \text{ otherwise}, \end{cases} \tag{18}$$

For binary functions defined on the real line, setting

$$h(\boldsymbol{x}) = \int_{-\infty}^{\infty} \int_{-\infty}^{\infty} f(\boldsymbol{x}) g(\boldsymbol{y} - \boldsymbol{x}) d\boldsymbol{x} \tag{19}$$

the relation

$$h(\boldsymbol{x}) = \begin{cases} 1, \text{ if } \boldsymbol{x} \in \mathbf{F} \oplus \mathbf{G}, \\ 0, \text{ otherwise}, \end{cases} \tag{20}$$

is satisfied, where integration is computed based on the relation

$$a + b = \max(a, b), \quad ab = \min(a, b), \quad a, b \in \{0, 1\} \tag{21}$$

For a binary functions, we have the relation

$$\Delta \mathbf{F} = \{\boldsymbol{x} | \nabla f \neq 0\}. \tag{22}$$

where the gradient is operated as a hyperfunction. Setting $\boldsymbol{x}(s) \in \delta \mathbf{F}$ for $0 \leq s \leq S$, we define the binary gradient as $\nabla f(\boldsymbol{x}s) = \dot{\boldsymbol{x}}(s)^\top$. Using this definition in $\mathbf{R}^2$ this equation is expressed as

$$\frac{\partial}{\partial \tau} \Delta \mathbf{F} = \frac{\nabla f}{|\nabla f|}. \tag{23}$$

Furthermore, for

$$S_i|_{\tau=0} = \{\boldsymbol{x} | |\boldsymbol{x} - \boldsymbol{c}_i| = r_i\}, \tag{24}$$

the curvature flow

$$\frac{\partial S}{\partial \tau} = \boldsymbol{n}, \tag{25}$$

where $\boldsymbol{n}$ is the unit outer normal of surface S describes the evolution of the boundary of $\mathbf{F}_i$, that is

$$S = \partial(\mathbf{F}_i \oplus \tau \mathbf{D}), \tag{26}$$

for $\mathbf{F}_i = \{\boldsymbol{x} | |\boldsymbol{x} - \boldsymbol{c}_i| \le r_i\}$.

The distance between a point and a set is defined as

$$d(\boldsymbol{x}, \mathbf{F}) = \min_{\boldsymbol{y} \in \mathbf{F}} d(\boldsymbol{x}, \boldsymbol{y}). \tag{27}$$

Furthermore, the distance between a pair of disjoint sets $\boldsymbol{F}$ and $\boldsymbol{G}$ in a space is defined as

$$d(\mathbf{F}, \mathbf{G}) = \min_{\boldsymbol{x} \in \mathbf{F}, \boldsymbol{y} \in \mathbf{G}} d(\boldsymbol{x}, \boldsymbol{y}), \tag{28}$$

for the distance $d(\boldsymbol{x}, \boldsymbol{y})$ in $\mathbf{R}^n$. For a collection of disjoint convex sets $\{\boldsymbol{F}_i\}_{i=1}^n$, such that

$$\mathbf{F}_{i\tau} \cap \mathbf{F}_{j\tau} = \emptyset, \tag{29}$$

using distance we define the Voronoi tessellation.

A point is a special convex set, since a point is a circle with zero radius. If each element of a collection of disjoint convex sets is a circle with finite radius, the Voronoi tessellation with this collection of sets preserves the same topology with the Voronoi tessellation with the centroid of each disc. This topological property is satisfied for the Voronoi tessellation with n-ball in n-dimensional Euclidean space. Furthermore, if each element of a collection of disjoint convex sets is convex set, this topological property is also preserved.

Using Voronoi tessellation with generators $\{\mathbf{F}_i\}_i^n$ such that $\mathbf{F}_i \cap \mathbf{F}_j = \emptyset$, we define and classify the singular points such that

$$\nabla f = 0, \quad det[\nabla \nabla^\top f] \ne 0. \tag{30}$$

The Voronoi tessellation in R^n corresponds to a convex n-polytope in $\mathbf{R}^{n+1}$. Setting $V(k)$ to be a k-dimensional facet of Voronoi tessellation, a generator defines a n-dimensional facet in $\mathbf{R}^{n+1}$. Using this convex polytope, we define the numbers of negative eigenvalue of the point such that $\nabla f = 0$.

Voronoi tessellation derives Delaunay triangulation defines the Delaunay facet using the Voronoi generator as 0-dimensional facets. a $(n-k)$ dimensional Voronoi facets for $k \ge 0$ shares points with k-dimensional facets. Using this geometric property, we define DSS in morphological scale space.

Definition 2. *A point in a k-dimensional Voronoi facet and $(n-k)$ dimensional Delaunay facet is a k-dimensional singular point.*

Definition 3. *At a singular point on a k-dimensional facet in* $\mathbf{R}^n$, $\nabla f = 0$ *and the number of negative eigenvalue of Hessian matrix at this point is n.*

It is possible to construct a conformal mapping to transform locally a collection of closed finite set $\{\mathbf{F}_i\}_{i=1}^n$ to a collection of n-disks $\{\mathbf{D_i}\}_{i=1}^n$. Using this conformal mapping, we can apply the disk-model of linear morphological scale space to general sets. In this case, mass conservation law described from from eq. (10) to eq. (16) is repressed by set preservation law

$$\mathbf{F}_{ij} \oplus \tau\mathbf{D} = (\mathbf{F}_{\cup}\mathbf{F}_j) \oplus \tau\mathbf{D}. \tag{31}$$

3   Scale Space Tree in 2D Linear Scale Space

Denoting the signs of the eigenvalues of the Hessian matrix of a function f which is expressed as $\boldsymbol{H} = \nabla\nabla^\top f$, as $(-,-)$, $(+,-)$ and $(+,+)$ in the linear scale space, these labels of points correspond to the local maximum points, the saddle points, and the local minimum points, respectively.

For two-dimensional positive functions with a finite number of extrema, we define labelling function such that

$$S(\boldsymbol{x},\tau) = \begin{cases} MM, r = 2, & \alpha_i < 0, \\ Mm, r = 2, & \alpha_1 \cdot \alpha_2 < 0, \\ mm, r = 2, & \alpha_i > 0, \\ sM, \quad r = 1, & \alpha_1 < 0, \\ sm, \quad r = 1, & \alpha_1 > 0, \\ m_\infty \quad |\boldsymbol{x}| = \infty, \end{cases} \tag{32}$$

for points $\nabla f = 0$. sM and sm correspond to s in 1D configurations.

Since Ms and ms correspond to s in 1D configurations, these configurations appear as $(MM)(Ms)^*(Mm)$ and $(mm)(ms)^*(mM)$ as 2D configurations. For these two configurations, we have local rewriting rules,

$$(MM)(Mm)(MM) = M(MmM) \rightarrow MM, \tag{33}$$
$$(Mm)(mm)(Mm) = (MmM)m \rightarrow Mm, \tag{34}$$
$$(mm)(Mm)(mm) = (mMm)m \rightarrow mm. \tag{35}$$

Using these rules, we have a simple example of transition such that

$$\begin{matrix} MM & MM \\ Mm \; mm \; mM & Mm \\ MM \; mM \; MM & MM \end{matrix} \rightarrow \cdots \rightarrow MM \rightarrow MMm_\infty \rightarrow \emptyset. \tag{36}$$

The intermediate configurations

$$M(MmM) \rightarrow M(sMM) \rightarrow MM, \tag{37}$$
$$(MmM)M \rightarrow (MsM)M \rightarrow MM, \tag{38}$$

$$m(MmM) \to m(sMM) \to Mm, \tag{39}$$
$$(MmM)m \to (MsM)m \to Mm, \tag{40}$$
$$m(mMm) \to m(smm) \to mm, \tag{41}$$
$$(mMm)m \to (msm)m \to mm, \tag{42}$$

define the order of the hierarchical expression same as the case of 1D functions. If all three transitions appear concurrently, the tree

$$MM\langle MM, Mm, MM, Mm, mm, Mm, MM\rangle \tag{43}$$

is derived. This transition appears if three isotropic Gaussian with the same variance are located at the three vertices of a regular triangle. For any triangles, the tree structures are derived as the rotations of the tree

$$MM\langle MM, sM\langle mM\langle mM, sm\langle mm, mM\rangle MM\langle MM, sM\langle mM, MM\rangle\rangle\rangle. \tag{44}$$

These two tree structures are the primitives of tree structures derived by the linear-scale-space singular points. In Figure 3, (a) and (b) show trees extracted from a regular triangle and a triangle, respectively. Therefore, combinations of these two primitives and three rewiring rules describe the transition of the configurations and derive tree structures from branching structures. This combinatorial structure is equivalent to the transition of the topology of a function with respect to the scale parameter τ on $\boldsymbol{x}$-τ-$f(\boldsymbol{x}, \tau)$ space. The branching geometry and the location of point $(\boldsymbol{x}_{s*}, \tau)^{\top}$, where $\nabla f(\boldsymbol{x}_{s*}) = 0$ and $rank \nabla \nabla^{\top} f(\boldsymbol{x}_{s*}) = 1$, completely and uniquely define the structure of tree in the linear scale space.

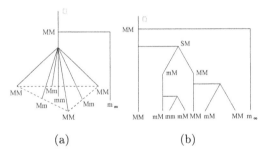

(a)   (b)

Fig. 3. Trees of triangle probes. (a) The tree extracted from a regular triangle. (b) The tree extracted from a triangle.

According to one-to-one mapping between function on Euclidean plane $\mathbf{R}^2$ and the unit sphere S^2, the scale space extrema, local maxima, saddle, local minima corresponds to vertices, edges, and faces on a polyhedron. Furthermore, these extrema are defined as the results of Voronoi tessellation and Delauney triangulation in the following manner.

1. Construct Voronoi tessellation form generators, such that,

$$\nabla f|_{\boldsymbol{x}} = 0, \ [\nabla \nabla^{\top} f]\boldsymbol{x} = \alpha \boldsymbol{x}$$

for $\alpha > 0$.

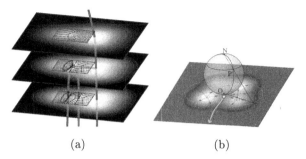

(a)   (b)

Fig. 4. Evolution of a function in the linear scale space. (a) Merging process of stationary curves for a two-dimensional function. (b) The spherical expression of the singular-point configuration.

2. Construct Delaunay triangulation from Voronoi tessellation.
3. Accept generators as local maxima, which become Delauney vertices.
4. Accept the common point between Voronoi edges and Delaunay edges as saddles.
5. Accept Voronoi vertices as local minima,
6. Accept the infinite point as a local minima.

Since a Delaunay graph of a finite number of generators on a plane is a finite graph, this Delaunay graph derives a polyhedron which is topologically equivalent to unit-sphere in the three-dimensional Euclidean space. In this transformation, the infinite face which is the infinite region separated by the boundary edges of the graph corresponds to the minimum at the infinity.

Same as the scale space evolution of one-dimensional functions, the evolution of the singular point configurations in the linear scale space of a two-dimensional function is the evolution of a polyhedral graph by the elimination of vertices.

Using one-to-one correspondences between a sphere and the Euclidean plane, for the numbers of the singular points, we have the next theorem.

Theorem 1. *Setting* $|MM|$, $Mm|$, $|mm|$, *and* $|m_\infty|$ *to be the numbers of singular points with symbols* MM, Mm, mm, *and* m_∞, *for* χ_2.

$$\chi_2 = |MM| - |Mm| + (|mm| + |m_\infty|) \tag{45}$$

the relation $\chi_2 = 2$ *is satisfied for* $0 \le \tau \le \infty$.

Figure 4 shows evolution of a function in the linear scale space. (a) Merging process of stationary curves for a two-dimensional function. (b) The spherical expression of the singular-point configuration.

4   Scale Space Tree in 2D Morphological Scape Space

For sets on the real plane such that

$$\boldsymbol{F}_\tau = \cup_{i=1}^n \boldsymbol{F}_{i\tau}, \quad \boldsymbol{F}_{i\tau} \cap \boldsymbol{F}_{j\tau} = \emptyset, \tag{46}$$

and each $\boldsymbol{F}_i$ is convex, we define three types of singular points

- The centroid of each set F_i where $f(x) = 1$ if $x \in F_i$.
- The Voronoi vertices.
- The Voronoi edges.

Using Voronoi tessellation, we define the singular points as followings.

1. Construct Voronoi tessellation form generators.
2. Construct Delaunay triangulation from Voronoi tessellation.
3. Accept generators as local maxima, which become Delaunay vertices.
4. Accept the common point between Voronoi edges and Delaunay edges as saddles.
5. Accept Voronoi vertices as local minima,
6. Accept the infinite point as a local minima.

Since a Delaunay graph of a finite number of generators on a plane is a finite graph, this Delaunay graph derives a polyhedron which is topologically equivalent to unit-sphere in the three-dimensional Euclidean space. In this transformation, the infinite face which is the infinite region separated by the boundary edges of the graph corresponds to the minimum at the infinity. The evolution of the singular point configurations in the linear scale space of a two-dimensional function is the evolution of a polyhedral graph by the elimination of vertices.

For two-dimensional positive functions with a finite number of extrema, we define labelling function such that

$$S_M(x, \tau) = \begin{cases} MM, \text{ local maxima,} \\ Mm, \text{ saddle,} \\ mm, \text{ local minima} \\ m_\infty \quad |x| = \infty. \end{cases} \tag{47}$$

According to the analysis in section 2.2, the branching points of the tree is determined by the minimum of the distances among discs. Furthermore, we have the next theorem for evolution of a tree in morphological scale space.

Theorem 2. *Tree evolution in morphological scale space satisfies the same relation with that in the linear scale space without symbol s.*

Therefore, we have local rewriting rules,

$$(MM)(Mm)(MM) = M(MmM) \to MM, \tag{48}$$
$$(Mm)(mm)(Mm) = (MmM)m \to Mm, \tag{49}$$
$$(mm)(Mm)(mm) = (mMm)m \to mm. \tag{50}$$

Moreover, using one-to-one correspondences between a sphere and the Euclidean plane, for the numbers of the singular points, we have the next theorem.

Theorem 3. *Setting $|MM|$, $Mm|$, $|mm|$, and $|m_\infty|$ to be the numbers of singular points with symbols MM, Mm, mm, and m_∞, for χ_2.*

$$\chi_2 = |MM| - |Mm| + (|mm| + |m_\infty|) \tag{51}$$

the relation $\chi_2 = 2$ is satisfied for $0 \leq \tau \leq \infty$.

5 Top Point Tree

We call the trees defined in the previous sections the stationary trees. The top points for two-variable functions are the points at which the rank of the Hessian matrix is one. Using singular points, we define the top-point.

Definition 4. *On the top points the rank of the Hessian matrix is less than the number of variable of the function.*

Therefore, a top point corresponds to the branch point of the stationary curves for two-valued functions. This geometric property derives an algorithm for the construction of the top-point tree from the stationary tree. Using the stationary trees, we define the hierarchical structure of the top points. On a stationary tree, a top point correspond to a node labelled s, $s*$, and $s**$ for 1- and 2-variable functions, respectively. This geometrical property permits us to mathematically define the top-point tree from the stationary tree.

Theorem 4. *Purring leaves without label s from the stationary tree derives the top-point tree.*

This definition implies there is one-to-one mapping between a stationary tree and a top point tree for a function. Figure 5 shows the extracted top-point tree from the stationary tree.

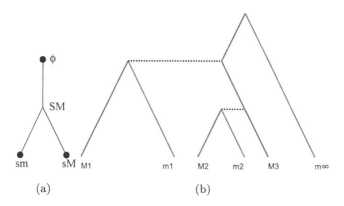

Fig. 5. Tree and top-point tree. (a) The top-point tree extracted from the tree of (b). (b) Merging process of stationary curves for a two-dimensional function.

6 Scale Space Tree in 3D

Symbolically it is possible to extend grammars proposed in the previous sections to 3D objects.

Denoting the signs of the eigenvalues of the Hessian matrix of a function f which is expressed as $\boldsymbol{H} = \nabla\nabla^{\top}f$, as $(-,-,-)$, $(+,-,-)$, $(+,+,-)$, and $(+,+,+)$ in the linear scale space, these labels of points correspond to the local maximum points, the negative saddle points, the positive saddle points and the local minimum points, respectively.

For three-dimensional positive functions with a finite number of extrema, we define a labelling function such that

$$S(\boldsymbol{x}, \tau) = \begin{cases} MMM, r = 3, & \alpha_i < 0, \\ MMm, r = 3, & \alpha_1 > 0 > \alpha_2 \geq \alpha_3, \\ Mmm, r = 3, & \alpha_1 \geq \alpha_2 > 0 > \alpha_3, \\ mmm, r = 3 & \alpha_i > 0, \\ sMM, r = 2, & \alpha_i < 0, \\ sMm, r = 2, & \alpha_1 > 0 > \alpha_2, \\ smm, r = 2, & \alpha_1 \geq \alpha_2 > 0, \\ ssM, r = 1, & \alpha_1 < 0, \\ ssm, r = 1, & \alpha_1 > 0, \\ m_\infty, & |\boldsymbol{x}| = \infty, \end{cases} \tag{52}$$

for points $\nabla f = 0$. Labels $s**$ and $ss*$ correspond to s in 1D configurations.

Since labels $s**$ and $ss*$ correspond to s in 1D configurations, we have the the rewriting rules.

$$(MMM)(MMm)(MMM) \rightarrow MM(MMm) \rightarrow MMM \tag{53}$$

$$(MMm)(Mmm)(MMm) \rightarrow Mm(MMm) \rightarrow MMm \tag{54}$$

$$(Mmm)(mmm)(Mmm) \rightarrow mm(MMm) \rightarrow Mmm \tag{55}$$

$$(MMm)(MMM)(MMm) \rightarrow MM(Mmm) \rightarrow Mmm \tag{56}$$

$$(Mmm)(MMm)(Mmm) \rightarrow Mm(MMm) \rightarrow MMm \tag{57}$$

$$(mmm)(Mmm)(mmm) \rightarrow mm(Mmm) \rightarrow mmm. \tag{58}$$

For these rules, the intermediate configurations are expressed as

$$(MMM)(MMm)(MMM) \rightarrow MM(MMm) \rightarrow MM(sMMm) \tag{59}$$

$$(MMm)(Mmm)(MMm) \rightarrow Mm(MMm) \rightarrow Mm(sMMm) \tag{60}$$

$$(Mmm)(mmm)(Mmm) \rightarrow mm(MMm) \rightarrow mm(sMMm) \tag{61}$$

$$(MMm)(MMM)(MMm) \rightarrow MM(Mmm) \rightarrow MM(sMmm) \tag{62}$$

$$(Mmm)(MMm)(Mmm) \rightarrow Mm(MMm) \rightarrow Mm(sMmm) \tag{63}$$

$$(mmm)(Mmm)(mmm) \rightarrow mm(Mmm) \rightarrow mm(sMmm). \tag{64}$$

These intermediate configurations define the order of branching structure of trees. This combinatorial structure is equivalent to the transition of the topology of a function with respect to the scale parameter τ on $\boldsymbol{x}$-τ-$f(x, \tau)$ space. The branching geometry and the location of point $(x_{s**}, \tau)^\top$, where $\nabla f(\boldsymbol{x}_{s**}) = 0$ and $rank\nabla\nabla^\top f(\boldsymbol{x}_{s**}) = 2$, completely and uniquely define the structure of a tree in the linear scale space.

Using one-to-one mapping from $\mathbf{R}^3$ to S^3 For the numbers of the singular points, we have the next theorem since singular points with labels (MMM), (MMm), (Mmm), and (mmm) correspond to the vertices, edges, faces, and volumes of polytopes in $\mathbf{R}^3$.

Theorem 5. *Setting* $| * * * |$ *to be the number of singular points with symbol* $* * *$*, for* χ_3*,*

$$\chi_3 = |MMM| - |MMm| + |Mmm| - (|mmm| + |m_\infty|), \qquad (65)$$

the relation $\chi_3 = 0$ *is satisfied.*

It is possible to deal with 4-dimensional polytopes as spatial graphs in $\mathbf{R}^3$, same as the case that we deal with a polyhedral without any holes as a planar graph.

For morphological scale space for 3D objects, the consideration of the symbols s and $s*$ is not required.

7   Conclusions

Singular points in the linear scale space provide fundamental features for the extraction dominant parts of an image. Employing the geometrical configuration of singular points, it is possible to construct a tree in scale space. This tree expresses a hierarchical structure of dominant parts. In this paper, we clarified the graphical grammar for the construction of this tree in the linear scale space. Furthermore, we show a combinatorial structure of singular points in morphological scale space using conformal mapping from Euclidean plane to the spherical surface.

We defined symbolic expression of the evolution of the configurations of singular points in the morphological scale space. These rules show that as the intermediate expression, the trajectory of saddle points defines the order of the branching geometry, that is, the saddle points define the local trunk as the main curve and local branch as the top point when a triplet of singular points merged to a singular point in the linear scale space of one-valuable functions. This geometry is first proposed by Zhao and Iijima [12, 13, 14]. For one-dimensional functions, in a successive triplet of singular points with a minimum and two maxima in both side, are merges to a saddle and a maximum and finally a maximum and a saddle point are merged to a local maximum. This local maximum is locally called the trunk. This transition of three singular points to a singular point is uniquely determined based on the configuration of singular points in the linear scale space. If the merge process of minimum and maximum appears for both maxima at the both ends of the configuration, we have the second branching geometry as trinary branching. In this case the main trunk after transition is the maximum. In higher dimensional functions, the same merging process appears at each branching point.

References

1. Bookstein, F. L., The line-skeleton, CVGIP, **11**, 1233-137, 1979
2. Rosenfeld, A., Axial representations of shapes, CVGIP, **33**, 156-173, 1986.
3. Serra, J., *mathematical Morphology*, Academic Press, 1982.
4. Iijima, T., *Pattern Recognition,* Corona-sha, Tokyo, 1974 (in Japanese).

5. Witkin, A.P., Scale space filtering, Pros. of 8th IJCAI, 1019-1022, 1993.

6. Lindeberg, T., *Scale-Space Theory in Computer Vision,* Kluwer, Boston 1994.

7. ter Haar Romeny, B.M. *Front-End Vision and Multi-Scale Image Analysis Multiscale Computer Vision Theory and Applications, written in Mathematica,* Springer 2003.

8. Lindeberg, T. Feature detection with automatic selection, International Journal of Computer Vision, **30**, 79-116, 1998.

9. Otsu, N., *Mathematical Studies on Feature Extraction in Pattern Recognition,* Researches of The Electrotechnical Laboratory, **818**, 1981 (in Japanese).

10. Weicker, J., *Anisotropic Diffusion in Image Processing,* Teubner, 1998.

11. Imiya, A., Sugiura, T., Sakai, T, Kato, Y., Temporal structure tree in digital linear scale space, LNCS, **2695**, 356-371, 2003.

12. Zhao, N.-Y., Iijima, T., Theory on the method of determination of view-point and field of vision during observation and measurement of figure IECE Japan, Trans. D., **J68-D**, 508-514, 1985 (in Japanese).

13. Zhao, N.-Y., Iijima, T., A theory of feature extraction by the tree of stable viewpoints. IECE Japan, Trans. D., **J68-D**, 1125-1135, 1985 (in Japanese).

14. Zhao, N.-Y., *A Study of Feature Extraction by the Tree of Stable View-Points,* Dissertation to Doctor of Engineering, Tokyo Institute of Technology, 1985 (in Japanese).

15. Pelillo, M., Siddiqi, K., Zucker, S.W., Matching hierarchical structures using Association graphs, IEEE, Trans, PAMI **21**, 1105-1120, 1999.

16. Yuille, A. L., Poggio, T., Scale space theory for zero crossings, IEEE PAMI, **8**, 15-25, 1986.

17. Kuijper, A., Florack, L.M.J., Viergever, M.A., Scale Space Hierarchy, Journal of Mathematical Imaging and Vision **18**, 169-189, 2003.

18. Kuijper, A.; Florack, L.M.J., The hierarchical structure of images, IEEE, Trans. Image Processing **12**, 1067- 1079, 2003.

19. Imiya, A., Katsuta, R. Extraction of a structure feature from three-dimensional objects by scale-space analysis, LNCS, **1252**, 353-356, 1997.

Additive Subsets

Yan Gerard

LLAIC, Auvergne University, Aubière, France
`gerard@llaic.u-clermont1.fr`

Abstract. Additive subsets have been introduced in the framework of discrete tomography with the underlying notion of x-rays. This notion can be defined from two different ways. We provide in the paper extensions of the two definitions and a proof of their equivalence in a framework where x-rays are replaced by any subsets. It results a pair of dual definitions of additivity cleared out from dispensable assumptions and a proof of their equivalence reduced to a separation theorem.

1 Introduction

Reconstruction of binary images from local information is an overall task which has been mainly investigated in the framework of discrete tomography. The problem introduced independently by H.J. Ryser [1] and D. Gale [2] in 1957 consisted in constructing a binary image (or a binary matrix) with prescribed numbers of 1s in each row and column. The generalization of the problem in dimension 3 (a kind of time-table problem) has been proved to be NP-hard in 1976 [3] and other generalizations have been considered since the nineties under influence of electron-microscopy or tomography. A new field called Discrete Tomography emerged in 1994 (DIMACS at Rutgers University) with the main purpose to provide solutions for reconstructing lattice sets with prescribed x-rays (historic details are developed in [4]).

The notion of x-ray (an x-ray of a lattice set S is the function which gives the cardinalities of the intersections of S with parallel lines or more generally with parallel affine spaces [5]) is in the center of the topic but it is also possible to consider intersections of a lattice set with other kinds of subsets than lines. Such an extension of the reconstruction problem has been done for instance in [6] where a rectangular window is translated all over the lattice. The combinatorial problem remains essentially the same: reconstructing a binary image with given numbers of 1s in some given subsets of the image. We can even disregard the lattice structure of the image and consider the problem of reconstruction of a subset S of a given unstructured set A with prescribed cardinalities y_i for its intersections $S \cap A^i$ with a family of given subsets $A^i \subset A$. This overall problem is purely combinatorial and it is clear that deciding the existence of a solution from the input is an overall class of problems known as NP-hard (even if the cardinalities given as input are restricted to be only 0s and 1s) since it covers tomographic problems known as NP-complete [3] [7].

We focus in this paper on a notion of discrete tomography called "additivity". This notion is of much interest for many questions regarding the uniqueness of a

U. Eckardt et al. (Eds.): IWCIA 2006, LNCS 4040, pp. 347–353, 2006.

solution or stability requirements [8]. It has been conjectured by A. Daurat and S. Brunetti in [9] that convex sets (and even Q-convex sets) are additive according to "good" sets of directions (see conjecture 11). Parallel conjectures have been formulated by L. Thorens or R. Gardner. We emphasize that if such results could be proved, they would generalize and extend some important theorems concerning for instance the uniqueness of a convex solution. Another framework in which additivity is used is the one of minimal matrices [10] [11].

The notion of additivity has been first introduced by Fishburn et al. in [12]. The tomographic framework of this work has leaded the authors to base their definition of additivity on a family of linear manifolds [13] while we could try to extend it to any family of subsets. The extension of tomography to non linear subsets as it is done for instance with rectangular subsets [6] leads to try to extend the notion of additivity in a more general framework and although previous authors did not mention it, this combinatorial extension is completely natural.

Thus the task of the paper is to show how the tomographic notion of additivity can be extended to the combinatorial framework of an unstructured set (we just assume that it is ordered to simplify notations). There exist two possible approaches, one related with linear programming and the other one related with existence of "generating" functions. The two corresponding definitions are equivalent in the original theory and the challenge is to prove that their extensions keep this property.

While we could think that the task could be harder in a general combinatorial framework, it appears in fact that the linearity of the subsets considered in tomography is useless for proving the equivalence of the two definitions. The proof that we present in the paper is a consequence of a classical geometrical theorem of separation of convex cones (or a Farkas lemma in linear programming theory). This approach is derived from a rewriting in a general combinatorial framework of a proof given by S. Onn and E. Vallejo in [10]. The use of the theorem of separation of convex cones leads to the conclusion that the two definitions of additivity are dual from each other.

Duality has leaded to present the result in two sections. First section presents the primal definiton of additivity while the second one is devoted to its transformation by duality in a second definition that we call "dual".

2   Primal Definition of Additivity

We start by introducing a combinatorial framework of a finite set of cardinality n. By assuming that its elements can be ordered, we identify it with the set of integers $\{1 \cdots n\}$. Any subset A of $\{1 \cdots n\}$ can be represented by its characteristic vector $\chi(A) \in \{0, 1\}^n$ that coordinates verify for any index j between 1 and n the equivalence between $\chi(A)_j = 1$ and $j \in A$. With this representation, the results of counting numbers of elements can be done by computing scalar products: the cardinality of the intersection $A \cap B$ of two subsets of $\{1 \cdots n\}$ is exactly $\chi(A).\chi(B)$.

We consider now the general combinatorial problem of computing a subset S of $\{1 \cdots n\}$ having fixed number of elements in given subsets of $\{1 \cdots n\}$.

Problem 1. Input: a list of m subsets $A^i \subset \{1 \cdots n\}$ and of m positive integers y_i (the index i going from 1 to m)
Output: a subset S of $\{1 \cdots n\}$ verifying for any index i from 1 to m the equalities $cardinality(S \cap A^i) = y_i$.

The vector of the cardinalities $cardinality(S \cap A^i) = \chi(S).\chi(A^i)$ can be expressed by matrix product $A\chi(S)$ where A is the (m, n)-matrix of entries $a_{i,j} = \chi(A^i)_j$. In this framework, conjunction of conditions $cardinality(S \cap A^i) = y_i$ is equivalent with $A\chi(S) = y$ where $y \in Z^m$ is the vector of coordinates y_i. Thus problem 1 is equivalent with solving a 0-1 integer programming instance $Ax = y$ where A is a given binary matrix and y a given vector. Problems of discrete tomography enter in this general framework. As already said, this class of problems is NP-hard (even in the case of a binary input y) since it covers some NP-complete classes of tomographic problems.

We consider now the relaxed problem where the binary constraint $x_j \in \{0, 1\}$ on each coordinate x_j of x becomes $x_j \in [0, 1]$. It means that x is requested to belong to unit hypercube $[0, 1]^n$ instead of being necessarily one of its vertices. Solving equation $Ax = y$ with x in the unit hypercube is a usual linear programming instance. Relaxing a binary constraint in a $[0, 1]$ interval is one of the main principles used in combinatorial optimization (see for instance branch and bound or branch and cut techniques).

We focus in this paper on the generalization of the notion of *additivity*.

Definition 1. *A subset S of $\{0, 1\}^n$ is said* additive *with respect to a finite family of subsets $A^i \subset \{1 \cdots n\}$ (with i between 1 and m) if the intersection between the unit hypercube $[0, 1]^n$ and the affine space $Ax = A\chi(S)$ (where entries of A are $a_{i,j} = \chi(A^i)_j$) does not contain any other point than $\chi(S)$.*

The points of the affine space $AX = A\chi(S)$ represent the multisets (with real multiplicities) of $\{1 \cdots n\}$ having the same number of points than S in each subset A^i (Fig. 1). The additivity of S with respect to A^i family means that S is

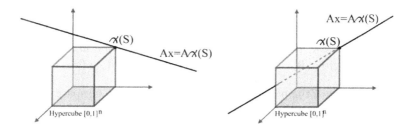

Fig. 1. We represent the hypercube $[0, 1]^n$. The point $\chi(S)$ is one of its vertices. The affine space $Ax = A\chi(S)$ represents the multisets that numbers of points in each A^j are the same than S. In the case of an additive set S (on the left) its intersection with the hypercube is reduced to $\chi(S)$ while in the case of a non additive set (on the right) the affine space contains other points than $\chi(S)$ in the hypercube.

unique among all multisets with multiplicities in $[0, 1]$ to share with all subsets A^i the same number of elements than S.

From a computational point of view, it means that S is the unique solution of the corresponding linear program $Ax = A\chi(S)$ with $x \in [0, 1]^n$. This condition guarantees that problem 1 with matrix A and vector $y = A\chi(S)$ as input can be solved quite easily because it can be obtained directly as unique solution of the linear program $Ax = A\chi(S)$ with $x \in [0, 1]^n$ without needing help of any integer or 0-1 linear programming technique.

3   Dual Definition of Additivity

This section requires to introduce some geometrical material. The duality that we mention is formalized by a separation theorem on polyhedral cones.

3.1   Separation Theorems

We mean by cones union of half-lines issued from a same vertex and by polyhedral cone the convex hull of a finite number of half-lines issued from a common vertex (see [14] for details). One of the main theorems of convex geometry says that convex cones issued from the origin have only the origin in common if and only if they can be "separated" by an hyperplane [15]:

Theorem 1. *Let K and M be closed convex cones in R^d issued from the origin. By assuming that K does not contain any line, we have the equivalence between*

(i) K and M intersect only in the origin.
(i) there is a non-zero vector $u \in R^d$ such that $u.x < 0$ for all $x \in K - \{O\}$ and $u.x \geq 0$ for all y in M.

By applying previous theorem on polyhedral cones we obtain next corollary:

Corollary 1. *Let C and C' be two polyhedral cones of R^d issued from a common vertex v. We assume also that C does not contain any line. Their intersection $C \cap C'$ is reduced to v if and only if there exists an affine hyperplane separating them (there exists a normal vector $u \in R^d$ verifying for any x in $C - \{v\}$ inequality $u.(x - v) < 0$ and for any x in C' the inequality $u.(x - v) \geq 0$).*

We can consider now the particular case where C' is an affine space containing v and of normal direction the linear space denoted $C'^\perp$. In this case the condition on u that for any x in C' we have $u.(x - v) \geq 0$ can be improved. If $u.(x - v) > 0$ then by taking as x' the symmetric point of x with respect to v (x' belongs also to C') we have $u.(x' - v) < 0$ which contradicts characterization of u. It follows that if C' is an affine space then the condition that for all x in C' we have $u.(x - v) \geq 0$ becomes that for any x in C' we have $u.(x - v) = 0$. It means exactly that u is in the normal linear space $C'^\perp$. Thus we obtain a new version of corollary 1 :

Corollary 2. *Let C be a polyhedral cone of R^d which does not contain any line and which has the point v as vertex. Let H be an affine space of R^d containing v. Their intersection $C \cap H$ is reduced to v if and only if there exists a vector $u \in H^\perp$ verifying for any x in $C - \{v\}$ inequality $u.(x - v) < 0$.*

We provide a last version of previous theorems by denoting that inequality $u.(x - v) < 0$ holds for all points x in polyhedral cone $C - \{v\}$ if and only if it holds for all the half-lines that polyhedral cone is the convex hull.

Corollary 3. *Let C be the convex hull of a finite number of half-lines $[v, v + y^k)$ of R^d. We also assume that C does not contain any line. Let H be an affine space of R^d containing v. Their intersection $C \cap H$ is reduced to v if and only if there exists a vector $u \in H^\perp$ verifying for any index k the inequality $u.y^k < 0$.*

3.2 Some Remarks

If we consider the details of primal definition of additivity (definition 1), it is question of the intersection between the affine space $Ax = A\chi(S)$ and hypercube $[0, 1]^n$. They both contain $\chi(S)$. The hypercube $[0, 1]^n$ is not a cone but we can introduce the cone containing all the half-lines issued from $\chi(S)$ and passing through one of its points (Fig. 2). We denote it $cone(S)$ (we have by definition $cone(S) = \{x \in R^n / \exists \lambda \in R^{*+}, \lambda x + (1 - \lambda)\chi(S) \in [0, 1]^n\}$).

Definition of $cone(S)$ leads to important remarks

1. The cone $cone(S)$ does not contain any line.
2. By construction, the cone of S $(cone(S))$ is the convex hull of the n half-lines issued from $\chi(S)$ and directed by the n oriented edges of the hypercube going out of $\chi(S)$. With an index j between 1 and n, these n vectors can be denoted $(-1)^{\chi(S)_j} e_j$ where e_j has null coordinates except the jth equal to 1 (the e_j are the canonical basis of R^n).

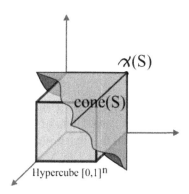

Fig. 2. We represent the hypercube $[0, 1]^n$, its vertex $\chi(S)$ and the polyhedral cone $cone(S)$

3. By linearity of the affine space $Ax = A\chi(S)$, its intersection with the hypercube $[0,1]^n$ is reduced to $\chi(S)$ if and only if its intersection with $cone(S)$ is also reduced to $\chi(S)$.

Remark 3 provides an equivalent definition of additivity in terms of cones: a set S is additive if and only if the intersection of the affine space $Ax = A\chi(S)$ with $cone(S)$ is reduced to $\chi(S)$. Corollary 3 associated to remarks 1 and 2 provides next intermediary lemma:

Lemma 1. *A subset S of $\{0,1\}^n$ is additive with respect to a finite family of subsets $A^i \subset \{1 \cdots n\}$ (with i between 1 and m) if and only if there exists a vector u in the orthogonal linear space $\{x \in R^n / Ax = A\chi(S)\}^{\perp}$ verifying for any index j from 1 to n inequality $u.(-1)^{\chi(S)_j} e_j < 0$.*

3.3 Dual Definition

To obtain the dual definition of additivity, it remains only to write what is the orthogonal to the affine space $Ax = A\chi(S)$ and to precise what means condition $u.(-1)^{\chi(S)_j} e_j < 0$ of lemma 1.

To answer first question we just recall that the n entries of ith line of matrix A are exactly the n coordinates of $\chi(A^i)$. It follows that the normal linear space to affine space $Ax = A\chi(S)$ is the linear space generated by m vectors $\chi(A^i)$. Thus existence of u in the normal space of Lemma 1 verifying for any index j from 1 to n inequality $u.(-1)^{\chi(S)_j} e_j < 0$ is equivalent with the existence of m coordinates λ_i verifying for any index j from 1 to n inequality $\sum_{i=1}^{m}(-1)^{\chi(S)_j} \lambda_i \chi(A^i).e_j < 0$. This last inequality can be rewritten $(-1)^{\chi(S)_j} \sum_{i=1}^{m} \lambda_i \chi(A^i)_j < 0$. Thus existence of u in Lemma 1 means exactly that there exists a linear combination $\sum_{i=1}^{m} \lambda_i \chi(A^i)$ of characteristic vectors $\chi(A^i)$ such that each coordinate $\sum_{i=1}^{m} \lambda_i \chi(A^i)_j$ has contrary sign from $(-1)^{\chi(S)_j}$. This formulation of existence of u in lemma 1 leads to a final dual definition of additivity:

Theorem 2. *A subset S of $\{0,1\}^n$ is additive with respect to a finite family of subsets $A^i \subset \{1 \cdots n\}$ (with i between 1 and m) if and only if there exists a linear combination of characteristics vectors $\chi(A^i)$ that j^{th} coordinate is strictly positive if $j \in S$ and strictly negative otherwise (for any index j between 1 and n).*

4 Conclusion

The first point of this paper is to break with the false idea that additivity is only related with "linear" x-rays: Other kinds of subsets can be introduced. It allows to generalize the notion of additivity in neighboring topics from discrete tomography. The second point is to show that the equivalence between the two possible definitions comes from duality.

These two ideas which have not been noticed in previous papers can be useful for a better understanding of this important notion. Additivity is the center of general conjectures related with convex sets. As other perspective than these hard theoretical problems of discrete tomography, we add a last question which is perhaps not completely disjoint: Given a subset, what are the different families of subsets which make it additive?

References

1. Ryser, H.J.: Combinatorial properties of matrices of zeros and ones. Canad. J. Math. **9** (1957) 371–377
2. Gale, D.: A theorem on flows in networks. Pacific J. Math. **7** (1957) 1073–1082
3. Even, S., Itai, A., Shamir, A.: On the complexity of timetable and multicommodity flow problems. SIAM J. Computing **5** (1976) 691–703
4. Herman, G.T., Kuba, A.: Discrete Tomography: A Historical Overview. Chapter 1 in Discrete Tomography, Foundations, Algorithms, and Applications. (1999) 3–34
5. Gardner, R.J.: Geometric Tomography. Encyclopedia of Mathematics and its Applications. **58**, Cambridge University Press (1995)
6. Frosini, A., Nivat, M.: Binary Matrices Under the Microscope: A Tomographical Problem. Lecture Notes in Computer Science. **3322** Proceedings of 10th International Workshop in Combinatorial Image Analysis (2004) 35–58
7. Gardner, R., Gritzmann, P., Prangenberg, D.: On the computational complexity of reconstructing lattice sets from their X-rays. Discrete Math. **202** (1999) 45–71.
8. Brunetti, S., Daurat, A.: Stability in Discrete Tomography: Linear Programming, Additivity and Convexity. Lecture Notes in Compter Science **2886** (2003) 398–407
9. Brunetti, S., Daurat, A.: Stability in Discrete Tomography: some positive results. Discrete Applied Mathematics. **147** (2005) 207–226
10. Onn, S., Vallejo, E.: Permutohedra and minimal matrices. Linear Algebra and its Applications. **412** (2006) 471–489
11. Vallejo, E.: Minimal matrices and discrete tomography. Electronic Notes in Discrete Mathematics. **20** (2005) 113–132
12. Fishburn, P.C., Lagarias, J.C., Reeds, J.A., Shepp, L.A.: Sets uniquely determined by projections on axes II. Discrete case. Discrete Mathematics. **91** (1991) 149–159
13. Fishburn, P.C., Shepp, L.: Sets of Uniqueness and Additivity in Integer Lattices. Chapter 2 in Discrete Tomography, Foundations, Algorithms, and Applications. (1999) 3–34
14. Ziegler, G.: Lectures on polytopes. Graduate Texts in Mathematics, Springer (1994)
15. Rockafellar, R.T.: Convex Analysis. Princeton Mathematical Series. **28** (1970)

Cooperating Basic Puzzle Grammar Systems

K.G. Subramanian[1], R. Saravanan[2], and P. Helen Chandra[3]

[1] Department of Mathematics,
Madras Christian College,
Chennai 600 059, India
kgsmani1948@yahoo.com
[2] Department of Mathematics,
Bharath Institute of Higher Education and Research,
Chennai 600 073, India
[3] Department of Mathematics,
Jayaraj Annapackiam College for Women
Periyakulam 625 601, India

Abstract. Nivat et al [3] introduced Context-free Puzzle grammars for generating connected picture arrays in the two-dimensional plane. Basic Puzzle grammars [6] constitute a subclass of these grammars. In this note we consider the Cooperating Array Grammar Systems introduced by Dassow et al [2] with Basic Puzzle grammar rules in the components instead of array grammar rules and examine the picture generating power of the resulting system, called, Cooperating Basic Puzzle Grammar System, in the maximal mode.

1 Introduction

The theory of Grammar systems was developed to provide a theoretical framework for modelling distributed computation at the symbolic level. A grammar system consists of several grammars or other language identifying mechanisms, that cooperate according to some well-defined protocol. The components of the system correspond to the agents, the current string(s) in generation to a symbolic environment, and the system's behaviour is represented by the language. A variety of string grammar system models have been introduced and studied [1].

On the other hand, in the study of generation and description of picture patterns considered as connected digitized, finite arrays of symbols, syntactic approaches have played a significant role on account of their structure-handling ability. Adapting the techniques of formal string language theory, various types of picture or array grammars have been introduced and investigated [4, 5]. Most of the array grammars developed to handle picture languages, are based on Chomskian string grammars. Context-free array grammars (CFAG) and Regular array grammars (RAG) are two such classes of array grammars well-studied for picture generation. Another model of grammars called Puzzle grammars has been introduced in [3]. It is known that context-free puzzle grammars (CFPG) and context-free array grammars (CFAG) coincide [3] whereas a subclass of CFPG, called Basic Puzzle grammars is known [6] to properly include the class of RAG's.

U. Eckardt et al. (Eds.): IWCIA 2006, LNCS 4040, pp. 354–360, 2006.

The power of the mechanism of cooperation in generating pictures by array grammars has been investigated in [2] by introducing Cooperating Array Grammar Systems and it is shown that cooperation increases the generative capacity even in the case of systems with regular array grammar components. Here we consider Basic Puzzle grammars in the components of Cooperating Array grammars. The resulting system, called Cooperating Basic Puzzle Grammar system, is examined, in this note, for its generative power in the maximal derivation mode.

2   Preliminaries

Let Σ be a finite alphabet. A picture over Σ is a finite connected array of symbols in the two-dimensional plane with the symbols belonging to Σ.

We refer to [3] for notions of Puzzle grammars and to [4] for array grammars. We recall only the definition of Basic Puzzle grammars [6].

Definition 1. *A Basic Puzzle Grammar (BPG) is a structure $G = (N, T, R, S)$ where N and T are finite sets of symbols; $N \cap T = \emptyset$. Elements of N are called non-terminals and elements of T, terminals. $S \in N$ is the start symbol or the axiom. R consists of rules of the following forms;*

$$A \longrightarrow \boxed{a}\, B \;, \quad A \longrightarrow a\boxed{B} \;, \quad A \longrightarrow B\boxed{a} \;, \quad A \longrightarrow \boxed{B}\, a \;,$$

$$A \longrightarrow \genfrac{}{}{0pt}{}{\boxed{a}}{B} \;, \quad A \longrightarrow \genfrac{}{}{0pt}{}{a}{\boxed{B}} \;, \quad A \longrightarrow \genfrac{}{}{0pt}{}{B}{\boxed{a}} \;, \quad A \longrightarrow \genfrac{}{}{0pt}{}{\boxed{B}}{a} \;, \quad A \longrightarrow \boxed{a}$$

where $A, B \in N$ and $a \in T$.

Derivations begin with S written in a unit cell in the two-dimensional plane, with all the other cells containing the blank symbol #, not in $N \cup T$. In a derivation step, denoted $\Rightarrow$, a non-terminal A in a cell is replaced by the right-hand member of a rule whose left-hand side is A. In this replacement, the circled symbol of the right-hand side of the rule used, occupies the cell of the replaced symbol and the non-circled symbol of the right side occupies the cell to the right or the left or above or below the cell of the replaced symbol depending on the type of rule used. The replacement is possible only if the cell to be filled in by the non-circled symbol contains a blank symbol.

The set of pictures or arrays generated by G, denoted $L(G)$, is the set of connected, digitized finite arrays over T, derivable in one or more steps from the axiom.

3   Cooperating Basic Puzzle Grammar Systems

Cooperating string grammar systems, [1] have been extended, for two-dimensional picture description, in [2] to Cooperating Array grammar Systems by taking the components as array grammars. We now introduce a variation in Cooperating Array Grammar system by considering Basic Puzzle Grammar rules instead of context-free or regular array rewriting rules. Also we consider here only the maximal derivation mode.

Definition 2. *Formally, a Cooperating Basic Puzzle grammar system (CBPGS) is $\Gamma = (N, T, S, P_1, P_2, \ldots, P_n)$, where N is the nonterminal alphabet; T is the terminal alphabet; $N \cap T = \emptyset; S \in N$ and $Pi, i = 1, \ldots, n$ are finite sets of basic puzzle grammar rules over $N \cup T$. For each $i, 1 \leq i \leq n, \Rightarrow_{P_i}$ is the usual derivation relation using the BPG rules of P_i. $\Rightarrow^*_{P_i}$ is the reflexive, transitive closure of $\Rightarrow_{P_i}$. For arrays X, Y the maximal derivation $\Rightarrow^t_{P_i}$ is defined by $X \Rightarrow^t_{P_i} Y$, if and only if $X \Rightarrow^*_{P_i} Y$ and there is no $Z \in (N \cup T)^{**}$ such that $Y \Rightarrow_{P_i} Z$. For CBPGS , the array language generated by Γ in the maximal derivation mode is*

$$L_t(\Gamma) = \{X \in T^{**}/S \Rightarrow^t_{P_{i_1}} X_1 \Rightarrow^t_{P_{i_2}} X_2 \Rightarrow \cdots \quad \Rightarrow^t_{P_{i_m}} X_m = X,$$
$$m \geq 1, 1 \leq i_j \leq n, 1 \leq j \leq m\}$$

The family of array languages generated by CBPGS with at most n components in the maximal derivation mode is denoted by $CD_n(BPG, t), n \geq 1$.

Remark 1. Every regular array grammar rule is indeed a *BPG* rule. Hence a cooperating regular array grammar system is a *CBPGS*.

Example 1. Consider the *CBPGS*
$\Gamma = (\{S, A, B, C, D, E, F, Y, Z\}, \{X\}, S, P_1, P_2)$ where

$$P_1 = \{ \quad S \overset{A}{\longrightarrow} \otimes \quad A \overset{B}{\longrightarrow} \otimes, \quad B \longrightarrow \otimes C, \quad C \longrightarrow \otimes D, \quad D \overset{A}{\longrightarrow} \otimes, \quad D \overset{A}{\underset{E}{\longrightarrow}} \otimes,$$

$$E \overset{}{\underset{E}{\longrightarrow}} \otimes, E \longrightarrow F \otimes, F \longrightarrow F \otimes, F \longrightarrow Y \otimes, Y \longrightarrow Z \otimes \}$$

$$P_2 = \{Y \longrightarrow \otimes\}$$

$$
\begin{array}{cccccccc}
 & & & & & x & x & x \\
 & & & & & x & & x \\
 & & & x & x & x & & x \\
 & & & x & & & & x \\
 & x & x & x & & & & x \\
 & x & & & & & & x \\
 & x & x & x & x & x & x & x \\
\end{array}
$$

Fig. 1.

Γ generates an array language that consists of "Staircases of x' s of fixed width"(Figure 1)

Note that Γ is in fact a cooperating regular array grammar system.

Proposition 1. *For* $n \geq 1$,

  i. $CD_n(BPG,t) \subseteq CD_{n+1}(BPG,t)$
  ii. $CD_n(REGA,t) \subseteq CD_n(BPG,t) \subseteq CD_n(CFA,t)$
 iii. $CD_1(BPG,t) = BPL$

Inclusions i) and ii) are obvious from definitions, as every regular array grammar rule is a Basic puzzle grammar rule, which in turn is a context - free array grammar rule. Equality iii) is due to the fact that a BPG can be considered as a system with one component.

Proposition 2. $REGA = CD_1(REGA,t) \subset CD_1(BPG,t) = BPL$
It is known [2] that $REGA = CD_1(REGA,t)$. *The strict inclusion is due to the fact that the set of pictures of "isosceles triangles of x 's" cannot be generated [7] by any regular array grammar but is generated [7] by a BPG.*

Proposition 3. $CD_2(BPG,t) - CFA \neq \emptyset$

Proof. The set R_{HF} *of hollow rectangular frames (Figure 2) of thickness one over one letter alphabet x can not be generated by any context - free array grammar, since even the set of hollow rectangles (Figure 3) is known [2] to be not generable by any context-free array grammar.*
   But the following Cooperating Basic Puzzle grammar system Γ *generates* R_{HF} *in the* $t-mode$

$$\Gamma = (\{S,A,B,C,D,E,I,J,K,X,Y,Z\}, \{x\}, S, P_1, P_2)$$

$$P_1 = \{S \longrightarrow \overset{S}{\boxed{x}}, \quad S \longrightarrow \overset{A}{\boxed{x}}, \quad A \longrightarrow x\,\boxed{B}, \quad B \longrightarrow \overset{x}{\boxed{C}},$$

$$C \longrightarrow \boxed{x}\,C, \quad C \longrightarrow \boxed{D}\,x, \quad D \longrightarrow \overset{x}{\boxed{E}}, \quad E \longrightarrow \boxed{x},$$

$$I \longrightarrow \boxed{J}\,x, \quad J \longrightarrow J\,\boxed{x}, \quad J \longrightarrow K\boxed{x}, \quad K \longrightarrow \overset{Z}{\underset{}{\boxed{x}}}^{F},$$

$$F \longrightarrow \underset{F}{\boxed{x}}, \quad F \longrightarrow \underset{x}{\boxed{I}} \,\}$$

$$P_2 = \{K \longrightarrow \underset{x}{\boxed{X}}, \quad X \longrightarrow x\,\boxed{Y}, \quad Y \longrightarrow \boxed{x} \,\}$$

$$\begin{array}{cccccccc} & x & & & & & x & \\ x & x & x & x & x & x & x & x \\ & x & & & & & x & \\ & x & & & & & x & \\ x & x & x & x & x & x & x & x \\ & x & & & & & x & \end{array}$$

Fig. 2.

$$\begin{array}{ccccc} x & x & x & x & x \\ x & & & & x \\ x & & & & x \\ x & x & x & x & x \end{array}$$

Fig. 3.

Proposition 4

i $CFA - CD_n(BPG, t) = \emptyset$, *for all* $n \geq 1$
ii CFA *and* $CD_n(BPG, t)$ *are incomparable for* $n \geq 2$.

Proof. The array language consisting of the only array M in Figure 4 cannot be generated by any $CBPGS$ as BPG rules can handle only one cell "protrusions" whereas in M, there are two-cell protrusions on all four directions (North, West, South, East) of the middle cell in M.

$$\begin{array}{ccccc} & & x & & \\ & & x & & \\ x & x & x & x & x \\ & & x & & \\ & & x & & \end{array}$$

Fig. 4. Array X

This proves (i)
Statement (ii) is consequence of (i) and proposition 3.

Proposition 5. $CD_n(BPG, t) = CD_2(BPG, t)$, *for* $n \geq 2$.
This result is analogous to and can be proved in a similar manner to the proof of the result $CD_n(REGA, t) = CD_2(REGA, t)$.

4 Parallel Basic Puzzle Grammars

An array with a "one-stroke path" [7] from a cell to another cell in the array and passing through all the cells in the array except for one-cell "protrusions" [6] can be generated by a sequence of Basic Puzzle grammar rules. For instance, the array X in Figure 5 has a one-stroke path from a to b and with a protrusions at c and d.

$$a$$
$$c\ x\ b$$
$$d$$

Fig. 5. Array X

So instead of a sequence of BPG rules generating the array X in Figure 5, we can have an equivalent single rule of the form $S \to X$ with the cell containing a being circled. With the modified BPG rules of this kind we can consider a variation of BPG that has an axiom array and tables of modified BPG rules that cooperate and rewrite an array in parallel, rewriting the nonterminals and terminals with applicable rules. The resulting grammar is called a Parallel Basic Puzzle Grammar.

We illustrate with an example to bring out the feature of increase in generative capacity when the rules of a table are applied in parallel.

Example 2. Consider the Parallel Basic Puzzle Grammar (PBPG) G with non-terminals A, D, terminal a and axiom array Z as in Figure 6 and tables t_1, t_2 of modified BPG rules:

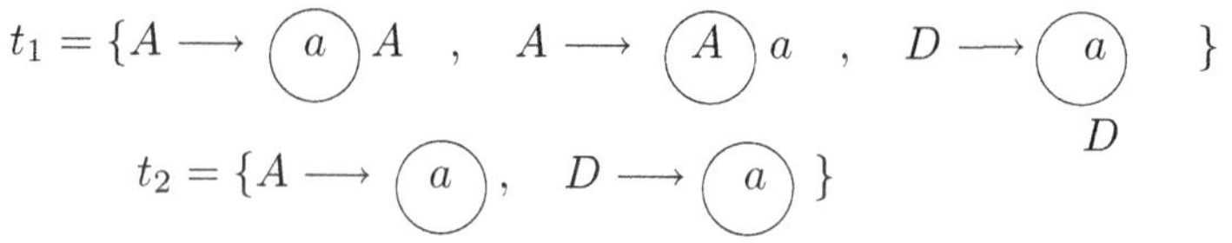

$$t_1 = \{A \longrightarrow \boxed{a} A \quad, \quad A \longrightarrow \boxed{A} a \quad, \quad D \longrightarrow \boxed{a} \quad \}$$
$$D$$
$$t_2 = \{A \longrightarrow \boxed{a}, \quad D \longrightarrow \boxed{a} \}$$

$$A\ a\ A$$
$$D$$

Fig. 6. Array Z

The PBPG G generates pictures describing token "T" with equal "arms" (Figure 7). These pictures cannot be generated by any BPG as equal "arms" cannot be maintained in a BPG.

$$a\ a\ a\ a\ a$$
$$a$$
$$a$$

Fig. 7. Array of TokenT

Although the example mentioned above has not really used the modified BPG rules, it can be seen that interesting picture classes such as sets of rectangles or squares can be generated by their use with a small number of rules.

5   Conclusion

Maximal derivation mode being an interesting mode, we have examined in this note, the picture description power of cooperation in grammar systems with

Basic puzzle grammar rules in the components. It remains to examine other modes of derivation in these systems.

Acknowledgement. The authors thank the referees for their useful comments.

References

1. E. Csuhaj-Varjú, J. Dassow, J. Kelemen and Gh. Păun: Grammar systems: A grammatical approach to distribution and cooperation, Gordon and Breach Science Publishers, 1994.
2. J. Dassow, R. Freund and Gh. Păun: Cooperating Array Grammar Systems, Int. Journal of Pattern Recognition and Artificial Intelligence, **9**, 1995, 1-25.
3. M. Nivat, A. Saoudi, K.G. Subramanian, R. Siromoney and V.R. Dare, Puzzle Grammars and Context-free Array Grammars, Int. Journal of Pattern Recognition and Artificial Intelligence, **5**, 1991, 663-676.
4. A. Rosenfeld, Picture Languages, Academic Press, 1979.
5. A. Rosenfeld and R. Siromoney: Picture languages - a survey, Languages of design, **1**, 1993, 229–245.
6. K.G. Subramanian, R. Siromoney, V.R. Dare, and A. Saoudi, Basic Puzzle Languages, Int.Journal of Pattern Recognition and Artificial Intelligence, **9**, 1995, 763-775.
7. Y. Yamamoto, K. Morita, and K. Sugata, Context-sensitivity of two-dimensional regular array grammars, Int.Journal of Pattern Recognition and Artificial Intelligence, **3,4**, 1989, 295-320.

Quasi-isometric and Quasi-conformal Development of Triangulated Surfaces for Computerized Tomography

Eli Appleboim[1], Emil Saucan[1], Yehoshua Y. Zeevi[1], and Ofir Zeitoun[2]

[1] Electrical Engineering Department, Technion, Haifa, Israel
{eliap, semil, zeevi}@tx.technion.ac.il
[2] FRS Ltd., Rosh HaAyin, Israel
ofir.zeitoun@gmail.com

Abstract. In this paper we present a simple method for minimal distortion development of triangulated surfaces for mapping and imaging. The method is based on classical results of F. Gehring and Y. Väisälä regarding the existence of quasi-conformal and quasi-isometric mappings between Riemannian manifolds. A random starting triangle version of the algorithm is presented. A curvature based version is also applicable. In addition the algorithm enables the user to compute the maximal distortion errors. Moreover, the algorithm makes no use to derivatives, hence it is suitable for analysis of noisy data. The algorithm is tested on data obtained from real CT images of the human brain cortex.

1 Introduction

Two-dimensional representation of three-dimensional object scan, are encountered in image processing. Medical imaging, computer aided design and reverse engineering are three of the important examples. For example, one should be able to present the three-dimensional MRI/CT scans of the brain cortex, as a (set of) two-dimensional images. Yet, in order to do so in a meaningful manner, so that diagnosis will be accurate, it is essential that the geometric distortion, in terms of change of angles and lengths, caused by this representation will be minimal. In computer graphic, this problem is sometimes refereed to as, surface/freeform parameterization. Applications are also found in texture mapping, surface re-meshing, surface compression, and others.

In most cases, since the surfaces considered are not isometric to the plane, one cannot expect a zero distortion solution. Yet, a reasonable solution to this problem is given by conformal maps (i.e. maps that preserve angles). This is done by mapping the surface conformally to the (complex) plane. Since this cannot be achieved in a global way, all solutions are local.

If one is willing to absorb some bounded amount of distortion then quasi-isometric/quasi-conformal maps (i.e. maps that are almost isometries/conformal – the precise definition will follow in Section 2) will also suffice.

As in many other cases, the tradeoff is between simplicity/cost of implementation on one hand and accuracy on the other. Common to all proposed solutions is

U. Eckardt et al. (Eds.): IWCIA 2006, LNCS 4040, pp. 361–374, 2006.

the fact, which really cannot be avoided because of the inevitable distortion, that the more locally one is willing to focus, the more accurate the results become.

1.1   Related Works

As stated above, the problem of minimal distortion flattening of surfaces attracted, in recent years, a great attention and interest, due to its wide range of applications.

In this section we briefly review some of the methods that were proposed for dealing with this problem.

Variational Methods. Haker et al. ([8], [9]) introduced the use of a variational method for conformal flattening of CT/MRI 3-D scans of the brain/colon for the purpose of medical imaging. The method is essentially based on solving Dirichlet problem for the Laplace-Beltrami operator $\triangle u = 0$ on a given surface Σ, with boundary conditions on $\partial \Sigma$. A solution to this problem is a harmonic (thus conformal) map from the surface to the (complex) plane. The solution suggested in [8] and [9] is a PL (piecewise linear) approximation of the smooth solution, achieved by solving a proper system of linear equations.

Circle Packing. Hurdal et al. ([10]) attempt to obtain such a conformal map by using circle packing. This relies on the ability to approximate conformal structure on surfaces by circle packings. The authors use this method for MRI brain images and conformally map them to the three possible models of geometry in dimension 2 (i.e. the 2-sphere, the Euclidian plane and the Hyperbolic plane). Yet, the method is applicable for surfaces which are topologically equivalent to a disk whereas the brain cortex surface is not. This means that there is a point of the brain (actually a neighborhood of a point), which will not map conformally to the plane, and in this neighborhood the dilatation will be infinitely large. An additional problem arises due to the necessary assumption that the surface triangulation is homogeneous in the sense that all triangles are equilateral. Such triangulations are seldom attainable.

Holomorphic 1-Forms. Gu et al. ([6], [7], [5]) are using holomorphic 1-forms in order to compute global conformal structure of a smooth surface of arbitrary genus given as a triangulated mesh. holomorphic 1-forms are differential forms (differential operators) on smooth manifolds, which among other things can depict conformal structures. The actual computation is done via computing homology/co-homology bases for the first homology/co-homology groups of the surface, H_1, H^1 respectively. This method indeed yields a global conformal structure hence, a conformal parameterization for the surface however, computing homology basis is extremely time consuming.

Angle Methods. In [12] Sheffer et al. parameterize surfaces via an angle based method in a way that minimizes angle distortion while flattening. However, the surfaces are assumed to be approximated by cone surfaces, i.e. surfaces that are composed from cone-like neighborhoods.

To summarize, all the methods described above compute only approximation to conformal mappings, therefore producing only quasi-conformal mappings, with no precise estimates on the dilatation.

In this paper we propose yet another solution to this problem. The proposed method relies on theoretical results obtained by Gehring and Väisälä in the 1960's ([4]). They were studying the existence of quasi-conformal maps between Riemannian manifolds. The basic advantages of this method resides in its simplicity, in setting, implementation and its speed. Additional advantage is that it is possible guarantee not to have distortion above a predetermined bound, which can be as small as desired, with respect to the amount of localization one is willing to pay (and, in the case of triangulated surfaces, to the quality of the given mesh). The suggested algorithm is best suited to cases where the surface is complex (high and non-constant curvature) such as brain cortex/colon wrapping, or of large genus such as skeleta, proteins, etc. Moreover, since toghether with the angular dilatation, both length and area distortions are readily computable, the algorithm is ideally suited for applications in Oncology, where such measurements are highly relevant.

The paper is organized as follows, in the next section we introduce the theoretical background, regarding the fundamental work of Gehring and Väisälä. Afterwards we describe our algorithm for surface flattening, based on their ideas. In Section 4 we present some experimental results of this scheme and in Section 5 we discuss possible extensions of this study. We include two appendices regarding some classical notions in quasi-conformal mapping theory and the definition of the *essential supremum*, respectively.

2   Theoretical Background

2.1   Basic Definitions

Definition 1. *Let $D \subset \mathbb{R}^3$ be a domain. A homeomorphism $f : D \to \mathbb{R}^3$ is called a* quasi-isometry *(or a* bi-lipschitz mapping*), if there exists $1 \leq C < \infty$, such that*

$$\frac{1}{C}|p_1 - p_2| \leq |f(p_1) - f(p_2)| < C|p_1 - p_2|, \text{ for all } p_1, p_2 \in D.$$

$C(f) = \min\{C \,|\, f \text{ is a quasi} - \text{isometry}\}$ *is called the* minimal distortion *of f (in D).*

Note. For the case of surface embedded in $\mathbb{R}^3$ distances are the induced intrinsic distances on the surfaces.

Remark 1. If f is a quasi-isometry then $K_I(f) \leq C(f)^2$ and $K_O(f) \leq C(f)^2$ where $K_I(f), K_O(f)$ represent the *inner*, respective *outer dilatation* of f, (see Appendix 1). It follows that any quasi-isometry is a quasi-conformal mapping (while – evidently – not every quasi-conformal mapping is a quasi-isometry). Quasi-conformal is the same as quasi-isometry where distances are replaced by angles between tangent vectors.

Definition 2. *Let $S \subset \mathbb{R}^3$ be a connected set. S is called* admissible *(see Fig. 1) iff for any $p \in S$, there exists a quasi-isometry i_p such that for any $\varepsilon > 0$ there exists a neighbourhood $U_p \subset \mathbb{R}^3$ of p, such that $i_p : U_p \to \mathbb{R}^3$ and $i_p(S \cap U_p) = D_p \subset \mathbb{R}^2$, where D_p is a domain and such that $C(i_p)$ satisfies:*

$$(i) \; \sup_{p \in S} C(i_p) < \infty$$

and

$$(ii) \; \text{ess} \sup_{p \in S} C(i_p) < 1 + \varepsilon \, .$$

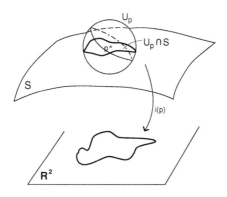

Fig. 1. An Admissible Surface

2.2   The Projection Map

Let S be a surface, $\bar{n}$ be a fixed unitary vector, and $p \in S$. Let $V \simeq D^2$, $D^2 = \{x \in \mathbb{R}^2 \,|\, ||x|| \leq 1\}$ be a disk neighbourhood of p. Moreover, suppose that for any $q_1, q_2 \in S$, the acute angle $\angle(q_1q_2, \bar{n}) \geq \alpha$ (see Figure 2). We refer to the last condition as *the Geometric Condition* or *Gehring Condition*.

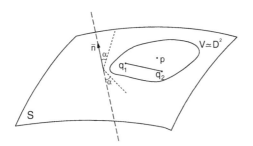

Fig. 2. The Geometric Condition

Then for any $x \in V$ there is a unique representation of the following form:

$$x = q_x + u\boldsymbol{n} \,;$$

where q_x lies on the plane through p which is orthogonal to $\boldsymbol{n}$ and $u \in \mathbb{R}$. Define:

$$Pr(x) = q_x \,.$$

Note: $\boldsymbol{n}$ need not be the normal vector to S at p.

By [4], Section 4.3 and Lemma 5.1, we have that for any $p_1, p_2 \in S$ and any $a \in \mathbb{R}_+$ the following inequalities hold:

$$\frac{a}{A} \,|p_1 - p_2| \leq |Pr(p_1) - Pr(p_2)| \leq A|p_1 - p_2| \,;$$

where

$$A = \frac{1}{2}[(a \csc \alpha)^2 + 2a + 1]^2 + \frac{1}{2}[(a \csc \alpha)^2 - 2a + 1]^2 \,.$$

In particular for $a = 1$ we get that

$$C(f) \leq \cot \alpha + 1$$

and

$$K(f) \leq \left(\left(\frac{1}{2}(\cot \alpha)^2 + 4 \right)^{\frac{1}{2}} + \frac{1}{2} \cot \alpha \right)^{\frac{3}{2}} \leq (\cot \alpha + 1)^{\frac{3}{2}} \,;$$

where

$$K(f) = \max \left(K_O(f), K_I(f) \right)$$

is the *maximal dilatation* of f.

Hence we have thus obtained the desired quasi-isometry, Pr, having maximal dilatation,

$$C(f) \leq \cot \alpha + 1 \,.$$

The Geometric Condition. From the discussion above we conclude that $S \subset \mathbb{R}^3$ is an admissible surface if for any $p \in S$ there exists $\boldsymbol{n}_p$ such that for any $\varepsilon > 0$, there exists $U_p \simeq D^2$, such that for any $q_1, q_2 \in U_p$ the acute angle $\angle(q_1 q_2, \boldsymbol{n}_p) \geq \alpha$, where

$$(i) \ \inf_{p \in S} \alpha_p > 0 \,;$$

and

$$(ii) \ \text{ess} \inf_{p \in S} \alpha_p < \frac{\pi}{2} - \varepsilon \,.$$

Example 1. Any surface in $S \in \mathbb{R}^3$ that admits a well-defined continuous turning tangent plane at any point $p \in S$ is admissible.

3   The Algorithm

We will present in this section the algorithm that is used for obtaining a quasi-isometric (flat) representation of a given surface.

First assume the surface is equipped with some triangulation T. Let N_p stand for the normal vector to the surface at a point p on the surface.

Second, a triangle Δ, of the triangulation must be chosen. We will project a patch of the surface quasi-isometrically onto the plane included in Δ. This patch will be called the patch of Δ, and it will consists of at least one triangle, Δ itself. There are two possibilities to chose Δ, one is in a random manner and the other is based on curvature considerations. We will refer to both ways later. For the moment assume Δ was somehow chosen. After Δ is (trivially) projected onto itself we move to its neighbors. Suppose Δ' is a neighbor of Δ having edges e_1, e_2, e_3, where e_1 is the edge common to both Δ and Δ'.

We will call Δ' *Gehring compatible w.r.t* Δ, if the maximal angle between e_2 or e_3 and N_Δ (the normal vector to Δ), is greater then a predefined measure suited to the desired predefined maximal allowed distortion, i.e. $\max\{\varphi_1, \varphi_2\} \geq \alpha$, where $\varphi_1 = \angle(e_2, N_\Delta)$, $\varphi_2 = \angle(e_3, N_\Delta)$.

We will project Δ' *orthogonally* onto the plane included in Δ and insert it to the patch of Δ, iff it is Gehring compatible w.r.t Δ.

We keep adding triangles to the patch of Δ moving from an added triangle to its neighbors (of course) while avoiding repetitions, till no triangles can be added.

If by this time all triangles where added to the patch we have completed constructing the mapping. Otherwise, chose a new triangle that has not been projected yet, to be the starting triangle of a new patch. A pseudocode for this procedure can be easily written.

Remark 2. There are two ways for choosing a base triangle for each patch. One is by taking a triangle which the sum of the (magnitude of) curvatures of its vertices is minimal, and the other one is by letting the user choose a triangle for each new patch.

Remark 3. One should keep in mind that the above given algorithm, as for any other flattening method, is local. Indeed, in a sense the (proposed) algorithm gives a measure of "globality" of this intrinsically local process.

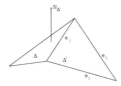

Fig. 3. Gehring Compatible Triangles

Remark 4. Our algorithm is best suited for highly folded surfaces, because of its intrinsic locality, on the one hand, and computational simplicity, on the other. However, on "quasi-developable" surfaces (i.e. surfaces that are almost cylindrical or conical) the algorithm behaves similar to other algorithms, with practically identical results).

4   Experimental Results

We now proceed to present some experimental results obtained by applying the proposed algorithm, both on synthetic surfaces and on data obtained from actual CT scans.

In each of the examples both the input surface and a flattened representation of some patch are shown. Details about mesh resolution as well as flattening distortion are also provided. The number of patches needed in order to flatten the surface is also given. In all images, the small rectangle shown on the surface represents a base triangle for the flattened patch.

The algorithm was implemented in two versions, or more precisely two possible ways of processing, automatic versus user defined.

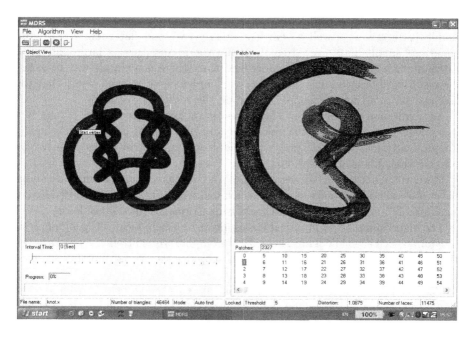

Fig. 4. Knotted Torus: The resolution of the model is of 46,464 triangles. The patch contains 11,475 triangles. The low selected value of α is $5°$, giving a dilatation equal to 1.0875. Due to the low Gauss curvature of the embedding of the knot, two such patches cover most of the flattened surface.

1. Automatic means that the triangles serving as base points for the patches to be flattened are chosen automatically according to curvature, as stated in Remark 3.1. The discrete curvature measure employed is that of *angular defect*, due to its simplicity and high reliability (see [13]).
2. User defined means that at each stage the user chooses a base triangle for some new patch.

The colored area in each of the images represents the patch being flattened.

Experiments have shown that results of the automatic process are similar, in terms of the dilatation, to those obtained from the user defined process yet, in order to flatten entire surface in the user defined method one needs in average 25 percent more patches.

5   Concluding Remarks and Future Study

In this paper we presented a new algorithm for flattened presentation of polyhedral meshes, with minimal dilatation while flattening is done. The algorithm is based upon the works of Gehring-Väisala and others concerning the existence of quasi-isometric/conformal/meromorphic mappings between Riemannian manifolds.

From the implementation results it is evident that this algorithm while being simple to program as well as efficient, also gives good flattening results and maintains small dilatations even in areas where curvature is large and good flattening is a challenging task. Moreover, since there is a simple way to assess the resulting dilatation, the algorithm was implemented in such a way that the user can set in advance an upper bound on the resulting dilatation.

An additional advantage of the presented algorithm is related to the fact that, contrary to some of the related studies, no use of derivatives is made. Consequently, the algorithm does not suffer from typical drawbacks of derivative computations like robustness, etc.

Moreover, since no derivatives are employed, no smoothness assumption about the surface to be flattened are made, which makes the algorithm presented herein ideal for use in cases where smoothness is questionable (to say the least).

The algorithm may be practical for applications where local yet, good analysis is required such as medical imaging with the emphasis on flattened representation of the brain and the colon (virtual colonoscopy) – see [1], [2]. Further study is currently undertaken.

The main issue for further investigation, remains the transition from local to global in a more precise fashion, i.e. how can one glue two neighbouring patches while keeping fixed bounded dilatation. (In more technical terms, this amounts to actually *computing* the *holonomy map* of the surface – see [14].)

Indeed, we may flatten the neighborhood of some vertex u obtaining the flat image I_u and the neighborhood of another vertex v obtaining the image I_v so that these two neighborhood have some intersection along the boundary yet, it will not be possible to adjust the resulting images to give one flat image $I_{u \cup v}$ of the union of these neighborhood, yet satisfy the quasi-isometric property. This too is also under current investigation.

Acknowledgment

Emil Saucan is supported by the Viterbi Postdoctoral Fellowship. Research is partly supported by the Ollendorf Minerva Center.

References

[1] Appleboim, E., Saucan E., and Zeevi, Y. *Minimal-Distortion Mappings of Surfaces for Medical Imaging*, Proceedings of VISAPP 2006, to appear.

[2] Appleboim, E., Saucan, E., and Zeevi, Y.Y. *On Sampling and Reconstruction of Surfaces*, Technion CCIT Report, 2006.

[3] Caraman, P. *n-Dimensional Quasiconformal (QCf) Mappings*, Editura Academiei Române, Bucharest, Abacus Press, Tunbridge Wells Haessner Publishing, Inc., Newfoundland, New Jersey, 1974.

[4] Gehring, W. F. and Väisälä, J. *The coefficients of quasiconformality*, Acta Math. **114**, pp. 1-70, 1965.

[5] Gu, X. Wang, Y. and Yau, S. T. *Computing Conformal Invariants: Period Matrices*, Communications In Information and Systems, Vol. 2, No. 2, pp. 121-146, December 2003.

[6] Gu, X. and Yau, S. T. *Computing Conformal Structure of Surfaces*, Communications In Information and Systems, Vol. 2, No. 2, pp. 121-146, December 2002.

[7] Gu, X. and Yau, S. T. *Global Conformal Surface Parameterization* , Eurographics Symposium on Geometry Processing, 2003.

[8] Haker, S. Angenet, S. Tannenbaum, A. Kikinis, R. *Non Distorting Flattening Maps and the 3-D visualization of Colon CT Images*, IEEE Transauctions on Medical Imaging, Vol. 19, NO. 7, July 2000.

[9] Haker, S. Angenet, S. Tannenbaum, A. Kikinis, R. Sapiro, G. Halle, M. *Conformal Surface Parametrization for Texture Mapping*, IEEE Transauctions on Visualization and Computer Graphics, Vol. 6, NO. 2, June 2000.

[10] Hurdal, M. K., Bowers, P. L., Stephenson, K., Sumners, D. W. L., Rehm, K., Schaper. K., Rottenberg, D. A. *Quasi Conformally Flat Mapping the Human Crebellum*, Medical Image Computing and Computer-Assisted Intervention - MICCAI'99, (C. Taylor and A. Colchester. eds), vol. 1679, Springer-Verlag, Berlin, 279-286, 1999.

[11] Stephenson, K. *personal communication*.

[12] Sheffer, A. de Stuler, E. *Parametrization of Faceted Surfaces for Meshing Using Angle Based Flattening*, Enginneering with Computers, vol. 17, pp. 326-337, 2001.

[13] Surazhsky, T., Magid, E., Soldea, O., Elber, G. and Rivlin E. *A Comparison of Gaussian and Mean Curvatures Estimation Methods on Triangular Meshes*, Proceedings of the IEEE International Conference on Robotics and Automation. Taipei, Taiwan, pp 1021-1026, September 2003.

[14] Thurston, W. *Three-Dimensional Geometry and Topology, vol.1, (Edited by S. Levy)*, Princeton University Press, Princeton, N.J. 1997.

[15] Väisälä, J. *Lectures on n-dimensional quasiconformal mappings*, Lecture Notes in Mathematics 229, Springer-Verlag, Berlin - Heidelberg - New-York, 1971.

Appendix 1

Let $D \subset \mathbb{R}^3$ be a domain, let $f : D \overset{\sim}{\to} f(D)$ and let $p \in D$.
We make the following notations:

$$L(p) = \limsup_{x \to p} \frac{|f(x) - f(p)|}{|x - p|} \; ;$$

$$l(p) = \liminf_{x \to p} \frac{|f(x) - f(p)|}{|x - p|} \; ;$$

$$J(p) = \limsup_{r \to 0} \frac{Vol\big(f(\mathbb{B}^3(p, r))\big)}{Vol\big(\mathbb{B}^3(p, r)\big)} \; .$$

$L(p)$, $l(p)$ are called the *maximal*, respective *minimal stretching*, of f. If f is differentiable then $J(p) = |Jacobian(f)(p)|$.

Then, if $J > 0$ and if f is ACL (see, e.g. [15]), then the maximal and minimal stretching can be defined as follows:

$$K_I(f) = \sqrt{\operatorname*{ess\,sup}_{p \in D} \frac{J(p)}{l^3(p)}} \; ;$$

$$K_O(f) = \sqrt{\operatorname*{ess\,sup}_{p \in D} \frac{L^3(p)}{J(p)}} \; .$$

Appendix 2

The *essential supremum* of f is the smallest number a for which f only exceeds a on a set of measure zero. More formally, we have the following definition:

Definition 3. *Let $(X, \mathcal{B}, \mu)$ be a measure space, let $f : X \to \mathbb{R}$, and let $a \in \mathbb{R}$. Define $M_a = \{x \,|\, f)x) > 0\}$ and $A_0 = \{a \in \mathbb{R} \,|\, \mu(x) = 0\}$. Then:*

$$\operatorname{ess\,sup} f = \inf A_0 \; .$$

(If $A_0 = \emptyset$, then we define $\operatorname{ess\,sup} f = \infty$.)

Remark 5. In our case μ is the 2-dimensional Hausddorff measure (see, e.g. [3]).

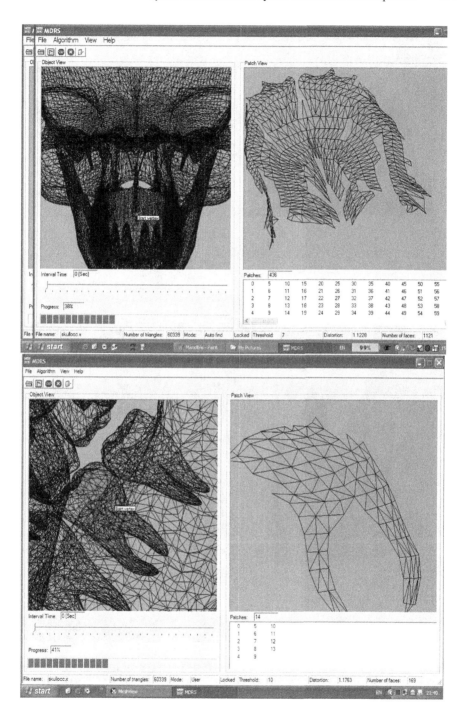

Fig. 5. Skull model: The resolution is of 60,339 triangles. Top: $\alpha = 7°$ and the dilatation is 1.1228. Bottom: $\alpha = 10°$ and the dilatation is 1.1763.

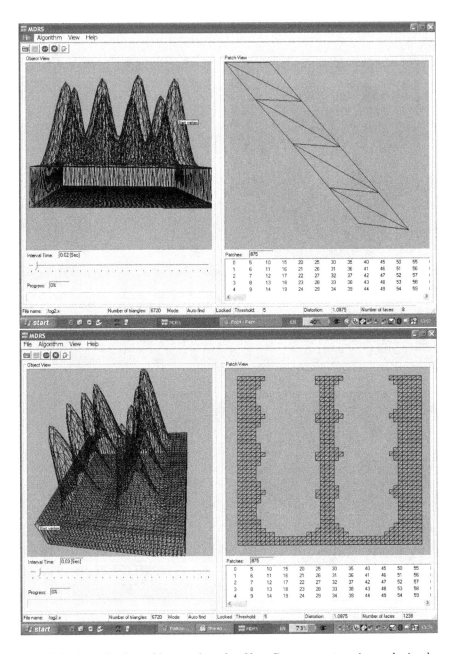

Fig. 6. A Analytic Surface: Observe the role of low Gauss curvature in producing large patches, even of genus higher then 0, (b). Here the resolution is 6720 triangles and $\alpha = 5°$.

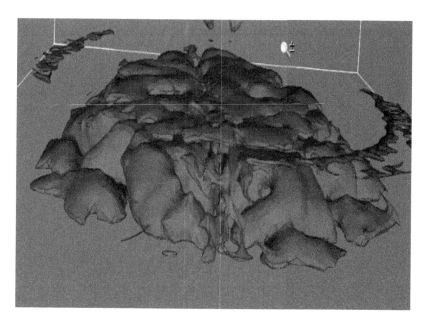

Fig. 7. Cerebral Cortex Flattening: The location of the cortical region selected for flattening in the previous figure

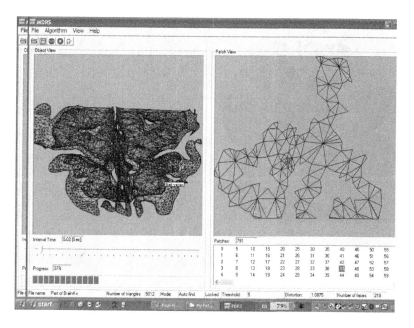

Fig. 8. Cerebral Cortex Flattening: Partial view of the parietal region. Observe that a non-simply connected patch is obtained. The resolution is 15.110 triangles, the angles chosen are 5°, producing dilatations of 1.0875.

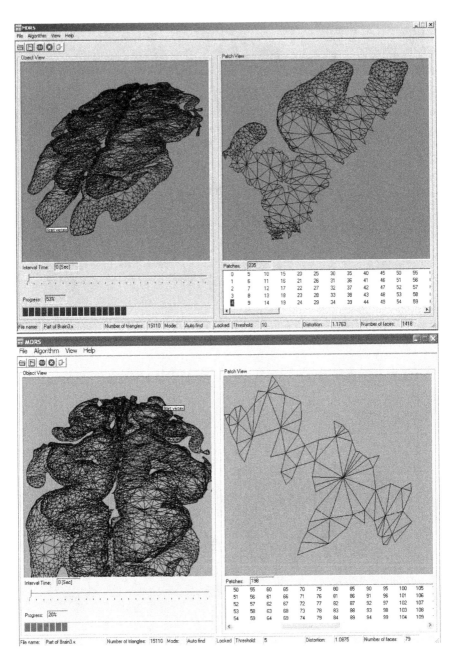

Fig. 9. Cerebral Cortex Flattening: Two patches obtained in the flattening of the parietal region. The resolution is 15.110 triangles, the angles chosen are 5° and 10°, producing dilatations of 1.0875 and 1.1763, respectively.

Binary Tomography with Deblurring

Stefan Weber[1], Thomas Schüle[1,3], Attila Kuba[2], and Christoph Schnörr[1]

[1] University of Mannheim,
Dept. of Mathematics and Computer Science, CVGPR-Group,
D-68131 Mannheim, Germany
{wstefan, schuele, schnoerr}@uni-mannheim.de
www.cvgpr.uni-mannheim.de
[2] University of Szeged,
Dept. of Image Processing and Computer Graphics,
H-6720 Szeged, Hungary
kuba@inf.u-szeged.hu,
www.inf.u-szeged.hu/~kuba
[3] Siemens Medical Solutions,
D-91301 Forchheim, Germany
www.siemensmedical.com

Abstract. We study two scenarios of limited-angle binary tomography with data distorted with an unknown convolution: Either the projection data are taken from a blurred object, or the projection data themselves are blurred. These scenarios are relevant in case of scattering and due to a finite resolution of the detectors. Assuming that the unknown blurring process is adequately modeled by an isotropic Gaussian convolution kernel with unknown scale-parameter, we show that parameter estimation can be combined with the reconstruction process. To this end, a recently introduced Difference-of-Convex-Functions programming approach to limited-angle binary tomographic reconstruction is complemented with Expectation-Maximization iteration. Experimental results show that the resulting approach is able to cope with both ill-posed problems, limited-angle reconstruction and deblurring, simultaneously.

1 Introduction

It is a general characteristic of imaging systems that the acquired images are some distorted versions of the ideal images of real objects. The distortion is due to physical limitations, e.g., finite resolution in space and time, non-uniform sensitivity in the field of view, etc. In many cases the distorted image can be modeled as the convolution of the ideal image with some function describing the distortion [1].

The situation is the same in tomography when the cross-sections of some 3D object are reconstructed from its projections. The pixel values in the projection images are usually only some approximations of the line integrals to be measured by a perfect imaging system in an ideal physical situation. In different application areas of tomography there are several correction methods to improve the

U. Eckardt et al. (Eds.): IWCIA 2006, LNCS 4040, pp. 375–388, 2006.

quality of the reconstructed images. The correction strategies can be divided into two classes roughly. The first class contains the methods aiming to correct the projection data before reconstruction (let us call them preprocessing) and then the reconstruction is performed from the corrected projection data. The second class is the family of special methods when the correction is included into the reconstruction process. We believe that both strategies can be useful. If the correction can be done as a preprocessing step before reconstruction then one of the methods from the first class is preferable. However, there are situations when the correction is impossible or too complicated before reconstruction, e.g., scatter correction in CT or in SPECT, then the correction during the reconstruction can still give a good solution.

The situation is very similar in the case of binary tomography, when the range of the function to be reconstructed is just the set $\{0, 1\}$ (as a summary of binary tomography see [2]). The known discrete range can be used in the reconstruction process as a kind of a priori information, and binary functions can be reconstructed effectively from very few projections (e.g., 2-5). As binary tomography is getting to be applied in several areas, the problem of distortion of such tomography images becomes an important problem to be studied. There are publications discussing different corrections in DT, e.g. in X-ray and neutron tomography [3, 4], and electron microscopy [5].

In this paper we deal with the general distortion model when the distortion can be described by the convolution with a Gaussian kernel $G_\sigma(\cdot)$. If the parameter σ is known in advance then the correction (deconvolution) can be done as a preprocessing step before the reconstruction. However, if the parameter is not known then we are going to show that there is still a way to binary tomography by including this parameter as an unknown value to be determined. To motivate our approach we present some reconstructions, see figure 1, performed without deblurring.

Section 2 shows the mathematical model of distorted DT and the reconstruction problem to be solved. Our reconstruction approach including the deconvolution adaptation is described in Section 3. The optimization algorithm is specified in section 4. Several experiments have been done to test our reconstruction procedure for both noiseless and noisy data. The corresponding results are presented in Section 5. We conclude and indicate further work in Section 6.

2   Problem Statement

2.1   Binary Tomography and Reconstruction by DC-Programming

We consider the reconstruction problem of transmission tomography for binary objects. As explained in figure 2, the imaging process is represented by the algebraic system of equations

$$Ax = b, \qquad A \in \mathbb{R}^{m \times n}, \; x \in \{0, 1\}^n, \; b \in \mathbb{R}^m, \qquad (1)$$

where A and b are given, and the binary indicator vector x representing the unknown object has to be reconstructed. To this end, we introduced in [6] the variational approach

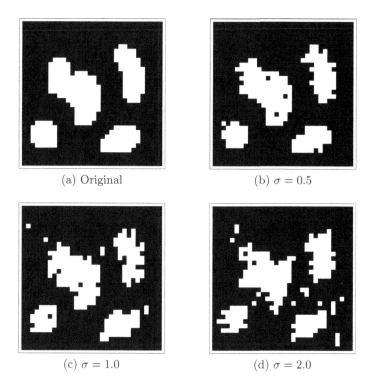

(a) Original

(b) $\sigma = 0.5$

(c) $\sigma = 1.0$

(d) $\sigma = 2.0$

Fig. 1. Reconstruction without deblurring fails. Panel (a) shows an object which was blurred with a Gaussian convolution kernel G_σ at three different scales $\sigma \in \{0.5, 1.0, 2.0\}$, and *then* projected along 5 directions $0°$, $22.5°$, $45°$, $67.5°, 90°$. Panels (b)-(d) show the reconstruction results *without* deblurring. The performance considerably deteriorates for increasing σ. Note that the original object (a) can be reconstructed without error from three projections only.

$$x_\mu^* = \operatorname*{argmin}_{x \in [0,1]^n} J_\mu(x) , \quad J_\mu(x) = D(x) + \alpha S(x) - \mu \frac{1}{2}\langle x, x - e \rangle \qquad (2)$$

where

$$D(x) = \frac{1}{2}\|Ax - b\|^2 \qquad (3)$$

and $S(x)$ is a convex smoothness prior (see section 3.2) which favors spatially homogeneous objects as reconstructions and $e := (1, 1, \ldots, 1)^\top \in \mathbb{R}^n$.

Problem (2) constitutes a numerically convenient relaxation of the combinatorial problem (1) because the set of feasible solutions $[0, 1]^n$ is convex. Starting with the global optimum x_0 of the *convex* functional J_0, the last term in (2) gradually enforces a locally optimal binary solution for increasing values of parameter μ. Although global optimality cannot be guaranteed, experimental results showed an excellent reconstruction performance [6, 7]. For further details of this framework and an overview from the optimization point of view, we refer to [8].

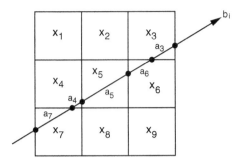

Fig. 2. Discretization model for transmission tomography. The measured projection data are given in terms of a vector $b \in \mathbb{R}^m$. Each component b_i corresponds to a projection ray measuring the absorption along the ray through the volume which is discretized into cells. The absorption a_j in each cell is assumed to be proportional to the density of the unknown object. $x_1, x_2, \ldots$ are binary variables indicating whether the corresponding cells belong to the object ($x_k = 1$) or not ($x_k = 0$). Assembling all projection rays into a linear system gives $Ax = b$, $x \in \{0, 1\}^n$, from which the unknown binary object, represented by x, has to be determined.

2.2   Binary Tomography with Blurred Data

Let G_σ denote the matrix that represents the linear mapping of some data by convolving it with an isotropic Gaussian kernel and scale-parameter σ. We generalize problem (1) along two directions:

Reconstruction from Projections of Blurred Objects
The corresponding generalization of the reconstruction problem (1) reads:

$$AG_\sigma x = b , \qquad A \in \mathbb{R}^{m \times n} , \; x \in \{0, 1\}^n , \; b \in \mathbb{R}^m \qquad (4)$$

Reconstruction from Blurred Projection Data
The corresponding generalization of the reconstruction problem (1) reads:

$$G_\sigma Ax = b , \qquad A \in \mathbb{R}^{m \times n} , \; x \in \{0, 1\}^n , \; b \in \mathbb{R}^m \qquad (5)$$

For notational simplicity, we used in both cases the same symbol G_σ, although G_σ denotes a block-circulant matrix in (4) corresponding to the convolution of multi-dimensional data x, whereas G_σ represents the one-dimensional convolution of the projection data in (5).

Accordingly, the variational approach (2) generalizes to

$$x_\mu^* = \operatorname*{argmin}_{x \in [0,1]^n} J_\mu(x; \sigma) , \quad J_\mu(x; \sigma) = D(x; \sigma) + \alpha S(x) - \mu \frac{1}{2} \langle x, x - e \rangle, \qquad (6)$$

where the data term $D(x; \sigma)$ indicates the dependency on the unknown convolution operator in (4) and (5), respectively.

3   Approach

3.1   Data Term and Scale Estimation

Optimization of J_μ in (6) is complicated through the unknown scale-parameter σ of the convolution operator G_σ. A common and natural approach to solve this problem is to apply the well-known Expectation-Maximization (EM) iteration (cf.,e.g. [9]) to the probabilistic interpretation of the data term $D(x;\sigma)$ as a likelihood term, provided this is computationally feasible. We elaborate this approach in this section.

We regard minimization of J_μ in (6) as Maximum-A-Posteriori (MAP) estimation of x, given the data b:

$$p(x|b) \propto \exp\left(-J_\mu(x;\sigma)\right) \propto p(b|x)p(x) \tag{7a}$$

$$p(x) \propto \exp\left(-\alpha S(x) + \mu\frac{1}{2}\langle x, x-e\rangle\right) \tag{7b}$$

Remark. The normalizing term missing in (7a) only depends on b and therefore it is unessential for estimating x.

The data likelihood $p(b|x)$ is unknown due to the dependency of the data term $D(x;\sigma)$ on the unknown parameter σ. Given some estimate $\hat{x}$, the standard EM-approach is then to maximize instead the following lower bound, commonly called Q-function, which does not depend on σ:

$$\log p(b|x) \geq \int_{\mathbb{R}_+} p(\sigma|b,\hat{x}) \log \frac{p(b,\sigma|x)}{p(\sigma|b,\hat{x})} d\sigma$$

Expanding the log-expression shows that only the first term, commonly called Q-function, depends on x and therefore is relevant:

$$Q(x|\hat{x},b) := \int_{\mathbb{R}_+} p(\sigma|b,\hat{x}) \log p(b,\sigma|x) d\sigma \tag{8}$$

To compute (8), the first term under the integral is evaluated via Bayes' rule

$$p(\sigma|b,\hat{x}) = \frac{p(b|\sigma,\hat{x})p(\sigma|\hat{x})}{p(b|\hat{x})}.$$

The denominator does not depend on σ and therefore it is unessential for marginalizing σ on the right in (8). The first term of the numerator is given by the data term $p(b|\sigma,\hat{x}) = Z^{-1}\exp(-D)$, where Z is a normalizing constant. Furthermore, it is reasonable to assume independency $p(\sigma|x) = p(\sigma)$. Thus, we obtain

$$p(\sigma|b,\hat{x}) \propto \frac{1}{Z}\exp\left(-D(\hat{x};\sigma)\right)p(\sigma). \tag{9}$$

For the second term under the integral in (8), we compute

$$\log p(b,\sigma|x) \propto \log p(b|\sigma,x) + \log p(\sigma) \propto -D(x;\sigma) + \log p(\sigma) \tag{10}$$

using again $p(\sigma|x) = p(\sigma)$, and dropping the normalizing constant of the first term on the right, as explained above in the remark after eqns. (7). Furthermore, we can drop the last term $\log p(\sigma)$ in (10) because it neither depends on x, nor does it contribute to the averaging of $D(x; \sigma)$ with respect to σ.

As a result, we insert the remaining term $-D(x; \sigma)$, together with (9), into (8) and denote the resulting expression again with Q:

$$Q(x|\hat{x}, b) := \int_{\mathbb{R}_+} \frac{1}{Z} \exp\big(-D(\hat{x}; \sigma)\big) p(\sigma)\big(-D(x; \sigma)\big) d\sigma \qquad (11)$$

This expression shows clearly how the unknown dependency on σ of the objective criterion (6) is dealt with: Given a current estimate $\hat{x}$ and a prior distribution $p(\sigma)$, the unknown data term $D(x; \sigma)$ is replaced by maximizing the average (11). Consequently, we replace the functional $J_\mu(x; \sigma)$ in (6) by the approximation

$$E_\mu(x; \hat{x}) := -Q(x|\hat{x}, b) + \alpha S(x) - \mu\frac{1}{2}\langle x, x - e\rangle. \qquad (12)$$

In practice, we choose the prior $p(\sigma)$ to be uniform within a reasonable interval $[\sigma_{\min}, \sigma_{\max}]$, and $\hat{x}$ is the current estimate on x. $Q(x|\hat{x}, b)$ is then evaluated by computing the one-dimensional integral (11) numerically using the trapezoidal rule.

3.2   Smoothness Term

As smoothness prior $S(x)$ in (12), we use a discrete approximation of the total-variation (TV) measure

$$\int_\Omega |\nabla x| d\Omega$$

of x (here temporarily regarded as a function), whose edge-preserving properties are well-known in image processing [10]. Recently, it has also been successfully used in connection with discrete tomography [11].

4   Optimization

The problem to minimize the functional $E_\mu(x; \hat{x})$ in (12) over the convex set of feasible solutions $B := [0, 1]^n$ can be written with a corresponding indicator function $I_B(x) = 0$ if $x \in B$ and $I_B(x) = +\infty$ if $x \notin B$, as

$$\inf_{x\in\mathbb{R}^n} E_\mu(x; \hat{x}) , \quad E_\mu(x; \hat{x}) = F(x; \hat{x}) - H_\mu(x), \qquad (13)$$

where

$$F(x; \hat{x}) := -Q(x|\hat{x}, b) + \alpha S(x) + I_B(x)$$

is a proper lower-semicontinuous convex functional, and where

$$H_\mu(x) := \mu\frac{1}{2}\langle x, x - e\rangle$$

is convex as well, thus concave when subtracted in (13). Therefore, a natural minimization approach is DC (Difference of Convex functions) programming [12, 13].

To specify the algorithm, recall the following definitions from convex analysis [14] for a function f:

$$\mathrm{dom}(f) := \{x \in \mathbb{R}^n \mid f(x) < +\infty\} \qquad \text{effective domain of } f$$

$$\partial f(\bar{x}) := \{v \mid f(x) \geq f(\bar{x}) + \langle v, x - \bar{x}\rangle\} \qquad \text{subdifferential of } f \text{ at } \bar{x}$$

$$f^*(y) := \sup_{x \in \mathbb{R}^n} \{\langle x, y\rangle - f(x)\} \qquad \text{conjugate function}$$

We apply to (13) the following algorithm adopted from [13]:

Subgradient Algorithm

Choose $x^0 \in \mathrm{dom}(F)$ arbitrary (this choice does not dependent on the second argument of F).

For $k = 0, 1, ...$ compute until convergence:

$$y^k \in \partial H_\mu(x^k) \tag{14}$$
$$x^{k+1} \in \partial F^*(y^k; \hat{x}) \tag{15}$$

The investigation of this algorithm in [13] includes the following results:

Proposition 1. *[13] Assume* $F(\cdot; \hat{x}), H_\mu : \mathbb{R}^n \to \overline{\mathbb{R}}$ *to be proper, lower-semicontinuous and convex, and* $\mathrm{dom}(F) \subset \mathrm{dom}(H_\mu)$, $\mathrm{dom}(H_\mu^*) \subset \mathrm{dom}(F^*)$. *Then*

(i) *the sequences* $\{x^k\}$, $\{y^k\}$ *according to the equations (15) and (14) are well-defined,*

(ii) $\{F(x^k; \hat{x}) - H_\mu(x^k)\}$ *is decreasing,*

(iii) *every limit point* x^* *of* $\{x^k\}$ *is a critical point of* $E_\mu(x; \hat{x}) = F(x; \hat{x}) - H_\mu(x)$.

Remarks

Concerning the full reconstruction algorithm, as listed on the subsequent page, we point out:

- Estimation of the unknown scale-parameter σ through the EM-iteration (cf. section 3.1) is done as part of step (15) – see lines 9-14 of the reconstruction algorithm listed on the following page.

- The global optimum of the convex optimization problem in line 11 of the reconstruction algorithm (cf. subsequent page) can be computed using any method. In our implementation, we used a dedicated algorithm [15] in view of the simple structure of the box-constraints $x \in [0, 1]^n$.

Reconstruction Algorithm

1 Choose x^0 arbitrary   (for example $x^0 := (\frac{1}{2}, ..., \frac{1}{2})^\top$)
2 Choose $\delta_\mu \in \mathbb{R}_+$ (our choice: $\delta_\mu \in (0, 0.5]$)
3 Choose $\epsilon > 0$ (our choice: $10^{-4} \leq \epsilon \leq 10^{-2}$)
4 Set $i := 0$, $\mu^0 := 0$
5 Do   (μ-loop)
6   Set $k := 0$
7   Do   (DC-loop)
8          $y^k := \nabla H_{\mu^i}(x^k)$
9   Set $l := 0$, $\hat{x}^0 := x^k$
10   Do   (EM-loop)
11             $\hat{x}^{l+1} := \underset{x \in [0,1]^n}{\operatorname{argmin}} \left\{ F(x; \hat{x}^l) - \langle y^k, x \rangle \right\}$
12             $l := l + 1$
13   while $||\hat{x}^l - \hat{x}^{l-1}||_2 > \epsilon$ (EM-loop)
14         $x^{k+1} := \hat{x}^l$
15         $k := k + 1$
15   while $||x^k - x^{k-1}||_2 > \epsilon$ (DC-loop)
16     $\mu^{i+1} := \mu^i + \delta_\mu$
17 while $\exists x_j^k \in [\epsilon, 1-\epsilon]$, $j = 1, \ldots, n$ (μ-loop)

5   Evaluation

5.1   Reconstruction from Projections of Blurred Objects

In figure 1, we showed that binary reconstruction fails in case of blurred objects. We repeated the experiment, however, this time taking deblurring into account. The results shown in figure 3 reveal that our novel reconstruction algorithm copes with both problems, deblurring by scale-parameter estimation and binary reconstruction, at the same time.

Further experiments showed, that the original object can be reconstructed even with four projections only ($0°$, $45°$, $90°$, and $135°$, for $\sigma = 1.0$).

5.2   Reconstruction from Blurred Projections

The upper-left image shown in figure 4 was projected along for directions $0°$, $45°$, $90°$, $135°$. Panel (b) shows these projections for illustration, and panel (d) the blurred version ($\sigma = 1.5$). The latter data was used to compute the reconstruction shown in panel (c). Panels (e) and (f) show the reconstructions for $\sigma = 1.0$ with and without deblurring, respectively. While the latter result clearly shows the ill-posedness of the combined deblurring-reconstruction problem, the results (c) and (e) demonstrate the stability of our new reconstruction algorithm even under such severe conditions.

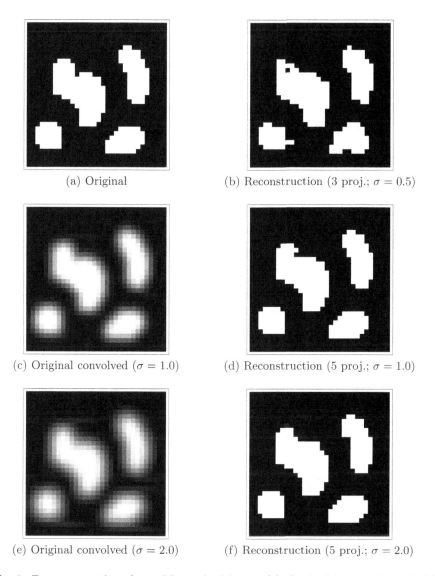

(a) Original

(b) Reconstruction (3 proj.; $\sigma = 0.5$)

(c) Original convolved ($\sigma = 1.0$)

(d) Reconstruction (5 proj.; $\sigma = 1.0$)

(e) Original convolved ($\sigma = 2.0$)

(f) Reconstruction (5 proj.; $\sigma = 2.0$)

Fig. 3. Reconstruction from blurred objects. (a) Original image, 32×32. (c) and (e): original image convolved with different Gaussian kernels, $\sigma \in \{1.0, 2.0\}$. 5 projections were taken for both images $0°$, $22.5°$, $45°$, $67.5°$, $90°$. Figures (d), and (f) show the corresponding results of our reconstruction algorithm. Since we obtained for $\sigma = 0.5$ the original image we present in this case the reconstruction from only three projections, $0°$, $45°$, and $90°$. Throughout the experiments the smoothing parameter α was set to 0.01.

To illustrate the deblurring process further, figure 5 depicts the expressions $\exp(-D(\hat{x}; \sigma))/Z$ and $D(\hat{x}; \sigma)$, respectively, as a function of σ during the reconstruction process. It can be clearly seen that the former expression peaks most

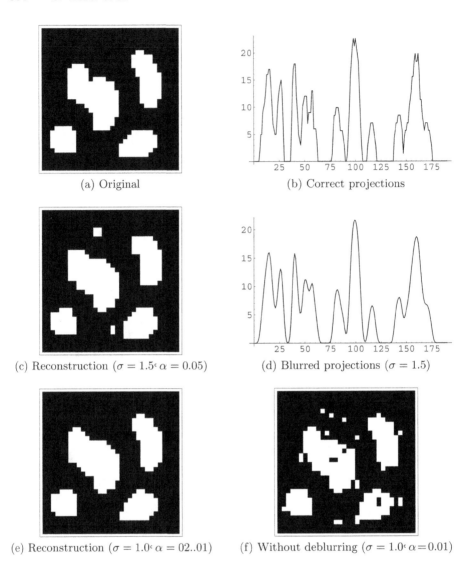

(a) Original

(b) Correct projections

(c) Reconstruction ($\sigma = 1.5$, $\alpha = 0.05$)

(d) Blurred projections ($\sigma = 1.5$)

(e) Reconstruction ($\sigma = 1.0$, $\alpha = 02..01$)

(f) Without deblurring ($\sigma = 1.0$, $\alpha = 0.01$)

Fig. 4. Reconstruction from blurred projections. Projections at $0°$, $45°$, $90°$, and $135°$ were taken from the image shown in panel (a) and convolved with a Gaussian kernel, $\sigma = 1.5$. Panels (b) and (d) show the correct projections and the blurred projections, respectively. Panel (c) shows the reconstruction result ($\alpha = 0.05$). Panel (e) shows the reconstruction from projection data that were blurred with $\sigma = 1.0$. Panel (f) shows the erroneous reconstruction without taking deblurring into account.

around the correct value $\sigma = 1.5$, whereas the latter term attains its global minimum there.

The experiments also revealed that reconstruction from blurred projections is more difficult that reconstruction from projections of blurred objects, as in the

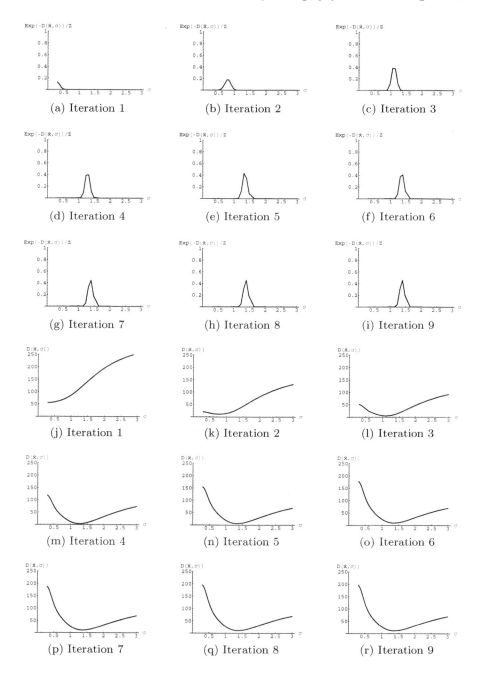

Fig. 5. Upper half: The term $\exp(-D(\hat{x};\sigma))/Z$ as a function of σ during the iteration. **Lower half:** The term $D(\hat{x};\sigma)$ as a function of σ during the iteration. While the former term peaks most near the correct value $\sigma = 1.5$, the latter attains its global minimum there. This illustrates that the inner EM-loop of the overall reconstruction algorithm is well-defined and robust.

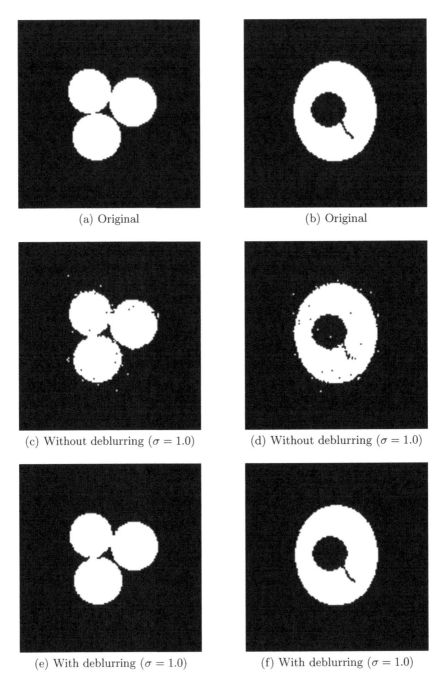

(a) Original

(b) Original

(c) Without deblurring ($\sigma = 1.0$)

(d) Without deblurring ($\sigma = 1.0$)

(e) With deblurring ($\sigma = 1.0$)

(f) With deblurring ($\sigma = 1.0$)

Fig. 6. Reconstruction from blurred projections. (a),(b) Original image, 128 × 128. For both images, reconstruction problems were set up using 5 projections, 0°, 36°, 72°, 108°, and 144°, and blurring these projections with a Gaussian kernel, $\sigma = 1.0$. (c),(d) Reconstruction without deblurring. (e),(f) Reconstruction with deblurring.

former section: Perfect reconstruction of the original object was possible for a smaller blurring scale only ($\sigma = 0.8$).

6   Conclusion and Further Work

We extended our reconstruction algorithm for binary tomography with an Expectation-Maximization (EM) step to improve its behavior in the presence of degradations during data acquisition. For evaluation purposes we defined two different degradation models. The same reconstruction algorithm can be applied to either of them which accurately estimates an unknown scale-parameter σ, during the reconstruction. Our results show that our approach stabilizes the reconstruction process in the presence of degradations.

Regarding the Q function in the EM-step, further work includes an adaptive sampling strategy of the supporting points. This is important for two reasons: First, it is expected to produce a more accurate approximation of the integral especially in areas where the true σ is suspected. Second, it should also reduce the number of supporting points since we can skip areas which are of low interest. The latter should further speed up our algorithm.

We suppose that our approach is sufficiently general to be applied to other combined reconstruction – missing parameter estimation scenarios as well. This will also be subject to our future work.

Acknowledgments

The third author was supported during his stay in Mannheim by the Alexander von Humboldt-Foundation. This work was partially supported by the NSF Grant DMS 0306215 and the OTKA Grant T048476.

The CVGPR-group gratefully ackowledges support from Siemens Medical Solutions, Forchheim, Germany.

References

1. Gonzales, R.C., Woods, R.E.: Digital Image Processing. 2 edn. Addison Wesley, Reading, MA (1992)
2. Herman, G.T., Kuba, A., eds.: Discrete Tomography: Foundations, Algorithms, and Applications. Birkhäuser Boston (1999)
3. Balaskó, M., Kuba, A., Nagy, A., Kiss, Z., Rodek, L., Ruskó, L.: Neutron-, gamma-, and X-ray three-dimensional computed tomography at the Budapest research reactor site. Nuclear Instruments and Methods in Physics Research A, **542** (2005) 22–27
4. Kuba, A., Ruskó, L., Rodek, L., Kiss, Z.: Preliminary sudies of discrete tomography in neutron imaging. IEEE Trans. Nucl. Sci. **NS-52** (2005) 380–385
5. Carazo, J.M., Sorzano, C.O., Rietzel, E., Schröder, R., Marabini, R.: Discrete tomography in electron microscopy. In: Discrete Tomography. Foundations, Algorithms, and Applications (Eds.: Herman, G. T., Kuba, A.). Birkhäuser, Boston, MA (1999) 405–416

6. Schüle, T., Schnörr, C., Weber, S., Hornegger, J.: Discrete tomography by convex-concave regularization and d.c. programming. Discrete Applied Mathematics **151** (2005) 229–243

7. Weber, S., Schüle, T., Schnörr, C., Hornegger, J.: A linear programming approach to limited angle 3d reconstruction from dsa projections. Special Issue of Methods of Information in Medicine **4** (2004) 320–326

8. Schnörr, C., Schüle, T., Weber, S.: Variational Reconstruction with DC-Programming. In Herman, G.T., Kuba, A., eds.: Advances in Discrete Tomography and Its Applications. Birkhäuser, Boston (2006) To appear.

9. MacLachlan, G., Krishnan, T.: The EM Algorithm and Extensions. Wiley (1996)

10. Rudin, L., Osher, S., Fatemi, E.: Nonlinear total variation based noise removal algorithms. Physica D **60** (1992) 259–268

11. Capricelli, T., Combettes, P.: Parallel block-iterative reconstruction algorithms for binary tomography. Electr. Notes in Discr. Math. **20** (2005) 263–280

12. Pham Dinh, T., Elbernoussi, S.: Duality in d.c. (difference of convex functions) optimization subgradient methods. In: Trends in Mathematical Optimization, Int. Series of Numer. Math. Volume 84. Birkhäuser Verlag, Basel (1988) 277–293

13. Pham Dinh, T., Hoai An, L.T.: A d.c. optimization algorithm for solving the trust-region subproblem. SIAM J. Optim. **8**(2) (1998) 476–505

14. Rockafellar, R.T.: Convex analysis. 2 edn. Princeton Univ. Press, Princeton, NJ (1972)

15. Birgin, E.G., Martínez, J.M., Raydan, M.: Algorithm 813: SPG - software for convex-constrained optimization. ACM Transactions on Mathematical Software **27** (2001) 340–349

A Neural Network Approach to Real-Time Discrete Tomography

K.J. Batenburg[1,2] and W.A. Kosters[1]

[1] Leiden University, Leiden, The Netherlands
kosters@liacs.nl
[2] CWI, Amsterdam, The Netherlands
K.J.Batenburg@cwi.nl

Abstract. *Tomography* deals with the reconstruction of the density distribution inside an unknown object from its projections in several directions. In *Discrete tomography* one focuses on the reconstruction of objects having a small, discrete set of density values. Using this prior knowledge in the reconstruction algorithm may vastly reduce the number of projections that is required to obtain high quality reconstructions.

Recently the first generation of *real-time* tomographic scanners has appeared, capable of acquiring several images per second. Discrete tomography is well suited for real-time operation, as only few projections are required, reducing scanning time. However, for efficient real-time operation an extremely fast reconstruction algorithm is also required.

In this paper we present a new reconstruction method, which is based on a feed-forward neural network. The network can compute reconstructions extremely fast, making it suitable for real-time tomography. Our experimental results demonstrate that the approach achieves good reconstruction quality.

1   Introduction

Tomography deals with the reconstruction of the density distribution inside an unknown object from its projections in several directions [8]. Figure 1 shows the basic principle. In this paper we look at *transmission tomography*, where the projections are obtained by sending a beam (e.g., X-rays, neutrons, etc.) through the object and measuring the attenuated beam that has passed through the object. Tomography is used extensively in medical imaging, industrial imaging and, more recently, in materials science and biology. Typically, a large number of projections is required (more than 100) to obtain good reconstruction quality.

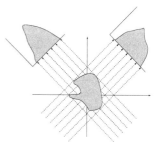

Fig. 1. Basic principle of tomography, 2 projections

When it is known in advance that the scanned object consists of only a few different materials, it may be possible to vastly reduce the number of required

U. Eckardt et al. (Eds.): IWCIA 2006, LNCS 4040, pp. 389–403, 2006.
© Springer-Verlag Berlin Heidelberg 2006

projections by using this prior knowledge in the reconstruction algorithm. *Discrete tomography* focuses on the tomographic reconstruction of objects for which the set of pixel values in the reconstructed image is discrete and small. In particular, the reconstruction of binary images (i.e., black-and-white images) has received considerable attention [6].

Most tomographic scanners acquire *static images*, i.e., a single image, either 2D or 3D, of the object is reconstructed from the projection data. Therefore it is not possible to do imaging of *dynamic processes*, where the scanned object changes significantly in a short period of time. Recently, real-time medical tomographic scanners have emerged [9], which are capable of acquiring several images per second. Besides medical imaging, real-time tomography could prove very useful in industrial imaging.

Discrete tomography is well suited for real-time imaging, since the small number of required projections results in a substantial reduction of the scanning time. However, to compute a long series of reconstructions in reasonable time, a very fast reconstruction algorithm is required. Several authors have proposed algorithms for discrete tomography, usually for the reconstruction of binary images. All these algorithms require at least several dozens of seconds to reconstruct a single 2D 256×256 image [1, 2, 12, 13].

In this paper we present a new reconstruction method, which is based on a feed-forward neural network. The neural network is first *trained* on a set of representative images, which may take a substantial amount of time. After the training phase, the network can be used to compute a reconstruction very fast. When implemented on a *Field Programmable Gate Array* (FPGA), a piece of computer hardware, frame rates of several hundreds per second are realistic.

We focus on the reconstruction of binary images from parallel projections. Additional prior knowledge other than the binary constraint that is present in the training set, is learned by the neural network during the training phase, so it does not have to be modelled explicitly. Besides real-time tomography, our approach can also be used to compute a good start solution for more accurate, time-consuming reconstruction algorithms.

Neural network reconstruction methods for other types of tomographic reconstruction have been considered in the literature, e.g., [10, 11]. Neural networks are not well suited for general transmission tomography from many projections, as the number of variables in the reconstruction problem is extremely large. Besides that, it is very difficult to outperform other available approaches. In discrete tomography the amount of projection data is much smaller, making the reconstruction problem underdetermined. Neural networks are well known for their ability to learn additional prior knowledge, which makes them suitable for discrete tomography.

Section 2 contains a short description of the tomographic reconstruction problem. In Section 3 we first propose a basic neural network approach and discuss its abilities and limitations. Subsequently we refine the approach, obtaining a so-called single-pixel neural network architecture that is capable of computing real-time reconstructions of large images. In Section 4 we provide experimental results of the neural network approach and a brief comparison with a continuous tomography algorithm. Section 5 concludes.

2   Reconstruction Problem

Figure 2 shows the main setting of the binary tomography problem. We assume that the object of interest is contained in the disc

$$A = \{(x, y) \in \mathbb{R}^2 : x^2 + y^2 \leq R^2\}$$

with radius R. We call this disc the *imaging area*. For the sake of convenience we assume that R is a positive integer.

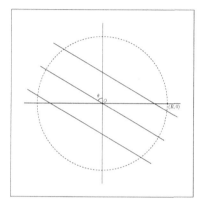

Fig. 2. Basic setting for the tomography problem in disc A; the angle between the parallel beam and the y-axis is denoted by θ

The *unknown binary image* that we would like to reconstruct is considered as a mapping $f : \mathbb{R}^2 \to \{0, 1\}$, where 0 is black and 1 is white. We assume that the support of f, i.e., the set $\{(x, y) \in \mathbb{R}^2 : f(x, y) = 1\}$, is a measurable set that is contained in A. Define the function $T_\theta : \mathbb{R}^2 \to \mathbb{R}$ as follows:

$$T_\theta(x, y) = x \cos \theta + y \sin \theta.$$

We call $T_\theta(x, y)$ the *point projection of* (x, y) *for angle* θ. Projections are measured along lines $L_{\theta,t}$ of the form

$$L_{\theta,t} = \{(x, y) \in \mathbb{R}^2 : T_\theta(x, y) = t\}.$$

The *Radon transform* P_f of f is defined (cf. [5], where also Radon's original paper is reproduced) as

$$P_f(\theta, t) = \int_{L_{\theta,t}} f(x, y) \, ds \qquad \text{for } \theta \in [0, 2\pi), \, t \in \mathbb{R}.$$

We can now formulate the main reconstruction problem, where the parameter n determines the number of angles:

Problem 1. Let $D = \{\theta_1, \theta_2, \ldots, \theta_n\}$ be a given set of projection angles with $0 \leq \theta_1 < \theta_2 < \ldots < \theta_n < \pi$, and let $\phi_1, \phi_2, \ldots, \phi_n$ be given functions (the measured projections) from $\mathbb{R}$ to $\mathbb{R}$. Construct a function $f : \mathbb{R}^2 \to \{0, 1\}$ such that $P_f(\theta_i, \cdot) = \phi_i(\cdot)$ for $i = 1, 2, \ldots, n$.

When the measured projections are obtained through physical measurements, Problem 1 usually does not have an exact solution. Even if the reconstruction problem has an exact solution theoretically, we need to approximate its solution by representing it on a pixel grid.

In practice, the function $P_f(\theta, \cdot)$ is usually not measured in single points t. Instead, the total projection in a strip, covering a small t-interval (t_ℓ, t_r), is measured as

$$S_{\theta,f}(t_\ell, t_r) = \int_{t=t_\ell}^{t_r} P_f(\theta, t)\, \mathrm{d}t.$$

Typically, the value $S_{\theta,f}(t_\ell, t_r)$ is measured for consecutive strips of fixed width. Without loss of generalization we assume that all these strips have width 1. For any angle θ, $2R$ strip projections are measured. The first strip corresponds to the t-interval $(-R, -R+1)$, the last strip to $(R-1, R)$. In our neural network approach we also need to evaluate $S_{\theta,f}(t_\ell, t_r)$ for other values of (t_ℓ, t_r), often with integer width $t_r - t_\ell$. These values are computed by linear interpolation of the measured projection data.

Although the reconstruction problem is defined using the projection data, the performance of reconstruction algorithms is often evaluated by considering a *known* image f and its projections $P_{\theta_1,f}, P_{\theta_2,f}, \ldots, P_{\theta_k,f}$, and comparing the reconstruction to the original image f. In practice, resemblance to the original image is often more important than perfect correspondence to the projection data. This is particularly important if the projection data by itself is not enough to determine the image f and additional prior knowledge must be used in the reconstruction algorithm, which is the case if the number of projection angles n is relatively small: the problem is then underdetermined.

3   Neural Network Approach

In this section we will discuss two neural network approaches to the discrete tomography problem. A *feed-forward neural network* consists of neurons, grouped in layers, where neurons from one layer can have a weighted connection to neurons from the next layer. The weights are trained simultaneously, hopefully toward optimal values, by presenting the network with correct input-output pairs.

Both proposed networks are feed-forward back-propagation networks (see, e.g., [4, 7]) with one input layer, one hidden layer and one output layer. The networks are fully connected. The first and probably most obvious version (referred to as a *full-image network*) has one output node for each pixel. The second version (a so-called *single-pixel network*) has only one output node, to reconstruct one pixel, but can be used — through appropriate adaptation — to reconstruct the whole image. This type of network has several advantages over the full-image network.

3.1 A Full-Image Network for Tomography

The input nodes of the network contain the values of the projections, the output nodes contain the pixel values. So the number of input nodes equals the number of projections, while the number of output nodes equals the size of the imaging area, i.e., approximately πR^2. The hidden nodes are connected to all input nodes and all output nodes. Output values are interpreted as gray values, yielding the gray level reconstruction. If necessary, these values can be rounded in a post-processing step to 0/1-values for a "crisp" or "rounded" reconstruction. These networks were first introduced and examined in [3].

Training is performed as follows. The input pattern, consisting of the projection values, is offered to the input nodes. Every connection has a real-valued weight, that is adapted during training. The hidden nodes receive the weighted sum of their incoming connections, and generate an output through the standard sigmoid $\sigma : x \mapsto 1/(1 + e^{-x})$. Output nodes operate in a similar way. In each epoch, a number (50,000, say) of random images (sampled from a certain distribution) with their projections are presented to the network; after each epoch the learning rate α is somewhat decreased. Note that samples are used only once, unless they are by chance regenerated.

The weights are adapted using the normal back-propagation rule. A weight w_{ji} from hidden node j to output node i is adapted through

$$w_{ji} \leftarrow w_{ji} + \alpha \cdot a_j \cdot \Delta_i, \quad \Delta_i = \sigma'(\mathrm{in}_i) \cdot (t_i - n_i).$$

Here a_j is the output of node j, $\mathrm{in}_i = \sum_j w_{ji} a_j$ is the weighted input to node i, $n_i = \sigma(\mathrm{in}_i)$ is its output (for output nodes this is the net output) and t_i is the desired target value, i.e., the true pixel value. The update rule for weight w_{kj} from input node k to hidden node j is a little more complicated (cf. [4, 7]):

$$w_{kj} \leftarrow w_{kj} + \alpha \cdot a_k \cdot \Delta_j, \quad \Delta_j = \sigma'(\mathrm{in}_j) \cdot \sum_i w_{ji} \Delta_i.$$

As usual, one extra input node and one extra hidden node clamped to -1 are added, the so-called *bias nodes*.

In [3] hidden nodes with only a small number of connections (so-called *local nodes*, as opposed to the more common *global nodes* mentioned above) are added. These local nodes are connected to a few input nodes and output nodes; they keep track of the constraints that affect a pixel and its immediate neighbours. Each local node corresponds with a unique pixel, receives input from the line projections that intersect with that pixel, and is connected to the 9 output nodes corresponding with the pixel and its immediate neighbours (6 or 4 near the boundaries). This general network architecture is depicted in Figure 3. Though this type of network was shown to perhaps have some advantages, in the sequel we will for comparison purposes just report on the version with only global nodes.

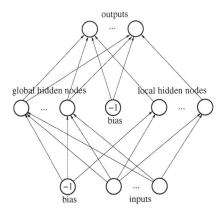

Fig. 3. General structure of the full-image network. Global hidden nodes are connected to all input nodes and all output nodes. Local hidden nodes are connected to only a few input and output nodes.

3.2   A More Efficient Architecture: The Single-Pixel Network

Although the network from Section 3.1 is suitable for real-time reconstruction once the training phase is complete, it has some disadvantages:

- The network contains a large number of input/output nodes; a large number of hidden nodes is required to obtain reasonable reconstructions. Due to the large number of nodes and connections between them, training the network takes a very long time.
- Millions of training images and their projections are required to train the network. In practical applications it is usually impossible to obtain such large data sets.

In the sequel we propose an improvement by focussing on the reconstruction of a single pixel, instead of the whole image. This vastly reduces the number of hidden to output connections.

Reconstructing a Single Pixel

One of the principal goals of our neural network design is a reduction of the number of input nodes in comparison to the network from Section 3.1. When reconstructing a single pixel $p = (x_p, y_p)$ within the imaging area A, it is clear that projected lines that pass through p are more important for determining its value than the other projected lines. Also, if we assume that the image is locally smooth, projected lines that pass near p are more important than lines that pass far away from this pixel, as neighbouring pixels of p are highly relevant to the value of p.

We use this intuitive notion of relative importance between projected lines to preprocess the projection data. The inputs of the neural network from Section 3.1

correspond directly to the measured projection values. In the new single-pixel network, each input corresponds to a *strip projection*. Strips that are far away from p are much broader than strips near p.

Let k be a positive integer. Let $0 < d_0 \leq d_1 \leq \ldots \leq d_k$ be real values, the *strip sizes*. We require that the strip sizes satisfy

$$d_0/2 + \sum_{i=1}^{k} d_i \geq 2R. \tag{1}$$

Define the *strip boundaries* $s_{-k}, \ldots, s_0, \ldots, s_{k+1}$ as follows:

$$s_0 = -d_0/2;$$
$$s_i = s_{i-1} + d_{i-1} \quad (i = 1, \ldots, k+1);$$
$$s_i = s_{i+1} - d_{-i} \quad (i = -k, \ldots, -1).$$

Put $\tau_\theta = T_\theta(x_p, y_p)$, the point projection of p for angle θ. The set I_θ of *input strips* for angle θ can now be defined as

$$I_\theta = \bigcup_{i=-k}^{k} \{(\theta, \tau_\theta + s_i, \tau_\theta + s_{i+1})\}.$$

Each element of I_θ is a 3-tuple, consisting of the angle θ and the left and right boundary of a t-interval, which jointly define a strip through the imaging area. The constraint in Equation (1) ensures that the strips in I_θ cover at least the entire imaging area A, independent of the position of p. Given the angles $\theta_1, \theta_2, \ldots, \theta_n$, define the set I of all input strips as $I = \bigcup_{i=1}^{n} I_{\theta_i}$.

For every triple $(\theta, t_\ell, t_r) \in I$ there is an input node in the neural network, giving a grand total of $(2k+1) \cdot n$ input nodes. The input for such a node is $S_{\theta, f}(t_\ell, t_r)$, the strip projection for angle θ in the t-interval (t_ℓ, t_r).

Figure 4 shows three possible choices for the set $\{d_0, d_1, \ldots, d_k\}$ of strip sizes. Setting $d_0 = d_1 = \ldots = d_k$ yields equally spaced strips (Figure 4a). Setting $d_0 = 1$ and $d_i = i$ for $i \geq 1$ yields a set of strips for which the size increases linearly as the distance from the pixel p increases (Figure 4b). In Section 4 we will show that even if we set $d_0 = 1$, $d_i = 2^{i-1}$ for $i \geq 1$ (Figure 4c), the results are still satisfactory. Using strip sizes that grow exponentially with the distance to p yields a large reduction in the number of inputs of the neural network.

Training the single-pixel network proceeds as in the case of the full-image network. However, training a separate network for each pixel would take a huge amount of time. Also, the problem that the training requires a large number of training examples remains. In the sequel we will show how both problems can be solved.

Reconstructing All Pixels

Although the binning approach from the previous section drastically reduces the number of inputs of the network, a large number of images is still required to

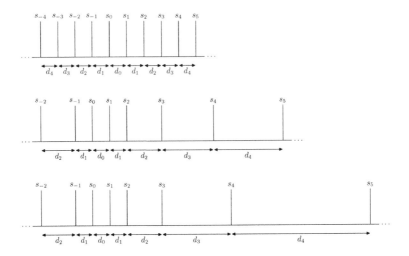

Fig. 4. Three different strip size configurations; a, top: constant, b, middle: linear, c, bottom: exponential

perform the training. If a separate network needs to be trained for each pixel in the image, the total training time may be even larger than for the basic network from Section 3.1.

Note that all inputs of the single-pixel network are *relative* to the projections of the center of the pixel. Therefore, there is no obvious reason why the network cannot be used to reconstruct a *different* pixel, elsewhere in the image. The only difference between two different pixels is that the relative position of the imaging area A is different. However, if we use a *varying* set of pixels from the training images, instead of using the same single pixel from each image, it may be possible to train the network *without* providing additional information on the relative position of the imaging area. This reconstruction task is harder, since less information is offered to the network. In Section 4 we show that one single-pixel network is capable of reconstructing arbitrarily positioned pixels. This offers some major advantages over the network from Section 3.1:

- If exponentially increasing strip sizes are used (see Figure 4c) both the number of inputs (logarithmic in R) and the number of outputs (constant, 1) is vastly reduced when compared to the network from Section 3.1, reducing training time for the single-pixel network.
- Only a single network has to be trained, instead of a new network for each pixel.
- Each training image, with its projections, now yields a new training example for each pixel in the image. The network from Section 3.1 requires a new image for each training example.

And finally, as we shall see in Section 4, the single-pixel networks seem more capable of reconstructing images from their projections, rather than learning images from certain classes.

3.3   Computational Complexity

For a feed-forward neural network having N_I input nodes, N_H hidden nodes and N_O output nodes, the time required to propagate an input pattern to the output nodes is $O(N_I N_H + N_H N_O)$. The time complexity for training a single input-output pattern is of the same order.

For the full-image network from Section 3.1 we have $N_I = O(nR)$ and $N_O = O(R^2)$, so that propagating a pattern takes $O(N_H(nR + R^2))$ time. (Remember that n is the number of projection angles or directions, and R is the radius of the imaging area.) Typically R is much larger than n. Once the training phase is complete, the time complexity of computing a reconstructed image from a set of projection data is of this same order.

For the single-pixel network with logarithmic strip sizes from Section 3.2 we have $N_I = O(n \log R)$ and $N_O = 1$, yielding a time complexity of $O(n N_H \log R)$ for propagating a pattern. To use the network for reconstruction after the training phase, a new input projection pattern has to be propagated through the network for every pixel in the image, yielding a time complexity of $O(n N_H R^2 \log R)$. Also, for this network the projection data must be preprocessed first to obtain the input data for the network, with time complexity $O(R)$. Note that the number N_H can vary heavily between different types of networks.

Since the value that is computed at each (non-input) node of a feed-forward network only depends on the values in the previous layer, the values for all nodes in a layer can be computed in parallel. Moreover, the computation that needs to be performed at each node is very simple. This allows for very efficient parallel implementations.

4   Experimental Results

In this section we present experimental results for the full-image network from Section 3.1 and the single-pixel network from Section 3.2. We restrict ourselves to two classes of synthetic images.

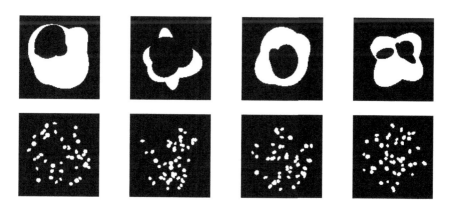

Fig. 5. Top: 128×128 images from the 7-class; bottom: images from the 50-class

The first class of examples consists of images with four large white ellipses and three smaller black ones inside, in a dark background. This class is referred to as the *7-class*, or just "7". The second class consists of images with fifty small white ellipses in a dark background, and is referred to as the *50-class*, or just "50". Figure 5 contains some examples.

Table 1. Results from experiments for the full-image network; 32 × 32 and 64 × 64 images; training runtime in hours

image width	angles	image class	hidden nodes	gray error	0–1 error	run-time
				average over 3 runs		
32	4	7	50	0.074	0.052	2.15
32	4	7	100	0.058	0.042	5.05
32	4	7	200	0.046	0.044	9.85
32	4	50	50	0.069	0.044	2.15
32	4	50	100	0.056	0.037	5.05
32	4	50	200	0.054	0.036	10.00
32	10	7	50	0.104	0.076	3.30
32	10	7	100	0.066	0.050	6.30
32	10	7	200	0.040	0.040	12.10
32	10	50	50	0.065	0.042	3.40
32	10	50	100	0.044	0.030	6.35
32	10	50	200	0.019	0.017	12.15
64	4	7	50	0.148	0.109	12.80
64	4	7	100	0.118	0.092	27.30
64	4	7	200	0.063	0.062	58.30
64	4	50	50	0.145	0.098	12.75
64	4	50	100	0.131	0.091	27.20
64	4	50	200	0.126	0.087	58.00
64	10	7	50	0.199	0.151	14.80
64	10	7	100	0.145	0.112	30.30
64	10	7	200	0.078	0.077	63.15
64	10	50	50	0.142	0.097	14.80
64	10	50	100	0.118	0.086	30.25
64	10	50	200	0.077	0.077	63.35

All experiments were repeated three times, and averages were taken over these three runs. Because all images are in fact located within a circle (the imaging area), we do not consider errors outside this area; therefore, mean error values are computed with respect to pixels within the imaging area. Average absolute errors are reported on independent test sets consisting of 1,000 images, both for gray level reconstruction and for rounded reconstruction (the 0–1 error). Table 1 shows results for the full-image network for two image sizes: 32 × 32 and 64 × 64. Table 2 gives results for the single-pixel network for three image sizes: 32 × 32, 64 × 64 and 128 × 128. The parameters are as follows: three sizes of the hidden layer: 50, 100 and 200 hidden nodes; and two different sets of projection angles,

Table 2. Results from experiments for the single-pixel network; 32×32, 64×64 and 128×128 images; training runtime in hours; (*) contains run with 0–1 error equal to 0

image width	angles	image class	hidden nodes	gray error	0–1 error	run-time
				average over 3 runs		
32	4	7	50	0.038	0.026	3.05
32	4	7	100	0.038	0.025	4.95
32	4	7	200	0.037	0.025	9.65
32	4	50	50	0.054	0.034	3.10
32	4	50	100	0.054	0.034	4.95
32	4	50	200	0.054	0.034	9.55
32	10	7	50	0.004	0.003	6.95
32	10	7	100	0.004	0.002(*)	12.20
32	10	7	200	0.004	0.003	21.65
32	10	50	50	0.016	0.011	6.95
32	10	50	100	0.015	0.011	12.10
32	10	50	200	0.015	0.011	21.50
64	4	7	50	0.046	0.029	3.70
64	4	7	100	0.044	0.032	5.80
64	4	7	200	0.044	0.029	11.10
64	4	50	50	0.116	0.083	3.70
64	4	50	100	0.114	0.083	5.80
64	4	50	200	0.118	0.082	11.15
64	10	7	50	0.006	0.005	8.95
64	10	7	100	0.006	0.005	14.75
64	10	7	200	0.006	0.005	25.20
64	10	50	50	0.051	0.034	9.00
64	10	50	100	0.052	0.033	14.65
64	10	50	200	0.051	0.033	25.20
128	4	7	50	0.042	0.028	4.75
128	4	7	100	0.041	0.029	7.15
128	4	7	200	0.039	0.028	13.35
128	4	50	50	0.130	0.094	4.70
128	4	50	100	0.129	0.092	7.10
128	4	50	200	0.129	0.100	13.40
128	10	7	50	0.005	0.003	12.20
128	10	7	100	0.005	0.002	18.30
128	10	7	200	0.005	0.003	30.50
128	10	50	50	0.057	0.038	12.20
128	10	50	100	0.056	0.039	18.15
128	10	50	200	0.055	0.040	30.45

consisting of 4 and 10 projections, equally spaced in the interval $[0°, 180°)$. In all experiments 200 epochs, each consisting of 50,000 examples (full-pixel network) or 2,300,000 examples (single-pixel network), were used for training. The number of examples per epoch was chosen so that the training of the two networks took the same amount of time for the case of 32×32 images from the 7-class using 4 projections and 100 hidden nodes. The reason for using more examples to train the single-pixel network is that each example only contains information on a single pixel in that case. When using a real measured dataset, a new training example can be constructed from each pixel in the dataset. Experiments were run on a single processor AMD Athlon 2.2 GHz PC. Runtimes in the tables refer to the *training times* of the networks; once trained, reconstruction of a new image is almost instantaneously. The learning rate α started at 0.5, and was multiplied by 0.99 after each epoch.

Experiments with the full-image network show that the network is capable of reconstructing small images with acceptable quality, e.g., 32×32 images. This requires a reasonably large number of hidden nodes, e.g., 200, giving a huge number of connections. For larger images however, quality drops down, while computing time increases heavily (therefore, no experiments on 128×128 images were performed). The results suggest that a further increase in the number of hidden nodes might improve reconstruction quality. Figure 6 shows some examples from a run of the full-image network for 10 projections on the 7-class for 64×64 images, using 100 hidden nodes. The average absolute pixel error for this particular run (using 1,000,000 training examples) was 0.133 (gray level reconstruction) and 0.102 (rounded reconstruction).

Fig. 6. From left to right: 64×64 original, gray level reconstruction and rounded reconstruction using a full-image network, with absolute total errors 413.7 and 291, respectively

As another example, we show results for 64×64 images within the 50-class, using 10 projections, and 100 hidden nodes, see Figure 7. The figure shows best and worst reconstruction from a random set of 25 images from the 50-class.

Fig. 7. From left to right: two pairs of 64×64 original and rounded reconstruction using a full-image network, with absolute total errors 235 and 310, respectively

Clearly, the results for the 50-class are not satisfactory. As we can see in Table 2, single-pixel networks give much better reconstructions. We now consider the single-pixel network exclusively.

For all test cases in Table 2 the number of hidden nodes has hardly any effect on the quality of the reconstructions. This suggests that a relatively small number of hidden nodes suffices. Structural Risk Minimization might be used to find a suitable size for the hidden layer. The 50-class is clearly more difficult to learn than the 7-class.

Figure 8 compares the single-pixel network and filtered backprojection (FBP), which is the fastest algorithm available for continuous tomography and the only one that can be used for real-time reconstruction. The FBP reconstructions were computed with the MATLAB Imaging Toolbox, using the Ram-Lak filter multiplied by a Hann window. The single-pixel reconstructions are clearly better. Other discrete tomography algorithms might be capable of producing more accurate reconstructions, perhaps even using fewer projections. However, to our knowledge these algorithms are far too slow for real-time reconstruction.

Fig. 8. From top to bottom: three 128 × 128 images and their reconstructions using filtered backprojection, rounded filtered backprojection and the single-pixel network, respectively; top and middle image from the 7-class, with 10 projections; bottom image from the 50-class, with 18 projections. Absolute 0–1 errors for the single-pixel network reconstructions are 101, 84 and 153, respectively.

A natural question to ask is whether the network is capable of reconstructing images outside the class it was trained on. Figure 9 shows the reconstruction results of two such images, which are clearly not in any of the training classes. Though not perfect, the results are surprisingly good.

Though results from [3] suggest that local nodes might give some improvement for the full-pixel network (putting more complexity into the network), this was only shown for relatively small images. In the current paper we have chosen

Fig. 9. From left to right: two pairs of a 128×128 image and its reconstruction; original images are *not* from the 7-class for which the single-pixel network was trained. Absolute 0–1 errors are 189 and 291, respectively.

Fig. 10. From left to right: 32×32 original and reconstruction using a full-image network and a single-pixel network, respectively; only one projection

for global nodes only, to allow for better comparison between full-image and single-pixel networks.

Another issue with the full-image networks is that they are more inclined to learn location dependent information, in particular when there are very few projections. In that extraordinary case they behave quite differently from the single-pixel networks. As an example, in Figure 10 we show some results for the reconstruction of a 32×32 image using only projection in one horizontal direction. The single-pixel reconstruction shows the density distribution of the 0–1 intensity in the projection direction, as any two pixels on the same horizontal line cannot be distinguished by the network. The imaging area is clearly visible.

As mentioned before, the final 0–1 image is generated from the gray level reconstruction by simply rounding, giving a crisp figure, cf. Figure 6. Experiments suggest that errors often occur for pixels that have a raw reconstruction value near 0.5. It is possible to slightly improve the final reconstructed image by (for those pixels, in parallel) trying both 0 and 1 as reconstruction value, meanwhile comparing with the projections. Time restrictions clearly allow just a few pixels to be toggled simultaneously. The quality of the reconstruction may also be improved by introducing a stochastic model of the image class and computing an image (in a postprocessing step) that corresponds well with both the output of the neural netwerk and the model of the image class.

5   Conclusions

We conclude that the single-pixel network from Section 3.2 is capable of generating very good quality reconstructions of images from both classes, given sufficiently many projection directions. Once trained, such a network can compute new reconstructions almost instantaneously, making it very suitable for real-time reconstruction. The full-image networks from Section 3.1 perform less satisfactory, but behave well if only very few projection directions are available and the images are small.

Although we use the networks for the reconstruction of *binary* images, there is no apparent reason why they could not be used for the reconstruction of images that contain a larger set of gray values, or even a continuous range. We intend to explore the possibility of using our neural network approach for continuous tomography in future research. For continuous tomography the size of the network will increase significantly, as the number of projections that is required to obtain a good reconstruction is typically much larger than for discrete tomography. This will increase the training time and the reconstruction time after training. For continuous tomography algorithms are already available that are both accurate and fast. For discrete tomography, however, the current approach seems to provide competitive algorithms, in particular for real-time reconstruction.

References

1. Batenburg, K.J.: Reconstructing Binary Images from Discrete X-rays, CWI Research Report PNA-E0418 (2004)
2. Batenburg, K.J.: An Evolutionary Algorithm for Discrete Tomography. Discrete Applied Mathematics **151** (2005) 36–54
3. Batenburg, K.J., Kosters, W.A.: Neural Networks for Discrete Tomography. Proceedings of the Seventeenth Belgium-Netherlands Conference on Artificial Intelligence (BNAIC), 21–27 (2005)
4. Bishop, C.M.: Neural Networks for Pattern Recognition. Oxford University Press (1995)
5. Helgason, S.: The Radon Transform. Birkhäuser, Boston (1980)
6. Herman, G.T., Kuba, A. (eds.): Discrete Tomography: Foundations, Algorithms and Applications. Birkhäuser, Boston (1999)
7. Hertz, J., Krogh, A., Palmer, R.G.: Introduction to the Theory of Neural Computation. Addison-Wesley (1991)
8. Kak, A.C., Slaney, M.: Principles of Computerized Tomographic Imaging. SIAM (2001)
9. Keat, N.: Real-time CT and CT Fluoroscopy. The British Journal of Radiology **74** (2001) 1088–1090
10. Kerr, J.P., Bartlett, E.B.: Neural Network Reconstruction of Single-photon Emission Computed Tomography Images. Journal of Digital Imaging **8** (1995) 116–126
11. Lampinen, J., Vehtari, A., Leinonen, K.: Application of Bayesian Neural Network in Electrical Impedance Tomography. Proceedings of the 1999 International Joint Conference on Neural Networks (1999)
12. Liao, H.Y., Herman, G.T.: A Coordinate Ascent Approach to Tomographic Reconstruction of Label Images from a Few Projections. Discrete Applied Mathematics **151** (2005) 184–197
13. Schüle, T., Schnörr, C., Weber, S., Hornegger, J.: Discrete Tomography by Convex-concave Regularization and D.C. Programming. Discrete Applied Mathematics **151** (2005) 229–243

A Novel Automated Hand-Based Personal Identification

Yinghua Lu[1,2], Yuru Wang[2,3], Jun Kong[2,3,*], and Longkui Jiang[2,3]

[1] Computer School, Jilin University, Changchun, China
[2] Computer School, Northeast Normal University Changchun, China
[3] Key Laboratory for Applied Statistics of MOE, China
{luyh, wangyr950, kongjun, jianglk701}@nenu.edu.cn

Abstract. A reliable and robust verification approach using hand-print features is presented in this paper. The characteristics of the proposed approach are that two hand-base features are employed, the palm-print and finger-print features. The system consists of two parts: a convenient device for hand-print image acquisition and an efficient algorithm for fast hand-print recognition. A robust and adaptive image coordinate system is defined to facilitate feature extraction. Discrete wavelet zero-crossing encoding scheme and 2-D Gabor filter is applied to hand-print feature extraction and representation. The experimental results demonstrate the effectiveness of the proposed system.

1   Introduction

Biometrics-based personal identification is regarded as an effective method for automatically recognizing, with a high confidence, a person's identity. Many biometric verification technique dealing with various human physiological features including facial images, hand geometry, fingerprint and iris have been proposed to improve the security of personal verification. Each technique has its strengths and limitations, and not being possible to determine which one is the best without considering the application environment. Nevertheless, it is known that, palm-print identification is regarded as one of the most promising and powerful means. Palm-print has received wide investigation from researchers lately [1]-[7]. There are two popular approaches. One is based on the statistical features, and works such as eighpalm [1], fisherpalms [2], Gabor filters [3], Fourier Transform [4], and local energy [5] appear. While the other approach is based on the structural features, the direction projection algorithm [6] and directional line detectors [7] are employed.

In this paper, a novel hand-print verification method for personal identification is presented. A hand-print identification approach, combining palm-print and finger-print, is proposed. 1-D finger-print feature represented using wavelet zero-crossing and palm-print features coded using wavelet zero-crossing and 2-D

---

* Corresponding author. This work is supported by science foundation for young teachers of Northeast Normal University, No. 20061002, China.

U. Eckardt et al. (Eds.): IWCIA 2006, LNCS 4040, pp. 404–414, 2006.

Gabor filter are applied. And the experimental results demonstrate the discriminative characteristic of the extracted 1-D features.

The rest of this paper is organized as follows: Section 2 gives a brief description of our hand-print identification approach. A hand-print coding scheme is described in Section 3. Model-based hand-print matching is detailed in Section 4. Section 5 reports our experimental results and presents the conclusions.

2   System Description

2.1   Hand-Print Images Acquisition

A feasible and convenient hand-print capture device is designed, a flatbed optical scanner is used. To facilitate the later processing, a case and a cover are used to form a semi-closed environment. It possesses the benefits of high availability,

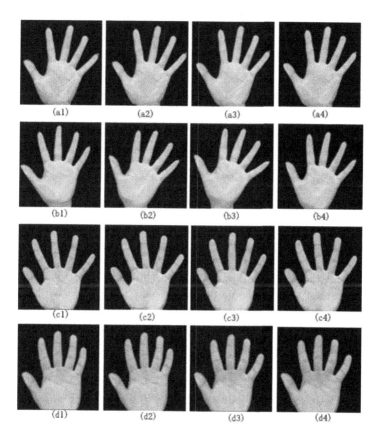

Fig. 1. The original gray-level images of hand-print captured from four different persons. (a1)−(a4) are captured from the same person. Similarly, (b1)−(b4), (c1)−(c4) and (d1)−(d4) are captured from three different persons, respectively. (a1)−(b4) are from female users, and (c1)−(d4)are from male.

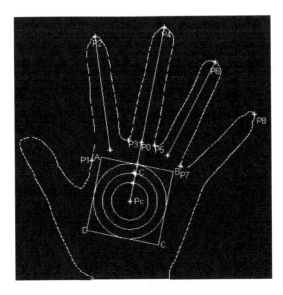

Fig. 2. The key-points detection and the hand-print feature generation

uniform and consistent good image quality, convenience and low cost. In our work, the resolution of the scanner is set at 150 *dpi*. What the users need to do is put their hands on the scanner platform. And the device avoid the limitation of guidance pegs. Fig. 1 shows the hand-print images captured in our work.

2.2   Adaptive Coordinate System

In method above, no guidance pegs are fixed on the scanner's platform and the users are allowed to place their hands freely on the platform of the scanner during capturing. Thus, hand images with different translation and rotations are produced. Therefore, it is important to define a robust coordinate system to provide the foundation for both reliable feature extracting, pattern matching in a certain degree of translation and rotation. To extract the central part of a palm-print, a Polar coordinate system on the palm is constructed, particularly, the pole is the palm center. (Fig. 2). The six major steps of coordinate system definition are:

Step 1. Obtain the hand contours $f(x, y)$ using the border tracing algorithm, and the coordinates of each traced pixel $(bx_i, by_i)(i = 1, 2, ...)$ should be maintained to represent the shape of the hand.

Step 2. Compute the coordinate (X_0, Y_0) of the centroid C of the hand shape, which is invariant to translation, using the regular moment m_{pq} [10].

$$m_{pq} = \int \int x^p y^q f(x, y) dx dy, p, q = 0, 1, 2, ...$$ (1)

$$X_0 = \frac{m_{10}}{m_{00}}, Y_0 = \frac{m_{01}}{m_{00}} .$$ (2)

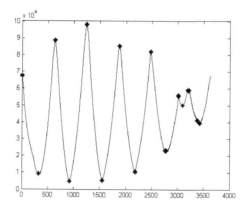

Fig. 3. Profile of Euclidean distance from hand contour pixels to the centroid C

Step 3. Transform the set of coordinates of contours of hand shape into the profile of Euclidean distance d_i to the centroid C as shown in Fig. 3.

$$d_i = \sqrt{(bx_i - X_0)^2 + (by_i - Y_0)^2} \,. \tag{3}$$

Step 4. Find the local points with extremum, which are regarded as the corner points.

Step 5. Line up P_0 and P_c to find the coordinate of center of the palm (cx, cy) based on the crucial points P_3, P_5 , and P_7. Where, $\overline{P_0 P_c}$ is the perpendicular and bisector line of $\overline{P_3 P_5}$, and $\overline{P_0 P_c}$ equals $\overline{P_3 P_7}$ in length (Fig. 2).

Step 6. Define a polar coordinate system based on pole P_c and polar axis $\overline{P_c P_0}$.

   Thus, to a specific sample image, an adaptive polar coordinate system is constructed.

3   Hand-Print Representation

In order to use the hand-print pattern for identification, it is important to define a representation that is well adaptive for extracting the hand-print information content for hand-print signatures. In this way, we introduce an algorithm for extracting unique features from hand-print signatures and representing these features using discrete dyadic wavelet transform [11] which can reflect the position and shape variance of signal and 2-D Gabor filter [3] which is an effective tool for texture analysis.

3.1   Wavelet Transform and Zero-Crossing Representation

When a signal includes important structures that belong to different scales, it is often helpful to reorganize the signal information into a set of "detail components" of varying size. The wavelet transform (WT) can decomposes a signal into components that appear at different resolutions.

Let $S(x) \in L^2(R)$ and $W_{2j}(S(x))_{j \in Z}$ be the original signal and its dyadic wavelet transform [9]. For constructing the zero-crossing representation, the mother wavelet $\psi(x)$ is defined here as the second derivative of a smoothing function θ_x.

$$\psi(x) = \frac{d^2 \theta(x)}{dx^2} . \tag{4}$$

Then, record the value of the integrals between any pair of consecutive zero-crossings of W_{2j} whose abscissae are (z_{n-1}, z_n) :

Then, record the value of the integrals between any pair of consecutive zero-crossings of whose abscissae are :

$$e_n = \int_{z_{n-1}}^{z_n} W_{2j} S(x) dx . \tag{5}$$

For any function $W_{2j} S(x)$, the position of the zero-crossings $(z_n)_{n \in Z}$ and the integral $(e_n)_{n \in Z}$ can be represented by a piece-wise constant function $Z_{2j} S(x)$ defined by:

$$Z_{2j} S(x) = \frac{e_n}{z_n - z_{n-1}}, x \in [z_{n-1}, z_n] . \tag{6}$$

The sequence of piece-wise constant functions $Zf = (Z_{2j} S(x))_{j \in Z}$ is referred to as the zero-crossing representation of $S(x)$.

In order to stabilize the zero-crossings, consideration can be restricted to dyadic scales, such as in the dyadic wavelet transform, instead of considering the zero-crossing on a continuous scale. In practical implementations, the input signal, in our case the hand-print signature is measured with both coarse and finite resolution that imposes both coarse and finite scale when computing the dyadic wavelet transform. Since information at fine resolution levels is strongly affected by noise, a few low resolution levels, excluding the coarsest one will be used in our work. Fig. 4 shows the wavelet transform and the zero-crossing representation at a particular scale 2^j .

3.2   2-D Gabor Filter

The general form of Gabor filter in spatial domain is :

$$G(x, y, \theta, u, \sigma) = \frac{1}{2\pi\sigma^2} exp-\frac{x^2 + y^2}{2\sigma^2} \times exp2\pi i(ux \cos\theta + uy \sin\theta) , \tag{7}$$

where $i = \sqrt{-1}$, u is the frequency of the sinusoidal wave, θ controls the orientation of the function and σ is the standard deviation of the Gaussian envelope. It has been widely used in iris recognition [11] and in palm-print recognition [3].

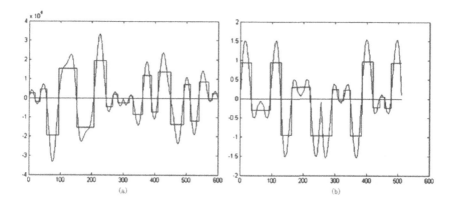

Fig. 4. Hand-print representation. (a)1-D Finger-print feature, (b)1-D palm-print feature.

3.3   Finger-Print Feature Coding

The lines on the fingers are important information used to guide a selection of a small set of similar candidates from the database, though is not discriminative enough to differentiate persons. The finger-print feature is coded in a plane rectangular coordinate system. The grey-level values on three center lines of three middle fingers (as shown in Fig. 2) are recorded as the finger-print signature S_f. In addition, we employ discrete wavelet zero-crossing representation to obtain finger-print pattern Zf_f (Fig. 4a).

$$Zf_f = (Z_{2j}S_f(x))_{j \in Z} \ . \tag{8}$$

3.4   Palm-Print Representing

In studying the characteristics of the palm-print, it is merely required to deal with samples of the grey-level profiles and use them to construct a representation. The main idea of the proposed technique is to represent the 1-D features of the palm-print by fine-to-coarse approximations at different resolution levels based on the WT zero-crossing representation, and to code the 2-D palm-print features by Gabor filter.

  As mentioned before, a polar coordinate system with pole P_c and polar axis $\overline{P_c P_0}$ has been defined, which ensures the representation be translation- and rotation-invariant. Thus, the palm-print signature S_p is the grey-level values on the contours of concentric circles, which is centered at the pole P_c (the center of the palm), with fixed radio and angular increments of $2\pi/L_s$, where L_s is the length of palm-print signature:

$$S_p = I(c_x + r\cos\theta, c_y + r\sin\theta), 0 \le \theta \le 2\pi/L_s \ , \tag{9}$$

where (cx, cy) is the coordinate of P_c. It should be pointed out that the start point of the signal is always on the polar axis defined above. pattern.

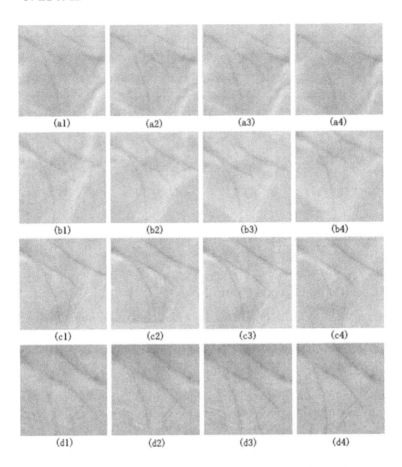

Fig. 5. The images cropped from original figures automatically to be used for 2-D palm-print representation

Then the zero-crossing representation Zf_p of palm-print feature is:

$$Zf_p = (Z_{2j}S_p(x))_{j\in Z} .\tag{10}$$

Fig. 2 shows the locations of these concentric circles on the palm. The extracted data from one of the circles are shown in Fig. 4b as the palm-print signature. Then its wavelet zero-crossing representation is obtained as the 1-D palm-print.

As shown in Fig. 2, the circumscribing square $ABCD$ of the maximal concentric circle can be found, and $\overline{AB}$ is perpendicular to the polar axis $\overline{P_cP_0}$. The 2-D palm images cropped from different individuals are shown in Fig. 5.

Then, according to formula (7) the 2-D palm-print feature coded by Gabor filter is:

$$G(x,y,\theta,u,\sigma) = \frac{1}{2\pi\sigma^2}exp-\frac{x^2+y^2}{2\sigma^2} \times exp2\pi i(ux\cos\theta + uy\sin\theta) .\tag{11}$$

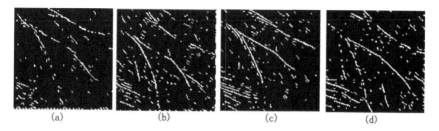

Fig. 6. The 2-D palm-print feature extracted using Gabor filter, from (a1), (b1), (c1), (d1) in Fig. 5 respectively

In our study, three parameters are selected to be $\theta = 3\pi/4, u = 1.1, \sigma = 0.36$, and the Gabor representation are shown in Fig. 6.

4   Identification

Identification is a process of comparing a query image against candidate images. In this section, we design a practical model-based technique to model the verifier of a specific person. Matching algorithm using the correlation function is a common and practical technique utilized in many pattern recognition applications. In the learning phase, the system will construct the model representations of the hand-print signature, and in the classification phase, a query signature will be matched with a specific model.

Considered a query signature q and candidate model t at a particular resolution level j of zero-crossing representation, in order to classify unknown signatures, the dissimilarity functions have been used as follows:

$$d_j^{(1)} = \min_m \sum_{n-1}^{N} |Z_j q(n) - Z_j t(n+m)|^2, m \in [0, N-1],\tag{12}$$

where N is the number of data points of the zero-crossing representation. In addition, the overall dissimilarity value of the query signature and the candidate model over the resolution interval $[K, L]$ will be the average of a certain dissimilarity function calculated at each resolution level in this interval.

The application of Hamming distance based matching is to compute the dissimilarity degree of 2-D palm-print features as follows.

$$d^{(2)} = \frac{1}{M \times M} \sum_{i=1}^{m} \sum_{j=1}^{m} q(i,j) \oplus t(i,j)),\tag{13}$$

where $\oplus$ denotes Exclusive-OR, and $M \times M$ is the size of 2-D palm-print feature.

There are three phases in identification, which are detailed as follows:

First of all, construct the finger-print and palm-print representation of the query sample.

Coarse-level identification based on finger-print representation:

Step 1. Calculate the dissimilarity degree between the unknown and the candidate models in the database.

Step 2. Provide the candidates whose dissimilarity degree $d_j^{(1)}$ is smaller than a pre-defined threshold value for the palm-print matching.

Fine-level identification based on 1-D and 2-D palm-print features:

Step 3. Compare the sample with the candidates selected in Step 2 in terms of 1-D palm-print feature to calculate their dissimilarity degree, for the palm-print pattern at a certain polar distance.

Step 4. Add the candidates whose dissimilarity degree $d_j^{(1)}$ is smaller than a pre-defined threshold to next level matching.

Step 5. Go to Step 3 and repeat the same procedure until all of the polar distances are considered or only one candidate remained.

Step 6. Compare the sample with the candidates selected in Step 2 in terms of 2-D palm-print feature, and add the candidates whose dissimilarity degree $d^{(2)}$ is smaller than a pre-defined threshold .

Step 7. Go to Step 6 and repeat the same procedure until all the candidates are considered or only one candidate remained.

Confirmation

Step 8. The candidate sample selected not only from step 5 but also step 7 is the final result.

5   Experimental Results

5.1   Hand-Print Database

We collect hand-print images from 50 individuals' right hand using our hand-print capture device. They are mainly from our students and teachers at the age of younger than 30 years old and 27 people are male. In addition, we collect the hand-print images at an interval of around 3 months. The captured images are of size 1000×1000, and resized to 500×500. Thus, the images used in the following experiments are 50×500 of 150 *dpi* resolution.

5.2   Speed

In our research, a standard PC with Intel Pentium processor (2.40GHZ) and 256 MB random access memorizes is used. The system is implemented using **Matlab** 6.5.1. The execution time in different phases is shown in Table. 1. In fact, we have not yet optimized the program code, so it is possible that the computation time could be reduced.

5.3   Identification Results

As mentioned above, coarse to fine scheme is employed in identification phase. This sub-section reports some experimental results obtained in each stages. In

Table 1. The average execution time in the system (in seconds(s))

	Preprocessing(s)	Center-location(s)	Feature extraction	Matching
Finger-print	0.5920	0.0320	0.1021	0.0290
Palm-print	0.5920	0.0320	0.0940	0.0014

Table 2. Comparison of different palm-print matching methods

	[8]	[5]	Ours
Database size	1500/50	200/	500/50
Device freedom	Pegs scanner	Prints	No pegs scanner
User comfort	No	Yes	Yes
Feature extraction	Wavelet/Pca	Texture/ feature points	WT zero-crossing/Gabor
Coarse matching	Hand geometry (1-D)	Texture energy (2-D)	Finger-print (1-D)
Fine matching	Palm-print (1-D)	Feature points (1-D)	Palm-print (1-D and 2-D)
Speed	Good	Limited	Good
Accuracy	FRR=5.3% FAR=3.7%	95%	FAR=FRR=2.3%

coarse level matching, we reduce the false rejection rate (FRR) to zero in order to increase the tolerance for similar features and the corresponding performance using finger-print features is evaluated by eliminating rate which is 87% at average. The final confirmation in fine level is operated on the candidate patterns selected from coarse level matching, the performance using 1-D zero-crossing features and 2-D Gabor features is 3.7% and 3.5% respectively (the value where the FAR and FRR are made equal by adjusting the decision threshold). Then, the coarse-level and the fine-level verification modules were sequentially combined to obtain the average performance of $FAR=FRR=2.3\%$.

5.4   Conclusions

In this paper, a novel approach is presented to authenticate individuals by using hand-print features. The hand images of low resolution are captured using a scanner without any fixed peg. This mechanism is very suitable and comfortable for all users. Compared with the current existing techniques for palm-print identification, our approach integrates finger-print and palm-print features represented by wavelet zero-crossing and 2-D Gabor. Table. 2 summarizes the major characteristics of our method and the other techniques [8] [5]. The experimental results presented above provide the basis for the further development of a fully automated hand-based security system.

References

1. Guangming Lu, David Zhang, Palm-print recognition using eigenpalms features, pattern recogni-tion leters, 24 (2003) 1463-1467.
2. Xiangqian Wu, David Zhang, Fisherpalms based palm-print recognition, Pattern recognition letters, 24 (2003) 2829-2838.

3. Wai Kin Kong, David Zhang, Palm-print feature extraction using 2-D Gabor filters,Pattern recog-nition, 36 (2003) 2339 - 2347.
4. W.Li,D.Zhang, Z.Xu, Palm-print identification by Fourier Transform, International Journal of Pattern Recognition and Artifical Intelligence, 16(4) (2003) 417-432.
5. Jane You, Wai Kin Kong, On hierarchical palm-print coding with multiple features for personal identification in large databases, IEEE Transactions on Circuits and Systems for Video Technology, vol. 14, no. 2, February 2004.
6. D.Zhang, W. Shu, Two novel characteristics in palmprint verification: datum point invariance and line feature matching, Pattern Recognition, 32 (1999)691-702.
7. Xiangqian Wu, David Zhang, Palm-print classification using principal lines, Pattern recognition, 37 (2004) 1987-1998.
8. Chin-Chuan Han, A hand-based personal authentication using a coarse-to-fine strategy, Image And Vision Computing, 22 (2004) 909-918.
9. S. Mallat, S. Zhong, Characterization of signals from multiscale edges, IEEE Trans. Pattern Anal. Mach. Intell. 14 (7) (1992) 710-732.
10. Dinggang Shen, Horace H.S. Ip, Discriminative wavelet shape descriptors for recognition of 2-D patterns, Pattern Recognition, 32 (1999) 151-165.
11. C. Sanchez-Avila, R. Sanchez-Reillo,Two different approaches for iris recognition using Gabor filters and multiscale zero-crossing representation, Pattern Recognition 38 (2005) 231-240.

Shortest Paths in a Cuboidal World

Fajie Li and Reinhard Klette

Computer Science Department, The University of Auckland
Auckland, New Zealand

Abstract. Since 1987 it is known that the Euclidean shortest path problem is NP-hard. However, if the 3D world is subdivided into cubes, all of the same size, defining obstacles or possible spaces to move in, then the Euclidean shortest path problem has a linear-time solution, if all spaces to move in form a simple cube-curve. The shortest path through a simple cube-curve in the orthogonal 3D grid is a minimum-length polygonal curve (MLP for short). So far only one general and linear (only with respect to measured run times) algorithm, called the *rubberband algorithm*, was known for an approximative calculation of an MLP. The algorithm is basically defined by moves of vertices along critical edges (i.e., edges in three cubes of the given cube-curve). A proof, that this algorithm always converges to the correct MLP, and if so, then always (provable) in linear time, was still an open problem so far (the authors had successfully treated only a very special case of simple cube-curves before). In a previous paper, the authors also showed that the original rubberband algorithm required a (minor) correction.

This paper finally answers the open problem: by a further modification of the corrected rubberband algorithm, it turns into a provable linear-time algorithm for calculating the MLP of any simple cube-curve.

The paper also presents an alternative provable linear-time algorithm for the same task, which is based on moving vertices within faces of cubes.

For a disticntion, we call the modified original algorithm now the *edge-based rubberband algorithm*, and the second algorithm is the *face-based rubberband algorithm*; the time complexity of both is in $\mathcal{O}(m)$, where m is the number of critical edges of the given simple cube-curve.

1   Introduction

A cube-curve g is a loop of face-connected grid cubes in the 3D orthogonal grid; the union **g** of those cubes defines the *tube* of g. The paper discusses Euclidean shortest paths in such tubes, which are defined by minimum-length polygonal (MLP) curves (see Figure 1).

The Euclidean shortest path problem is as follows: Given a Euclidean space which contains (closed) polyhedral obstacles; compute a path which *(i)* connects two given points in the space, *(ii)* does not intersect the interior of any obstacle, and *(iii)* is of minimum Euclidean length. This problem (starting with dimension 2) is known to be NP-hard [2].

U. Eckardt et al. (Eds.): IWCIA 2006, LNCS 4040, pp. 415–429, 2006.
© Springer-Verlag Berlin Heidelberg 2006

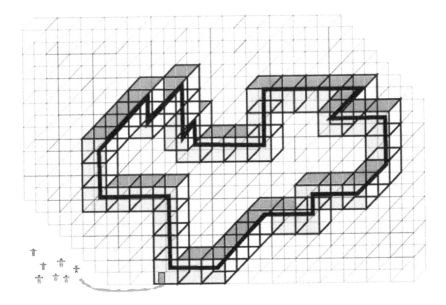

Fig. 1. A cuboidal world: seven robots at the lower left corner, and the bold curve is an initial guess for a 3D walk (or flight) through the given loop of shaded cubes. The length of the 3D walk needs to be minimized, which defines the MLP.

There are algorithms solving the approximate Euclidean shortest path problem in 3D in polynomial time, see [3]. Shortest paths or path planning in todays 3D robotics (see, for example, [15], or the annual ICRA conferences in general) seems to be dominated by heuristics rather than by general geometric algorithms. If a cuboidal world can be assumed then this paper provides two general and linear-time shortest path algorithms.

3D MLP calculations generalize MLP computations in 2D; see, for example, [8, 17] for theoretical results and [5, 21] for 2D robotics scenarios. Shortest curve calculations in image analysis also use graph metrics instead of the Euclidean metric; see, for example, [20].

Interest in 3D MLPs was also raised by the issue of multigrid-convergent length estimation for digitized curves. The length of a simple cube-curve in 3D Euclidean space can be defined by that of the MLP; see [18, 19], which can be characterized as a 'global approach'. A 'local approach' for 3D length estimation, allowing only weighted steps within a restricted neighborhood, was considered in [7]. Alternatively to the MLP, the length of 3D digital curves can also be measured (within time, linear in the number of grid points on the curve) based on DSS-approximations [4].

The computation of 3D MLPs was first time published in [1], proposing a 'rubberband' algorithm[1]. This iterative algorithm was experimentally tested and showed linear run-time behavior. It also was correct for all the tested inputs

---

[1] Not to be confused with a 2D image segmentation algorithm of the same name [16].

(where correctness was tested manually!). However, in this publication, no mathematical proof was given for linear run time or general correctness (i.e., that its solution iterates, for any simple cube-curve, to the MLP). This *original rubberband algorithm* is also published in the monograph [10]. Recent applications of this algorithm are in 3D medical imaging; see, for example, [6, 22].

The authors approached the correctness and linearity problem of the rubberband algorithm along the following steps:

[11] only considered a very special class of simple cube-curves and developed a provable correct MLP algorithm for this class. The main idea was to decompose a cube-curve of that class into arcs at "end angles" (see Definition 3 in [11]), that means, the cube-curves have to have end-angles, that the algorithm can be applied.

[12] constructed an example of a simple cube-curve whose MLP does not have any of its vertices at a corner of a grid cube. It followed that any of cube-curve with this property does not have any end angle, and this means that we cannot use the MLP algorithm as proposed in [11]. This was the basic importance of the result in [12]: we showed the existence of cube-curves which require further algorithmic studies.

[14] showed that the original rubberband algorithm requires a modification (in its Option 3) to guarantee that calculated curves are always contained in the tube **g**. This *corrected rubberband algorithm* achieves (as the original rubberband algorithm) minimization of length by moving vertices along *critical edges* (i.e., grid edges incident with three cubes of the given simple cube-curve).

This paper now (finally) extends the corrected rubberband algorithm into the *edge-based rubberband algorithm* and shows, that it is correct for any (!) simple cube-curve. The paper also presents a totally new algorithm, the *face-based rubberband algorithm*, and shows that it is also correct for any simple cube-curve. We prove that both, the edge-based and the face-based rubberband algorithm, have time complexity in $\mathcal{O}(m)$ time, where m is the number of critical edges in the given simple cube-curve.

Further (say, 'more elegant') algorithms for calculating MLPs in simple cube-curves may exist; this way this article may be just the starting point for more detailed performance evaluations. Also, the given modifications of the original rubberband algorithm might be not always necessary, or the simplest ones.

The paper is organized as follows: Section 2 describes the concepts used in this paper. Section 3 provides mathematical fundamentals for our two algorithms. Section 4 describes the edge-based and face-based rubberband algorithm, and discusses their time complexity. Section 5 presents an example illustrating how the edge-based and face-based rubberband algorithms are converging to identical results (i.e., to the MLP). Section 6 gives our conclusions.

2   Definitions

Following [10], a grid point $(i, j, k) \in \mathbb{Z}^3$ is assumed to be the center point of a *grid cube* with *faces* parallel to the coordinate planes, with *edges* of length 1, and *vertices* at its corners. *Cells* are either cubes, faces, edges, or vertices. The intersection of two cells is either empty or a joint *side* of both cells. A *cube-curve* is an alternating sequence $g = (f_0, c_0, f_1, c_1, \ldots, f_n, c_n)$ of faces f_i and cubes c_i, for $0 \leq i \leq n$, such that faces f_i and f_{i+1} are sides of cube c_i, for $0 \leq i \leq n$ and $f_{n+1} = f_0$. It is *simple* iff $n \geq 4$ and for any two cubes $c_i, c_k \in g$ with $|i - k| \geq 2$ (mod $n + 1$), if $c_i \bigcap c_k \neq \phi$ then either $|i - k| = 2$ (mod $n + 1$) and $c_i \bigcap c_k$ is an edge, or $|i - k| \geq 3$ (mod $n + 1$) and $c_i \bigcap c_k$ is a vertex.

A *tube* **g** is the union of all cubes contained in a cube-curve g. A tube is a compact set in $\mathbb{R}^3$; its frontier defines a polyhedron. A curve in $\mathbb{R}^3$ is *complete* in **g** iff it has a nonempty intersection with every cube contained in g. Following [18, 19], we define:

Definition 1. *A* minimum-length curve *of a simple cube-curve g is a shortest simple curve P which is contained and complete in tube $\mathbf{g}$. The length $\mathcal{L}(g)$ of g is defined to be the length $\mathcal{L}(P)$.*

It turns out that such a shortest curve P is always a polygonal curve, called MLP for short; it is uniquely defined if the cube-curve is not contained in a single layer of cubes of the 3D grid (see [18, 19]). If it is contained in just one layer then the MLP is uniquely defined up to a translation orthogonal to that layer. We speak about *the* MLP of a simple cube-curve.

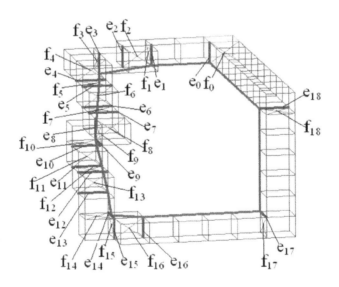

Fig. 2. A simple cube-curve and its MLP (see also Table 1)

Figure 2 shows a simple-cube curve and (as bold polygonal curve) its MLP; grid edges containing vertices of the MLP are also shown in bold.

Definition 2. *A critical edge of a cube-curve g is a grid edge which is incident with exactly three different cubes contained in g. If e is a critical edge of g and l*

Table 1. Coordinates of endpoints of critical edges shown in Figure 2 (also used later in an experiment)

Critical edge	x_{i1}	y_{i1}	z_{i1}	x_{i2}	y_{i2}	z_{i2}
e_0	-0.5	1	-0.5	-0.5	1	0.5
e_1	-0.5	2	-0.5	-0.5	2	0.5
e_2	-1.5	3	-0.5	-1.5	3	0.5
e_3	-2.5	3	-0.5	-2.5	4	-0.5
e_4	-3.5	3	-0.5	-3.5	4	-0.5
e_5	-3.5	3	-1.5	-3.5	4	-1.5
e_6	-4.5	3	-1.5	-4.5	4	-1.5
e_7	-5.5	4	-2.5	-5.5	4	-1.5
e_8	-6.5	4	-2.5	-5.5	4	-2.5
e_9	-6.5	4	-2.5	-6.5	5	-2.5
e_{10}	-6.5	4	-3.5	-6.5	5	-3.5
e_{11}	-7.5	4	-3.5	-7.5	5	-3.5
e_{12}	-7.5	4	-4.5	-7.5	5	-4.5
e_{13}	-8.5	4	-5.5	-7.5	4	-5.5
e_{14}	-8.5	4	-6.5	-8.5	4	-5.5
e_{15}	-8.5	3	-6.5	-8.5	3	-5.5
e_{16}	-9.5	-1	-5.5	-8.5	-1	-5.5
e_{17}	-8.5	-2	-0.5	-8.5	-1	-0.5
e_{18}	-0.5	-1	-0.5	-0.5	-1	0.5

is a straight line such that $e \subset l$, then l is called a critical line of e in g or critical line for short. If f is a face of a cube in g and one of f's edges is a critical edge e in g then f is called a critical face of e in g or critical face for short.

Definition 3. A simple cube-curve g is called first-class iff each critical edge of g contains exactly one vertex of the MLP of g.

Figure 3 shows a first-class simple cube-curve. The cube-curve shown in Figure 2 is not first-class because there are no vertices of the MLP on the following critical edges: e_1, e_4, e_5, e_6, e_8, e_9, e_{10}, e_{11} and e_{14} (will be later shown in the experiments, summarized in Table 4).

Unfortunately, we need also a few rather technical definitions:

Definition 4. Let e be a critical edge of a simple cube-curve g and f_1, f_2 be two critical faces of e in g. Let c_1, c_2 be the centers of f_1, f_2 respectively. Then a polygonal curve can go in the direction from c_1 to c_2, or from c_2 to c_1, to visit all cubes in g such that each cube is visited exactly once. If e is on the left of line segment c_1c_2, then the orientation from c_1 to c_2 is called counter-clockwise orientation of g. f_1 is called the first critical face of e in g. If e is on the right of line segment c_1c_2, then the direction from c_1 to c_2 is called clockwise orientation of g.

Figure 2 shows all critical edges (e_0, e_1, e_2, ..., e_{18}) and their first critical faces (f_0, f_1, f_2, ..., f_{18}) of a simple cube-curve g.

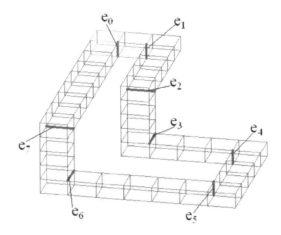

Fig. 3. A first-class simple cube-curve

Definition 5. *A* minimum-length pseudo polygon *of a simple cube-curve g, de-noted by $MLPP$, is a shortest curve P which is contained and complete in tube g such that each vertex of P is on the first critical face of a critical edge in g.*

From results in [19] it follows that the $MLPP$ of a simple cube-curve g is unique. The number of vertices of an $MLPP$ is the number of all critical edges of g. $p_{4_0}p_{4_1}\cdots p_{4_{18}}$ (see Table 3) is the $MLPP$ of g as shown in Figure 2.

Let f_{i1}, f_{i2} be two critical faces of e_i in g, $i = 1$, 2. Let c_{i1}, c_{i2} be the centers of f_{i1}, f_{i2} respectively, for $i = 1, 2$. Obviousely, the counter-clockwise orientation of g defined by c_{11}, c_{12} is identical to the one defined by c_{21}, c_{22}.

Definition 6. *Let e_0, e_1, e_2, ... e_m and e_{m+1} be all consecutive critical edges of g in the counter-clockwise orientation of g. Let f_i be the first critical face of e_i in g, and p_i be a point on f_i, where $i = 0, 1, 2, ..., m$ or $m + 1$. Then the polygonal curve $p_0p_1\cdots p_mp_{m+1}$ is called an* approximate minimum-length pseudo polygon *of g, denoted by $AMLPP$.*

The polygonal curve $p_{1_0}p_{1_1}\cdots p_{1_{18}}$ (see Table 2) is an $AMLPP$ of g shown in Figure 2.

Definition 7. *Let p_1, p_2 and p_3 be three consecutive vertices of an $AMLPP$ of a simple cube-curve g. If p_1, p_2 and p_3 are colinear, then p_2 is called a* trivial *vertex of the $AMLPP$ of g. p_2 is called a* non-trivial *vertex of the $AMLPP$ of g if it is not a trivial vertex of that $AMLPP$ of g.*

A simple *cube-arc* is an alternating sequence $a = (f_0, c_0, f_1, c_1, \ldots, f_k, c_k, f_{k+1})$ of faces f_i and cubes c_i with $f_{k+1} \neq f_0$, denoted by $a = (f_0, f_1, \ldots, f_{k+1})$ or $a(f_0, f_{k+1})$ for short, which is a consecutive part of a simple cube-curve. A subarc of an arc $a = (f_0, f_1, \ldots, f_{k+1})$ is an arc $(f_i, f_{i+1}, \ldots, f_j)$, where $0 \leq i \leq j \leq k$.

Definition 8. *Let a polygonal curve $P = p_0p_1\cdots p_mp_{m+1}$ be an $AMLPP$ of g and $p_i \in f_i$, where f_i is a critical face of g, $i = 0, 1, 2, ..., m$ or $m + 1$. A cube-arc $\rho = (f_i, f_{i+1}, \ldots, f_j)$ is called*

- a $(2,3)$-cube-arc *with respect to* P *if each vertex* p_k *is identical to* p_{k-1} *or* p_{k+1}, *where* $k = i+1, \ldots, j-1$, [2]
- a maximal $(2,3)$-cube-arc *with respect to* P *if it is a $(2,3)$-cube-arc and* p_i *is not identical to* p_{i+1} *and* p_{i-1}, *and* p_j *is not identical to* p_{j-1} *and* p_{j+1},
- a 3-cube-arc unit *with respect to* P *if it is a $(2,3)$-cube-arc such that* $j = i + 4 \pmod{m+2}$ *and* $p_{i+1}, p_{i+2}, p_{i+3}$ *are identical.*
- a 2-cube-arc *with respect to* P *if it is a $(2,3)$-cube-arc and no three consecutive vertices of* P *on* a *are identical,*
- a maximal 2-cube-arc *with respect to* P *if it is both a maximal $(2,3)$-cube-arc and a 2-cube-arc as well,*
- a 2-cube-arc unit *with respect to* P *if it is a 2-cube-arc such that* $j = i + 3 \pmod{m+2}$ *and* p_{i+1} *is identical to* p_{i+2},
- a regular cube-arc unit *with respect to* P *if* $a = (f_i, f_{i+1}, f_j)$ *such that* p_i *is not identical to* p_{i+1} *and* p_j *is not identical to* p_{i+1},
- a cube-arc unit *with respect to* P *if* a *is a regular cube-arc unit, 2-cube-arc unit or 3-cube-arc unit, or*
- a regular cube-arc *with respect to* P *if no two consecutive vertices of* P *on* a *are identical.*

Let $P_{18_i} = p_{i_0} p_{1_1} \cdots p_{1_{18}}$ (see Table 2), where $i = 1, 2, 3, 4$. Then there are four maximal 2-cube-arcs with respect to P_{18_i}: $(p_{i_{18}}, p_{i_0}, p_{i_1}, p_{i_2})$, $(p_{i_2}, p_{i_3}, p_{i_4}, p_{i_5})$, $(p_{i_7}, p_{i_8}, p_{i_9}, p_{i_{10}})$ and $(p_{i_{12}}, p_{i_{13}}, p_{i_{14}}, p_{i_{15}})$ in total, where $i = 1, 2, 3$. They are also maximal 2-cube-arcs and 2-cube-arc units with respect to P_{18_i}, where $i = 1, 2, 3$. There are no 3-cube-arc units with respect to P_{18_i}, where $i = 1, 2, 3$. $(p_{i_1}, p_{i_2}, p_{i_3})$ is a regular cube-arc unit with respect to P_{18_i} and $(p_{i_4}, p_{i_5}, p_{i_6}, p_{i_7}, p_{i_8})$ is a regular cube-arc with respect to P_{18_i}, where $i = 1, 2, 3$.

There are three maximal 2-cube-arcs with respect to P_{18_4}: $(p_{4_{18}}, p_{4_0}, p_{4_1}, p_{4_2})$, $(p_{4_2}, p_{4_3}, p_{4_4}, p_{4_5})$, and $(p_{4_{12}}, p_{4_{13}}, p_{4_{14}}, p_{4_{15}})$ in total. They are also maximal 2-cube-arcs and 2-cube-arc units with respect to P_{18_4}. $(p_{4_6}, p_{4_7}, p_{4_8}, p_{4_9}, p_{4_{10}}, p_{4_{11}}, p_{4_{12}})$ is a $(2,3)$-cube-arc with respect to P_{18_4}. $(p_{4_6}, p_{4_7}, p_{4_8}, p_{4_9}, p_{4_{10}})$ is a unique 3-cube-arc unit with respect to P_{18_4}.

Definition 9. *Let* $\rho = (f_i, f_{i+1}, \ldots, f_j)$ *be a simple cube-arc and* $p_k \in f_k$, *where* $k = i, j$. *A* minimum-length arc *with respect to* p_i *and* p_j *of* ρ, *denoted by* $MLA(p_i, p_j)$, *is a shortest arc (from* p_i *to* p_j*) which is contained and complete in* ρ *such that each vertex of* $MLA(p_i, p_j)$ *is on the first critical face of a critical edge in* ρ.

3 Basics

We provide mathematical fundamentals to be used in the following. We start with citing a theorem from [9]:

Theorem 1. *Let* g *be a simple cube-curve. Critical edges are the only possible locations of vertices of the MLP of* g.

---

[2] Note that it is impossible that four consecutive vertices of P on ρ are identical.

Let $d_e(p, q)$ be the Euclidean distance between points p and q.

Let $e_0, e_1, e_2, \ldots, e_m$ and e_{m+1} be $m+2$ consecutive critical edges in a simple cube-curve g, and let $l_0, l_1, l_2, \ldots, l_m$ and l_{m+1} be the corresponding critical lines. We express a point $p_i(t_i) = (x_i + k_{x_i} t_i, y_i + k_{y_i} t_i, z_i + k_{z_i} t_i)$ on l_i in general form, with $t_i \in \mathbb{R}$, where $i = 0, 1, \ldots$, or $m+1$.

Let e_i, e_j, and e_k be three (not necessarily consecutive) critical edges in a simple cube-curve.

Lemma 1. ([11], Lemma 1) *Let $d_j(t_i, t_j, t_k) = d_e(p_i, p_j) + d_e(p_j, p_k)$. It follows that $\frac{\partial^2 d_j}{\partial t_j^2} > 0$.*

By elementary geometry, we also have:

Lemma 2. *Let P be a point in $\triangle ABC$ such that P is not on any of the three line segments AB, BC and CA. Then $d_{PA} + d_{PB} < d_{CA} + d_{CB}$.*

The following Lemma is straightforward but useful in our description of the edge-based rubberband algorithm in Section 4.

Lemma 3. *Let p_i and p_{i+1} be two consecutive vertices of an AMLPP of g. If p_i is identical to p_{i+1} then p_i and p_{i+1} are on a critical edge of g.*

Lemma 4. ([13], Lemma 4) *The number of MLPPs of a first-class simple cube-curve g is finite.*

Let $p_i \in f_i$, where f_i is the first critical face of e_i in g, $i = 0, 1, 2, \ldots, m$ or $m + 1$. Let P be a polygonal curve $p_0 p_1 \cdots p_m p_{m+1}$.

Corollary 1. *The number of MLPPs of a simple cube-curve g is finite.*

Proof. If there is a vertex $p_i \in f_i$ such that p_i is not on an edge of f_i (i.e, p_i is a trivial vertex of $MLPP$), then p_{i-1}, p_i and p_{i+1} are colinear. In this case, p_i can be ignored because it is defined by p_{i-1} and p_{i+1}. Therefore, without loss generality, we can assume that each p_i is on one edge of f_i, where $i = 0, 1, 2, \ldots, m$ or $m + 1$. In this case, the proof of this lemma is exactly the same as that of Lemma 4. □

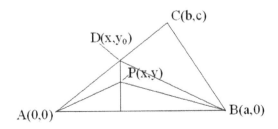

Fig. 4. Illustration for the proof of Lemma 2

Lemma 5. ([13], Lemma 14) *Each first-class simple cube-curve g has a unique MLPP.*

Analogous to the proof of Lemma 5, we also have

Lemma 6. *Each simple cube-curve g has a unique MLPP.*

Theorem 2. *P is an MLPP of g iff for each cube-arc unit $a = (f_i, f_{i+1}, \ldots, f_j)$ with respect to P, the arc $(p_i, p_{i+1}, \ldots, p_j)$ is equal to $MLA(p_i, p_j)$.*

Proof. The necessity is straightforward. The sufficiency is by Lemma 6.   □

Analogously to the proof of Theorem 1 we also obtain

Lemma 7. *If a vertex p of an AMLPP of g is on a first critical face f but not on any edge of it, then p is a trivial vertex of the AMLPP.*

4   Algorithms

We present two algorithms which are both linear-time and provable convergent to the *MLP* of a simple cube-curve. We start to describe some useful procedures which will be used in those two algorithms (as subroutines).

4.1   Procedures

Given a critical e in g, and two points p_1 and p_3 in g, by Procedure 1, we can find a unique point p_2 in f such that $d_{p_1 p_2} + d_{p_3 p_2} = \min\{d_{p_1 p} + d_{p_3 p} : p \in e\}$.

Procedure 1
Let the two endpoints of e be a and b. Then by Lemma 6 of [13], $p_2 = a + t * (b - a)$, where $t = -(A_1 B_2 + A_2 B_1)/(B_2 + B_1)$, A_1, A_2, B_1 and B_2 are functions of the coordinates of p_1, p_3, a and b.

Given a critical face f of a critical edge in g, and two points p_1 and p_3 in g, by Procedure 2, we can find a point p_2 in f such that $d_{p_1 p_2} + d_{p_3 p_2} = \min\{d_{p_1 p} + d_{p_3 p} : p \in f\}$.

Procedure 2
Case 1. $p_1 p_3$ and f are on the same plane. *Case 1.1.* $p_1 p_3 \cap f \neq \phi$. In this case, $p_1 p_3 \cap f$ is a line segment. Let p_2 be the end point of this segment such that it is close to p_1. *Case 1.2.* $p_1 p_3 \cap f = \phi$. By Lemmas 2, p_2 must be on the edges of f. By Lemmas 1, p_2 must be uniquely on one of the edges of f. Apply Procedure 1 on the four edges of f, denoted by e_1, e_2, e_3 and e_4, we get p_{2i} such that $d_{p_1 p_{2i}} + d_{p_3 p_{2i}} = \min\{d_{p_1 p} + d_{p_3 p} : p \in e_i\}$, where $i = 1, 2, 3, 4$. Then we can find a point p_2 such that $d_{p_1 p_2} + d_{p_3 p_2} = \min\{d_{p_1 p_{2i}} + d_{p_3 p_{2i}} : i = 1, 2, 3, 4.\}$.

Case 2. $p_1 p_3$ and f are not on the same plane. *Case 2.1.* $p_1 p_3 \cap f \neq \phi$. It follows that $p_1 p_3 \cap f$ is a unique point. Let p_2 be this point. *Case 2.2.* $p_1 p_3 \cap f = \phi$. In this case, p_2 can be found exactly the same way as in Case 1.2.

The following procedure is used to convert an $MLPP$ into an MLP.

Procedure 3
Given a polygonal curve $p_0 p_1 \cdots p_m p_{m+1}$ and three pointers addressing vertices at positions i - 1, i, and $i + 1$ in this curve. Delete p_i if p_{i-1}, p_i and p_{i+1} are colinear. Next, the subsequence (p_{i-1}, p_i, p_{i+1}) is replaced in the curve by (p_{i-1}, p_{i+1}). Then, continue with vertices $(p_{i-1}, p_{i+1}, p_{i+2})$ until $i + 2$ is $m + 1$.

Let $p_i \in l_i \subset f_i$, ..., $p_j \in l_j \subset f_j$ be some consecutive vertices of the $AMLPP$ of g, where f_i, ..., f_j are some consecutive critical faces of g, and l_k is a line segment on f_k, $k = i, i+1, ..., j$. Let $\epsilon = 10^{-10}$ (this value defines the accuracy of the output of this algorithm). We can apply the method of Option 3 of rubberband algorithm (page 967, [1], and see correction in [14]) on cube-arc $\rho = (f_i, f_{i+1}, ..., f_j)$ to find an approximate $MLA(p_i, p_j)$ as follows:

Procedure 4

1. Calculate the length of arc $p_i p_{i+1} \cdots p_{j-1} p_j$, denoted by L_1;
2. Let $k = i+1$;
3. Take two points $p_{k-1} \in f_{k-1}$ and $p_{k+1} \in f_{k+1}$;
4. For line segment l_k on a critical face f_k in g, and points p_{k-1} and p_{k+1} on l_{k-1} and l_{k+1}, respectively, apply Procedure 1 to find a point $q_k \in l_k$ such that $d_{p_{k-1} q_k} + d_{p_{k+1} q_k} = \min\{d_{p_{k-1} p} + d_{p_{k+1} p} : p \in l_k\}$. Let $p_k = q_k$.
5. $k = k + 1$;
6. If $k = j$, calculate the length of arc $p_i p_{i+1} \cdots p_{j-1} p_j$, denoted by L_2.
7. If L_1 - $L_2 > \epsilon$, let $L_1 = L_2$ and go to Step 2. Otherwise, output the arc $p_i p_{i+1} \cdots p_{j-1} p_j$.

Let e_0, e_1, e_2, ... e_m and e_{m+1} be all consecutive critical edges of g in the counter-clockwise orientation of g. Let f_i be the first critical face of e_i in g, and c_i be the center of f_i, where $i = 0, 1, 2, ..., m$ or $m + 1$. All indices of points, edges and faces are taken $mod\ m + 2$. Let $\epsilon = 10^{-10}$. By Procedure 5, we can compute an $AMLPP$ of g and its length.

Procedure 5

1. Let P be a polygonal curve $p_0 p_1 \cdots p_m p_{m+1}$;
2. Calculate the length of P, denoted by L_1;
3. Let $i = 0$;
4. Take two points $p_{i-1} \in f_{i-1}$ and $p_{i+1} \in f_{i+1}$;
5. For the critical face f_i of an critical edge e_i in g, and points p_{i-1} and p_{i+1} in f_{i-1} and f_{i+1}, respectively, apply Procedure 2 to find a point q_i in f_i such that $d_{p_{i-1} q_i} + d_{p_{i+1} q_i} = \min\{d_{p_{i-1} p} + d_{p_{i+1} p} : p \in f_i\}$. Let $p_i = q_i$.
6. $i = i + 1$;
7. If $i = m + 3$, calculate the length of the polygonal curve $p_0 p_1 \cdots p_m p_{m+1}$, denoted by L_2.
8. If L_1 - $L_2 > \epsilon$, let $L_1 = L_2$ and go to Step 2. Otherwise, output the polygonal curve $p_0 p_1 \cdots p_m p_{m+1}$ as an $AMLPP$ of g and its length L_2.

Given an n-cube-arc unit $(f_i, \ldots, f_j)$ with respect to a polygonal curve P of g, where $n = 2$ or 3. Let $p_i \in f_i$ and $p_j \in f_j$. We can find an $MLA(p_i, p_j)$ by the following procedure.

Procedure 6

1. Compute the set $E = \{e: e$ is an edge of f_k, $k = i + 1, \ldots, j - 1\}$;
2. Let $I = 1$ and $L = 100$;
3. Compute the set $SE = \{S: S \subseteq E$ and $|S| = I \}$;
4. Go through each $S \in SE$, input $p_i, e_1, \ldots, e_l, p_j$ to Procedure 4 to compute an approximate $MLA(p_i, p_j)$ such that it has minimal length with respect to all $S \in SE$, denoted by $AMLA(I, SE)$, where $e_k \in S$, $k = 1, 2, \ldots, l$ and $l = |S|$. If the length of $AMLA(I, SE) < L$, let $MLA(p_i, p_j) = AMLA(I, SE)$ and $L =$ the length of $AMLA(I, SE)$;
5. Let $I = I + 1$.
6. If $I < n$ then go to Step 3. Otherwise, stop.

Lemma 8. *For each cube-arc unit* $\rho = (f_i, f_{i+1}, \ldots, f_j)$ *with respect to* P, $MLA(p_i, p_j)$ *can be computed in* $\mathcal{O}(1)$.

Proof. If ρ is a regular cube-arc unit, then $MLA(p_i, p_j)$ can be found by Procedure 2, which has complexity $\mathcal{O}(1)$. Otherwise, ρ is an n-cube-arc unit, where $n = 2$ or 3. Then, by Lemma 7, $MLA(p_i, p_j)$ can be found by Procedure 6, which can be computed in $\mathcal{O}(1)$ because $n = 2$ or 3. □

4.2   Algorithms

The original rubberband algorithm was published in [1] and slightly corrected in [14]. We now extend this corrected rubberband algorithm into the following (provable correct) algorithm.

The Edge-Based Rubberband Algorithm

1. Let P_0 be the polygon obtained by the (corrected) rubberband algorithm;
2. Find a point $p_i \in f_i$ such that p_i is the intersection point of an edge of P_0 with f_i, where $i = 0, 1, 2, \ldots, m$ or $m + 1$. Let P be a polygonal curve $p_0 p_1 \cdots p_m p_{m+1}$;
3. Apply Procedure 6 to all cube-arc units of P. If for each cube-arc unit $\rho = (f_i, f_{i+1}, \ldots, f_j)$ with respect to P, the arc $(p_i, p_{i+1}, \ldots, p_j) = MLA(p_i, p_j)$, then P is the $MLPP$ of g (by Theorem 2), and go to Step 4. Otherwise, go to Step 3.
4. Apply Procedure 3 to obtain the final MLP.

The Face-Based Rubberband Algorithm

1. Take a point $p_i \in f_i$, where $i = 0, 1, 2, \ldots, m$ or $m + 1$;
2. Apply Procedure 5 to find an $AMLPP$ of g, denoted by P;

3. Find all maximal 2-cube-arcs with respect to P, apply Procedure 4 to update the vertices of the $AMLPP$, which are on one of the 2-cube-arcs. (By Lemma 3, the input line segments of Procedure 4 are critical edges.) Repeat this step until the length of the updated $AMLPP$ is sufficiently accurate (i.e., previous length minus current length $< \epsilon$);

4. Apply Procedure 5 to update the current $AMLPP$;

5. Find all maximal (2,3)-cube-arcs with respect to the current P, apply Procedure 4 to update the vertices of the current $AMLPP$, which are on one of the (2,3)-cube-arcs. The input line segments of Procedure 4 can be found such that they are on the critical face and parallel or perpendicular to the critical edge of the face. Repeat this step until the length of the updated $AMLPP$ is sufficiently accurate;

6. Apply Procedure 5 to update the current $AMLPP$.

7. Apply Procedure 6 to all cube-arc units of P. If for each cube-arc unit $\rho = (f_i, f_{i+1}, \ldots, f_j)$ with respect to P, the arc $(p_i, p_{i+1}, \ldots, p_j) = MLA(p_i, p_j)$, then P is the $MLPP$ of g (by Theorem 2), and go to Step 8. Otherwise, go to Step 3.

8. Apply Procedure 3 to obtain the final MLP.

4.3 Computational Complexity

It is obvious that Procedures 1 and 2 can be computed in $\mathcal{O}(1)$, and Procedure 3 can be computed in $\mathcal{O}(m)$, where m is the number of critical edges of g. The

Table 2. Comparison of results of steps of the face-based rubberband algorithm. p_{1_0}, $p_{1_1}, \ldots, p_{1_{18}}$ are the results of Step 2, and $p_{2_0}, p_{2_1}, \ldots, p_{2_{18}}$ are the results of Step 3.

p_{1i}	x_{1i}	y_{1i}	z_{1i}	p_{2i}	x_{2i}	y_{2i}	z_{2i}
p_{1_0}	-0.5	1	0	p_{2_0}	-0.5	1	-0.21
p_{1_1}	-0.5	1	0	p_{2_1}	-0.5	1	-0.21
p_{1_2}	-1.5	3	-0.34	p_{2_2}	-1.5	3	-0.34
p_{1_3}	-2.5	3.29	-0.5	p_{2_3}	-2.5	3.23	-0.5
p_{1_4}	-2.5	3.29	-0.5	p_{2_4}	-2.5	3.23	-0.5
p_{1_5}	-3.5	3.5	-1.11	p_{2_5}	-3.5	3.45	-1.11
p_{1_6}	-4.15	3.64	-1.5	p_{2_6}	-4.15	3.64	-1.5
p_{1_7}	-5.5	3.94	-2.32	p_{2_7}	-5.5	3.94	-2.32
p_{1_8}	-5.8	4	-2.5	p_{2_8}	-5.69	4	-2.5
p_{1_9}	-5.8	4	-2.5	p_{2_9}	-5.69	4	-2.5
$p_{1_{10}}$	-6.5	4	-3.32	$p_{2_{10}}$	-6.5	4	-3.32
$p_{1_{11}}$	-6.65	4	-3.5	$p_{2_{11}}$	-6.65	4	-3.5
$p_{1_{12}}$	-7.5	4	-4.5	$p_{2_{12}}$	-7.5	4	-4.5
$p_{1_{13}}$	-7.95	4	-5.5	$p_{2_{13}}$	-8	4	-5.5
$p_{1_{14}}$	-7.95	4	-5.5	$p_{2_{14}}$	-8	4	-5.5
$p_{1_{15}}$	-8.5	3	-5.5	$p_{2_{15}}$	-8.5	3	-5.5
$p_{1_{16}}$	-8.5	-1	-5.5	$p_{2_{16}}$	-8.5	-1	-5.5
$p_{1_{17}}$	-8.5	-1	-0.5	$p_{2_{17}}$	-8.5	-1	-0.5
$p_{1_{18}}$	-0.5	-1	-0.1	$p_{2_{18}}$	-0.5	-1	-0.1

Table 3. Comparison of results of steps of the face-based rubberband algorithm. p_{3_0}, p_{3_1}, ..., $p_{3_{18}}$ are the results of Step 4, and p_{4_0}, p_{4_1}, ..., $p_{4_{18}}$ are the results of Step 7.

p_{3i}	x_{3i}	y_{3i}	z_{3i}	p_{4i}	x_{4i}	y_{4i}	z_{4i}
p_{3_0}	-0.5	1	-0.21	p_{4_0}	-0.5	1	-0.5
p_{3_1}	-0.5	1	-0.21	p_{4_1}	-0.5	1	-0.5
p_{3_2}	-1.5	3	-0.41	p_{4_2}	-1.5	3	-0.5
p_{3_3}	-2.5	3.23	-0.5	p_{4_3}	-2.5	3.22	-0.5
p_{3_4}	-2.5	3.23	-0.5	p_{4_4}	-2.5	3.22	-0.5
p_{3_5}	-3.5	3.47	-1.13	p_{4_5}	-3.5	3.48	-1.17
p_{3_6}	-4.09	3.62	-1.5	p_{4_6}	-4	3.61	-1.5
p_{3_7}	-5.5	3.95	-2.38	p_{4_7}	-5.5	4	-2.5
p_{3_8}	-5.69	4	-2.5	p_{4_8}	-5.5	4	-2.5
p_{3_9}	-5.69	4	-2.5	p_{4_9}	-5.5	4	-2.5
$p_{3_{10}}$	-6.5	4	-3.4	$p_{4_{10}}$	-6.5	4	-3.5
$p_{3_{11}}$	-6.59	4	-3.5	$p_{4_{11}}$	-6.5	4	-3.5
$p_{3_{12}}$	-7.5	4	-4.5	$p_{4_{12}}$	-7.5	4	-4.5
$p_{3_{13}}$	-8	4	-5.5	$p_{4_{13}}$	-8	4	-5.5
$p_{3_{14}}$	-8	4	-5.5	$p_{4_{14}}$	-8	4	-5.5
$p_{3_{15}}$	-8.5	3	-5.5	$p_{4_{15}}$	-8.5	3	-5.5
$p_{3_{16}}$	-8.5	-1	-5.5	$p_{4_{16}}$	-8.5	-1	-5.5
$p_{3_{17}}$	-8.5	-1	-0.5	$p_{4_{17}}$	-8.5	-1	-0.5
$p_{3_{18}}$	-0.5	-1	-0.27	$p_{4_{18}}$	-0.5	-1	-0.5

main operation of Procedure 4 is Step 4, which can be computed in $\mathcal{O}(n)$, where n is the number of points of the arc. Analogously, Procedure 5 can be computed in $\mathcal{O}(m)$, where m is the number of points of the polygonal curve.

[14] has proved that the (corrected) rubberband algorithm can be computed in $\mathcal{O}(m)$, where m is the number of critical edges of g. The main additional operation of the edge-based rubberband algorithm is Step 3 which can be computed in $\mathcal{O}(m)$, where m is the number of critical edges of g (by Lemma 8). It follows that the edge-based rubberband algorithm can be computed in $\mathcal{O}(m)$, where m is the number of critical edges of g.

For the face-based rubberband algorithm, Step 1 is trivial; Steps 2, 4, 6 have the same complexity as Procedure 5. Step 3 can be computed in $N_1(\epsilon)\mathcal{O}(m)$, where $N_1(\epsilon)$ depends on the accuracy ϵ, where m is the number of points of the polygonal curve. Analogously, Step 5 can be computed in $N_2(\epsilon)\mathcal{O}(m)$, where $N_2(\epsilon)$ depends on the accuracy ϵ (Note that there is a constant number of different combinations of input line segments of Procedure 4). By Lemma 8, Step 7 can be computed in $\mathcal{O}(m)$, where m is the number of critical edges of g. Therefore, the face-based rubberband algorithm can be computed in $\mathcal{O}(m)$, where m is the number of critical edges of g.

5   An Example

We approximate the MLP of the simple cube-curve g, shown in Figure 2. Table 1 lists all coordinates of critical edges of g. We take the centers of the first critical

Table 4. Results of the edge-based rubberband algorithm. p_{4_0}, p_{4_1}, ..., $p_{4_{18}}$ are the vertices of the MLP of the simple cube-curve shown in Figure 2.

$final_{p_i}$	x_i	y_i	z_i
p_{4_0}	-0.5	1	-0.5
p_{4_2}	-1.5	3	-0.5
p_{4_3}	-2.5	3.22	-0.5
p_{4_7}	-5.5	4	-2.5
$p_{4_{12}}$	-7.5	4	-4.5
$p_{4_{13}}$	-8	4	-5.5
$p_{4_{15}}$	-8.5	3	-5.5
$p_{4_{16}}$	-8.5	-1	-5.5
$p_{4_{17}}$	-8.5	-1	-0.5
$p_{4_{18}}$	-0.5	-1	-0.5

Table 5. Lengths of calculated curves at different steps of the face-based rubberband algorithm, compared with the length calculated by the edge-based rubberband algorithm

step	initial	2	3	4	8	edge-based rubberband algorithm
length	35.22	31.11	31.08	31.06	31.01	31.01

faces of g to produce an initial polygonal curve for the face-based rubberband algorithm. The updated polygonal curves are shown in Tables [3] 2 and 3. We take the middle points of each critical edge of g for the initialization of the polygonal curve of the (corrected) rubberband algorithm. The resulting polygon is shown in Table 4. Table 5 shows that the edge-based and face-based rubberband algorithms converge to the same MLP of g.

6   Conclusions

We have presented an edge-based and a face-based rubberband algorithm and have shown that both are provable correct for any simple cube-curve. We also have proved that their time complexity is $\mathcal{O}(m)$, where m is the number of critical edges of g. The presented algorithms followed the basic outline of the original rubberband algorithm [1].

References

1. T. Bülow and R. Klette. Digital curves in 3D space and a linear-time length estimation algorithm. *IEEE Trans. Pattern Analysis Machine Intell.*, **24**:962–970, 2002.
2. J. Canny and J.H. Reif. New lower bound techniques for robot motion planning problems. In Proc. *IEEE Conf. Foundations Computer Science*, pages 49–60, 1987.

---

[3] Two digits are used only for displaying coordinates. Obviously, in the calculations it is necessary to use higher precision.

3. J. Choi, J. Sellen, and C.-K. Yap. Approximate Euclidean shortest path in 3-space. In Proc. *ACM Conf. Computational Geometry*, ACM Press, pages 41–48, 1994.
4. D. Coeurjolly, I. Debled-Rennesson, and O. Teytaud. Segmentation and length estimation of 3D discrete curves. In Proc. *Digital and Image Geometry*, pages 299–317, LNCS 2243, Springer, 2001.
5. M. Dror, A. Efrat, A. Lubiw, and J. Mitchell. Touring a sequence of polygons. In Proc. *STOC*, pages 473–482, 2003.
6. E. Ficarra, L. Benini, E. Macii, and G. Zuccheri. Automated DNA fragments recognition and sizing through AFM image processing. *IEEE Trans. Inf. Technol. Biomed.*, **9**:508–517, 2005.
7. A. Jonas and N. Kiryati. Length estimation in 3-D using cube quantization, *J. Math. Imaging and Vision*, **8**: 215–238, 1998.
8. M. I. Karavelas and L. J. Guibas. Static and kinetic geometric spanners with applications. In Proc. *ACM-SIAM Symp. Discrete Algorithms*, pages 168–176, 2001.
9. R. Klette and T. Bülow. Critical edges in simple cube-curves. In Proc. *Discrete Geometry Comp. Imaging*, LNCS 1953, pages 467–478, Springer, Berlin, 2000.
10. R. Klette and A. Rosenfeld. *Digital Geometry: Geometric Methods for Digital Picture Analysis*. Morgan Kaufmann, San Francisco, 2004.
11. F. Li and R. Klette. Minimum-length polygon of a simple cube-curve in 3D space. In Proc. *Int. Workshop Combinatorial Image Analysis*, LNCS 3322, pages 502–511, Springer, Berlin, 2004.
12. F. Li and R. Klette. The class of simple cube-curves whose MLPs cannot have vertices at grid points. In Proc. *Discrete Geometry Computational Imaging*, LNCS 3429, pages 183–194, Springer, Berlin, 2005.
13. F. Li and R. Klette. Minimum-Length Polygons of First-Class Simple Cube-Curve. In Proc. *Computer Analysis Images Patterns*, LNCS 3691, pages 321–329, Springer, Berlin, 2005.
14. F. Li and R. Klette. Analysis of the rubberband algorithm. Technical Report CITR-TR-175, Computer Science Department, The University of Auckland, Auckland, New Zealand, 2006 (www.citr.auckland.ac.nz).
15. T.-Y. Li, P.-F. Chen, and P.-Z. Huang. Motion for humanoid walking in a layered environment. In Proc. *Conf. Robotics Automation*, Volume 3, pages 3421–3427, 2003.
16. H. Luo and A. Eleftheriadis. Rubberband: an improved graph search algorithm for interactive object segmentation. In Proc. *Int. Conf. Image Processing*, Volume 1, pages 101–104, 2002.
17. J. Sklansky and D. F. Kibler. A theory of nonuniformly digitized binary pictures. *IEEE Trans. Systems Man Cybernetics*, **6**:637–647, 1976.
18. F. Sloboda, B. Zaťko, and R. Klette. On the topology of grid continua. In Proc. *Vision Geometry*, SPIE 3454, pages 52–63, 1998.
19. F. Sloboda, B. Zaťko, and J. Stoer. On approximation of planar one-dimensional grid continua. In R. Klette, A. Rosenfeld, and F. Sloboda, editors, *Advances in Digital and Computational Geometry*, pages 113–160. Springer, Singapore, 1998.
20. C. Sun and S. Pallottino. Circular shortest path on regular grids. CSIRO Math. Information Sciences, CMIS Report No. 01/76, Australia, 2001.
21. M. Talbot. A dynamical programming solution for shortest path itineraries in robotics. *Electr. J. Undergrad. Math.*, **9**:21–35, 2004.
22. R. Wolber, F. Stäb, H. Max, A. Wehmeyer, I. Hadshiew, H. Wenck, F. Rippke, and K. Wittern. Alpha-Glucosylrutin: Ein hochwirksams Flavonoid zum Schutz vor oxidativem Stress. *J. German Society Dermatology*, **2**:580–587, 2004.

Surface Registration Markers
from Range Scan Data

John Rugis[1,2] and Reinhard Klette[1]

[1] CITR, Dep. of Computer Science, The University of Auckland
Auckland, New Zealand
[2] Dep. of Electrical & Computer Engineering, Manukau Institute of Technology
Manukau City, New Zealand
john.rugis@manukau.ac.nz

Abstract. We introduce a data processing pipeline designed to generate registration markers from range scan data. This approach uses *curvature maps* and *histogram-templates* to identify local surface features. The noise associated with real-world scans is addressed using a (common) *Gauss filter* and *expansion-segmentation*. Experimental results are presented for data from The Digital Michelangelo Project.

1   Introduction

There are a number of challenges associated with 3D range scan digitizations (or 3D surface reconstructions in general) of real-world objects. At first the captured data should be accurate, and second the (in general) various data sets need to be unified into one consistent surface model. In general, only a partial section of an object surface is acquired with each scan, and a number of scans need to be aligned and subsequently merged together. When each scan is taken from a different uncalibrated viewpoint, aligning the scans can be time consuming [11, 15]. There exists a number of algorithms [2, 3, 14], including the widely used Iterative Closest Point algorithm (ICP), that can refine a given rough alignment of scan pairs, two at a time. Initial matching and alignment of multiple (possible many) 3D scans is still largely an open problem. Since range scans normally already each have the same scaling, alignment to some fixed reference points involves a linear transformation consisting of a 3D translation and a 3D rotation. Exact alignment is in practice not possible due to the noise in real-world data. Input data uncertainty is also present when surface patches have been generated via less reliable computer vision 3D surface recovery techniques such as photometric stereo or structure from motion (see, e.g., [10] for 3D surface recovery techniques).

Early work used 2D features, such as contour based grouping using relaxation methods [13]. For the visualization of implicit surfaces we cite [1]. More recent work in [5] uses (what they refer to) a "curvature map method" to characterize a local signature for every point in a scan. This can be seen as a continuation of surface characterizations by Gauss maps in differential geometry, or by extended Gaussian images (see, e.g., [17]) in computer vision.

U. Eckardt et al. (Eds.): IWCIA 2006, LNCS 4040, pp. 430–444, 2006.

In this paper we propose a matching and initial alignment approach based on identifying a small number of registration markers. The registration markers that we generate are based on local surface curvature features. An advantage to using surface curvature is that it is rotation and translation invariant. However, one problem that needs to be overcome is that curvature, being a second derivative property [4, 9], is very sensitive to local noise.

The sequence of steps in the processing pipeline that we use for surface marker generation is to (i) calculate (noisy) curvatures from the point data, (ii) filter and segment the curvature data, and then (iii) identify and mark local features using the curvature data. We illustrate the application of this sequence of steps to multiple scans by an example. For each step in this processing pipeline, we select and define appropriate methods and algorithms.

As a particular difficulty, in practice, scan overlap regions can be small relative to the scan size, and thus our registration markers need to be smaller than the scan overlap. We keep, from the outset, the demands of extensive real-world datasets in mind. Finally we demonstrate our approach on data from The Digital Michelangelo Project [11, 12].

2   Curvature and Curvature Estimators

Studies on surface curvature can be traced back to original work by Gauss [6]; see books on differential geometry (e.g., [4]) or a discussion of surface curvature in the context of 3D image analysis in [9]. The surface curvature of continuous smooth surfaces is a well-defined property, however, when working with point set data, triangulated surfaces or 3D digital images, the surface curvature can only be estimated. There are a number of different existing curvature estimators for such situations [9]. In this paper, we use an uncompensated orthogonal cut method to calculate a mean curvature as described in [16].

The curvature estimation approach that we use, for each scan point, is to firstly identify the four nearest neighbor points associated with two orthogonal planar cuts. Next, for each of the two cuts, calculate an estimated signed planar line curvature. Finally, take the mean of these two curvatures as an estimate of the mean surface curvature.

With reference to Figure 1, the planar line curvature (in each planar cut) is estimated as the incremental angular advance divided by the incremental change in length

$$\kappa = \frac{\alpha}{(d_1 + d_2)/2} \qquad (1)$$

where d_1 is the length of the line segment from $\mathbf{p}_1$ to $\mathbf{p}_2$ and d_2 is the length of the line segment from $\mathbf{p}_2$ to $\mathbf{p}_3$. Consequently, the curvature can be calculated as

$$\kappa = \left(\frac{2}{||\mathbf{v}_1|| + ||\mathbf{v}_2||} \right) \cos^{-1} \left(\frac{\mathbf{v}_1 \cdot \mathbf{v}_2}{||\mathbf{v}_1|| \, ||\mathbf{v}_2||} \right) \qquad (2)$$

where $\mathbf{v}_1 = \mathbf{p}_2 - \mathbf{p}_1$ and $\mathbf{v}_2 = \mathbf{p}_3 - \mathbf{p}_2$.

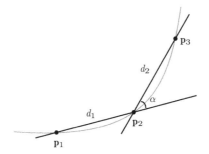

Fig. 1. Planar line curvature estimation

In practice this results in a (noisy) mean curvature value associated with each 3D scan point. We also tested further curvature measures (as described in [9]); however, the mean curvature estimate gives results that correspond to actual image appearance of the object and thus is the most relevant feature for our purposes.

3   Curvature Maps, Filtering and Segmentation

In this section, we convert the mean curvature data at surface scan points into a (2D) *curvature map*, which is an array of the same dimension (ignoring squashing) as the given 2D scan array, where values are mean curvatures at scan points [16]. There are a number of advantages when using such curvature maps, including the possibility of visualization and processing using 2D image processing techniques. Pixels in 2D images are, by standard convention, stored in 2D arrays and we apply adjacency definitions based on the related orthogonal grid. If the 3D point data has been acquired in a 3D orthogonal grid, then the curvature mapping is straightforward (defined by orthogonal cuts parallel to coordinate planes, see [7]).

Data acquired in the Michelangelo project uses a hexagonal adjacency (i.e., 6-adjacency in the image plane) in a virtual projection plane, defined by the order and geometry of scan acquisition. In such a case we can use a special squashed dot mapping [16].

Mappings for both cases, either orthogonal or hexagonal, are shown in Figure 2, with the reverse mapping as shown in Figure 3. The second mapping has the expense of quadrupling the number of pixels. (We do not discuss further scan acquisition geometries in this paper).

Fig. 2. Mappings of orthogonal (on the left) or hexagonal (on the right) grids into an orthogonal grid

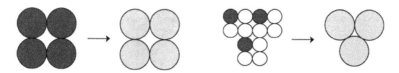

Fig. 3. Reverse mappings of orthogonal (on the left) or squashed hexagonal (on the right) grids into the original grid

We then apply a 2D Gaussian filter to the curvature map to reduce noise. As an illustration of this effectiveness of this technique, consider noisy data from a sampled planar surface. The noisy data points will (incorrectly) exhibit a symmetrical distribution of positive and negative curvatures centered around zero, with the average value being (correctly) zero. The Gaussian filter performs exactly the desired spatial averaging. Note that we are addressing the noisy data problem in 2D rather than in the original 3D domain. The terms in an $n_1 \times n_2$ Gaussian convolution kernel centered at $(0, 0)$ are determined (as usual) using the formula

$$h(n_1, n_2) = \frac{h_g(n_1, n_2)}{\sum_{n_1} \sum_{n_2} h_g} \quad \text{with} \quad h_g(n_1, n_2) = e^{-(n_1^2 + n_2^2)/2\sigma^2}$$

where σ is effectively (in our case) a smoothing area factor.

The filtering process correctly reduces the extremes in the distribution of mean curvature values. The resulting mean curvatures are more representative of the actual surface curvature, which is concentrated in a much smaller range around zero. Now, in order to reveal previously obscured detail, we perform a linear expansion centered around this zero-curvature region of interest.

Multi-threshold segmentation of the expanded data is then performed to partition the data into a number of *curvature bins*. This data reduction enables the approach presented in the next section. An example of how we perform a combined expansion-segmentation operation is given in the experiment section of this paper.

Also, in preparation for local feature identification, we reverse the 3D to a 2D curvature mapping process, by selecting pixels, depending on the original adjacency as illustrated in Figure 3, and update the mean curvature associated with each of the 3D scan points with its filtered and expansion-segmented value.

4   Local Feature Identification

We use *histogram-templates* to search through the filtered and segmented curvature data for local surface features. A histogram-template consists of (i) a sliding window that is segmented into subsets and (ii) a set of histograms, one for each subset. Each histogram accumulates curvature counts into a number of curvature bins. The number of histogram curvature bins is assigned to be exactly the same as the number of segmentation curvature bins in the previous section.

Fig. 4. Subsets for the definition of a histogram template (here: four histograms for indicated positions) for a surface pit; the individual histograms are not shown in this figure

Note that, even though we are now working within the 3D dataset, we do the histogram-template search in a 2D fashion using the 2D orthogonal grid.

The histogram-templates are carefully designed (at the given scale of scan data) to characterize local surface features; a multi-scale approach would be relevant if uncertainty increases for the given scale of scan data. An example histogram template is shown in Figure 4.

In the example, the histogram-template is designed to characterize pit-like surface defects which have radial symmetry. The template indicates that four histograms, indexed zero through three, are tabulated for each selected 3D data point [1]. Histogram two is the center region which should contain negative curvature values, histogram one is the pit perimeter which should contain positive curvature values, histogram zero is a guard ring which should contain low curvature magnitude values, and histogram three is a "don't care" region. Individual histogram bins are assigned such that first bin accumulates the count of largest negative magnitude curvatures and the last bin accumulates the count of largest positive magnitude curvatures.

The histogram-template is used to search through the curvature values looking for the desired feature match.

5   Performance Evaluation

The value of algorithm performance evaluation in the field of computer vision has been discussed in [8]. A possible starting point for performance evaluation is

---

[1] Excluding those close to the array's border.

to use a data set having known indisputable characteristics. These indisputable characteristics can be used to establish what is referred to as a *ground truth*.

One way to insure the existence of ground truth in a data set is to model and generate synthetic data having known properties. In our case we have chosen to synthesize data using a hexagonal adjacency orthogonal projection grid. Because curvature is scaling dependent, we needed to establish numerical distance values for our grid. Figure 5 shows the values that we used[2]. We will refer to the illustrated vertical dimension as the *resolution* of the scan.

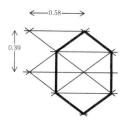

Fig. 5. Scanning grid

The synthetic objects that we "scanned" consisted of a planar surfaces having spherical indentations and spherical bumps. The indentation model is shown in Figure 6. Note that the sphere cuts into the plane by a distance of one half the radius. Spherical bumps are modeled similarly.

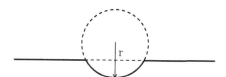

Fig. 6. The indentation model

5.1   Synthetic Data

5.2   Performance Measures

We evaluated our curvature estimator in the presence of varying levels of Gaussian noise. We defined noise level by a scale factor relative to the scan resolution. A noise scale factor of one means that the noise sigma value is equal to the resolution. A factor of two means that the noise sigma value is equal to the resolution dived by two, and so on. Of course, the actual curvature of a planar surface is zero and the curvature at any point on a sphere is a constant equal to the reciprocal of its radius.

---

[2] These values closely match those used in the real-world data that we explore later in this paper.

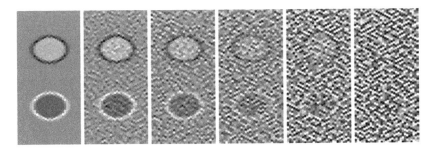

Fig. 7. Curvature estimation: no noise, sf:16, sf:8, sf:4, sf:2, sf:1

Figure 7 illustrates the performance of the 2-cut curvature estimator in the presence of varying levels of noise. Positive curvature is shading coded as white. Zero curvature is shading coded as medium gray and maximum negative curvature is encoded as black. Thus, bumps are shown in the top half of the figure and indentations in the bottom half of the figure.

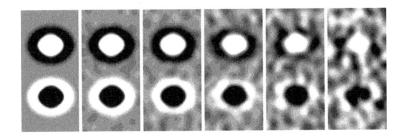

Fig. 8. Filtered and segmented: no noise, sf:16, sf:8, sf:4, sf:2, sf:1

Figure 8 illustrates the performance of the filtering and segmentation process, again in the presence of varying noise levels.

5.3   Feature Identification Evaluation

We applied the complete feature identification process to a scan of a more comprehensive synthetic object. The object has sixteen indentations associated with spheres having radii which vary from 0.1 to 2.0. Figure 9 illustrates the results pictorially. Identified features are each marked with five red pixels. Note that because the feature identification process searches through (nearly) every image pixel, the same feature is usually hit and marked more than once.

The results from Figure 9 are summarized in Table 1. We can make a number of observations. Increasing noise levels tend to cause the process to associate a smaller size to the feature. The fixed size matching template does in fact identify a range of indentation feature sizes. There are no false positives, even in the presence of rather high noise.

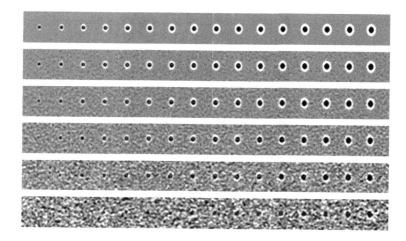

Fig. 9. Feature identification results: synthetic data and six noise levels

Table 1. Summary of feature marker hits from Figure 9

noise factor	marker hits: small radius ———————— large radius indentation														features
none	3	9	10	8	4	3	2	1								8
16		9	10	12	8	6	5	3	3							8
8		3	9	10	12	10	5	5	3	2						9
4			7	10	10	8	12	6	6	5	3					9
2				2	8	10	13	3	9	7	11		2		2	10
1									1	3	9	12		2	4	6

6 Experiments and the David Dataset

We have also tested our approach using the extensive David dataset from the University of Stanford Digital Michelangelo Project [11, 12]. The David data set consists of 1.93 giga-bytes of data which has been made available in nine

Fig. 10. The scanner acquiring 3D data [11] using structured lighting (known as light plane projection [10])

Fig. 11. Points from two scans ("blue" and "red scan") rendered with lighting

compressed files. The uncompressed data set represents approximately 1.1 billion 3D space points. There are a total of 6,540 raw scan files collected into 515 groupings. Each scan was acquired over a fixed width of approximately 140mm

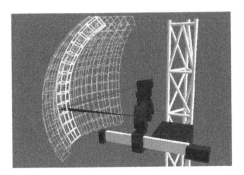

Fig. 12. The scanner's imaging volume [11]

and a height of generally no larger than 600mm. The David statue is over five meters tall.

The scanner is shown acquiring data in Figure 10, and a rendered image of the points from two adjacent overlapping scans is shown in Figure 11. It is not uncommon for a scan to contain upwards of 800,000 points. The scanning system's physical geometry is illustrated in Figure 12. The cyan colored boxes represent the volume regions associated with possible individual scans.

The University of Stanford group responsible for the project undertook a rather time consuming initial manual alignment of the individual scans. They reported both on this process and the need for an automated global matching method[12]. The surface identification markers described in this paper are candidates for a feature based global matching technique.

6.1   Data Structure

Each scan is acquired in a regular sweeping pattern of scan lines, left-to-right, and top-to-bottom. Each scan line consists of a zig-zag pattern of 486 points as shown in Figure 13. The scanning pattern results in a hexagonal grid. Of course, the scanner does not find the reflective surface of an object at all points, and this is illustrated, for example, in Figure 13 where the white dots represent the surface of an object and the darker dots mean that nothing was found in that direction.

The regular scanning pattern suggests 6-adjacency (based on the hexagonal grid) shown in Figure 14. A '-1' entry in the array means that no surface point was found. A non-negative integer entry is an index value into another array of surface point data including its 3D location and a not yet calculated mean curvature value.

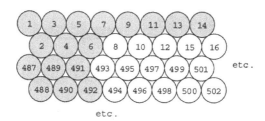

Fig. 13. Top left corner of the scanning sequence

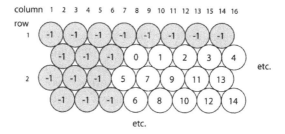

Fig. 14. The hexagonal grid of a scan

6.2   Curvature Estimators, Filtering and Segmentation

Since the 3D data points have 6-adjacency within each scan, the nearest neighbor orthogonal planar cut points [3] are selected as shown in Figure 15 (dark squares), and the squashed dot mapping is used as discussed above.

Fig. 15. 6-adjacency and orthogonal cut points

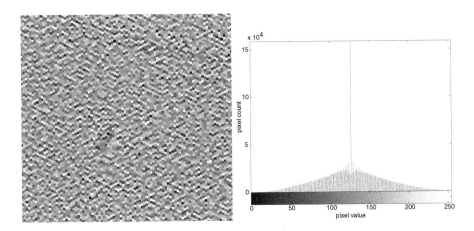

Fig. 16. Noisy mean curvature map (on the left) and its histogram

An eight-bit precision mean curvature map of a section of a scan of David's face is shown in Figure 16 along with a histogram of that entire scan. Positive surface curvature is defined as that which bends away from the scanning source or, equivalently in this case, as that curvature associated with viewing a convex hull from the outside. Maximum positive curvature is shading coded as white. Zero curvature is shading coded as medium gray and maximum negative curvature is encoded as black. Surface detail in this image is obscured by noise.

For Gaussian filtering of the David data, a sigma of four is large enough to smooth noise but not so large as to remove surface detail. A Gaussian filtered version of the previous curvature map is shown in Figure 17. Note that there

---

[3] Note that, particularly in the case of data having 6-adjacency, a three-cut mean calculation is also possible. Experiments using this more expensive method gave similar results to those presented in this paper.

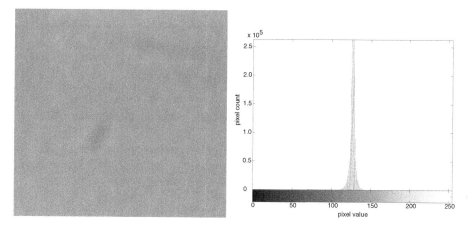

Fig. 17. Processed map (on the left) and used Gaussian filter

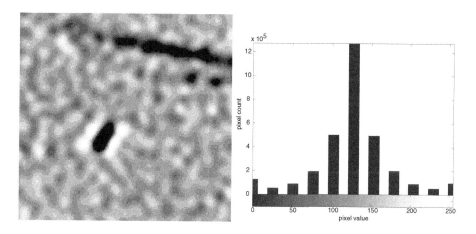

Fig. 18. Processed map: multi-level segmentation

is now not much visible detail and that the levels are now mostly concentrated around the middle medium gray as shown in the associated histogram.

As mentioned earlier in this paper, we need to expand the middle histogram region to recover detail. Because of the eight-bit precision used, the middle region contains only a small discrete range of integer values. We perform a linear expansion around the center (medium gray) and clip the results (to full-black and full-white). The result, in this example, is an image containing pixels having only eleven different integer intensity values.

In general, this limited precision expansion will always also simultaneously multi-level segment the data into a limited number of segmentation bins.

An *expansion segmented* version of the previous mean curvature map is shown in Figure 18. Curvature detail is now readily apparent. The histogram shows the range of eleven segmentation values.

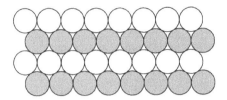

Fig. 19. Every-other-row pixel subset for feature searching (shown as white)

We now reverse the squashed dot mapping process, by selecting pixels as illustrated in Figure 3, and update the mean curvature associated with each 3D point with its filtered and expansion-segmented value.

6.3   Feature Identification

There are numerous small (approximately 2-4mm diameter) surface 'pits', due to weathering and abuse, found scattered on the surface of the David statue[4] that are ideal candidates for surface markers. The pit-like features certainly meet the requirement of being smaller than scan overlap regions. One of these pits is clearly visible in Figure 18. A pit can be characterized by its center of negative curvature and a rim of positive curvature.

Experimental comparison showed that it was possible to reduce the histogram template search time by employing data thinning, as illustrated in Figure 19, where only every other row is examined[5]. This thinning also results in pixels defined in an orthogonal grid, which, rather conveniently, means that we can use a rectangular search template.

We used precisely the histogram template given earlier as an example in Figure 4 to search through the curvature data for pit surface features. In this particular experiment, the non-square dimensions of the template compensate for the unequal aspect ratio of the rectangular data pixel subset. In fact, this template maps back to a (nearly) square region in the 3D dataset.

The identification test for pit features that we used with the David dataset is a minimum of twelve black (fully negative curvature) values in the center region, a minimum of four white, or nearly white (very positive curvature) values in the rim, a maximum of four black or nearly black (very negative curvature) values in the guard ring, and a maximum of two white (fully positive curvature) values in the guard ring. This identification test was able to identify pit surface features as reported in the next section.

6.4   Some Feature Mapping Results

We illustrate the application of the example histogram template to four different, but overlapping, scans of David. The results are shown in Figure 20. The first

---

[4] This can be verified by rendering images of triangle mesh versions of the scan data with specular reflections and appropriately placed lighting.

[5] The sampling frequency appears to be at least double that of the Nyquist frequency associated with the smallest underlying surface features, and thus, there is no aliasing associated with data thinning by a factor of two.

and third scans where acquired from the same view-point. The second and forth scans where acquired from different view-points. Note that there is at least one feature that links each scan to at least two other scans. For example, there is a feature just below the left eye that links scan one to each of scans two and three.

Comparison, by the authors, with visual observations of the 3D triangle mesh models mentioned previously, confirms that common features in different scans have been successfully identified and marked using our processing pipeline. Because these registration markers are associated with the 3D dataset, additional 3D marker characteristics, such as the size of the pits, can be calculated.

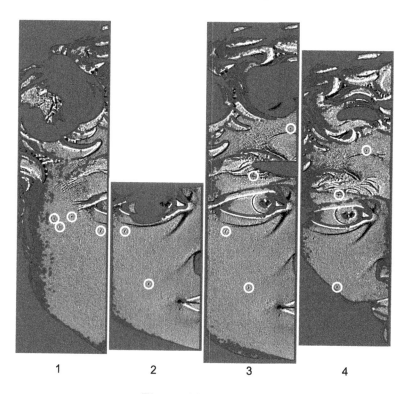

Fig. 20. Marker results

7   Conclusions

We have demonstrated the possibility and effectiveness of data smoothing *after* a shading encoded mean curvature calculation on noisy data. In general, this approach reduces complexity in that it moves the smoothing process from the 3D domain into the 2D domain. We have also demonstrated the effectiveness of using 2D (rotation invariant) radially symmetric matching templates to identify 3D surface features.

Further work is anticipated to include additional marker characterizations and using curvature based registration markers to automatically generate an initial alignment of diverse overlapping scans of the same object. Results using different curvature measures (besides the mean curvature) will be published in a forthcoming report.

Acknowledgements. The first author acknowledges the financial support from The Department of Electrical and Computer Engineering at Manukau Institute of Technology. The authors thank Stanford University for access to the Digital Michelangelo Project archive.

References

1. Balsys, R.J. and K.G. Suffern: Visualisation of implicit surfaces. *Computers & Graphics,* **25**:89–107, 2001.
2. Brown, B.J. and S. Rusinkiewicz: Non-rigid range-scan alignment using thin-plate splines. In Proc. *3D Data Processing Visualization Transm.,* pages 759–765, 2004.
3. Chang, M.C., F.F. Leymarie, and B.B. Kimia: 3D shape registration using regularized medial scaffolds. In Proc. *3D Data Processing Visualization Transmission,* pages 987–994, 2004.
4. Davies, A., and P. Samuels: *An Introduction to Computational Geometry for Curves and Surfaces.* Oxford University Press, Oxford, 1996.
5. Gatzke, T., C. Grimm, M. Garland, and S. Zelinka: Curvature maps for local shape comparison. In Proc. *Shape Modeling and Applications,* pages 244–253, 2005.
6. Gauss, C.F.: *General Investigations of Curved Surfaces* (Reprint of publications from 1825 and 1827). Dover Publications, New York, 2005.
7. Hermann, S. and R. Klette: Multigrid analysis of curvature estimators. In Proc, *Image and Vision Computing New Zealand,* pages 108–112, 2003.
8. Klette, R., Stiehl, H.S., Viergever. M.A, and Vincken, K.L. (eds.): *Performance Characterization in Computer Vision.* Kluwer, Dordrecht, 2000.
9. Klette, R. and A. Rosenfeld: *Digital Geometry.* M. Kaufmann, San Francisco, 2004.
10. Klette, R., K. Schlüns, and A. Koschan. *Computer Vision - Three-Dimensional Data from Images.* Springer, Singapore, 1998.
11. Levoy, M.: The digital Michelangelo project. In Proc. *3D Digital Imaging Modeling,* pages 34–43, 1999.
12. Levoy, M., K. Pulli, B. Curless, S. Rusinkiewicz, D. Koller, L. Pereira, M. Ginzton, S. Anderson, J. Davis, J. Ginsberg, J. Shade, D. Fulk: The digital Michelangelo project: 3D scanning of large statues. In Proc. *SIGGRAPH,* pages 131–144, 2000.
13. Parent, P. and S. Zucker: Trace inference, curvature consistency, and curve detection. *IEEE Trans. Pattern Analysis and Machine Intelligence,* **11**:823–839, 1989.
14. Pulli, K.: Multiview registration for large data sets. In Proc. Int. Conf. *3D Digital Imaging and Modeling,* pages 160–168, 1999
15. Pulli, K., B. Curless, M. Ginzton, S. Rusinkiewicz, L. Pereira, and D. Wood: Scanalyze v1.0.3: A computer program for aligning and merging range data. Stanford Computer Graphics Laboratory, Stanford, 2002.
16. Rugis, J.: Surface curvature maps and Michelangelo's David. In Proc. *Image and Vision Computing New Zealand,* pages 218–222, 2005.
17. Sun, C. and J. Sherrah. 3-D symmetry detection using the extended Gaussian image. *IEEE Trans. Pattern Analysis and Machine Intelligence,* **19**:164–168, 1997.

Two-Dimensional Discrete Shape Matching and Recognition

Isameddine Boukhriss, Serge Miguet, and Laure Tougne

Université Lyon 2, Laboratoire LIRIS
Bâtiment Europe, 5 av. Pierre Mendès-France
69 676 Bron Cedex, France
{iboukhri, smiguet, ltougne}@liris.univ-lyon2.fr
http://liris.cnrs.fr

Abstract. We present a 2D matching method based on corresponding shape outlines. By working in discrete space, our study is done by using discrete operators and avoids interpolations and approximations. To encode shapes, we polygonalize their contours and we proceed by the extraction of intrinsic properties namely length, curvature and normal vectors. We optimize then a measure of similarity controlled by weight parameters over a dynamic programming process. The approach is not sensitive to sampling errors and affine transformations. We validate our approach on simple and complex forms, we made tests also to recognize shapes. The weight parameters could be interactively modified by an end-user to customize the matching.

1 Introduction

The object matching is of great interest in computer vision operation like registration, segmentation and shape recognition. The methods applied can be further classified as boundary, region and model based methods. Any of these applications may require correspondence between pairs of regions, curves or points. In our case, we deal with 2D shapes and we try to match their contours. The curve-based methods suffer from problems related to sampling, affine transformations and shapes' articulations. A good shape description and a pre-processing stage could solve some of these problems.

We present in this paper a curve-based matching method based on finding an optimized correspondence. This optimization is done by minimizing a similarity function over the whole set of points. The criteria are based on distance, curvature and normals properties and are controlled by weight parameters. The order in which points are treated is taken into account in the optimization process. We describe first of all, in section 3 a pre-processing stage to encode our contours and compute our discrete parameters. In section 4 we optimize the matching with a dynamic programming approach. Results and shape recognition extension are given in section 5. But first of all, let us give a little state of the art about shape matching in the following section.

U. Eckardt et al. (Eds.): IWCIA 2006, LNCS 4040, pp. 445–452, 2006.

2   State of the Art

Many approaches are proposed in literature. Deformation-based methods try to find a mapping between two curves that minimizes an elastic function [You98]. Other approaches choose to match shapes basing on their medial axis or skeletons [ZY96]. Some of them try to resolve a part of the problem. For Instance Wolfson tries to find the longest common sub-curve of two curves to assemble them like in a puzzle [Wol90]. This approach was motivated by the algorithm of Schwartz and Sharir [SS87] consisting of finding the translation and the rotation of a sub-curve giving its best least squares fit to a longer curve.

More general techniques use geodesic paths [CH98]. The curves are defined as a source area S and destination area D. The matching is done through the computation of paths connecting these two areas. Geodesic distance maps are computed for each area in order to quantify the similarity. Parametric solutions could also resolve matching problem [WT04]. Authors propose an approach which combines the use of B-spline and Curvature Scale Space (CSS) matching. It generates the CSS image of the smoothed B-Spline of the original curve and then matches an input image with the generated model following the CSS matching algorithm [MaK96].The CSS matching technique suffers from noise sensibility due to curvature computation. A similar approach was also used in [AG03]. Authors try, starting from a given pair of curves (C_1, C_2), to find a re-parametrization such that the pair is better aligned with respect to a cost function of curvature conservative variational energy.

In [SKKC00], computing a curve atlas, based on deriving a correspondence between two curves, is studied. The optimal correspondence is found by a dynamic programming method based on a measure of similarity between the intrinsic properties of curves. Our method can be considered as a discrete version of this approach but with more robust description of curves by avoiding an uniform subdivision of contours, we do not interpolate nor approximate. We have furthermore the use of normals criteria and the possibility to optimize results interactively if the end-user approves the need by introducing a control coefficient for each parameter. Let us now explain it more precisely.

3   Pre-processing

Let us consider two binary figures F and F', each of them contains only one connected component, without hole, called shape in the following and denoted respectively by f and f'. We proceed by a registration and scaling process. f and f' are aligned according to their maximal elongation (principal component associated to the first eigen vector) with maximum interior intersection. Let us now consider the respective borders of each shape, that is to say the set of points that belongs to the shape and that have at least one 4−background neighbor. Just remark that we consider then 8−connected borders. Let us denote respectively by C and C' the borders of f and f'. We suppose that the border C contains N pixels and the border C' contains M pixels with N not necessary equals to M.

In order to accelerate the process, we do not consider all the points of the borders but just some of them: the points which are the extremities of the segments obtained by polygonalization. We explain more precisely this extraction in the following subsection. We also precise in this subsection the parameters we compute for each point.

The subsection 3.2 deals with the weights we affect to each parameter and how we combine them to obtain a cost for each association between one point of C and one of C'.

3.1   Discrete Parameters

Just remark that we could have consider all the points of the two borders and compute parameters for each of them. But, in order to be more efficient, we first make a polygonalization of the borders and compute parameters on the extremities of the segments extracted from the polygonalization. More precisely, we use Debled's linear polygonalization algorithm[DR94] which divides C and C' into digital line segments. Just remark that, as C and C' are closed curves, the number of obtained segments depends on the starting point and we could have used Feschet and Tougne's algorithm [FT05] in order to minimize the number of obtained segments. But, as this number only differs by one, we thought it was negligible in our context. Let us denote by P_i with i from 0 to n the extremities of segments extracted from C, remark that $P_0 = P_n$, and P'_j with j from 0 to m the extremities of segments extracted from C' with $P'_0 = P'_m$. On each point P_i and P'_j we compute parameters that will help to associate a cost to each pair of points (P_i, P'_j).

The first parameter that seems obvious to compute is the distance between the points of one pair. So, we consider the Euclidean distance and we compute :

$$d(P_i, P'_j) = \sqrt{(x_{P_i} - x_{P'_j})^2 + (y_{P_i} - y_{P'_j})^2}$$

Remark that this parameter may dominate the other parameters because intuitively we would like to associate nearest points. Nevertheless, this parameter is not sufficient because sometimes it will be better to associate points that are not the nearest but those which are not too moved away and such that the borders on these points are locally similar.

So, the two other parameters we compute compare locally the two borders C and C'. One measures the difference between the curvatures in the points P_i and P'_j and the second compares the normal vectors.

We obtain very easily the curvature on each point P_i (respectively P'_j) by considering the points P_{i-1} and P_{i+1} (respectively P'_{j-1} and P'_{j+1}). As a matter of fact, we compute the radius R_{P_i} (respectively $R_{P'_j}$) of the osculator circle $\mathcal{C}(P_i)$ (respectively $\mathcal{C}(P'_j)$) going through the three points P_i, P_{i-1} and P_{i+1} (respectively P'_j, P'_{j-1} and P'_{j+1}). We associate to the point P_i (respectively P'_j) the curvature $\kappa_{P_i} = \frac{1}{R_{P_i}}$ (respectively $\kappa_{P'_j} = \frac{1}{R_{P'_j}}$).

Just remark that if we have considered all the points of the borders, we could have used an efficient and similar algorithm to compute the curvature in each point. As a matter of fact, we could have compute for each point its right tangent and its left tangent and compute the circle going through the point and its semi-tangent extremities. Such computation has a complexity in $O(n)$ [Tou04] but obviously the number of computations would have be more important.

So, to the pair of points (P_i, P'_j) we associate the difference of curvatures :

$$\kappa(P_i, P'_j) = \mid \kappa_{P_i} - \kappa_{P'_j} \mid$$

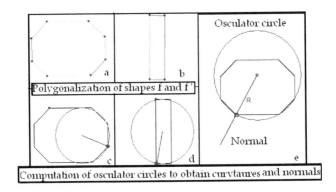

Fig. 1. Discrete parameters computation: polygonalization, curvature and normals. Figures a and b illustrate the polygonalization process. Figures c and d show oscula-tor circles used in figure e to compute curvature=1/R and extract the normal as an extension of the line handling the radius.

In order to avoid to associate points that would be near, with a quasi-similar local curvature but with inverse curve orientation, we compute a last parameter that compares the normal vectors at the points P_i and P'_j. Such vectors are also obtained very easily. The direction of the normal vector associated to the point P_i (respectively P'_j) is given by the direction of the line going through the center $c(P_i)$ (respectively $c(P'_j)$) of the circle $\mathcal{C}(P_i)$ (respectively $\mathcal{C}(P'_j)$) and the point P_i (respectively P'_j). Its orientation is given by the orientation of the vector $\overrightarrow{c(P_i)P_i}$ (respectively $\overrightarrow{c(P'_j)P'_j}$). Hence, we associate to the pair of points (P_i, P'_j), the angle made by the two normal vectors :

$$\alpha(P_i, P'_j) = \widehat{\overrightarrow{c(P_i)P_i}, \overrightarrow{c(P'_j)P'_j}}$$

Figure 1 illustrates the different parameters computation. Let us now explain how we combine such parameters in order to obtain a cost associated to each pair (P_i, P'_j).

3.2   Parameters Weighting

We suppose, at this stage, that for each point P_i, P'_j we know its curvature, its curvilinear coordinates and its normal. We will now quantify curves matching.

For this reason, lets consider a mapping M of the two curves:

$$M : [0, n] \longrightarrow [0, m], M(P_i) = P'_j.$$

Our goal is to minimize a measure of similarity on this mapping. Lets define this measure by $\phi[M] = \sum F(P_i, P'_j)$ where F is similarity cost function. F is computed by considering the distance parameter between each pair of points, the curvature difference and respective normal vectors angle. The question is how to try all combinations of matching points and determine the best one. This is ensured by dynamic programming process. Knowing that we registrated our shapes before, it would be easy to locate and predict the best matching between points of each curve. We just have to precise how should F be quantified:

$$F[0, n] \times [0, m] \longrightarrow R+$$

$$F(P_i, P'_j) = d(P_i, P'_j) + \kappa(P_i, P'_j) + \alpha(P_i, P'_j)$$

Note that in order to make more robust and efficient optimization, we chose to add to coefficients k_1 and k_2 $\epsilon[0, 1]$ in order to control the influence of $d(P_i, P'_j)$ and $\kappa(P_i, P'_j)$ respectively in the optimization. By considering $k_1 = 1 - k_2$: $F(P_i, P'_j) = k_1 d(P_i, P'_j) + k_2 \kappa(P_i, P'_j) + \alpha(P_i, P'_j)$. This can give more flexibility to the end-user to control matching especially if there are shapes disturbed or rather different. The normal vectors angle criteria should have a constant influence in order to make convergence fast. It should be noticed also that all parameters are normalized in order to be in the same scale of comparison. They can also be determined automatically: we have just to vary them in a complementary way such that we obtain the minimum of the cost function. It is a standard minimization process.

4   Matching Optimization

Before proceeding with the optimization process, based on dynamic programming, lets define the structure which will be used. This structure, as shown in figure 2 is a grid of intersections expressing the cost $F(P_i, P'_j)$. So we can consider that our grid is $n \times m$ elements where each intersection of the axes joining P_i and P'_j is the cost of matching P_i to P'_j. This structure contains twice the same matrix for circular matching reason. When we suppose for example that P_0 is matched to P'_0, we update all other costs according to this choice. It is important to notice that the global cost function is monotic in the sense that we could not match point P_{i+1} with P'_{j-1} if we matched before P_i to P'_j as illustrated in the right part of figure 2. That is to say that the update for a cost between a pair of points (P_i, P'_j) could only be done by adding $min(cost(P_{i-1}, P'_{j-1}), cost(P_{i-1}, P'_j), cost(P_i, P'_{j-1}))$. This is the application of the direct acyclic graph (DAG). The idea is to find a path which permits the matching of all chosen points of polygonalization with the minimum cost [DAG]. With this technique, we could find a combination that minimize the whole cost of matching. The combination could match one point of a curve to more than one in the other curve and *vice versa*. The minimal cost path determines the pairs of points to be matched while going up and picking the positions i and j.

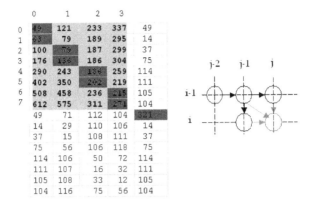

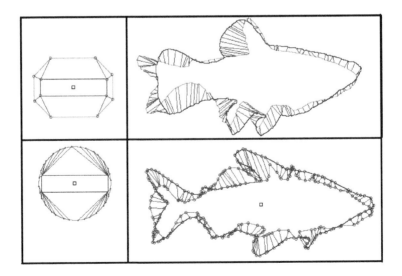

Fig. 2. Matching optimization based on directed acyclic graph

5   Results and Application to Shape Recognition

We made tests on simple and complex shapes. For each pair of shapes, we proceed by a registration process, then we compute the intrinsic properties and finally we search the best path candidate to join all points forming the corresponding contours. We can modify the final result, by modifying the associated weighted parameters to curvature and curvilinear coordinates as shown in figure 3. The complexity is linear for both the registration and parameters computing processes contrary to the optimization stage which presents a logarithmic complexity. Figure 3(left) illustrates the result we obtain on the examples of

Fig. 3. Simple and complex shape matching

figure 1 and a more complex shape matching is shown in figure 3(right). We should obtain best results if we take into account noise factor specially in the polygonalization process [RRR03]. The final result could be a pre-processing stage to 2D atlas construction.

We extended our tests to shape recognition application. Starting for a given form and a database of different shapes, we try to measure the similarity between the input shape and the base. This measure is based on the same intrinsic properties: curvature, euclidian distance and normals angle. The shape which has the maximum (the minimum) value of resemblance (difference) is the best candidate. Figure 4 shows a comparative study between different shapes and gives the final choice for each input image. The last row of the figure shows the best candidate selected, the result is satisfying.

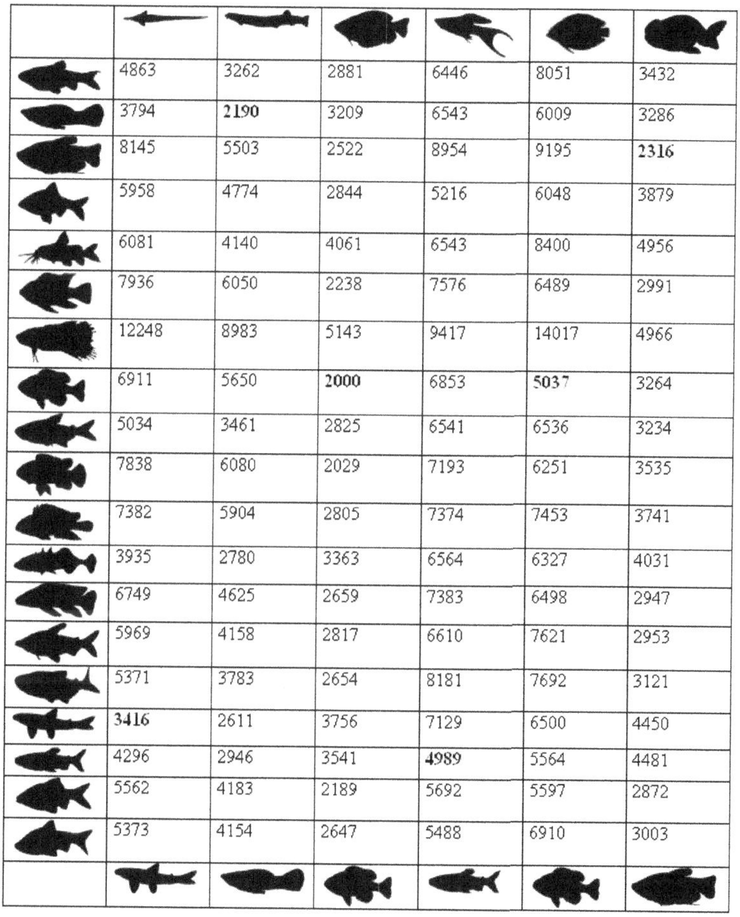

	4863	3262	2881	6446	8051	3432
	3794	**2190**	3209	6543	6009	3286
	8145	5503	2522	8954	9195	**2316**
	5958	4774	2844	5216	6048	3879
	6081	4140	4061	6543	8400	4956
	7936	6050	2238	7576	6489	2991
	12248	8983	5143	9417	14017	4966
	6911	5650	**2000**	6853	**5037**	3264
	5034	3461	2825	6541	6536	3234
	7838	6080	2029	7193	6251	3535
	7382	5904	2805	7374	7453	3741
	3935	2780	3363	6564	6327	4031
	6749	4625	2659	7383	6498	2947
	5969	4158	2817	6610	7621	2953
	5371	3783	2654	8181	7692	3121
	3416	2611	3756	7129	6500	4450
	4296	2946	3541	**4989**	5564	4481
	5562	4183	2189	5692	5597	2872
	5373	4154	2647	5488	6910	3003

Fig. 4. Shape recognition: best candidate selection

6   Conclusion and Future Works

We have presented a matching method between two 2D curves. Our approach is totally discrete and could be extended to shape recognition by a good parametrization of the optimization process. It should be noticed also that the method is independent of scaling and initial positions of shapes, but it depends on a course of contours in the same direction. Noise also could be a source of inefficiency specially in polygonalization, but this can be avoided by fuzzy polygonalization [RRR03]. The proposed approach could be a pre-processing stage to compute 2D statistical atlases based on principal component analysis [CT01]. Extension to 3D is under study in spite of problem of landmarks order that will not be determined in a similar way between shapes.

References

[AG03]   B. Avants and J. Gee. Matching with Scale-Space Curvature and Extrema-Based Scale Selection. *Lecture Notes in Computer Science*, 2695:798–813, 2003.

[CH98]   I. Cohen and I. Herlen. Curves matching using geodesic paths. *IEEE Computer Society Conference on Computer Vision and Pattern Recognition*, pages 741–46, 1998.

[CT01]   T.F. Cootes and C.J. Taylor. Statistical models of appearance for medical image analysis and computer vision. *SPIE Medical Imagin*, 2001.

[DAG]   Directed acyclic graph. http://mathworld.wolfram.com/AcyclicDigraph.html.

[DR94]   I. Debled and J. P. Reveilles. A linear algorithm for segmentation of digital curves. *Third International Workshop on Parallel image analysis*, 1994.

[FT05]   F. Feschet and L. Tougne. On the min dss problem of the closed discrete curves. *Discrete Applied Mathematics*, 2005. Accepted for publication.

[MaK96]   F. Mokhtarian, S. abbasi, and J. Kittler. Robust and Efficient shape indexing through Curvature Scale Space. *BMVC*, 1:53–62, 1996.

[RRR03]   I. D. Rennesson, J. L. Remy, and J. Rouyer. Segmentation of discrete curves into fuzzy segments. *IWCIA*, 12, 2003.

[SKKC00]   B. Sebastian, P. N. Klein, B. B. Kimia, and Joseph J. Crisco. Constructing 2D Curve Atlases Thomas. *IEEE Workshop on Mathematical Methods in Biomedical Image Analysis*, pages 70–77, 2000.

[SS87]   J. T. Schwartz and M. Sharir. Identification of partially obscured objects in two dimensions by matching of noisy characteristics curves. *International Journal of Robotics Research*, 6:29–44, 1987.

[Tou04]   L. Tougne. Descriptions robustes de formes en géométrie discrète. Habilitation à diriger des recherches de l'université Lyon 2, december 2004.

[Wol90]   Haim J. Wolfson. On curve matching. *Transactions on pattern analysis and machine intelligence*, 12:483–89, 1990.

[WT04]   Y. Wang and E. K. Teoh. A novel 2d shape matching algorithm based on B-Spline Modeling. *International Conference on Image Processing*, pages 409–12, 2004.

[You98]   L. Younes. Computable elastic distance between shapes. *SIAM Journal on Applied Mathematics*, 58:565–86, 1998.

[ZY96]   S. S. Zhu and A. L. Yuille. Forms: A flexible object recognition and momdeling system. *International Journal of Computer Vision*, 20:187–212, 1996.

Hierarchical Tree of Image Derived by Diffusion Filtering

Haruhiko Nishiguchi[1], Atsushi Imiya[2], and Tomoya Sakai[2]

[1] School of Science and Technology, Chiba University, Japan
[2] Institute of Media and Information Technology, Chiba University, Japan
Yayoi-cho 1-33, Inage-ku, Chiba, Japan, 263-8522
halu@graduate.chiba-u.jp
{imiya, tsakai}@faculty.chiba-u.jp

Abstract. This paper aims to introduce a class of non-linear diffusion filterings based on deep structure analysis in scale space. In linear scale space, the trajectory of extrema is called stationary curves. This curves provides deep structure analysis and hierarchical expression of signals. The motion of extrema in linear scale space is controlled by a function of the higher derivatives of the signals. We introduce a non-linear diffusion filterings based on the absolute values of second derivative of signals.

1 Introduction

In this paper, we show an algorithm for the construction of hierarchical tree of an image using diffusion filtering. The singular points in the scale space derived by diffusion filtering define the deep structures. This structure yields hierarchical structure of images.

The deep structures of the linear scale space are defined as the trajectories of extrema of topographic map of diffused images derived from diffusion filtering based on a linear diffusion equation. These trajectories are called critical curves [5] and stationary-curves [2, 3, 4] in the context of linear scale space analysis. Singular points on the curves derive a collection of feature points for the expression of hierarchical properties of images. In this paper, we call these trajectories stationary-curves. In the linear scale space, the deep structures provide features for the hierarchical description of images, since the linear scale space analysis was first introduced for extraction of the hierarchy of dominant parts as the attention field [2, 3, 4, 6].

Non-linear diffusion filtering performs region segmentation employing the extraction of boundaries of segments. The segmented regions yield features for the hierarchical expression of these extracted regions in an image if we extract regions for many scales in the scale space. Existing non-linear diffusion filtering control smoothness of images using the gradient map of an image, since the gradient map indicates boundaries of regions as the steepest points at appropriate neighbourhoods in an image for each scale. However, the deep structures of the non-linear scale space have not analysed yet. In this paper, we first show properties of 1D deep structure of the non-linear scale space. Constructing 1D basic theory, we discuss properties of 2D deep structure of the non-linear scale space.

U. Eckardt et al. (Eds.): IWCIA 2006, LNCS 4040, pp. 453–465, 2006.

2   1D Filtering

2.1   Linear Diffusion Filtering

On the real line **R**, the linear scale-space transform for function $f(x)$, such that

$$f(x, \tau) = \frac{1}{(\sqrt{2\pi\tau})} \int_{-\infty}^{\infty} f(y) \exp(-\frac{(x-y)^2}{2\tau}) dy, \tag{1}$$

defines the general function of function $f(x)$. Therefore, function $f(x, \tau)$ is defined in $\mathbf{R} \times \mathbf{R}_+$ [1]. The function $f(x, \tau)$ is the solution of the linear diffusion equation

$$\frac{\partial}{\partial \tau} f(x, \tau) = \frac{\partial^2}{\partial x^2} f(x, \tau), \ \tau > 0, \ f(x, 0) = f(x). \tag{2}$$

The solution of eq. (2) is formally expressed

$$f(x, \tau) = \exp(\tau \frac{d^2}{dx^2}) f(x) \tag{3}$$

using the theory of Lie group. Stationary points for the topographical maps in the scale space [1, 2] are defined as the solutions of the equation.

2.2   Non-linear Diffusion Filtering

On the real line **R**, we deal with the operations in the form

$$\frac{\partial}{\partial \tau} f(x, \tau) = \frac{\partial}{\partial x} \left(g(x) \frac{\partial}{\partial x} f(x, \tau) \right), \ f(x, 0) = f(x), \tag{4}$$

for a positive function $g(x)$, which is the diffusivity of the diffusion equation.

The diffusivity of the Perona-Malik type [7] and the Weickert type [8] are, for a non-zero constant λ,

$$g(|f_x|) = \frac{1}{1 + \left(\frac{|f_x|}{\lambda}\right)^2}, \tag{5}$$

and

$$g(|f_x|) = \begin{cases} 1 & \text{if } |f_x| = 0 \\ 1 - \exp\left(\frac{-3.31488}{\left(\frac{|f_x|}{\lambda}\right)^4}\right) & \text{otherwise}, \end{cases} \tag{6}$$

respectively.

2.3   Linear Scale Space Hierarchy

The stationary-curves in the scale space are the collections of the stationary points. We denote the trajectories of the stationary points as $x(\tau)$. Setting $\boldsymbol{H}$ to be the Hessian matrix of $f(\boldsymbol{x}, \tau)$, Zhao and Iijima [2] showed that the stationary-curves for a two-dimensional image are the solution of

$$\frac{\partial^2}{\partial^2 x} f(x, \tau) \frac{d}{d\tau} x(\tau) = -\frac{\partial^3}{\partial x^3} f(x, \tau) \tag{7}$$

and clarified topological properties of the stationary-curves for signals.

The point x_∞ for $\lim_{\tau \to \infty} x(\tau) = x_\infty$ is uniquely determined for any $f(x)$.

We call a curve on which point x_∞ lies and a curve which is open to the direction of $-\tau$ the trunk and branch, respectively. On the top of each branch, a singular point exists. For the construction of unique hierarchical expression of stationary points, Zhao and Iijima [2] considered that the sub-root of a branch is the stationary point of the top of the branch curve and that a sub-root and the trunk are connected by a line segment parallel to the x axis.

2.4   Non-linear Scale Space Hierarchy

The stationary-curve is the solution of a system of equations

$$\frac{\partial}{\partial \tau} f(x, \tau) = \frac{\partial}{\partial x} \left(g(x) \frac{\partial}{\partial x} f(x, \tau) \right), \quad \frac{\partial}{\partial x} f(x, \tau) = 0. \tag{8}$$

The derivative of the system of equations eq. (8) is expressed in the form

$$\frac{d}{d\tau} x(\tau) = G(g, f_x, f_{xx}, f_{xxx}, \cdots f^{(n)} \cdots). \tag{9}$$

If $g = 1$ for G, following theorem is proven.

Theorem 1. *If $g = 1$,*

$$G(g, f_x, f_{xx}, f_{xxx}, \cdots f^{(n)} \cdots) = -f_{xx}^{-1} f_{xxx}. \tag{10}$$

This theorem implies that for the linear diffusion filtering, the speed of motion of extrema in the linear scale space is

$$\frac{\partial^2 f}{\partial x^2} \frac{dx}{d\tau} = -\frac{\partial^3 f}{\partial x^3}. \tag{11}$$

Therefore, in this paper, for $g \neq 1$, we deal with the case

$$G(g, f_x, f_{xx}, f_{xxx}, \cdots f^{(n)} \cdots) = -\gamma g f_{xx}^{-1} f_{xxx}, \tag{12}$$

for a positive constant γ, since G is the speed of the trajectory of the extrema. Therefore, if the relation of eq. (12) is fulfilled, the extrema move in a similar manner with the motion of extrema in the linear scale space.

For G, we have the next theorems.

Theorem 2. *For $g(x) > 0$, if $g_x(0) = 0$, eq. (12) is fulfilled.*

Corollary 1. *The Perona-Malik type diffusivity fulfils the condition of the theorem 2.*

Corollary 2. *The Weickert type diffusivity fulfils the condition of the theorem 2.*

Theorem 3. *For a positive constant α, if $\alpha g f_{xxx} = 2 g_x f_{xx}$, eq. (12) is fulfilled.*

Corollary 3. *$g = |f_{xx}|$ fulfils the condition of the theorem 2.*

Proofs of Theorems and Corollaries. For a one dimensional non-linear diffusion equation in the form eq. (4), considering $f_x(x, \tau) = 0$, we have the relation

$$f_{xx}\frac{dx}{d\tau} = -g f_{xxx} - 2 g_x f_{xx}. \tag{13}$$

Therefore, we have theorems 2 and 3.

For the Perona-Malik type diffusivity defined by eq. (5), we have the relation

$$g_x = -2 f_x f_{xx}\left\{\lambda\left(1 + \frac{f_x^2}{\lambda^2}\right)\right\}^{-2}. \tag{14}$$

Since for extrema $f_x = 0$, we have the relation

$$\frac{d}{dx}\left(1 + \frac{|f_x|}{\lambda^2}\right)^{-1} = 0. \tag{15}$$

For the Weickert type diffusivity defined by eq. (6), we have the relation

$$g_x = -4 \times 3.31488 \times \lambda^4 \exp\left(-\frac{3.31488}{(|f_x/\lambda|^4)}\right)\frac{f_{xx}}{f_x^5}. \tag{16}$$

Since $\lim_{f_x \to 0+} g_x = 0$, we have the corollary.

Since, for $g = |f_{xx}|$, we have the relation

$$\frac{\partial^2 f}{\partial x^2}\frac{dx}{d\tau} = \begin{cases} -3\frac{\partial^2 f}{\partial x^2}\frac{\partial^3 f}{\partial x^3} & \text{if } f_{xx} > 0, \\ 3\frac{\partial^2 f}{\partial x^2}\frac{\partial^3 f}{\partial x^3} & \text{if } f_{xx} < 0. \end{cases} \tag{17}$$

this relation derives the equation,

$$\frac{\partial^2 f}{\partial x^2}\frac{dx}{d\tau} = -3\left|\frac{\partial^2 f}{\partial x^2}\right|\frac{\partial^3 f}{\partial x^3}. \tag{18}$$

2.5   Stable View-Points

Setting $S(x, \tau) = |\frac{dx(\tau)}{d\tau}|$, Zhao and Iijima [2, 3] defined the stable view-points. The stable view-points are the points which satisfy $S(x, \tau) = 0$, or are the isolated points with the conditions $\frac{dS(x,\tau)}{d\tau} = 0$, and $\frac{d^2 S(x,\tau)}{d^2\tau} > 0$. They also developed an algorithm to define a stable view-point tree whose nodes are the stable view-points on the stationary-curves, and introduced a hierarchical expression of a signal using this tree. From the stable view-points on the stationary-curves, the tree is constructed according to the order of the stable view-points. For the stable view-points on the stationary-curves, the order of the stable view-points is defined as

$$x(\tau) \succ x(\tau') \text{ if } \tau > \tau'. \tag{19}$$

3   2D Filtering

3.1   Linear Scale Space

In the two-dimensional Euclidean space $\mathbf{R}^2$, for an orthogonal coordinate system x-y defined in $\mathbf{R}^2$, a vector in $\mathbf{R}^2$ is expressed by $\boldsymbol{x} = (x, y)^\top$, where $\cdot^\top$ is the transpose of a vector.

For the original function $f(\boldsymbol{x})$, the general function $f(\boldsymbol{x}, \tau)$ is the solution of the equation

$$\frac{\partial}{\partial \tau} f(\boldsymbol{x}, \tau) = \Delta f(\boldsymbol{x}, \tau), \ \tau > 0, \ f(\boldsymbol{x}, 0) = f(\boldsymbol{x}). \tag{20}$$

The space $\mathbf{R}^2 \times (0 \leq \tau < \infty)$ is called the scale space. Points $\boldsymbol{x}$ which satisfies the condition $\nabla f(\boldsymbol{x}, \tau) = 0$ with the classification based on the signs of the eigenvalues of Hessian matrix $\boldsymbol{H}$ expresses the topological structure of image $f(\boldsymbol{x}, \tau)$. We extend this idea to the general images. The solution of eq. (20) is expressed as

$$f(\boldsymbol{x}, \tau) = \frac{1}{4\pi\tau} \int_{-\infty}^{\infty} f(\boldsymbol{y}) \exp(-\frac{|\boldsymbol{x} - \boldsymbol{y}|^2}{4\tau}) d\boldsymbol{y}. \tag{21}$$

Zhao and Iijima [2] showed that the stationary-curves for a function are the solution of

$$\boldsymbol{H} \frac{d}{d\tau} \boldsymbol{x}(\tau) = -\nabla \Delta f(\boldsymbol{x}(\tau), \tau). \tag{22}$$

For fixed τ, since the Hessian matrix is always singular for singular points, this equation is valid for nonsingular points. Using the second derivations of $f(\boldsymbol{x}, \tau)$, we classify the topological properties of the stationary points on the topographical maps. The relation $\frac{d^2 f}{dn^2} = \boldsymbol{n} \cdot \nabla(\boldsymbol{n} \cdot \nabla f) = \boldsymbol{n}^\top \boldsymbol{H} \boldsymbol{n}$ means that the eigenvectors of Hessian matrix of $f(\boldsymbol{x}, \tau)$ gives the extrema of $D^2 f(\boldsymbol{x}, \tau)$ and that the extremal are achieved by the eigenvectors of the Hessian of $f(\boldsymbol{x}, \tau)$, since $\alpha_1 \geq \boldsymbol{n}^\top \boldsymbol{H} \boldsymbol{n} \geq \alpha_2$ for $|\boldsymbol{n}| = 1$, where $\alpha_1 \geq \alpha_2$ are two eigenvalues of the 2×2 Hessian matrix $\boldsymbol{H}$. Therefore, we have the relations $\max(D^2_{\boldsymbol{x}(\tau)}) = \alpha_1$ and $\min(D^2_{\boldsymbol{x}(\tau)}) = \alpha_2$.

Denoting the signs of the eigenvalues of the minus of the Hessian matrix as $(+, +)$, $(+, -)$ and $(-, -)$ in the linear scale space, these labels of points correspond to the local maximum points, the saddle points, and the local minimum points, respectively for fixed τ.

In [2, 3], they pay attention to the maximum and minimum points. In this paper, we deal with all of three classes of extrema. The saddle points in the scale space appear on walls and valley which connect maximum points and minimum points, respectively, Therefore, the motion of the saddle points in the scale space corresponds to the changes of the topology of images in the scale space. According to the second directional derivation, we can define three types of stationary-curves: maximum curves, minimum curves, and saddle curves.

3.2 Non-linear Scale Space

In the two-dimensional Euclidean space $\mathbf{R}^2$, For the original function $f(\boldsymbol{x})$, we deal with the diffusion filtering in the form

$$\frac{\partial}{\partial \tau} f(\boldsymbol{x}, \tau) = \nabla \cdot \boldsymbol{F}(\nabla f(\boldsymbol{x}, \tau)), \quad \boldsymbol{x} \in \mathbf{R}^n, \quad \tau > 0. \tag{23}$$

If $\boldsymbol{F}(\nabla f(\boldsymbol{x}, \tau)) = \nabla f(\boldsymbol{x}, \tau)$, that is, $\boldsymbol{F}$ is the identity operation, eq. (23) coincides with the linear diffusion filtering. Next, we take a more precise form of $\boldsymbol{F}$ as

$$\boldsymbol{F}(\nabla f(\boldsymbol{x}, \tau)) = g(|\nabla f|)\nabla f(\boldsymbol{x}, \tau). \tag{24}$$

This class of $\boldsymbol{F}$ means that diffusion is controlled by the gradient map $|\nabla f|$ of f at each τ.

The stationary points for the non-linear diffusion filtering satisfy the system of equations

$$\frac{\partial}{\partial \tau} f(\boldsymbol{x}, \tau) = \nabla \cdot \boldsymbol{F}(\nabla f(\boldsymbol{x}, \tau)), \quad \boldsymbol{x} \in \mathbf{R}^n, \quad \tau > 0 \ \nabla f(\boldsymbol{x}, \tau) = 0 \tag{25}$$

From eq. (25), the stationary-curves in the non-linear scale space are the solution of

$$\boldsymbol{H}_f \frac{d\boldsymbol{x}(\tau)}{d\tau} = -g\nabla \Delta f - (\Delta f \boldsymbol{I} + \boldsymbol{H}_f)\nabla g. \tag{26}$$

Theorem 4. If $g = 1$,

$$\boldsymbol{H}\frac{d}{d\tau}\boldsymbol{x}(\tau) = -\nabla \Delta f(\boldsymbol{x}, \tau). \tag{27}$$

Therefore, in this paper, for $g \neq 1$, we deal with the case

$$\boldsymbol{H}\frac{d}{d\tau}\boldsymbol{x}(\tau) = -\gamma g\nabla \Delta f(\boldsymbol{x}, \tau) \tag{28}$$

for a positive constant γ, since

$$\frac{d}{d\tau}\boldsymbol{x}(\tau) = -\gamma g\boldsymbol{H}^{-1}\nabla \Delta f(\boldsymbol{x}, \tau) \tag{29}$$

is the speed of the trajectory of the extrema for $det(\boldsymbol{H}) \neq 0$. Therefore, if the relation of eq. (28) is fulfilled, the extrema move in a similar manner with the motion of extrema in the linear scale space. We have the next theorem.

Theorem 5. *For $g(x) > 0$, if $\nabla g(0) = 0$, eq. (28) is fulfilled.*

The diffusion functions of the Perona-Malik type [7] and the Weickert type [8] are, for a positive constant λ,

$$g(|\nabla f|) = \frac{1}{1 + \left(\frac{|\nabla f|}{\lambda}\right)^2},$$ (30)

and

$$g(|\nabla f|) = \begin{cases} 1 & \text{if } |\nabla f| = 0 \\ 1 - \exp\left(\frac{-3.31488}{\left(\frac{|\nabla f|}{\lambda}\right)^8}\right) & \text{otherwise,} \end{cases}$$ (31)

respectively. These diffusion functions are basically and historically introduced to the edge detection for segmentation. For the Perona-Malik type diffusion function, we can derive the relation,

$$\nabla g(|\nabla f|) = \frac{-2\Delta f \nabla f}{\lambda^2 (1 + |\nabla f/\lambda|^2)^2}.$$ (32)

Therefore, the Perona-Malik type diffusion function satisfies the relations $g = 1$ and $\nabla g = 0$ for $\nabla f = 0$. Furthermore, from eq. (31), the Weickert diffusion function satisfies the relation $\nabla g = 0$ for $\nabla f = 0$. From these analysis on the non-linear diffusion filterings, we have the following corollaries.

Corollary 4. *The Perona-Malik type diffusion fulfils the condition of the theorem 5.*

Corollary 5. *The Weickert type diffusion fulfils the condition of the theorem 5.*

Additionally, we have the next theorem.

Theorem 6. *If the diffusivity g which satisfies a condition*

$$\alpha g \nabla \Delta f = (\Delta f \boldsymbol{I} + \boldsymbol{H}) \nabla g$$ (33)

is exist, eq. (28) is fulfilled.

Since for one-dimensional signals, $\boldsymbol{H} = \frac{\partial^2 f}{\partial x^2}$, theorem 5 derives theorem 3 for one-dimensional signals.

As an analogous to corollary 3, it is possible to deal with diffusion controlled by Laplacian $|\Delta f|$,

$$\boldsymbol{F}(\nabla f(\boldsymbol{x}, \tau)) = g(|\Delta f|) \nabla f(\boldsymbol{x}, \tau),$$ (34)

for

$$g(x)(1 + \frac{x}{\lambda}), \tag{35}$$

since for $g(\Delta f) = c \exp\left(\frac{|\Delta f|}{\lambda}\right)$

$$(1 + \frac{|\Delta f|}{\lambda}) \simeq \exp\left(\frac{|\Delta f|}{\lambda}\right). \tag{36}$$

However, this function does not satisfy theorem 6.

The Laplacian of an image generally approaches to zero with increasing τ. Therefore, the diffusion based on eq. (35) converges to the linear diffusion with the constant diffusivity 1. This mathematical property means that the diffusion is controlled by the Laplacian map for small values of τ, and the diffusion behaves as the linear diffusion filtering for large values of τ. In the sense of geometry, the diffusivity of the form $(1 + \frac{|\Delta f|}{\lambda})$ accelerates the diffusion term in small values of τ.

3.3   Trees of Scale Space Hierarchy

The Structure Tree. Since the stationary-curves consist of many curves for $\tau > 0$, we call each curve a branch curve. The point $\boldsymbol{x}_\infty$ for

$$\lim_{\tau \to \infty} \boldsymbol{x}(\tau) = \boldsymbol{x}_\infty \tag{37}$$

is uniquely determined for any image. We call a curve on which point $\boldsymbol{x}_\infty$ lies and a curve which is open to the direction of $-\tau$ the trunk and branch, respectively.

On the top of each branch, a singular point exists. Therefore, for the construction of a unique hierarchical expression of stationary points, Zhao and Iijima [2] proposed the following rules.

1. *The sub-root of a branch is the singular point, such that $\det \boldsymbol{H} = 0$, of the top of the branch curve and a sub-root.*
2. *The sub-root is connected to the trunk by a line segment parallel to x-y plane.*

Using these rules, the structure tree is constructed. The nodes of structure tree are $\boldsymbol{x}_\infty$ (root), singular points (sub-root) and stationary points at the finest scale (leaf).

The Stable View-point Tree. Setting $S(\boldsymbol{x}, \tau) = |\frac{d\boldsymbol{x}(\tau)}{d\tau}|$, Zhao and Iijima [2,3] defined the stable view-points. The stable view-points are the points which satisfy $S(\boldsymbol{x}, \tau) = 0$, or are the isolated points with the conditions $\frac{dS(\boldsymbol{x},\tau)}{d\tau} = 0$ and $\frac{d^2 S(\boldsymbol{x},\tau)}{d^2\tau} > 0$. They also developed an algorithm to define a unique tree whose nodes are the stable view-points on the stationary-curves, and introduced a unique hierarchical expression of an image using this tree. For the stable view-points on the stationary-curves, the order of the stable view-points is defined as

$$\boldsymbol{x}(\tau) \succ \boldsymbol{x}(\tau') \text{ if } \tau > \tau'. \tag{38}$$

From the stable view-points on the stationary-curves, the tree is constructed according to the order of the stable view-points.

4   Numerical Experiments

4.1   Numerical Scheme

The stability of the schemes based on the explicit finite difference schemes such as forward time centred space (FTCS) depends on the magnitude of the diffusivity g.

Since the diffusivity depends on the values of f at each points, we cannot pre-determinate the diffusivity of filtering. This mathematical structure of non-linear diffusion might cause instability of computation. Therefore, implicit schemes or backward in the iteration steps have been adopted to generate the nonlinear scale space in a number of applications [9, 12].

4.2   1D Filtering

Using $f(x) = (6x-1)^2 \exp(-x^2/2)$, we show 1D numerical examples. The structure trees corresponding to these stationary-curves are equivalent as shown in Fig 1. In Fig. 1, (e), (f), (g) and (h) show the stationary-curves yielded by the linear, Perona-Malik type, Weickert type and second-derivative-based diffusion filtering operations, respectively.

There are many stable view-points, which are not exist on the trunk of stationary-curve yielded by the linear diffusion filtering operation. According to this consequence, it might be said that the linear and the second-derivative-based diffusion are better than the Perona-Malik type and the Weickert type diffusion for constructing the structure tree based on stable view-point as the points of attention for the extraction of global features of signals. For 1D signals, eq. (12) indicated that the direction of the motion of stationary point is determined by the sign of f_{xxx}/f_{xx} for nonsingular points. Second derivative f_{xx} is negative (positive) at the local maximum (minimum) points. Therefore, the direction of motion is controlled by the sign of f_{xxx}.

4.3   2D Filtering

In Fig. 2, (b), (e) and (h) show the stationary-curves yielded by the linear, Perona-Malik and Weickert type diffusion filtering operations, respectively. We set $\lambda = 7.5$ for the Perona-Malik type diffusion function, $\lambda = 25.0$ for the Weickert type diffusion function. (e) and (h) show topologically similar curves. These figures show that the Perona-Malik type and the Weickert type diffusions derive similar stationary-curves as shown in Figs. 2 (e) and (h). In these two stationary-curves, there exist side-branches in the left of main trunks. Furthermore, trees of (c) and (l) posses the same structure in the large scale.

Since in 2D, we extracted points such that $\nabla f = 0$, for $(+, +)$, $(+, -)$, and $(-, -)$, the extracted trees by the linear and non-linear scale spaces are different. In 2D scale space, we deal with a class of trajectory of stationary point which

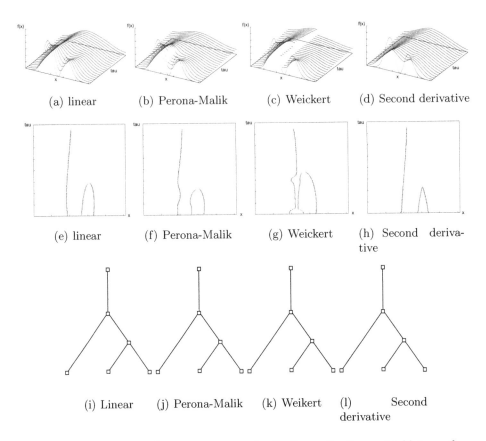

(a) linear   (b) Perona-Malik   (c) Weickert   (d) Second derivative

(e) linear   (f) Perona-Malik   (g) Weickert   (h) Second deriva-
                                                                     tive

(i) Linear   (j) Perona-Malik   (k) Weikert   (l)   Second
                                                        derivative

Fig. 1. Diffused signals and stationary-curves for the linear, the Perona-Malik type, the Weickert type, and the second-derivative-based diffusion filtering operations. (a)(e)(i) Signals in the scale space, the stationary-curve, and the structure tree of the linear diffusion. (b)(f)(j) Signals in the scale space, the stationary-curve, and the structure tree of the Perona-Malik type diffusion, (c)(g)(k) Signals in the scale space, the stationary-curve, and the structure tree of the Weickert type diffusion, (d)(h)(l) Signals in the scale space, the stationary curve, and the structure tree of $g = |f_{xx}|$ diffusion. As shown in the fourth column, the structure trees derived from these stationary-curves are equivalent.

satisfy eq. (28). Let $\boldsymbol{A} = diag(\alpha_1, \alpha_2)$ for the eigenvalues of the Hessian matrix $\boldsymbol{H}$ and $\boldsymbol{R}$ be the eigenmatrix of $\boldsymbol{H}$, that is, $\boldsymbol{H} = \boldsymbol{RAR}^\top$. From eq. (29)

$$\frac{d\boldsymbol{y}}{d\tau} = -\boldsymbol{K}\nabla\Delta f(\boldsymbol{y}, \tau), \quad \boldsymbol{K} = \begin{pmatrix} \gamma g \alpha_1^{-1} & 0 \\ 0 & \gamma g \alpha_2^{-1} \end{pmatrix}, \quad \boldsymbol{y} = \boldsymbol{Rx}, \qquad (39)$$

since $\boldsymbol{R}^\top \nabla f(\boldsymbol{Rx}) = \nabla f(\boldsymbol{x})$. Therefore, the direction of the velocity is controlled by the signs of the eigenvalues of $\boldsymbol{H}$ locally in the coordinate system whose axes

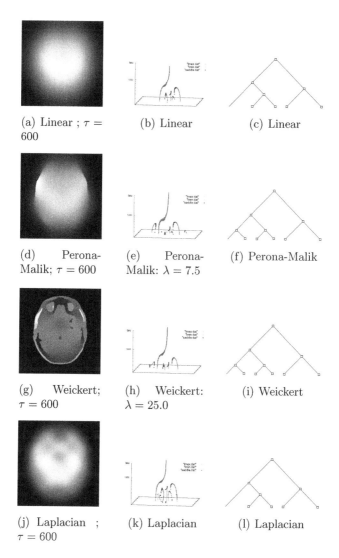

(a) Linear ; $\tau = 600$

(b) Linear

(c) Linear

(d) Perona-Malik; $\tau = 600$

(e) Perona-Malik: $\lambda = 7.5$

(f) Perona-Malik

(g) Weickert; $\tau = 600$

(h) Weickert: $\lambda = 25.0$

(i) Weickert

(j) Laplacian ; $\tau = 600$

(k) Laplacian

(l) Laplacian

Fig. 2. Input image and Stationary-curves yielded by the three types of diffusion filtering. (a) The linear diffusion, (d) the Perona-Malik type diffusion, $\lambda = 7.5$, (g) the Weickert type diffusion, $\lambda = 25.0$. (j) Laplacian Diffusion (b), (e), (h), and (k) are stationary curve in scale space. (c) Structure tree derive by linear diffusion. (f), (i),and (l) show the Perona-Malik type, the Weickert type and he Laplacian-controlled diffusions derive equivalent structure trees. In two structure trees in (e) and (h), there exist side-branches in the left of main trunks.

depend on the direction of the eigenvectors of the matrix $\boldsymbol{H}$. The direction of the motion of singular points in the scale space is not usually parallel to the coordinate axes of the image expression.

5   Concluding Remarks

In this paper, we analysed local geometric properties of the deep structures yielded by non-linear scale space analysis. We first showed properties of 1D deep structure of the non-linear scale space. Constructing 1D basic theory, we discuss properties of 2D deep structure of the non-linear scale space, that is, we described the local properties of the deep structures of the Perona-Malik and Weickert types diffusion filtering. Furthermore, we introduce the Laplacian-map-based diffusion filtering. Comparisons of deep structure of linear and non-linear diffusion filterings, we found that difference of diffusivity functions affects to discrete properties of deep structure.

Setting

$$h(\lambda) = \left\{ \sum_{n=0}^{\infty} a_n \lambda^n \right\}^{-1}, \quad a_0 = 1, \quad a_n \geq 0 \tag{40}$$

$$\sum_{n=0}^{\infty} a_n < \infty, \quad \sum_{n=0}^{\infty} a_n^2 < \infty, \quad a_1 = 0 \tag{41}$$

the diffusion functions of Perona-Malik and Weickert types are expressed as

$$g(|\nabla f|) = h(\lambda)\big|_{\lambda = |\nabla f|} \tag{42}$$

for specified coefficient set $\{a_n\}_{n=0}^{\infty}$ for small $|a_n|$, $n \geq 1$ and $|\nabla f|$. These properties of the diffusion functions, which control diffusivity, suggest that it is possible to derive a class of diffusions for non-linear diffusion filtering operations substituting the gradient map $|\nabla f|$ in a class of filter functions. As the Laplacian map version of the diffusion kernel we deal with filtering operations expressed as a series of $|\Delta f|$ such that,

$$g(|\Delta f|) = h(z)\big|_{z = |\Delta f|} \tag{43}$$

for a specified coefficient set.

References

1. Iijima, T., *Pattern Recognition*, Corona-sha, Tokyo, 1974 (in Japanese).
2. Zhao, N.-Y., Iijima, T., "Theory on the method of determination of view-point and field of vision during observation and measurement of figure," *IEICE Japan, Trans. D.*, Vol. J68-D, pp. 508-514, 1985 (in Japanese).
3. Zhao, N.-Y., Iijima, T., "A theory of feature extraction by the tree of stable view-point," *IEICE Japan, Trans. D.*, Vol. J68-D, pp. 1125-1132, 1985 (in Japanese).
4. Imiya, A., Sugiura, T., Sakai, T., Kato, Y., "Temporal structure tree in digital linear scale space," *Scale Space Methods in Computer Vision 2003*, Lecture Notes in Computer Science, Vol. 2695, pp. 356-371, 2003.
5. Kuijper, A., *The Deep Structure of Gaussian Scale Space Images*, Ph.D. thesis, Utrecht University, 2002.

6. Sakai, T., Imiya, A., "Hierarchical analysis of low-contrast temporal images with linear scale space," *ECCV CVAMIA-MMBIA 2004*, Lecture Notes in Computer Science, Vol. 3117, pp. 145-156, 2004.
7. Perona, P., Malik, J., "Scale space and edge detection using anisotropic diffusion," *IEEE Transaction on Pattern Analysis and Machine Intelligence*, Vol. 12, No. 7, pp. 629-639, 1990.
8. Weickert, J., *Anisotropic Diffusion in Image Processing*, Teubner, Stuttgart, 1998.
9. Weickert, J., "Applications of nonlinear diffusion in image processing and computer vision," *Acta Mathematica Universitatis Comenianae*, Vol. 70, No. 1, pp. 33-50, 2001.
10. Bracewell, R. N., *Two-Dimensional Imaging*, Prentice-Hall, New Jersey, 1995.
11. Kimmel, R., *Numerical Geometry of Images*, Springer, New York, 2004.
12. Barash, D., Kimmel, R., "An accurate operator splitting scheme for nonlinear diffusion filtering," *Scale-Space and Morphology in Computer Vision*, Lecture Notes in Computer Science, Vol. 2106, pp. 281-289, 2001.
13. Schroeder, M., R., *Number Theory in Science and Communication*, Springer, 1997 (Third Edition).
14. Neelin, P.,"McConnell Brain Imaging Centre," http://www.bic.mni.mcgill.ca/

Object Tracking Using Genetic Evolution Based Kernel Particle Filter

Qicong Wang[1], Jilin Liu[1], and Zhigang Wu[2]

[1] Departmant of Information and Electronics Engineering , Zhejiang University, Hangzhou,
Zhejiang 310013, China
{wang_qi_cong, liujl}@zju.edu.cn
[2] College of Information & Electronics Engineering, Taizhou University, Linhai,
Zhejiang 317000, China
wuzg@163.com

Abstract. A new particle filter, which combines genetic evolution and kernel density estimation, is proposed for moving object tracking. Particle filter (PF) solves non-linear and non-Gaussian state estimation problems in Monte Carlo simulation using importance sampling. Kernel particle filter (KPF) improves the performance of PF by using density estimation of broader kernel. However, it has the problem which is similar to the impoverishment phenomenon of PF. To deal with this problem, genetic evolution is introduced to form new filter. Genetic operators can ameliorate the diversity of particles. At the same time, genetic iteration drives particles toward their close local maximum of the posterior probability. Simulation results show the performance of the proposed approach is superior to that of PF and KPF.

1 Introduction

Object tracking is required by many vision applications such as human-computer interfaces, video communication, or surveillance [1]. Particle filtering has proven very successful for solving non-linear and non-Gaussian state estimation problems and managing multi-modal density function effectively [4]. So PF is widely applied in visual tracking. In Monte Carlo simulation, the posterior density is approximated by a weighted sum based on the discrete grid sequentially chosen by the importance sampling. Generally, uniform re-sampling is employed in particle filtering, which results in the particle impoverishment problem, i.e., the loss of diversity for the particles. A large number of particles can be sampled to deal with this problem in the filtering. However, this produces large computational costs.

In order to reduce the impoverishment effect, layered sampling [6], annealed importance sampling [7], and partitioned sampling [5] are proposed. They employ complex sampling strategies or specific prior knowledge about the objects. In [9], kernel particle filter which includes an inherent re-sampling step at every iteration can approximate the posterior with smaller number of kernel. However, the particle impoverishment problem still exists. In [10], a mean shift based particle filter is proposed to obtain robust tracking in which mean shift is applied to each particle independently.

In this paper we present genetic evolution based kernel particle filter (GEKPF) for visual tracking. Introducing optimization procedures into particle filtering is similar to

U. Eckardt et al. (Eds.): IWCIA 2006, LNCS 4040, pp. 466 – 473, 2006.
© Springer-Verlag Berlin Heidelberg 2006

the mean shift based particle filter. However, it is derived in a different manner that uses evolutionary computation. Genetic evolution is incorporated into KPF. In GEKPF, kernel density estimation can form continuous estimate of the posterior. The iterative procedure of genetic evolution can redistribute particles to the local modes of the observation effectively. At the same time, this procedure can perform implicitly iterative sampling to alleviate the particle impoverishment phenomenon greatly through genetic operator. The experiments on visual tracking demonstrate that compared to PF and KPF, the performance of the proposed approach is superior.

This paper is organized as follows. Section 2 presents the kernel density estimation of the posterior density. In section 3, we discuss the classical particle filter and KPF respectively. We propose the new filter which incorporates genetic evolution into KPF in section 4. Some results of experiments on object tracking are showed in section 5. Finally conclusions are briefly drawn in section 6.

2   Kernel Density Estimation of the Posterior Density

In [2], through non-linear and time-varying functions f and h particle filter solves the problem based on the system model

$$\mathbf{x}_t = f\left(\mathbf{x}_{t-1}, \mathbf{w}_t\right),$$ (1)

and on the observation model

$$\mathbf{y}_t = h\left(\mathbf{x}_t, \mathbf{v}_t\right),$$ (2)

where $\mathbf{x}_t$ is the state at time t, $\mathbf{y}_t$ the observation at time t, and both $\mathbf{w}_t$ and $\mathbf{v}_t$ the independent white noises. $\mathbf{y}_{0:t}$ is defined as the history sequence of the random variables. Our problem consists in computing the posterior density $p\left(\mathbf{x}_t \mid \mathbf{y}_{0:t}\right)$ of the state $\mathbf{x}_t$ at each time t, which can be obtained through prediction and update recursively. By Eq. (1), we realize prediction according to the following equation

$$p(\mathbf{x}_t \mid \mathbf{y}_{0:t-1}) = \int p\left(\mathbf{x}_t \mid \mathbf{x}_{t-1}\right) p\left(\mathbf{x}_{t-1} \mid \mathbf{y}_{0:t-1}\right) d\mathbf{x}.$$ (3)

To obtain the posterior density, we update this prediction with the observation $\mathbf{y}_t$ in terms of the Bayes' rule

$$p\left(\mathbf{x}_t \mid \mathbf{y}_{0:t}\right) = \frac{p\left(\mathbf{y}_t \mid \mathbf{x}_t\right) p\left(\mathbf{x}_t \mid \mathbf{y}_{0:t-1}\right)}{\int p\left(\mathbf{y}_t \mid \mathbf{x}\right) p\left(\mathbf{x} \mid \mathbf{y}_{0:t-1}\right) d\mathbf{x}}.$$ (4)

For particle set $\left\{\left(\mathbf{s}_t^{(n)}, q_t^{(n)}\right)\right\}_{n=1,2,\dots,N}$, where $\mathbf{s}_t$ is the particle state, q_t the weight associated to the particle, and n the number of particles, we approximate the posterior with the following weighted sum on the discrete grids.

$$p\left(\mathbf{x}_t \mid \mathbf{y}_{0:t}\right) \approx \sum_{n=1}^{N} q_t^{(n)} \delta\left(\mathbf{x}_t - \mathbf{s}_t^{(n)}\right), \tag{5}$$

where $\delta\left(\bullet\right)$ is Dirac's delta function.

Kernel density estimation is one of the most popular nonparametric methods of probability density estimation [3]. It is used here to form a continuous estimate of the posterior. Given a set $\left\{\mathbf{x}_i, i = 1, 2, ..., N\right\}$ in the d-dimensional space $\mathbf{R}^d$, kernel density estimation with kernel $k\left(\mathbf{x}\right)$ and bandwidth h is defined as

$$p_k\left(\mathbf{x}\right) = \frac{1}{Nh^d} \sum_{i=1}^{N} k\left(\frac{\mathbf{x} - \mathbf{x}_i}{h}\right), \tag{6}$$

where $k\left(\mathbf{x}\right)$ is the Gaussian kernel. So Gaussian kernel density estimation of the posterior $p\left(\mathbf{x}_t \mid \mathbf{y}_{0:t}\right)$ is obtained by the formulation

$$p\left(\mathbf{x}_t \mid \mathbf{y}_{0:t}\right) = \frac{1}{Nh^d} \sum_{n=1}^{N} q_t^{(n)} k\left(\frac{\mathbf{x}_t - \mathbf{s}_t^{(n)}}{h}\right). \tag{7}$$

3 Kernel Particle Filter

Particle filter solves non-linear and non-Gaussian state estimation problems in Monte Carlo simulation using importance sampling, in which the posterior density is approximated by the relative density of particles in a neighborhood of state space. Generally, the SIR algorithm [11] is composed of three steps: sampling, weighting, and re-sampling. In the sampling step, new particles are generated by drawing from the importance distribution. In the importance step, the weights associated to particles are evaluated by means of the observation model. Then the weights are normalized so that the weights add up to unity. In the re-sampling step, new particles are drawn from the distribution representation by the previous particle set, and all weights associated to the particles are set to be equal weight $1/N$. The re-sampling step is crucial in the implementation of particle filtering because without it, the variance of the particle weights quickly increases, i.e., very few normalized weights are substantial. If uniform re-sampling is used in particle filtering, it will result in the particle impoverishment problem, i.e., the loss of diversity for the particle set.

The kernel based particle filter improves the performance of the traditional algorithm by using density estimation of broader kernel [9]. It has three merits. Firstly, it does not need optimization of kernel parameters at every step. Secondly, the approximation of the integral includes an inherent re-sampling step, which allows the particle filter accuracy to survive longer than the standard version. Finally, using kernels with nonzero width can realize the continuous density estimation of the posterior. Each kernel can be propagated through the mapping $p\left(\mathbf{x}_t \mid \mathbf{x}_{t-1}\right)$ by using a local linearization, yielding a continuous output distribution $p\left(\mathbf{x}_t \mid \mathbf{y}_{0:t}\right)$, this is again a sum of kernels but the kernels are no longer identical. The prediction step is defined by the kernel representation equation

$$p\left(\mathbf{x}_t \mid \mathbf{y}_{0:t-1}\right) = \sum_{n=1}^{N} q_{t-1}^{(n)} \int p\left(\mathbf{x}_t \mid \mathbf{x}_{t-1}\right) k\left(\mathbf{A}_{t-1}^{(n)}(\mathbf{x}_{t-1} - \mathbf{s}_{t-1}^{(n)})\right) d\mathbf{x}_{t-1} , \qquad (8)$$

where $\mathbf{A}_t$ is a transformation matrix used to keep track of distortions of the kernel in each iteration. To avoid too much distortion, a re-sampling schema can be applied. All kernels in Eq. (8) are assumed that they are small compared to the dynamic in the non-linearity such that f can be locally linearized. Then the Jacobian $\mathbf{J} \mid_{\mathbf{s}_{t-1}^{(n)}} = \dfrac{\partial f}{\partial \mathbf{x}} \mid_{\mathbf{s}_{t-1}^{(n)}}$ is obtained by linearizing f around $\mathbf{s}_{t-1}^{(n)}$, and $\mathbf{A}_t^{(n)}$ is updated by the equation

$$\mathbf{A}_t^{(n)} = \mathbf{A}_{t-1}^{(n)} \mathbf{J} \mid_{\mathbf{s}_{t-1}^{(n)}}^{-1} . \qquad (9)$$

If we consider only Gaussian kernel, the transformation matrix is the covariance matrix. So the update of the transformation matrix $\mathbf{A}_t^{(n)}$ can be replaced with an update of the covariance matrix Σ as follows:

$$\Sigma_t^{(n)} = \mathbf{J} \mid_{\mathbf{s}_{t-1}^{(n)}} \Sigma_{t-1}^{(n)} \mathbf{J} \mid_{\mathbf{s}_{t-1}^{(n)}}^{T} . \qquad (10)$$

For given particle set, the posterior density $p\left(\mathbf{x}_t \mid \mathbf{y}_{0:t}\right)$ can be estimated by the formulation:

$$p\left(\mathbf{x}_t \mid \mathbf{y}_{0:t}\right) \approx \sum_{n=1}^{N} q_t^{(n)} k\left(\Sigma_t^{(n)} \left(\mathbf{x}_t - \mathbf{s}_t^{(n)}\right)\right) . \qquad (11)$$

And the weight update is obtained as follow:

$$q_t^{(n)} = q_{t-1}^{(n)} p\left(\mathbf{y}_t \mid \mathbf{s}_t^{(n)}\right) \mid \mathbf{J} \mid_{\mathbf{s}_{t-1}^{(n)}} \mid^{-1} . \qquad (12)$$

4   Genetic Evolution Based Kernel Particle Filter

Kernel particle filter can approximate the posterior with smaller number of kernel than PF, which produces smaller computation cost. However, the posterior density is estimated by a single kernel after several updates. This problem is similar to the impoverishment phenomenon of PF. We presented a novel approach to overcome this problem by incorporating genetic evolution into KPF.

Genetic evolution is stochastic global optimization approach and the solution of problem is a searching process in which the chromosomes are then evolved through iteratively performed selection and genetic operators [8]. At the beginning of each update of GEKPF, all particles are taken as the initial population of genetic operation. We only utilize the genetic evolution to achieve sub-optimization. Then genetic iteration can realize two functions. One is an iterative seeking procedure and the other is the implicitly iterative sampling. The iterative seeking procedure can redistribute particles to the local maxima of the posterior density, which produces the proper local representation of the posterior density. During implicitly iterative sampling, genetic operation can work as the re-sampling step in which mutation and crossover operator which can ameliorate the diversity of particles. If the iteration number of genetic operation, crossover probability and mutation probability are set properly, the consequential particles set will not include too many repeated points, so the impoverishment problem is efficiently overcome. Moreover, the proposed algorithm requires fewer particles to maintain multiple modes, because particles can be redistributed to their close local maxima actively after genetic iteration. Even though we sample more particles, most particles may converge to the same local maximum due to genetic iteration. Thus GEKPF can maintain multiple modes using fewer particles than KPF.

We can divide the proposed tracker into the following steps [12]. The first step is initialization. The particles are generated by using Eq. (1) and code the state of each particle to be a binary chromosome. So a particle set is mapped to a chromosome set. Initialize the values of crossover rate and mutation rate. The second step evaluates the fitness of each chromosome in the chromosome set, in which the weights, i.e. the fitness of chromosomes, are recomputed by using the observation model. The third step selects chromosomes to perform crossover and mutation operation from the chromosome set according to the fitness, the crossover ratio and mutation ratio. New chromosomes are evaluated by their fitness function value in the fourth step. In the final step new chromosomes are inserted into chromosome set and bad chromosomes in fitness are eliminated through selection. The above steps iterate several times. When iteration stops, particles can reach their close local maxima and the weighted average of all chromosomes is calculated to obtain the output state of the tracked object. At time t, we can define the procedure of one genetic iteration as operation $\mathbf{GE}: R^d \rightarrow R^d$, where d is the state space dimension. So we can rewrite Eq. (11) as

$$p\left(\mathbf{x}_t \mid \mathbf{y}_{0:t}\right) \approx \sum_{n=1}^{N} q_t^{(n)} k\left(\Sigma_t^{(n)}\left(\mathbf{x}_t - \mathbf{GE}\left(\mathbf{s}_t\right)\right)\right), \tag{13}$$

where $\mathbf{GE}\left(\mathbf{s}_t\right)$ iterates several times.

5   Experiment

In this section we present the comparison among PF, KPF, and GEKPF. All algorithms are applied to track moving human head which include horizontal acceleration, changes of direction, and self-occlusion and moving face whose motion is rapid. The test sequences of moving human which are downloaded from [3] are a resolution of 384x288 pixels and a slow motion. The real video of moving face with noises is a resolution of 320x240 pixels. Trackers are initialized manually. The object is modeled as a rectangle. So we define the state of the tracked object using four

components $\mathbf{s} = \left(x, y, \overset{\bullet}{x}, \overset{\bullet}{y} \right)^{T}$, where (x, y) is the coordinates of the rectangle

center, and $\overset{\bullet}{x}$, $\overset{\bullet}{y}$ the velocity. We only focus on the dynamics of (x, y). In the experiments, we use a simple second-order AR process as the system model

$$\mathbf{s}_t - \mathbf{s}_{t-1} = \mathbf{s}_{t-1} - \mathbf{s}_{t-2} + \mathbf{w}_t .$$ (14)

We adopt color distribution based the method in the observation model [1]. Then the particles are measured on the color probability distribution image. So the weights associated to particles, which represent the fitness of chromosomes, can be approximated using a Gaussian distribution of the Bhattacharyya distance Bd as follow:

$$q_t^{(n)} = \frac{1}{\sqrt{2\pi}\sigma} \exp\left(-\frac{Bd^{(n)2}}{2\sigma^2} \right),$$ (15)

where variance σ is empirically set as 0.2.

The best state at each time is derived based on the approximation of the expectation

$$E\left(\mathbf{x}_t \mid \mathbf{y}_{0:t} \right) \approx \frac{1}{N} \sum_{n=1}^{N} \mathbf{s}_t^{(n)} q_t^{(n)} .$$ (16)

A comparison of the tracking results is shown in Fig. 1 and Fig. 2. In Fig. 1, the white rectangles represent the mean states; the tracking results of PF with 220 particles are shown in the top row; the tracking results of KPF with 65 particles are shown in the middle row; the tracking results of GEKPF with 25 particles and 4 iterations are shown in the bottom row. In Fig. 2, the red rectangles represent the mean states; the first row is the tracking results using PF with 250 particles; the second row is the tracking results using KPF with 60 particles; the final row is the tracking results using GEKPF with 25 particles and 4 iterations.

All results show PF even with a lot of particles produces unstable results. For slow motion, KPF can achieve good tracking using fewer particles than PF. However, KPF loses tracking because of rapid motion of the object. GEKPF can obtain robust tracking with fewer particles than the conventional algorithms for both slow and rapid motion.

GEKPF improves the performance of particle filtering at the cost of introducing extra complexity. Due to reduce the total number of particles, GEKPF can alleviate the computation costs.

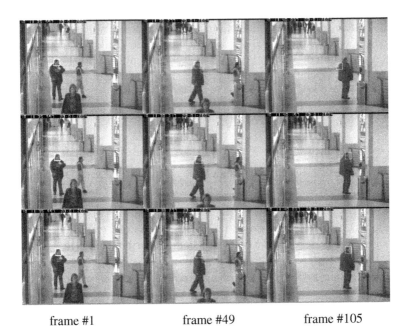

frame #1   frame #49   frame #105

Fig. 1. Some frames illustrate the tracking of three algorithms in slow motion including self-occlusions and changes of direction

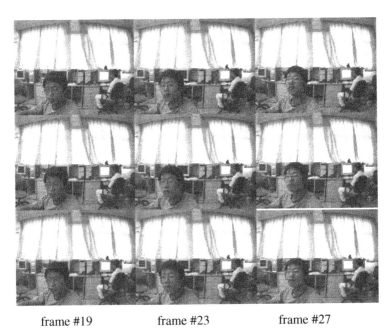

frame #19   frame #23   frame #27

Fig. 2. Some frames of tracking rapid motion by using three algorithms respectively

6   Conclusion

For robust tracking, we present a novel particle filter, i.e., genetic evolution based kernel particle filter. It incorporates genetic evolution and kernel method in particle filtering. Genetic evolution can shift particles to the local maxima of the posterior density and reduce the particle impoverishment problem. Kernel density estimation can realize a continuous estimate of the posterior. The experiments on visual tracking demonstrate that the proposed approach outperforms particle filter and kernel particle filter while using fewer particles than the conventional algorithms. Our future work is focused on analyzing and comparing the performance of other optimization and swarm intelligence approaches in particle filtering.

Acknowledgements

This work is supported by Grant No.60534070 from the National Science Foundation of China.

References

1. Nummiaro, K., Koller-Meier, E., Gool, L.V.: An Adaptive Color-Based Particle Filter. Image and Vision Computing, 21 (2003) 99-110
2. Isard, M., Blake, A.: CONDENSATION--Conditional Density Propagation for Visual Tracking. International Journal on Computer Vision, 29 (1998) 5-28
3. http://groups.inf.ed.ac.uk/vision/CAVIAR/CAVIARDATA
4. Arulampalam, M. S., Maskell, S., Gordon, N., Clapp T.: A Tutorial on Particle Filters for Online Nonlinear/Non-gaussian Bayesian Tracking. IEEE Transactions on Signal Processing, 50 (2002) 174-188
5. MacCormick, J., Blake, A.: Partitioned Sampling, Articulated Objects and Interface-quality Hand Tracking. Lecture Notes in Computer Science, 1843 (2000) 3-19
6. MacCormick, J., Blake, A.: A Probabilistic Exclusion Principle for Tracking Multiple Objects. In: Proceedings of International Conference on Computer Vision. (1999) 572-578
7. Deutscher, J., Blake, A., Reid, I.: Articulated Body Motion Capture by Annealed Particle Filtering. In: Proceedings of Computer Vision and Pattern Recognition. (2000) 126-133
8. Hwang, S.W., Kim, E.Y., Park, S.H.: Object Extraction and Tracking Using Genetic Algorithms. In: Proceedings of International Conference on Image Processing. (2001) 383-386
9. Lehn-Schioler, T., Erdogmus, D., Principe, J.C.: Parzen Particle Filters. In: Proceedings of International Conference on Acoustics, Speech, and Signal Processing. (2004) 781-784
10. Chang, C., Ansari, R.: Kernel Particle Filter: Iterative Sampling for Efficient Visual Tracking. In: Proceedings of International Conference on Image Processing. (2003) 977-980
11. Doucet, A., Freitas, N.de, Gordon, N.: Sequential Monte Carlo Methods in Practice. Springer-Verlag, New York (2001)
12. Hu, C.B., Yu, Q.F., Li Y., Ma, S.D.: Extraction of Parametric Human Model for Posture Recognition Using Genetic Algorithm. In: Proceedings of International Conference on Automatic Face and Gesture Recognition. (2000)

An Efficient Reconstruction of 2D-Tiling with $t_{1,2}$, $t_{2,1}$, $t_{1,1}$ Tiles

Masilamani Vedhanayagam and Kamala Krithivasan

Dept of Computer Science and Engineering
Indian Institute of Technology Madras
Chennai India 600036
jvmiitkgp@yahoo.com, kamala@iitm.ernet.in

Abstract. We define the projection of a tiling as a matrix $P = (p_{ij})$ where p_{i1} is number of $t_{1,2}$ tiles in row i and p_{i2} is the number of $t_{2,1}$ tiles in row i. We give an efficient algorithm to tile a 2D-square grid with only $t_{1,2}$, $t_{2,1}$, $t_{1,1}$ tiles such that the projection of this tiling is the same as the given projection.

1 Introduction

The area of *discrete tomography* is concerned about reconstruction of a discrete object or its geometrical properties from its projections or some other information. This has application in fields such as: computer vision, VLSI design, image processing, statistical data security, biplane angiography, graph theory, crystallography etc. [4] gives the fundamentals related to this topic.

Here we consider the reconstruction of tilings. We are given a collection of tiles where each tile can have different shapes. A tiling is a placement of non overlapping copies of the tiles in a $n \times n$ grid, where each copy is obtained by translating one of the tiles. [2, 3] deal with several problems related to this. [1] gives many results relating to the complexity of problems for different types of tiles. In [1], reconstruction of the tiling is considered where both row and column projection are given. If the tiles are ▯▯ , ⊟ and ▯ the problem is NP complete. But with one projection the problem can have a solution. In fact the solution need not be unique. Usually, there exist several possibilities of tiling for the same projection. This is shown in [1]. In [1] it is shown that the problem is NP complete for 3 atoms.

Considering the aim of tomography to reconstruct objects, it should be noted that objects are in general three dimensional. Here our tiling problem is restricted to two dimensions. It would be interesting to extend the reconstruction of tiling problems to three dimensions. In 3D-tiling, 3D tiles such as $t_{2,1,1}$, $t_{1,2,1}$, $t_{1,1,2}$ are to considered . For example, the 2 atom problem and 3 atom problem can be considered in three dimensions also. When the problem is NP complete in two dimensions, it would naturally be NP complete for three dimensions. In a plane, we consider row and column projections. In three dimensions, we have to consider the projections along CY plane, YZ plane and XZ plane.

U. Eckardt et al. (Eds.): IWCIA 2006, LNCS 4040, pp. 474–480, 2006.

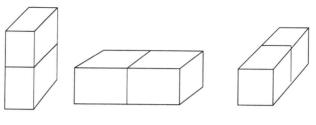

Fig. 1.

Extending the problem of 2D-tiling to three dimensions, we consider 3 types of dominoes shown in figure below.

Considering projections in 3 planes the problem will be NP complete. As a first step, we can consider reconstruction from projection on one plane. Projecting on to XY plane we will have three types of tiles in one plane ☐, ⊟ and ☐. Hence as a first step of solving this three dimensional problem, we consider the following two dimensional problem in this paper: Construct a tiling with the given projection in one direction. We have given an algorithm in [5] for this problem using backtracking. Here we give an algorithm without backtracking. The algorithm works in polynomial time. It is to be noted that we are considering "complete" tiling and not "partial" tiling.

In the next section, we give the algorithm and illustrate with a few examples. In section 3, we briefly discuss the complexity and correctness and also the possible solutions one can get from a particular solution using switching components. The paper concludes with a brief remark in section 4.

2   Reconstruction of Tiling from One Projection

In this section, we give an algorithm to reconstruct a tiling whose tiling has the same projection as the given projection. Let $\tau_{n \times n}$ be a family of tilings of $n \times n$ grid with $t_{1,2}$ (☐☐), $t_{2,1}$ (⊟) and $t_{1,1}$ (☐)   Let $T \in \tau_{n \times n}$ and

$$P = \begin{matrix} r_1^h & r_1^c \\ r_2^h & r_2^c \\ \vdots & \vdots \\ r_n^h & r_n^c \end{matrix}$$

Let $A_{i,j}^l = \{i' | i' < i \text{ and } t_{i',j} \text{ is covered with } t_{1,2} \text{ and } t_{1,1}\}$

and $A_{i,j}^r = \{i' | i' < i \text{ and } t_{i',j+1} \text{ is covered with at } t_{1,2} \text{ or } t_{1,1} \}$.

We define the $t_{i,j}$ of T to be *possible starting position* of $t_{1,2}$ (*PSP*). Let T be consistent with the vector P. Consider the following properties for *PSP* computation : $t_{i,j}$ of T is *PSP* if any of the following cases are true:

$$Let\ i^l_m = \begin{cases} max\ \{i'|i' \in A^l_{i,j}\}\ if\ A^l_{i,j} \neq \phi \\ 0 \qquad\qquad\quad Otherwise \end{cases}$$

$$and\ i^r_m = \begin{cases} max\ \{i'|i' \in A^r_{i,j}\}\ if\ A^r_{i,j} \neq \phi \\ 0 \qquad\qquad\quad Otherwise \end{cases}.$$

$$Let\ l_{diff} = i - i^l_m\ and\ r_{diff} = i - i^r_m.$$

$Case(i)$: Both l_{diff} and r_{diff} are odd.
$Case(ii)$: If l_{diff} is odd and r_{diff} is even then there should be a row j such that $j - i^r_m$ is odd and $r^c_j \geq 1$
If r_{diff} is odd and l_{diff} is even then there should be a row j such that $j - i^l_m$ is odd and $r^c_j \geq 1$
$Case(iii)$: Both l_{diff} and r_{diff} are even and there exist two rows j and k such that $j - i^l_m$ and $k - i^r_m$ are odd, and $r^c_j + r^c_k \geq 2$

A pair of sub tilings is said to be switching components if one component can be replaced by other component of the pair without affecting projection.

If the switching component does not have any non trivial switching component (switching component with more then one tile) as its sub tiling, then the switching component is called as elementary switching component. The possible elementary switching components with respect to projection onto column are given below:

1. Switching components involving only $t_{1,2}$ and $t_{2,1}$ tiles

$$\left\{ \begin{array}{|c|c|c|} \hline 1 & 0 & 0 \\ \hline 1 & 0 & 0 \\ \hline \end{array}\ , \begin{array}{|c|c|c|} \hline 0 & 0 & 1 \\ \hline 0 & 0 & 1 \\ \hline \end{array} \right\}$$

2. Switching components involving only $t_{1,1}$ and $t_{2,1}$ tiles

$$\left\{ \begin{array}{|c|c|} \hline 1 & 2 \\ \hline 1 & 2 \\ \hline \end{array}\ , \begin{array}{|c|c|} \hline 2 & 1 \\ \hline 2 & 1 \\ \hline \end{array} \right\}$$

3. Switching components involving only $t_{1,2}$ and $t_{1,2}$ tiles

$$\left\{ \begin{array}{|c|c|c|} \hline 2 & 0 & 0 \\ \hline \end{array}\ , \begin{array}{|c|c|c|} \hline 0 & 0 & 2 \\ \hline \end{array} \right\}$$

4. Switching components involving $t_{1,2}$, $t_{1,2}$ and $t_{1,2}$ tiles

$$\left\{ \begin{array}{|c|c|c|} \hline 0 & 0 & 1 \\ \hline 2 & 2 & 1 \\ \hline \end{array}\ , \begin{array}{|c|c|c|} \hline 1 & 0 & 0 \\ \hline 1 & 2 & 2 \\ \hline \end{array} \right\}$$

$$and\ also \left\{ \begin{array}{|c|c|c|} \hline 2 & 2 & 1 \\ \hline 0 & 0 & 1 \\ \hline \end{array}\ , \begin{array}{|c|c|c|} \hline 1 & 2 & 2 \\ \hline 1 & 0 & 0 \\ \hline \end{array} \right\}$$

Algorithm: Tiling_Reconstruction ($\tau_{n \times n}$, P)
Input:

$$A\ matrix\ P\ =\ \begin{bmatrix} r_1^h & r_1^c \\ r_2^h & r_2^c \\ \vdots & \vdots \\ r_n^h & r_n^c \end{bmatrix}$$

Output: A tiling $T \in \tau_{n \times n}$ such that, for each $1 \leq i \leq m$, r_i^h counts the number of $t_{1,2}$ in the i-th row of T and r_i^c counts the number of $t_{1,1}$ in the i-th row of T, if it exists else give FAILURE.

Step1 :
   If $2r_1^h \leq n$ then place r_1^h number of $t_{1,2}$ from position (1,1) to position
   $(1, 2r_1^h)$ in the first row of T and set $r_1^h = 0$, else give FAILURE.
   If $r_1^c > n - 2r_1^h$ Give FAILURE.
   Step2 :
   for each $2 \leq i \leq n$
      If $r_i^h \neq 0$ then
         for each $1 \leq k \leq r_i^h$
            (i) Find the leftmost PSP cell of the i-th row of T if possible,
            else give FAILURE
            (ii) Check whether or not $P[i][1]$ horizontal dominoes can be placed
            in row i using even number of clear cells.
            if possible, Place an $t_{1,2}$ starting from leftmost PSP
            if (i, j) is the PSP cell and $(i - 1, j)$ is not tiled with clear cell
            otherwise Place an $t_{1,2}$ starting from leftmost PSP.
            (iii) $k = k + 1$
   Step3 :
   for each incorrect tiled column j
   / * *Here incorrect tiled column means the odd number of consecutive*
   vacant cells * /
      (i) Find i such that $r_i^c > 0$
      (ii) Place clear cell at $t_{i,j}$ if correct tiling is possible.
   Step4 :
   for each pair (i, j), $1 \leq i \leq n$, $1 \leq j \leq n$ such that $r_i^c \neq 0$ and $r_j^c \neq 0$ and
   $j - i$ is odd,
      find column k such that placing clear cells at $t_{i,k}$ and $t_{j,k}$ gives correctly
   tiled column k.
   Step5 :
   Fill the empty cells with vertical dominoes
      If not possible give FAILURE.

EXAMPLE 1
The above algorithm is illustrated with an example, Let us represent horizontal
domino by $\boxed{0\,0}$, vertical domino by $\boxed{\begin{smallmatrix}1\\1\end{smallmatrix}}$ and clear cell $\boxed{2}$.

$$Input: \quad P = \begin{bmatrix} 3 & 2 \\ 1 & 3 \\ 1 & 1 \\ 1 & 3 \\ 1 & 2 \\ 1 & 1 \\ 1 & 1 \\ 0 & 5 \end{bmatrix}$$

*Step*1 : Tile first row with h-dominoes and even part of clear cells

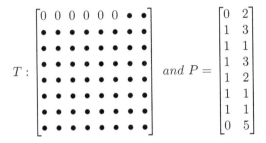

$$T: \begin{bmatrix} 0 & 0 & 0 & 0 & 0 & 0 & \bullet & \bullet \\ \bullet & \bullet & \bullet & \bullet & \bullet & \bullet & \bullet & \bullet \\ \bullet & \bullet & \bullet & \bullet & \bullet & \bullet & \bullet & \bullet \\ \bullet & \bullet & \bullet & \bullet & \bullet & \bullet & \bullet & \bullet \\ \bullet & \bullet & \bullet & \bullet & \bullet & \bullet & \bullet & \bullet \\ \bullet & \bullet & \bullet & \bullet & \bullet & \bullet & \bullet & \bullet \\ \bullet & \bullet & \bullet & \bullet & \bullet & \bullet & \bullet & \bullet \\ \bullet & \bullet & \bullet & \bullet & \bullet & \bullet & \bullet & \bullet \end{bmatrix} \quad and\ P = \begin{bmatrix} 0 & 2 \\ 1 & 3 \\ 1 & 1 \\ 1 & 3 \\ 1 & 2 \\ 1 & 1 \\ 1 & 1 \\ 0 & 5 \end{bmatrix}$$

*Step*2 : Tiling with h-bars starting from 2nd row

$$T: \begin{bmatrix} 0 & 0 & 0 & 0 & 0 & 0 & \bullet & \bullet \\ 0 & 0 & \bullet & \bullet & \bullet & \bullet & \bullet & \bullet \\ 0 & 0 & \bullet & \bullet & \bullet & \bullet & \bullet & \bullet \\ 0 & 0 & \bullet & \bullet & \bullet & \bullet & \bullet & \bullet \\ 0 & 0 & \bullet & \bullet & \bullet & \bullet & \bullet & \bullet \\ 0 & 0 & \bullet & \bullet & \bullet & \bullet & \bullet & \bullet \\ 0 & 0 & \bullet & \bullet & \bullet & \bullet & \bullet & \bullet \\ \bullet & \bullet & \bullet & \bullet & \bullet & \bullet & \bullet & \bullet \end{bmatrix} \quad and\ P = \begin{bmatrix} 0 & 2 \\ 0 & 3 \\ 0 & 1 \\ 0 & 3 \\ 0 & 2 \\ 0 & 1 \\ 0 & 1 \\ 0 & 5 \end{bmatrix}$$

*Step*3 : Correct all incorrect columns with available clear cells.

$$T: \begin{bmatrix} 0 & 0 & 0 & 0 & 0 & 0 & \bullet & \bullet \\ 0 & 0 & 2 & 2 & 2 & \bullet & \bullet & \bullet \\ 0 & 0 & \bullet & \bullet & \bullet & \bullet & \bullet & \bullet \\ 0 & 0 & \bullet & \bullet & 2 & \bullet & \bullet & \bullet \\ 0 & 0 & \bullet & \bullet & \bullet & \bullet & \bullet & \bullet \\ 0 & 0 & \bullet & \bullet & \bullet & \bullet & \bullet & \bullet \\ 0 & 0 & \bullet & \bullet & \bullet & \bullet & \bullet & \bullet \\ 2 & 2 & \bullet & \bullet & \bullet & \bullet & \bullet & \bullet \end{bmatrix} \quad and\ P = \begin{bmatrix} 0 & 2 \\ 0 & 0 \\ 0 & 1 \\ 0 & 2 \\ 0 & 2 \\ 0 & 1 \\ 0 & 1 \\ 0 & 3 \end{bmatrix}$$

Step4 : Place all possible pairs of remaining clear cells with even distance apart

$$
T: \begin{bmatrix}
0 & 0 & 0 & 0 & 0 & 0 & 2 & 2 \\
0 & 0 & 2 & 2 & 2 & \bullet & \bullet & \bullet \\
0 & 0 & 2 & \bullet & \bullet & \bullet & \bullet & \bullet \\
0 & 0 & \bullet & \bullet & \bullet & 2 & 2 & 2 \\
0 & 0 & \bullet & 2 & 2 & \bullet & \bullet & \bullet \\
0 & 0 & 2 & \bullet & \bullet & \bullet & \bullet & \bullet \\
0 & 0 & \bullet & \bullet & \bullet & 2 & \bullet & \bullet \\
2 & 2 & \bullet & 2 & 2 & 2 & \bullet & \bullet
\end{bmatrix}
\quad and \ P = \begin{bmatrix}
0 & 0 \\
0 & 0 \\
0 & 0 \\
0 & 0 \\
0 & 0 \\
0 & 0 \\
0 & 0 \\
0 & 0
\end{bmatrix}
$$

Step5 :

$$
T: \begin{bmatrix}
0 & 0 & 0 & 0 & 0 & 0 & 2 & 2 \\
0 & 0 & 2 & 2 & 2 & 1 & 1 & 1 \\
0 & 0 & 2 & 1 & 1 & 1 & 1 & 1 \\
0 & 0 & 1 & 1 & 1 & 2 & 2 & 2 \\
0 & 0 & 1 & 2 & 2 & 1 & 1 & 1 \\
0 & 0 & 2 & 1 & 1 & 1 & 1 & 1 \\
0 & 0 & 1 & 1 & 1 & 2 & 1 & 1 \\
2 & 2 & 1 & 2 & 2 & 2 & 1 & 1
\end{bmatrix}
\quad and \ P = \begin{bmatrix}
0 & 0 \\
0 & 0 \\
0 & 0 \\
0 & 0 \\
0 & 0 \\
0 & 0 \\
0 & 0 \\
0 & 0
\end{bmatrix}
$$

3 Correctness

Let us denote a cell (i, j) as $E - cell$ or $O - cell$ if l_{diff} is odd or even respectively. We denote sequence of consecutive $E - cells$ as $E - block$ and sequence of consecutive $O - cells$ as $O - block$. An $E - block$ or $O - block$ is said to be odd or even if number of cells in the block is odd or even respectively.

Lemma 1. *In any row i, the number of odd $E - blocks$ is at most one.*

Let us prove the lemma by induction on row i. Since all the cells in row 1 is $E - cell$, the number of odd $E - blocks$ is one if n is odd, zero otherwise. Assume that the lemma is true for all all $i < k$. If each row $i < k$ has even number of clear cells, then each horizontal dominoes should have been placed starting from odd column. Hence odd $E - block$ is possible in row k only when no cells in A is tiled by any tile.

where $A = \{(i, j) \mid 1 \leq i \leq k$ and $(i, j) \in$ E-block $\}$.

If some rows have odd number of clear cells, then consider the largest row i such that $1 \leq i < k$, and row i has odd number of clear cells, and consider column j such that j is maximum and (i, j) is tiled by clear cell. clearly there is no odd$E - blocks$ among cells (k, j') such that $1 \leq j' \leq j + 2 * h$ where h is number of horizontal dominoes placed in row $i + 1$ starting from column j . Hence placement of clear cell in position (i, j) will not make new odd $E - blocks$ in row $i + 1$ and hence in remaining rows. Hence there can't be more than one $E - blocks$ in any row.

Theorem 1. *If the algorithm could not place a horizontal domino in any row i, then there exists no tiling consistent with given projection.*

By lemma 1, algorithm places maximum number of horizontal dominoes in each row. Hence the Theorem.

Theorem 2. *If the algorithm could not place any clear cell in any row i then there exists no tiling consistent with given projection.*

Theorem 3. *If the algorithm could not place any vertical domino in any row i then there exists no tiling consistent with given projection.*

As the placement of horizontal and clear cells are made such that there is no odd length of vertical blocks of free cells, theorem is evident.

4 Complexity

Step 1 uses $\mathbf{O(n)}$ operations. In Step 2, PSP is called $\mathbf{O}(n^2)$ time, PSP finds l_{diff} and r_{diff} in $\mathbf{O(n)}$ time and then checks whether or not there exist j and k such that $j - i_m^l$ and $k - i_m^r$ are odd in $\mathbf{O(n)}$ time. Hence step 2 takes $\mathbf{O}(n^3)$ time. Step 3 and 4 take $\mathbf{O}(n^4)$ time.

5 Conclusion

In this paper we have considered a particular case of tiling problem with 3 tiles, with one projection. We have also shown that another solution can be obtained from an existing one by switching the switching components. We are working on improving the algorithm. The extension of this problem to three dimensions where we consider dominoes in three directions is a topic for further exploration.

References

1. Frosini Andrea. *Complexity Results and Reconstruction Algorithms for Discrete Tomography.* PhD thesis, 2002.
2. M.Chrobak P.Couperus C.Durr and G.Woeginger. A note on tiling under tomographic constraints. *Theoretical Computer Science*, 290(3):2125–2136, 2003.
3. C.Durr E.Goles I.Rapaport E.Remila. Tiling with bars under tomographic constraints. *Theoretical Computer Science*, 290(3):1317–1329, 2003.
4. Gabor T. Herman, Attila Kuba, editor. *DISCRETE TOMOGRAPHY Foundations, Algorithms, and Applications.* Birkhauser, 1999.
5. V. Masilamani, K. Krithivasan, K. Nalinadevi. Reconstruction of 2D-tiling with horizontal, vertical dominoes and a clear cell. In *ADCOM*, 2005. to appear.

Author Index

Lecture Notes in Computer Science

For information about Vols. 1–3933

please contact your bookseller or Springer

Vol. 3982: M. Gavrilova, O. Gervasi, V. Kumar, C.J. K. Tan, D. Taniar, A. Laganà, Y. Mun, H. Choo (Eds.), Computational Science and Its Applications - ICCSA 2006, Part III. XXV, 1243 pages. 2006.

Vol. 3981: M. Gavrilova, O. Gervasi, V. Kumar, C.J. K. Tan, D. Taniar, A. Laganà, Y. Mun, H. Choo (Eds.), Computational Science and Its Applications - ICCSA 2006, Part II. XXVI, 1255 pages. 2006.

Vol. 3980: M. Gavrilova, O. Gervasi, V. Kumar, C.J. K. Tan, D. Taniar, A. Laganà, Y. Mun, H. Choo (Eds.), Computational Science and Its Applications - ICCSA 2006, Part I. LXXV, 1199 pages. 2006.

Vol. 3979: T.S. Huang, N. Sebe, M.S. Lew, V. Pavlović, M. Kölsch, A. Galata, B. Kisačanin (Eds.), Computer Vision in Human-Computer Interaction. XII, 121 pages. 2006.

Vol. 3978: B. Hnich, M. Carlsson, F. Fages, F. Rossi (Eds.), Recent Advances in Constraints. VIII, 179 pages. 2006. (Sublibrary LNAI).

Vol. 3976: F. Boavida, T. Plagemann, B. Stiller, C. Westphal, E. Monteiro (Eds.), Networking 2006. Networking Technologies, Services, and Protocols; Performance of Computer and Communication Networks; Mobile and Wireless Communications Systems. XXVI, 1276 pages. 2006.

Vol. 3975: S. Mehrotra, D.D. Zeng, H. Chen, B.M. Thuraisingham, F.-Y. Wang (Eds.), Intelligence and Security Informatics. XXII, 772 pages. 2006.

Vol. 3973: J. Wang, Z. Yi, J.M. Zurada, B.-L. Lu, H. Yin (Eds.), Advances in Neural Networks - ISNN 2006, Part III. XXIX, 1402 pages. 2006.

Vol. 3972: J. Wang, Z. Yi, J.M. Zurada, B.-L. Lu, H. Yin (Eds.), Advances in Neural Networks - ISNN 2006, Part II. XXVII, 1444 pages. 2006.

Vol. 3971: J. Wang, Z. Yi, J.M. Zurada, B.-L. Lu, H. Yin (Eds.), Advances in Neural Networks - ISNN 2006, Part I. LXVII, 1442 pages. 2006.

Vol. 3970: T. Braun, G. Carle, S. Fahmy, Y. Koucheryavy (Eds.), Wired/Wireless Internet Communications. XIV, 350 pages. 2006.

Vol. 3968: K.P. Fishkin, B. Schiele, P. Nixon, A. Quigley (Eds.), Pervasive Computing. XV, 402 pages. 2006.

Vol. 3967: D. Grigoriev, J. Harrison, E.A. Hirsch (Eds.), Computer Science - Theory and Applications. XVI, 684 pages. 2006.

Vol. 3966: Q. Wang, D. Pfahl, D.M. Raffo, P. Wernick (Eds.), Software Process Change. XIV, 356 pages. 2006.

Vol. 3965: M. Bernardo, A. Cimatti (Eds.), Formal Methods for Hardware Verification. VII, 243 pages. 2006.

Vol. 3964: M. Ü. Uyar, A.Y. Duale, M.A. Fecko (Eds.), Testing of Communicating Systems. XI, 373 pages. 2006.

Vol. 3963: O. Dikenelli, M.-P. Gleizes, A. Ricci (Eds.), Engineering Societies in the Agents World VI. XII, 303 pages. 2006. (Sublibrary LNAI).

Vol. 3962: W. IJsselsteijn, Y. de Kort, C. Midden, B. Eggen, E. van den Hoven (Eds.), Persuasive Technology. XII, 216 pages. 2006.

Vol. 3960: R. Vieira, P. Quaresma, M.d.G.V. Nunes, N.J. Mamede, C. Oliveira, M.C. Dias (Eds.), Computational Processing of the Portuguese Language. XII, 274 pages. 2006. (Sublibrary LNAI).

Vol. 3959: J.-Y. Cai, S. B. Cooper, A. Li (Eds.), Theory and Applications of Models of Computation. XV, 794 pages. 2006.

Vol. 3958: M. Yung, Y. Dodis, A. Kiayias, T. Malkin (Eds.), Public Key Cryptography - PKC 2006. XIV, 543 pages. 2006.

Vol. 3956: G. Barthe, B. Grégoire, M. Huisman, J.-L. Lanet (Eds.), Construction and Analysis of Safe, Secure, and Interoperable Smart Devices. IX, 175 pages. 2006.

Vol. 3955: G. Antoniou, G. Potamias, C. Spyropoulos, D. Plexousakis (Eds.), Advances in Artificial Intelligence. XVII, 611 pages. 2006. (Sublibrary LNAI).

Vol. 3954: A. Leonardis, H. Bischof, A. Pinz (Eds.), Computer Vision – ECCV 2006, Part IV. XVII, 613 pages. 2006.

Vol. 3953: A. Leonardis, H. Bischof, A. Pinz (Eds.), Computer Vision – ECCV 2006, Part III. XVII, 649 pages. 2006.

Vol. 3952: A. Leonardis, H. Bischof, A. Pinz (Eds.), Computer Vision – ECCV 2006, Part II. XVII, 661 pages. 2006.

Vol. 3951: A. Leonardis, H. Bischof, A. Pinz (Eds.), Computer Vision – ECCV 2006, Part I. XXXV, 639 pages. 2006.

Vol. 3950: J.P. Müller, F. Zambonelli (Eds.), Agent-Oriented Software Engineering VI. XVI, 249 pages. 2006.

Vol. 3948: H.I Christensen, H.-H. Nagel (Eds.), Cognitive Vision Systems. VIII, 367 pages. 2006.

Vol. 3947: Y.-C. Chung, J.E. Moreira (Eds.), Advances in Grid and Pervasive Computing. XXI, 667 pages. 2006.

Vol. 3946: T.R. Roth-Berghofer, S. Schulz, D.B. Leake (Eds.), Modeling and Retrieval of Context. XI, 149 pages. 2006. (Sublibrary LNAI).

Vol. 3945: M. Hagiya, P. Wadler (Eds.), Functional and Logic Programming. X, 295 pages. 2006.

Vol. 3944: J. Quiñonero-Candela, I. Dagan, B. Magnini, F. d'Alché-Buc (Eds.), Machine Learning Challenges. XIII, 462 pages. 2006. (Sublibrary LNAI).

Vol. 3943: N. Guelfi, A. Savidis (Eds.), Rapid Integration of Software Engineering Techniques. X, 289 pages. 2006.

Vol. 3942: Z. Pan, R. Aylett, H. Diener, X. Jin, S. Göbel, L. Li (Eds.), Technologies for E-Learning and Digital Entertainment. XXV, 1396 pages. 2006.

Vol. 3941: S.W. Gilroy, M.D. Harrison (Eds.), Interactive Systems. XI, 267 pages. 2006.

Vol. 3940: C. Saunders, M. Grobelnik, S. Gunn, J. Shawe-Taylor (Eds.), Subspace, Latent Structure and Feature Selection. X, 209 pages. 2006.

Vol. 3939: C. Priami, L. Cardelli, S. Emmott (Eds.), Transactions on Computational Systems Biology IV. VII, 141 pages. 2006. (Sublibrary LNBI).

Vol. 3936: M. Lalmas, A. MacFarlane, S. Rüger, A. Tombros, T. Tsikrika, A. Yavlinsky (Eds.), Advances in Information Retrieval. XIX, 584 pages. 2006.

Vol. 3935: D. Won, S. Kim (Eds.), Information Security and Cryptology - ICISC 2005. XIV, 458 pages. 2006.

Vol. 3934: J.A. Clark, R.F. Paige, F.A. C. Polack, P.J. Brooke (Eds.), Security in Pervasive Computing. X, 243 pages. 2006.